W0246509

APPROPRIATING INNOVATIONS

Entangled Knowledge in Eurasia, 5000–1500 BCE

Edited by

PHILIPP W. STOCKHAMMER AND JOSEPH MARAN

OXBOW | books

Oxford & Philadelphia

Published in the United Kingdom in 2017 by
OXBOW BOOKS
The Old Music Hall, 106–108 Cowley Road, Oxford OX4 1JE

and in the United States by
OXBOW BOOKS
1950 Lawrence Road, Havertown, PA 19083

© Oxbow Books and the individual contributors 2017

Hardback Edition: ISBN 978-1-78570-724-7
Digital Edition: ISBN 978-1-78570-725-4 (epub)

A CIP record for this book is available from the British Library and the Library of Congress

All rights reserved. No part of this book may be reproduced or transmitted in any form or by any means, electronic or mechanical including photocopying, recording or by any information storage and retrieval system, without permission from the publisher in writing.

Printed in Malta by Melita Press Ltd

For a complete list of Oxbow titles, please contact:

UNITED KINGDOM
Oxbow Books
Telephone (01865) 241249, Fax (01865) 794449
Email: oxbow@oxbowbooks.com
www.oxbowbooks.com

UNITED STATES OF AMERICA
Oxbow Books
Telephone (800) 791-9354, Fax (610) 853-9146
Email: queries@casemateacademic.com
www.casemateacademic.com/oxbow

Oxbow Books is part of the Casemate Group

Front cover: "Wheels of Innovation", Jelena Radosavljević

Contents

Chapter 1

Introduction

Joseph Maran and Philipp W. Stockhammer

The question of how to conceptualise the role of technological innovations over the last 12,000 years is of crucial importance in understanding the mechanisms and rhythms of long-term cultural change in prehistoric and early historic societies. Although the accelerating force of the advent of agriculture and sedentary village life during the early Holocene is widely acknowledged, the changes that have come about since then have often been modelled as gradual and linear. Already back in the 1920s such an approach was challenged by Vere Gordon Childe (1925; 1929), who insisted on the importance of technological and economic innovation coupled with human mobility and communication between societies in Asia and Europe in triggering periods of upheaval that he envisaged as 'revolutionary' in their consequences. Childe was rightly criticised for his belief in teleological progress and his oversimplified idea of diffusion as an almost natural force spreading from a few 'civilizational cores' in the Near East and Egypt to 'peripheral areas'. Yet, his ideas were groundbreaking in their emphasis on coupling societal change with interaction and technological innovation.

It is not the aim of this volume to develop a non-linear perspective for the large number of technological innovations that have shaped human existence since Childe's 'Neolithic Revolution'. Instead, we focus on two major clusters of innovation, namely the Secondary Products Revolution (Sherratt 1981) and bronze casting. What the introduction of the Secondary Products Revolution and of bronze casting have in common is that, roughly until the 1980s, they were mostly addressed with explanatory frameworks that relied on the concept of diffusion. In the early 20th century, diffusionist thought was a very potent factor in the emergence of what was later called '*Kulturkreislehre*' and soon became the most influential current of culture–historical ethnography in German-speaking academia (Trigger 1996, 235–241;

Rebay-Salisbury 2011). Its proponents aimed to distinguish cultural 'circles' by charting the distribution of certain aspects of social structure, material culture, technology, language, religion, music and myth. These circles were believed to be so definitive that even similarities in the form of specific cultural traits between extremely distant regions were regarded as meaningful and used to reconstruct cultural origins by unravelling sequences of diffusion or migration in space and time. The *Kulturkreislehre* introduced the concept of diffusion to archaeology. Its most influential champion was Vere Gordon Childe.

In the 1970s, archaeology understandably tried to rid itself of the influence of diffusionism and to develop new approaches explaining the appearance of similar cultural traits in different areas. Unfortunately, this quest for new approaches sometimes led to a wholesale abandonment of the study of intercultural contacts, replacing it by explanations rooted in the notion of independent, autochthonous development in various regions. Only in the last decade have new approaches emerged that aim to overcome the division between diffusionism and autochthonism. It was in line with this recent line of thought that we designed the conference 'Appropriating Innovations' (15–17 January 2015, Internationales Wissenschaftsforum Heidelberg) on which the present volume is based. The aim of the conference was to map out alternatives to the impasse that research on the introduction of innovations in early societies had manoeuvred itself into by opting either for diffusionism or autochthonism, that is either for unidirectional flow from a few 'civilizational cores' towards peripheral areas or polycentric invention occurring independently in different areas.

One of the most critical flaws of diffusionism was its obsession with questions of origin. It emphasised the importance of clarifying when and where an innovation was first discovered. The question of why such innovations were taken up by other societies never occurred to proponents of diffusionism, who held it to be self-evident that, once the practicability of an innovation was proven, it would automatically be adopted by other societies because it was so useful. In contrast to previous research, we set out to shift our perspectives away from diffusion and towards the concept of translation. To accomplish this, the discussion of innovations in archaeology needs to gravitate away from a focus on when and where they were developed and towards an investigation of how these innovations were ingested by societies and how this affected the lives and worldviews of the people constituting them. In this context, it is particularly interesting not only to address the question of 'what societies did with innovations', but especially to deal with the dialectically related question of 'what innovations did with society' (Maran and Kostoula 2014).

During the first period of the Heidelberg Cluster of Excellence 'Asia and Europe in a Global Context' from 2008 to 2012, our research focused on the local appropriation of objects coming from faraway places. At the 'Materiality and Practice' conference in March 2010 our chief concern was with the transformation of functions and meanings of objects and related practices in contexts of intercultural encounter (Maran and Stockhammer 2012). Since 2012 our focus has shifted from objects to innovations, but it is still the process of translation and appropriation of what we archaeologists define as foreign that lies at the heart of our research. Nevertheless, the transfer and translation of innovations requires different forms of intercultural contact than the transfer of objects: objects can be exchanged without any instructions for use – and sometimes develop their transformative potential especially when the recipient lacks any further information. The objects are appropriated, which is why our analytical focus has been on the transformative powers triggered by individual actors' interaction with objects of distant provenance. From a conceptual point of view, however, such objects can hardly be 'translated' as the notion of translation implies and even emphasises 'reaching out' (Fuchs 2009) in the sense of acquiring knowledge about, and from, actor(s) from a long way off and creating local knowledge within this process of negotiation with the other. Innovation transfer requires personal contact and an exchange of ideas and knowledge in the framework of encounters with people from faraway places. As every technology is socially constructed, not only technological knowledge is exchanged in the context of these personal encounters, but also non-technological knowledge, *i.e.* ideas and world-views connected with the innovation in question. Unlike the exchange of objects, the exchange of innovations is always connected with an overspill of information about foreign world-views that hinders rather than enhances acceptance and appropriation, as not only technological but

also non-technological knowledge has to be translated into individual world-views.

The particular kind of personal interaction typical for the exchange of innovations also entails another factor not necessarily involved in the transfer of objects in situations of intercultural encounter, namely the possibility of communicating with each other at least at a basic level of language. To put it differently, learning language is part of learning the practices necessary to deal with a given technology. When it comes to innovation transfer, language and hence translation, in the truest sense of the word, becomes a much more important factor.

We understand technologies as networks where the different nodes (*i.e.* human and non-human actors and actants such as world-views) are linked with each other. The ability to appropriate and translate a new technology is also conditioned by the previous existence of these nodes, whether they have to do with the ability to produce a certain temperature in ovens or with the tradition of exchanging knowledge and practices between different crafts. This relates to what Avinoam Meir (1988) has defined as the 'adoption environment', the analysis of which is crucial in understanding the appropriation of innovations.

For a long time, archaeologists have attributed an interest in foreign knowledge and technologies to male individuals of high status, *i.e.* those actors who have long been taken as the guiding factors in historical developments. Without further reflecting on this issue, archaeologists have generally assumed that at a given time foreign male actors met local male elites and provided them with the knowledge necessary to handle a particular technology. However, Science and Technology Studies (STS) have shown that male actors of high status are in fact among those with relatively little interest in innovation, as they consider every potential change of the existing system a danger for their position in the hierarchy. On the other hand, individuals with low or very flexible status positions – for reasons associated with migration, mobility or gender – seem to be much more open to innovation, as their lack of power and status means there is less danger of their forfeiting influence (*cf.* De Laet and Mol 2000). Can we archaeologists imagine that the knowledge required to cast bronze was not transmitted by a male smith to a male elite but by mobile women? What prerequisites have to be taken into consideration if we want to find arguments linking the acceptance of a certain innovation with what we perceive to be a local elite? Which actors were actually interested in new technologies? How can we identify and trace the process of translation in the archaeological record? And how can we describe the transformative dynamics triggered in the process? Questions such as these are addressed in the present volume, which assembles an international group of scholars drawing upon their specialist expertise to address a wide array of questions that are relevant to the overarching topic of the book.

Acknowledgements

This volume would not have been possible without generous financial support from the Heidelberg Cluster of Excellence 'Asia and Europe in a Global Context: The Dynamics of Transculturality' as part of the Excellence Initiative of the German Research Foundation. Our research has been undertaken as part of the Cluster project 'Appropriating Innovations: Entangled Knowledge in Late Neolithic and Early Bronze Age Eurasia', and we would like to thank the members of this project for their intellectual input over the years. With regard to the present publication, we would like to thank Maleen Leppek for taking charge of the overall preparation of the publication, Andrew, Jennifer and Eva-Maria Jenkins for their English language editing and Jelena Radosavljević for the editing of the illustrations. Their careful work has made this publication possible. Moreover, we would like to thank Oxbow Books for their continuous collaboration with us and especially Julie Gardiner as chief editor. Last but not least, we would like to thank all the contributors for enriching this volume with their carefully considered remarks on the appropriation of prehistoric innovations.

References

Childe, V. G. (1925) *The Dawn of European Civilization*. London, K. Paul.

Childe, V. G. (1929) *The Danube in Prehistory*. London, K. Paul.

De Laet, M. and Mol, A. (2000) The Zimbabwe Bush Pump: Mechanics of a Fluid Technology. *Social Studies of Science* 30/2, 225–263.

Fuchs, M. (2009) Reaching Out; or, Nobody Exists in One Context Only. Society as Translation. *Translation Studies* 2, 21–40.

Maran, J. and Kostoula, M. (2014) The Spider's Web: Innovation and Society in the Early Helladic 'Period of the Corridor Houses'. In Y. Galanakis, T. Wilkinson and J. Bennet (eds), *AΘYPMATA. Critical Essays on the Archaeology of the Eastern Mediterranean in Honour of E. Susan Sherratt*, 141–158. Oxford, Archaeopress.

Maran, J. and Stockhammer, P. W. (eds) (2012) *Materiality and Social Practice. Transformative Capacities of Intercultural Encounters*. Oxford, Oxbow Books.

Meir, A. (1988) Adoption Environment and Environmental Diffusion Processes: Merging Positivistic and Humanistic Perspectives. In P. J. Hugill and D. B. Dickson (eds), *The Transformation and Transfer of Ideas and Material Culture*, 233–347. College Station, TX, Texas A&M University Press.

Rebay-Salisbury, K. C. (2011) Thoughts in Circles: Kulturkreislehre as a Hidden Paradigm in Past and Present Archaeological Interpretations. In B. W. Roberts and M. Vander Linden (eds), *Investigating Archaeological Cultures: Material Culture, Variability, and Transmission*, 41–59. New York *et al.*, Springer.

Sherratt, A. (1981) Plough and Pastoralism: Aspects of the Secondary Products Revolution. In I. Hodder, G. Isaac and N. Hammond (eds), *Pattern of the Past*, 261–306. Cambridge, Cambridge University Press.

Trigger, B. G. (1996) A History of Archaeological Thought. 2nd ed. Cambridge *et al.*, Cambridge University Press.

Chapter 2

Innovation Minus Modernity? Revisiting Some Relations of Technical and Social Change

Cornelius Schubert

Introduction

Most innovation studies are concerned with distinctly modern phenomena. Modern societies are seen as driven by manifold innovation dynamics that evolve at the interstices between science, technology, politics and the economy. Sociology itself was born of the urge to understand the societal transformations at the end of the 19th century and it is still fuelled today by the unceasing dynamics of social change. One of society's most obvious dynamics simultaneously poses one of the discipline's most vexing conceptual problems: the relation between technical and social change. Despite the almost 150 years since the publication of *Das Kapital* (Marx 1867), the intricate relations between technology and society still form a 'moving target' that sparks continued interest among the social sciences. But we have also come a long way since then. In the course of this paper I shall revisit some classical approaches and engage with more recent developments in order to outline what insights archaeology might stand to gain from the sociology of innovation. Put briefly, my approach is to subtract modernity from innovation studies and then ask how what remains might be used to conceptualise 'non-modern' relations between technical and social change.

Recent conceptual developments in particular might prove difficult to transfer because they have emphasised analyses of interrelated technical and social dynamics *in practice*. The main gist of these studies is that technology (and society) are never ready-made. They are continually in the making. The new studies reject linear models of progress in favour of emergent entanglements of actors and artefacts. Without access to the concrete practices of design, manufacture, use and appropriation, both technology and society remain – at least partially – obscured.[1] Sociology and archaeology thus approach the problem from different directions. Mainstream sociology has long neglected the material constitution of society by focussing on idealist or immaterial norms and values, emphasising human interpretation over material arrangements. Sociological conceptualisations often dematerialise tools and devices into abstract interests, mechanisms or power relations. This bias has been thoroughly criticised in the last 30 years as the unabating interest in questions of materiality continues to fuel discussion (*cf.* Knappett and Malafouris 2008). Archaeology, on the other hand, has little but the durable material artefacts from which to reconstruct prehistoric societies. The question is not how to (re)integrate materiality into social theory but how to extract social relations in the past from their material remains. Indeed, archaeology is praised by one of the first sociologists of invention (Gilfillan 1952, 192) for refuting hero-centric views of history and acknowledging the significance of technical inventions in work and war.

Given their sheer number, it is a formidable challenge, perhaps even a futile endeavour, to select and sort out in the confines of one chapter even a fraction of the studies concerned with the myriad relations of technical and social changes. Accordingly, my approach will be to outline a perspective from the sociology of technology and innovation, drawing on its central tenets and elaborating possible contributions it may be able to make to archaeology. My argument is divided into two parts. First, I revisit evolutionary models of innovation that emphasise the emergent and complex nature of social and technical change in contrast to simplified models of linear innovation or progress. Second, I specifically address the creative adaptation of technologies in innovation processes by looking at the concept of diffusion and by considering the role of users. This shifts the burden of explanation away from design issues towards application contexts and essentially breaks up the innovation process into multiple local, and sometimes also contradictory, reinterpretations or reinventions along the assumed pathway of diffusion.

Evolutionary models of social and technical change

First and foremost, evolutionary models of innovation attack two prominent myths of social and technical change: technological determinism and linear progression (*cf.* Basalla 1988; Ziman 2000).[2] Both these myths represent simplistic reductions of the complex and emergent processes constituting socio-technical change. A closer look at the literature, however, reveals that ideas strongly associated with technological determinism and linear progress are actually quite hard to find in innovation studies. Rather than *bona fide* positions, they are more often used to describe tendencies in the development of innovations or to provide a rhetorical foil. In other words, both terms figure mainly as a kind of shorthand for denouncing reductionism in the study of technology and society. In combination, technical determinism and linear progress portray social transformation as an inevitable process prompted by uncontrollable technological developments. These ideas have been prominently refuted by sociologists since the dawn of the discipline. One excellent example is the heated debate that broke out at the first Congress of the German Sociological Association in 1910 after Werner Sombart delivered a talk on the interrelation between technology and culture. Max Weber closed his reply by protesting strongly against the idea that anything, be it technology or economics, could be the last or final cause of something else: 'If we look at the causal chain, it sometimes runs from the technical to the economic to the political. At other times, the progression runs from the political to the religious to the economic [...]. Nowhere do we find a final cause (*Ruhepunkt*) [C.S.]' (Deutsche Gesellschaft für Soziologie 1911, 101).

Colum Gilfillan, referred to above, stressed this point in a similar manner by drawing on the insights of archaeology 40 years later: 'The vast development of archaeology and ethnology since Morgan has proved that all sorts of social systems, religions, art, *etc.* can coexist with all sorts of economic systems and all sorts of technology, in preliterate cultures, and presumably in civilizations too' (Gilfillan 1952, 194). Hence, no causal connections between technology and society are likely to be found. Gilfillan was also one of the first scholars to link technical invention with biological evolution by highlighting the conviction that technological invention should be seen as a contingent series of small steps rather than clearly identifiable breakthroughs: 'An invention is an evolution, rather than a series of creations, and much resembles a biological process [...]' (Gilfillan

1935, 5). What is more, the classic linear model of innovation itself has a history that is in no way linear. Godin (2006) recently traced the twisted evolution of the concept and its continual reshaping by divergent interests between science, politics and the economy since the early 20th century. The powerful image of innovation as following a more or less straightforward sequence, from basic to applied research and then to development and finally to diffusion, can be seen as a rhetorical device that only crystallised into a specific meaning by way of its application in different contexts over time. Unlike technological (or social) determinism and models of linear progress, evolutionary understandings of innovation emphasise its contingent nature, arguing that 'it could have been otherwise'. Inventions are not simply adopted; complex and situated judgements often lead to 'retarded' appropriation and local reconfiguration by adopters. I will go into more detail on some of these points below.

Approaches to evolutionary thinking

An evolutionary perspective in innovation studies cannot be described as a uniform approach. Rather, it draws on different lines of thought and shares some basic assumptions. It is prominent in fields such as economics, history and sociology. The lowest common denominator is – as I have just indicated – a firm rejection of deterministic and linear models of innovation.

Evolutionary thinking in economics has been used especially to counter simplistic neo-classical notions of profit-maximising actors and economic equilibria. In 'search of a useful theory of innovation', Nelson and Winter (1977) developed an 'evolutionary theory of economic change' (1982). Their approach underscores the fact that innovation necessarily entails uncertainty, meaning that actors do not have clear criteria by which to evaluate their choices. Moreover, actors do not simply orient their behaviour towards maximising profit but have muddled interests and bounded rationalities. Rather than being rational agents of change, actors in organisations tend to stick to organisational routines – or as Nelson and Winter put it, organisations are 'much better at changing in the direction of "more of the same" than they are at any other kind of change' (1982, 10). Their argumentation, in a nutshell, is that economic organisations and their tendency for self-preservation are best suited to stable environments, ones that are not constantly perturbed by innovation. Yet the economic environment of modern societies is one of constant change, driven by technical invention and economic competition. Evolutionary economists thus target the theoretical shortcomings of neoclassical economics under the conditions of industrial capitalism. Their understanding of innovation emphasises three key issues: first, the continuous disequilibrium of modern economic environments, second, the tenacity of organisational routines and the development of technologies according to 'natural trajectories' (1977, 56), and third, a heterogeneous institutional selection environment that is difficult, if not impossible, to predict. Even though this concept is geared to understanding modern economic organisations, *i.e.* firms in capitalist dynamics, the evolutionary perspective holds a more general insight, indicating that (more radical) technical change is not readily adopted but tends to be resisted, making for a considerable time lag between invention and use, if an invention 'catches on' at all.

One of the most salient concepts for understanding the tenacity of previous technical choices in evolutionary economics is that of path dependence (David 1985; Arthur 1989). Drawing on the historical example of the QWERTY keyboard layout, David argues that suboptimal technological arrangements can emerge and stabilise to the point of near irreversibility. This happens not despite, but because of single optimising choices of economically oriented actors, *i.e.* the sum of individual optimisations does not necessarily lead to collective optimisation. The *de facto* standard of the QWERTY layout is 'locked-in'. It is held in place by self-reinforcing dynamics such as technical interrelatedness (between technical artefact and human skill), economies of scale (decreasing costs for standardised products) and the quasi-irreversibility of investments (costs associated with switching systems) despite the fact that supposedly better keyboard layouts like the Dvorak Simplified Keyboard have been available since the early 1930s. David and Arthur emphasise that markets do not automatically select the best technological option (as neoclassical reasoning suggests) but one that suffices at the time and directs the course of further progress. To unfold its full analytical potential, the original notion of path dependence essentially requires modern market conditions; however, the concept has also been transferred and adapted to issues of institutional continuity and change (Streeck and Thelen 2005). Without going into greater detail, it suffices for our present purposes to establish that, in line with evolutionary economics, path dependence mainly counters voluntarist notions of technical and social change. Change largely happens 'behind the backs' of actors, despite their efforts to exercise control. From the perspective of evolutionary economics, technical and social transformations are emergent phenomena that cannot be reduced to rational actors, technical effectiveness or economic efficiency; they unfold from historical and cultural contingencies that slowly stabilise over time. In evolution and in economics, entities compete for scarce resources and this competition is analysed on the level of populations, not individuals.

Other fields of research, such as the history and sociology of technology, have also borrowed evolutionary perspectives to understand socio-technical innovations in ways that reject the linear model. These approaches do not use the term 'evolution' as prominently as evolutionary economics, but there are some resemblances.[3]

Historical research on large technological systems like electricity has emphasised their evolution as an interdependent interlocking of technical and social components (Hughes 1987). The systems perspective highlights the fact that (modern) technical progress emerges from a near 'seamless web' (Hughes 1986) that cuts across the supposed dichotomies between science, technology, economics and politics, connecting heterogeneous entities such as 'physical artefacts, mines, manufacturing firms, utility companies, academic research and development laboratories, and investment banks' (Hughes 1986, 287). In order to manage such a system, entrepreneurs – Hughes chooses to focus on Thomas Edison – have to weave together all these entities into a functioning whole. Like evolutionary economics, Hughes explicitly targets modern industrial change. The term 'evolution' thus resonates strongly with more general ideas of the contingent emergence, successive expansion and increasing irreversibility of socio-technical systems. As systems mature and expand, they lose some – or most – of their initial flexibility and acquire momentum (Hughes 1986, 76) to the point of near irreversibility or lock-in. If we look at Hughes's argument from a more theoretical angle, he seeks to refute one-sided conceptions of either technological or social reductionism. The evolution of the system does not strictly follow technical requirements, nor is it shaped by social concerns alone. Instead, technology and society co-evolve (and expand) by mutually stabilising each other.

Last but not least, evolutionary thinking has been employed in the sociology of technology. One prominent approach is the social construction of technology (SCOT, Pinch and Bijker 1984). Again, these scholars reject a linear model of innovation in favour of an evolutionary or multidirectional model of variation and selection (Pinch and Bijker 1984, 411). Their main criticism is levelled at a functionalist bias arising from the study of mainly successful innovations. This leads to a distorted perception in which the individual stages of technological development appear in retrospect to follow naturally from one another. In short, *explanans* and *explanandum* are collapsed into the seemingly self-explanatory category of 'success'. Like other studies regarding the 'social construction of X', the SCOT approach seeks to open up narratives of normalisation by showing how things could have turned out differently. Its main emphasis lies, first, on demonstrating the 'interpretative flexibility' (Pinch and Bijker 1984, 421) of a technological solution and, second, on showing how closure is brought about through social negotiations. SCOT thus deflates functionalist notions of success achieved via technical effectiveness and brings social processes into the picture. Such processes become especially evident when one looks at technical controversies about what counts as a 'successful' design. Pinch and Bijker use the historical example of the bicycle to show how previously marginalised social groups (such as women) gain relevance in the transition from the high-wheeled penny-farthing or ordinary bicycle to the low-wheeled safety bicycle in the late 19th century. It is not only the shape of the bicycle that is at stake but the social position of women and other groups that were effectively excluded from riding high-wheeled ordinaries.

The evolutionary approaches in economics and the history and sociology of technology that I have briefly revisited indicate that evolutionary concepts employed to understand technical and social change are not strictly modelled on biological evolution. Instead, they draw on a general notion of biological evolution to illustrate the fact that technical change and social change are inherently unpredictable and uncontrollable. Like co-evolving species, technological solutions and social structures emerge out of complex interdependencies in processes of variation, selection and retention. They are not created *ex nihilo* by omniscient actors in a linear fashion but follow from a steady sequence of contingent mutual adaptations. Of course, the dynamics of variation, selection, and retention differ significantly for biological and socio-technical evolution and this is where the analogy ends. But this abstract conception does provide us with some insights as to what remains when we subtract the distinctly modern elements from evolutionary innovation studies.

First, technical progress should not be seen as self-evident. The advantage of one technology over another is typically conflated with, and contested in, countless social interests and negotiations. Like biological mutations, radically new technologies may be conceived of as 'hopeful monsters' (Goldschmidt 1933, 547), *i.e.* their survival is far from certain and depends on finding a suitable niche in the selection environment.

Second, technological change more often than not equals cultural change. With each new technology come new skills and social structures that co-evolve over time. What comes first is still a matter of debate and a co-evolutionary perspective refutes both social and technical determinism. However, we can observe a tendency to emphasise the relative rigidity and durability of existing socio-technical structures in the face of technical change.

Third, the origins of novelty are difficult, if not impossible, to isolate. Rarely can it be attributed to a single act of creation; it is much more aptly described as a sequence of interdependent variations that slowly stabilises over time. In the course of this process, social identities and technical functionalities are mutually configured until a (temporarily) stable inter-definition – or translation (Callon 1986) – of all involved entities has been achieved. Finding a clear distinction between the old and the new is thus a matter of scale, because the closer we look, the more continuity we are likely to find.

Fourth, as technical evolution does not follow a simple linear pattern from creative invention to social diffusion, the later stages of selection and retention acquire greater explanatory power in the understanding of innovations.

The fate of a new technology lies only partially with its creators, more with its users and the way in which it is appropriated and adapted, redefined and domesticated. Innovation studies have thus highlighted not only the successful diffusion, but also the disuse, discontinuation, repurposing, or abandonment of novel technologies (*cf.* Law and Callon 1992).

Do these four insights hold when we subtract modernity from the innovation-equation, or are they still tied to modes of industrial production and abundant technological choices? It is indeed difficult to jettison the modernist legacies they carry, but what are the indicators that non-modern socio-technical change was less complex? Which assumptions do we have to make in order to argue for a more functionalist mode of innovation? Why would we assume that people in prehistoric societies were more receptive to technological change than their modern counterparts? Where do we situate the sources of novelty if we cannot easily fall back on institutionalised settings such as industrial laboratories? I must admit that as a sociologist who normally studies modern high-tech settings, I do not feel qualified to reply to these questions. To avoid being completely lost for an answer, I will draw on some additional literature to argue for possible continuities between non-modern and modern forms of social and technical change.

Revisiting models of diffusion

Two of the main questions to be answered concern, first, the source of novelty and, second, the modes of diffusion. Because the first question is very tricky, I will start by addressing the second, which also bears on a fundamental sociological question: the propagation of social change. How do new ideas, artefacts and practices spread and find acceptance? On the one hand, innovations seem to diffuse at an increasing pace in modern societies. On the other, each innovation disrupts existing habits and practices.

The single most prominent concept in studying the diffusion of innovations is the so-called S-curve model of diffusion. In his influential book *Diffusion of Innovations*, Rogers (2003[1962]) refers to the early French sociologist Gabriel Tarde (1903[1890]) as one of the first proponents of an S-curve to describe adoption processes for innovations. Put briefly, the S-shaped diffusion curve denotes a relatively slow rate of adoption in the early phases of an innovation. This rate becomes more rapid as the majority of people follow suit, only to decelerate again in later stages. Social explanations for this pattern typically invoke an innovative elite (Tarde called them 'opinion leaders') with a capacity to induce imitative behaviour in others. For Tarde, this was the most fundamental process at the core of all societal order. Some individuals gain superiority over others, not necessarily through force, but by prestige (Tarde 1903, 78) and are subsequently imitated by others, as children imitate their parents. The diffusion of novelty, then, not only has an internal dynamic (slow–fast–slow) but also a specific direction (from prestigious positions into the mainstream) and a fundamental mechanism (imitation). The adoption of a novel idea, artefact or practice is explained not by the novelty itself but by the social asymmetry inherent in society and a desire to copy those who are 'better off'. The S-curve neatly depicts this top-down model of diffusion through imitation. As Tarde points out (Tarde 1903, 127), it also guards against a misreading of diffusion as a constantly accelerating process. In its basic form, S-curve diffusion can be 'viewed thus as an expansion of waves issuing from distinct centres' (Tarde 1903, 109) which expand over time. It is important to note that this description of distinct wave-emitting centres does not specify where these centres can actually be found. One of Tarde's main examples concerns the diffusion of religious ideas through prophets and apostles, who initially tend to be in marginal rather than central positions. Centrality can thus be more an effect of diffusion rather than its primary cause, which brings us closer to answering the first question about what can be understood as the source of a novelty.

In his overview of diffusion studies, Rogers (2003[1962]) builds upon Tarde's early ideas and, among other insights that cannot be reported here, argues that more radical novelties are more likely to stem from marginal groups and especially from risk-taking individuals: 'The salient value of the innovator is venturesomeness. He or she desires the hazardous, the rash, the daring, and the risky' (Rogers 2003[1962], 283). This figure stands in contrast to the early adopters, who are likely to be better integrated into local social structures. As Rogers puts it: 'Whereas innovators are cosmopolites, early adopters are localites' (Rogers 2003[1962]), 248). In general terms, we could say that social pressure to conform rises with a more central position in a given social group, so marginal figures or positions are more likely to champion radical novelties. This argument has been strongly supported by subsequent research, for instance by Granovetter's (1973) emphasis of the 'strength of weak ties' or Tushman and Anderson's (1986) explanation of technological evolution and the preference among companies for technological change that is 'competence-enhancing' rather than 'competence-destroying'. Rogers, Granovetter, Tushman and Anderson, however, all explicitly target modern innovations. Rogers' innovator, except maybe for the prophet, is a distinctly modern figure. Granovetter is concerned with the diffusion of innovation in more or less close-knit societies, while Tushman and Anderson argue from the perspective of the capitalist firm.

Central to their arguments is the uncertainty that attends the introduction of novelty and the question of who is willing to risk a major change and how much they are prepared to put at stake. This would not be such a problem if the technology's superiority were evident. But, as many

innovation studies have shown, this is very difficult to judge in advance. It depends on the specific trials of strength, *i.e.* criteria of judgement and valuation that are employed to determine superiority: exchange, competition, conflict, cooperation, *etc.* One of the main insights from evolutionary innovation studies is that technical superiority is hardly ever self-evident, *i.e.* it cannot be reduced to technical criteria. Numerous considerations play their part and choices often only become seemingly obvious in retrospect. We can thus assume that uncertainty was a function of novelties before modern innovation dynamics arrived, and that novel artefacts, for instance, are essentially indeterminate until stabilised in practices of use.

In short, diffusion models highlight a specific dynamic of innovations that can be represented by an S-shaped curve to describe the spread of novel artefacts or ideas. In addition, they emphasise that if a novel artefact or idea is considered more radical with respect to the existing arrangements, it is more likely to start in marginal positions and then diffuse inwards towards the centre, where the social pressure for conformity is higher. This may subsequently cause a previously marginal position to become mainstream, as Tarde pointed out for the diffusion of religious ideas.

Diffusion research thus gives two short answers to the questions posed above. The source of novelty depends upon its perceived radicality. The more radical it is, the more likely a novel idea or artefact will be found on the margins due to social conformity in the centre and a subsequent preference for competence-enhancing rather than competence-destroying socio-technical change. This boils down to the idea of a conservative convergence within central positions and the related observation in evolutionary economics that, to quote again, organisations are 'much better at changing in the direction of "more of the same" than they are at any other kind of change' (Nelson and Winter 1982, 10). The mode of diffusion, however, is largely independent of the source of novelty. It is often understood as a form of imitation through which novel ideas and artefacts are adopted by an increasing number of people – or not, if an innovation fails. In either event, the idea that a novel idea or artefact remains unchanged in the process of innovation has been criticised for some time (*e.g.* by Latour 1986). I will use this criticism to highlight the last step in my argumentation and especially to the role users play in innovation processes.

Considering users: adoption, adaptation, appropriation or domestication?

The question of how much agency should be assigned to users over designers defines most of the discussion about the role users play in innovation processes. The linear model of innovation proceeds from a passive user who readily adopts a technical change because of its inherent advantages. Such an understanding of adoption can be largely equated with imitation, as proposed by Tarde's diffusion framework. Users typically acquire a more active role within the concept of adaptation, where novelties are not simply accepted but altered to fit the respective local conditions. Two additional concepts, appropriation and domestication, extend this line of thought. Like adoption and adaptation, these terms are not based on clear-cut definitions and in the literature tend to overlap. Appropriation in its strongest sense means that a novel artefact or idea is integrated into the established order without necessarily changing the existing socio-material structures. In other words, its inherent strangeness is transformed into a set of characteristics deemed appropriate within a pre-existing framework. The notion of domestication runs along similar lines. It was first used in media studies (Silverstone and Hirsch 1992) and was soon taken up within science and technology studies (Sørensen and Lie 1996) to highlight the ways in which users 'tame' new technologies to fit their everyday lives.[4] The most radical version, of course, is the active rejection of a novel artefact or idea; cases of non-adoption and non-use are by no means rare in innovation studies. Plenty of inventions fail, especially when there are various solutions competing for supremacy.

Once again, it becomes apparent that the way we view adoption, adaptation, appropriation, domestication and non-use is largely determined by modern conditions of innovation. The main argument is that novel media and technologies are not passively consumed but actively recreated by the users. This essentially presupposes a society in which designers and users are separated by an advanced division of labour and linked only through anonymous market transactions – to the point of mutual unawareness or ignorance.[5] If the modern distinction between design and use becomes less rigid, however, design, production and use start to merge into each other and the notion of a primarily passive user becomes questionable in itself. But we can also try to rescue some of the more general insights to be gained through a more active conception of the user. One quite important insight is that societies are not homogeneous but consist of a plurality of 'social worlds' (Strauss 1978). Users and designers do not share the same set of assessment criteria. In other words, innovations have to travel across social or cultural boundaries. This was probably also true of non-modern innovations that asserted themselves across long time-spans and distances (Frachetti 2012). And in each social world, a novel artefact or idea would have been interpreted differently. I will try to make this point clearer by briefly revisiting some empirical cases.

Kline and Pinch (1996) use the example of the diffusion of the automobile in rural parts of the United States. Early cars were mostly used in cities and only slowly became accepted outside urban settings. The authors argue that even though the motorcar had acquired its characteristic shape and specific practices of use, it was subject to interpretative flexibility once it travelled outside its established terrain. First of all, rural Americans did not simply welcome the motorcar. They complained about its noise and smell, its high cost and equally high unreliability. Compared to a horse, the early motorcars were essentially useless to them. Only after the affordable and durable Ford Model T became available did the farmers start to become more interested. But they still did not simply adopt the car. Instead of making use of its most obvious feature, mobility, many farmers put the vehicle on jacks, removed the tyres, and used it as a stationary motor for powering appliances such as washing machines (Kline and Pinch 1996, 775). Other conversions included removing the rear seats and turning the back section of the car into a loading space for farm goods. In terms of the relation between designers and users, car designers lost a sizeable portion of control over the vehicle's usage.

A similar case to the conversion of the car in the rural U.S. is made by de Laet and Mol (2000). They analyse a modern-day technology transfer project between Europe and Africa, in which a supposedly robust water pump for Africa is developed in Europe but is not adopted by users in Africa. De Laet and Mol argue that the first version of the Zimbabwe Bush Pump essentially excluded local users from maintenance and repair. Pumps that broke down or got clogged with sand could not be repaired. Later pump versions were crafted with more readily available parts and their nuts and bolts were accessible for local maintenance and repair. De Laet and Mol and call this later bush pump version a 'fluid' technology, because it can be easily adapted to the diverse local needs and conditions of Zimbabwean villages. Even though we might call all bush pumps by a single name, *i.e.* the B-type Zimbabwe Bush Pump, the individual artefacts are likely to differ somewhat in construction and use. It is precisely this indeterminate fluidity that has made the B-type pump a success in contrast to its more closed or rigid predecessors.

Both these cases highlight the fact that technical artefacts are not immutable, invariable entities prescribing specific forms of usage. The more open and flexible a technology is, the more easily it can cross social and cultural boundaries and be appropriated by heterogeneous social worlds. This point helps us to give a more detailed answer to the questions we were looking at earlier about the origin of novelty and modes of diffusion. In a very basic sense, novelty and creativity are not confined to the early phases of an innovation but stretch out along the entire diffusion process. With each step in the direction of the users, we are likely to find some creative appropriation or domestication that involves the mutual transformation of artefacts and established social structures. Instead of a diffusion model based on imitation and originating from a single innovative source, this research suggests that diffusion is an active process of appropriation that extends creative agency to a variety of users or social groups that further the spread of a specific novel artefact by adapting it to their local needs (Latour 1986). Innovation researchers have thus started to take a closer look at seemingly marginal user groups and settings, from amateur clubs to housewives or rural communities. In contrast to the dominant models and their focus on innovative engineers in company laboratories, these groups show how inventive creativity is born of local contingencies through tinkering or bricolage, which is not necessarily inferior to planned engineering but involves sophisticated physical and technical skills and considerable know-how. This becomes especially evident in cases where so-called 'high-tech' is transferred to 'low-tech' settings that often point up the limitations of the technical design by placing it in a different social, technical and natural environment (Akrich 1992).

Conclusion: evolution, innovation, modernity

The concepts and cases encountered in innovation studies repeatedly demonstrate a strong affinity with modernity. As I have tried to show, however, it is possible to extract some more general insights that may also apply to non-modern conditions.

In the studies I have revisited in this article, the main aim is to point out variations of technical and social change by making strong claims against reductionist or deterministic conceptualisations of innovation. Taken alone, neither technical nor social aspects suffice to explain the dynamics of change in modern societies. Modern societies are often described as 'technical civilisations', but it is difficult to imagine that non-modern societies or communities were less dependent on technical applications. Innovation studies, especially in the tradition of actor-network theory, have strongly emphasised the mutual constitution of humans and non-humans on a fundamental level that extends well beyond modernity (Webmoor 2007). But the actor-network approach requires detailed empirical material to trace the socio-technical connections; hence it can only follow these assemblages as far as its methodology allows. What can be preserved is an understanding of innovation that does not rely on functionalist or reductionist concepts but that remains sensitive to heterogeneous configurations of technical and social structures that may co-exist and continue to vary over time. Such localised configurations can be imagined as 'nonuniform institutional alignments' (Frachetti 2012) extending back much further than modernity.

We may also assume that technical and social change have always been a matter of controversy. If we are prepared to accept a second cue from evolutionary economics and innovation sociology, it would probably be that incumbents, *i.e.* the established and powerful actors in a given field, are more likely to be conservative and opposed to radical change than those in marginal positions. Gradual change then happens along the 'technological

trajectories' contained within a 'technological paradigm' (Dosi 1982) and maintains the status quo, whereas large-scale transformations – or revolutions – are typically not initiated by those in power. A sensitivity to marginal and subordinate positions in the innovation process helps to uncover the manifold material and social negotiations, contestations, adaptations, meanderings and choices, made or abandoned, that essentially constitute an innovation in the making.

I would like to end this paper on a third note: the relations between stability and change. Evolutionary thinking conceives of the world as caught up in a continuous flux, as a place that never stands still. In all cases of innovation, however, we find stabilising forces in technical artefacts, human actors and social structures. The difficult challenge that remains is to conceive of the interrelations between social and material configurations on the one hand and the dynamics of stability and change on the other. Technological novelty is not the primary motor of change, just as social structures are not the primary agents of stability, or vice versa. Only rigorous empirical work can disentangle the intimate ties binding technical and social change. This will provide a basis for further conceptual elaboration and perhaps also constitute another arena in which sociology and archaeology can join forces.

Notes

1 A similar tendency can be observed in the anthropological and ethnological study of material culture. Lemonnier (1986), for example, argued in favour of increased observation of the use and the socio-material interrelations of technical systems rather than mere descriptions of technical objects on their own.

2 The following evolutionary approaches must not be misunderstood as an 'evolutionist' perspective, which presupposes a linear unfolding along an inherent progress logic, *e.g.* a linear progression from rudimentary forms to more evolved, refined and effective types in the development from stone to metal axes (Montelius 1903). In contrast, evolutionary approaches in innovation studies seek to account for situated appropriations, diverse interpretations and multiple uses of novel artefacts as well as unexpected continuity (lock-in) in apparently sub-optimal solutions.

3 There are also marked differences between the large technical systems approach in the history of technology and the social constructivist approach in the sociology of technology. I will not address them in greater detail here. Suffice it to say that the social constructivist approach tends to emphasise social factors in the shaping of technologies, whereas the systems approach (much like actor-network theory) seeks to privilege neither society nor technology in its analysis.

4 Rogers (2003[1962], 180) uses the term 're-invention' to point out that the diffusion of an innovation should not be reduced to exact copying or imitation, but that novel ideas and artefacts are likely to change as they spread across time and space.

5   Since the 1980s, an increasing number of user-centred or participatory design approaches have been formulated in order to bridge the gap between design and use (*cf.* Oudshoorn and Pinch 2003).

References

Akrich, M. (1992) The De-Scription of Technical Objects. In W. E. Bijker and J. Law (eds), *Shaping Technology – Building Society. Studies in Sociotechnical Change*, 205–224. Cambridge, MA, MIT Press.

Arthur, W. B. (1989) Competing Technologies, Increasing Returns, and Lock-In by Historical Events. *Economic Journal* 99, 116–131.

Basalla, G. (1988) *The Evolution of Technology.* Cambridge, Cambridge University Press.

Callon, M. (1986) Some Elements of a Sociology of Translation. Domestication of the Scallops and the Fishermen of Saint Brieuc Bay. In J. Law (ed.), *Power, Action and Belief: A New Sociology of Knowledge?* 196–233. London, Routledge.

David, P. A. (1985) Clio and the Economics of QWERTY. *American Economic Review* 75/2, 332–337.

Deutsche Gesellschaft für Soziologie (ed.) (1911) *Verhandlungen des Ersten Deutschen Soziologentages vom 19. bis 22. Oktober 1910 in Frankfurt a. M.* Tübingen, Mohr Siebeck.

Dosi, G. (1982) Technological Paradigms and Technological Trajectories. *Research Policy* 11, 147–162.

Frachetti, M. D. (2012) Multiregional Emergence of Mobile Pastoralism and Nonuniform Institutional Complexity Across Eurasia. *Current Anthropology* 53/1, 2–38.

Gilfillan, S. C. (1935) *The Sociology of Invention.* Chicago, IL, Follett Publishing Co.

Gilfillan, S. C. (1952) Social Implications of Technical Advance. *Current Sociology* 1/2–3, 191–207.

Godin, B.(2006) The Linear Model of Innovation. The Historical Construction of an Analytical Framework. *Science, Technology & Human Values* 31/6, 639–667.

Goldschmidt, R. (1933) Some Aspects of Evolution. *Science* 78/2033, 539–547.

Granovetter, M. (1973) The Strength of Weak Ties. *American Journal of Sociology* 78/6, 1360–1380.

Hughes, T. P. (1986) The Seamless Web. Technology, Science, etcetera, etcetera. *Social Studies of Science* 16/2, 281–292.

Hughes, T. P. (1987) The Evolution of Large Technological Systems. In W. E. Bijker, T. P. Hughes and T. J. Pinch (eds), *The Social Construction of Technological Systems,* 51–82. Cambridge, MA, MIT Press.

Kline, R. and Pinch, T. J. (1996) Users as Agents of Technological Change. The Social Construction of the Automobile in the Rural United States. *Technology and Culture* 37/4, 763–795.

Knappett, C. and Malafouris, L. (ed.) (2008) *Material Agency. Towards a Non-Anthropocentric Approach.* New York, Springer.

Laet, M. de and Mol, A. (2000) The Zimbabwe Bush Pump. Mechanics of a Fluid Technology. *Social Studies of Science* 30/2, 225–263.

Latour, B. (1986) The Powers of Association. In J. Law (ed.), *Power, Action and Belief: A New Sociology of Knowledge?* 264–280. London, Routledge.

Law, J. and Callon, M. (1992) The Life and Death of an Aircraft. A Network Analysis of Technical Change. In W. E. Bijker and J. Law (ed.), *Shaping Technology/Building Society. Studies in Sociotechnical Change*, 21–52. Cambridge, MA, MIT Press.

Lemonnier, P. (1986) The Study of Material Culture Today. Toward an Anthropology of Technical Systems. *Journal of Anthropological Archaeology* 5/2, 147–186.

Marx, K. (1867) *Das Kapital. Kritik der politischen Ökonomie 1. Der Produktionsprozess des Kapitals.*1st ed. Hamburg, Otto Meissner.

Montelius, O. (1903) *Die älteren Kulturperioden im Orient und in Europa 1. Die Methode.* Stockholm, self-published.

Nelson, R. R. and Winter, S. (1977) In Search of Useful Theory of Innovation. *Research Policy* 6/1, 36–76.

Nelson, R. R. and Winter, S. G. (1982) *An Evolutionary Theory of Economic Change.* Cambridge, MA, Belknap Press.

Oudshoorn, N. and Pinch, T. J. (ed.) (2003) *How Users Matter. The Co-Construction of Users and Technology.* Cambridge, MA, MIT Press.

Pinch, T. J. and Bijker, W. E. (1984) The Social Construction of Facts and Artefacts. Or How the Sociology of Science and the Sociology of Technology Might Benefit Each Other. *Social Studies of Science* 14/3, 399–441.

Rogers, E. M. (2003) *Diffusion of Innovations.* 5th ed. New York, Free Press.

Silverstone, R. and Hirsch, E. (ed.) (1992) *Consuming Technologies. Media and Information in Domestic Spaces.* London, Routledge.

Sørensen, K. H. and Lie, M. (ed.) (1996) *Making Technology Our Own. Domesticating Technology into Everyday Life.* Oslo, Scandinavian University Press.

Strauss, A. L. (1978) A Social World Perspective. In N. K. Denzin (ed.), *Studies in Symbolic Interaction* 1, 119–128. Greenwich, Jai Press.

Streeck, W. and Thelen, K. (2005) Introduction. Institutional Change in Advanced Political Economies. In W. Streeck and K. Thelen (eds), *Beyond Continuity. Institutional Change in Advanced Political Economies*, 3–39. Oxford, Oxford University Press.

Tarde, G. (1903) *The Laws of Imitation.* New York, Henry Holt.

Tushman, M. L. and Anderson, P. (1986) Technological Discontinuities and Organizational Environments. *Administrative Science Quarterly* 31/3, 439–465.

Webmoor, T. (2007) What About 'One More Turn After the Social' in Archaeological Reasoning? Taking Things Seriously. *World Archaeology* 39/4, 563–578.

Ziman, J. (ed.) (2000) *Technological Innovation as an Evolutionary Process.* Cambridge, Cambridge University Press.

Chapter 3

From Counting to Writing: The Innovative Potential of Bookkeeping in Uruk Period Mesopotamia

Kristina Sauer[1]

[Die Schrift] zu entziffern heißt vielmehr, ihre sozialen Funktionen zu rekonstruieren, die Dynamik des Entwicklungsprozesses der Schrift zu studieren und die Folgen auf das Denken und den Umgang mit Informationen zu ermitteln, die mit der Schriftentwicklung verbunden waren.

(Nissen *et al.* 2004, X).

The Uruk Period: a brief sketch

The Late Chalcolithic in the Near East, the so-called Uruk Period (*c.* 3800–3000 BC), sees the formation of previously unknown urban complexity in southern Mesopotamia and the adjacent Susiana, involving intricate, extensive settlement and communication networks (Fig. 3.1). These networks were not only limited to the south, they also extended northwards along the major river courses into the Syrian Plain and the Taurus Piedmont and eastwards into the Iranian Plateau (*cf.* Adams 1981; Algaze 1993). They are understood as the expression of cataclysmic societal processes which probably led to the consolidation of an early state society by the end of the Uruk Period (most recently Algaze 2013; furthermore Algaze 1993; Rothman 2001; Stein and Özbal 2007). Not only the dissemination of material culture and technical and economic improvements but also the transmission of knowledge and cultural practices created an extensively cross-linked transregional cultural landscape (*cf.* Algaze 2013; Rothman *et al.* 2001; Stein 1999; Stein and Özbal 2007).

The Uruk society is characterised by social stratification, economic and technological differentiation and specialisation, plus institutionalised resource management and the development of elaborate administrative structures (*cf.* Algaze 2001; Wright 2001; Nissen *et al.* 2004). These aspects are reflected in the creation of public spaces and monumental architecture (*i.e.* in Uruk; *cf.* Eichmann 2013 for a recent discussion and further references) as well as in the deployment of bureaucratic mechanisms, *i.e.* the emergence of writing and the introduction of the cylinder seal as a means of supervision (*cf.* Englund 1998 for a comprehensive study on the emergence of writing). Increasing specialisation and standardisation is, for instance, reflected in the ceramic repertoire, which is distinguished by mostly mass-produced wares (*cf. inter alia* Helwing 2002; Bachmann 2011; Sürenhagen 2014). The social, religious, and ritual practices of that time can be traced through artistic products such as statues, carved stone vessels, the images on cylinder seals, and the earliest known texts.

The archaeological complex of the Uruk culture was first identified in the 1930s during excavations in the ancient city of Uruk (*cf.* Butterlin 2014–2016). Soon it became clear that its sphere of influence extended far beyond southern Mesopotamia into the north and also had strong parallels in the Susiana material, a phenomenon for which the term 'Uruk expansion' was coined, postulating an expansion of the technologically superior, dominant, and colonialist Uruk culture (*cf.* Algaze 1989; 1993).

In the past decades, research in Syria, the Taurus Piedmont, and Iran has led to a more sophisticated

Figure 3.1: Map of Mesopotamia illustrating sites mentioned in the text (© K. Sauer).

understanding of the complex processes taking place in late Chalcolithic Mesopotamia (*cf.* Rothman *et al.* 2001; Algaze 2013 with further references). It has become clear that the process of so-called 'Uruk expansion' was much lengthier and more diversified than initially assumed, commencing in the Middle Uruk Period around 3700 BC (*cf.* Table 3.1) and with origins reaching back as far as the Ubaid Period in the 5th millennium BC (*cf.* Sürenhagen 1986; Wengrow 1998; Wright and Rupley 2001; Algaze 2013).

Analysis of these highly interactive and interconnected cultural spheres reveals paths facilitating the mediation and appropriation of innovations and knowledge. But the opposite can also be observed: rejection of innovations. In contradiction to Algaze's initial assumption, ongoing research has revealed that in some centres – independently of, and prior to, contact with the Uruk horizon – local complex societies emerged with administrative systems of their own and a high degree of receptiveness to innovation (*cf.* Frangipane 2007 for an extensive survey of the administrative mechanisms encountered in Arslantepe). These societies interacted with the Uruk communities but did not necessarily adopt 'Urukean' innovations such as preliterate accounting devices or writing (*e.g.* Arslantepe (*cf.* Frangipane 1997; 2002; 2007) or Tall Brak (*cf.* Oates 2002; Oates *et al.* 2007; McMahon 2009).

Table 3.1: Chronological chart of Mesopotamia in the 4th and 3rd millennia BC.

Date	Period	Administrative Device
4000	Terminal Ubaid	Stamp Seals (StS), Tokens (T)
3900	Early Uruk	StS, T
3700		StS, T, Cylinder Seals (CS),
	Middle Uruk	Hollow Clay Balls, Numerical
3500		Tablets
3300	Late Uruk	StS, T, CS, numero-ideographic tablets, archaic texts (Uruk IV)
3100	Ǧamdat Naṣr	StS, T, CS, archaic texts (Uruk III)
2900		
		Archaic texts from Ur
2700		
	Early Dynastic	Early Dynastic texts from Fāra
2500		and Abu Ṣalabīḫ
2300	Dynasty of Akkad	Old Akkadian texts
2100	Ur III Dynasty	Neo-Sumerian texts

Revolutionary innovations: Information technologies on the verge of literacy

This paper focuses on a specific 'Urukean' innovation – the deployment of intricate bureaucratic mechanisms, such as tokens, bullae, hollow clay balls, and numerical tablets as novel means of supervision, culminating in the emergence of writing – one of the most intriguing and far-reaching innovations of the Uruk Period.

The ability to transform thoughts into images and symbols can be traced back over thousands of years (*cf.* White 1989; Renfrew 2001), although the interpretation of the content they may have transmitted tends to be purely speculative. The situation is different, however, when it comes to inscribed artefacts. The emergence of writing – understood here in its broadest sense as a system of 'codes for visual representation' retaining information (*cf.* Assmann and Assmann 2003, 394) – represents a milestone in the cognitive development of mankind. Script and inscribed artefacts serve as instruments of memory expansion and enable the communication of content through space and time.

Nevertheless, understanding meaning and content as well as language in the form both of image and text presupposes the presence of a recipient (*cf.* Hilgert 2010, 2–3). Writing is used for communication, thus one could say 'where there is writing, there is a reader' (Powell 2009, 13). Accordingly, writing is the product of agreement, the shared assignment of meaning (*cf.* Ott and Kiyanrad 2015, 160). Communication can only take place if the signs used are known, regardless of whether they are letters or other symbols. It is initially irrelevant whether the signs actually represent phonetic elements of a language. For example, the symbol 🚭 'no smoking' is understood almost everywhere in the world, whoever sees it (*cf.* Powell 2009, 19).

Writing systems are therefore cultural artefacts that are not based in nature but in the human mind (Powell 2009, 11). At the same time, they consist of markings on a physical medium, so they are part of the material culture. Script depends on its material basis, its carrier – without the latter, it cannot communicate, it is useless. Writing is also fundamentally material (*cf.* Powell 2009, 13, 18; Piquette and Whitehouse 2013, 1). An analysis of inscribed artefacts as part of the material culture must therefore also extend to the materiality of these artefacts (*cf.* Piquette and Whitehouse 2013). Consideration of the materiality of inscribed artefacts should thus also encompass questions about the nature of script carriers, their characteristics and their origins and about the communicating agents and their knowledge and skills (*cf.* Hilgert 2010, 116; Piquette and Whitehouse 2013, 6).

The material evidence: Early means of administrative control

The invention of writing is a quantum leap in the evolution of mankind. But from an Urukean perspective, it was first and foremost an essential step in assuring the much-desired efficiency of administrative systems (*cf.* Nissen *et al.* 2004, 55).

The oldest known texts noted down on clay tablets were discovered in the ancient city of Uruk in present-day southern Iraq, most of them in the temple precinct of Eanna (Fig. 3.2) dedicated to the goddess Inanna/Ištar (*cf.* Falkenstein 1936; Englund 1998). Dating them accurately is difficult, since these early specimens all stem from debris and none have been found *in situ* (for detailed discussion of the find-spots see Englund 1998; Sauer and Sürenhagen 2016). The *communis opinio* favours a dating to the Eanna IV level, somewhere around 3300 BC (*cf.* Sallaberger and Schrakamp 2015, 55).

The homogeneity of the archaic corpus and its conventionalised form is remarkable, both in terms of the writing medium and of the script itself. This has prompted historians to assume that there must have been precursors that have either not survived or have yet to be discovered (*cf.* Green 1981; Postgate *et al.* 1995). But precursors of the earliest texts have been found in the archaeological record from Uruk itself and from numerous other Late Uruk sites. They take the form of an ever-increasing number of involved administrative tools, namely cylinder seals, tokens, bullae, hollow clay balls, and numerical tablets. These seem to have been utilised not only for the notation of internal procedures, but possibly also for the control of commodity exchange over long distances. They all share common features such as the use of clay for their manufacture or techniques for marking symbols by means of grooving and by impressing symbols with a tool.

The fact that writing appears to have developed within a relative short time span points to a degree of urgency in

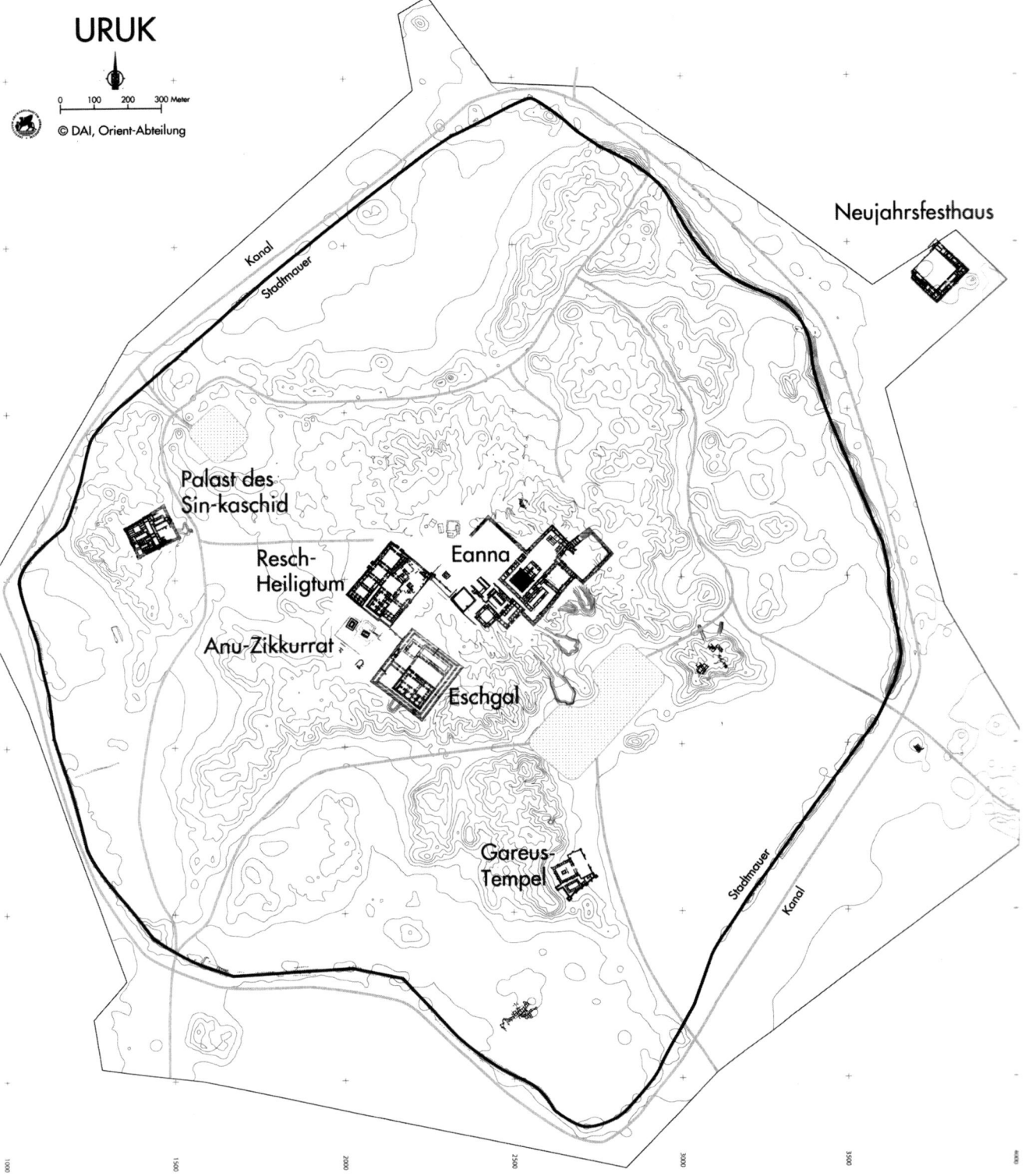

Figure 3.2: The ancient city of Uruk (© DAI Orient-Abteilung).

the need for such a medium (*cf.* Michalowski 1993, 54, 56; Glassner 2003, 4).

Preliterate accounting techniques

Whereas the sealing of objects is not an innovation from the Uruk Period – stamp seals as a means of verification had been in use for a long time (the oldest known seals and sealings date back to the aceramic Neolithic in the 8th millennium BC, *cf.* Collon 1987; von Wickede 1990) – the cylinder seal definitely is. Appearing around 3500 BC, the origins of the cylinder seal still remain unknown (*cf.* Pittman 2013, 324). Up to now, the earliest known evidence for the use of cylinder seals, on sealings made out of clay, has not been found in southern Mesopotamia but in Iran (Tepe Sharafabad: Wright *et al.* 1981, 279, figs 6–8; Susa: Le Brun 1999, 140) and Syria (Tall Brak: Oates and Oates 1993, 176, figs 31, 44; Tall Sheikh Hassan: Boese 1995, 95, fig. 8b–d).

The new cylindrical shape of the seal offered greater communication potential since there was now room for broader narratives on the surface (*cf.* Ross 2012, 305; Pittmann 2013, 324–325.). As a bureaucratic tool, the cylinder seal, or more precisely the image of the seal, could thus transport 'literal messages' (Pittmann 2013, 325) pertaining to economic units involved in transactions and the institutions participating. However, some important information is absent on cylinder seals: the objects concerned and the actual quantities involved. This information was provided and stored by means of the different kinds of preliterate accounting tools referred to earlier.

Tokens

Tokens (Fig. 3.3) are small objects mostly 1 to 2 cm in size, mainly made of clay, occurring in a wide range of different forms and ornamentation (*cf.* Schmandt-Besserat 1992 for the most extensive investigation on these objects; for critical reviews of her work and the objects themselves, *cf. inter alia* Oates and Jasim 1986; Strommenger *et al.* 2014; Sauer and Sürenhagen 2016). Like seals, tokens are not innovations from the Uruk Period itself. They have also been found in Neolithic contexts, the earliest known examples dating back to the 8th millennium BC (*cf.* Schmandt-Besserat 1992, 17). However, in the Late Chalcolithic and especially in the Uruk Period, this instrument becomes more sophisticated. Denise Schmandt-Besserat (1992) distinguishes 'plain' and 'complex' tokens, 'plain tokens' referring to objects with simple geometric forms such as spheres, squares, cones, and the like (*cf.* Fig. 3.3a–d), and 'complex tokens' exhibiting the same forms but now featuring grooves, punctuation, and appliques (*cf.* Fig. 3.3e–h). It has proved useful to distinguish one other group of tokens, the so-called 'naturalistic' variety, (*cf.* Fig. 3.3i–k), which appear to represent actual objects such as vessels, animals, tools *etc.* These first occur during the Uruk Period (*cf.* Sauer and Sürenhagen 2016).

For their manufacture, tokens required no a great knowledge. The simplest forms are generated by messing around with clay (Schmandt-Besserat 1992, 30). More precise forms were obtained by kneading and rolling small pieces of clay between the palms of the hands or compressing clay between the fingertips. Finally, the pieces were air-dried or baked (the latter mostly the case with differentiated tokens, *cf.* Sauer and Sürenhagen 2016).

The interpretation of these small implements as counting devices has been widely accepted. This goes hand in hand with the way they were stored. Perforated tokens could be tied with string or enclosed in a container (Schmandt-Besserat 1992, 109). Both alternatives ensured that groups of tokens could be securely held together and the transaction in question sealed.

Bullae

Bullae are another instrument of administrative control (Fig. 3.4). Like tokens, bullae are not an innovation from the Uruk Period, they can be traced back to the Ḥalāf Period in the 6th millennium BC (*cf.* Mallowan and Rose 1935, 98f., pl. IXb: the specimens being called 'clay lumps'). The term bulla denotes mostly oblong biconoid solid clay objects grouped around a string or knot (see Fig. 3.4b; Schmand-Besserat 1992, 109; Rittig 2014, 347). In most cases, the surfaces of these clay implements were covered with sealings and, rarely, notational marks (see Fig. 3.4a; further, in Ḥabūba Kabīra-South: M II:164a–b, Rittig 2014, 200.10, 209.7; notations on bullae have also been observed in Susa: Amiet 1972, no. 599).

These implements are thought to have been tags tied to commodities (*cf.* Mallowan and Rose 1935; Amiet 1972, 70; Otto 2009–2011, 470–471), but since they are frequently found together with pierced tokens, Schmandt-Besserat (1992, 109–110) has proposed that they may have secured strings of tokens (Fig. 3.4c).

Hollow Clay Balls

Hollow clay balls (Fig. 3.5), occasionally also called clay envelopes (*cf.* Schmandt-Besserat 1980, 1992; Englund 1998), were another possibility for storing tokens. In this case, the counting devices were embedded in hollow clay balls which were then closed and (sometimes) sealed (*cf.* Fig. 3.6b–d). These devices first appear in the archaeological record shortly before first instances of archaic writing and represent a true Urukean innovation (Nissen *et al.* 2004, 48.). The interpretation of these objects as purely administrative was first advanced by Pierre Amiet (1966, 70). He interpreted the '*bulles sphériques*' known from Susa and Uruk as instruments serving the need to control trade and the exchange of commodities. His theory was further substantiated by a find from Nuzi (modern Yorgan Tape in the eastern Tigris region of Iraq, *cf.* Fig. 3.1). It was a hollow clay ball – albeit considerably younger in date – containing 48 tokens pertaining to a sheep transaction. The clay ball is covered

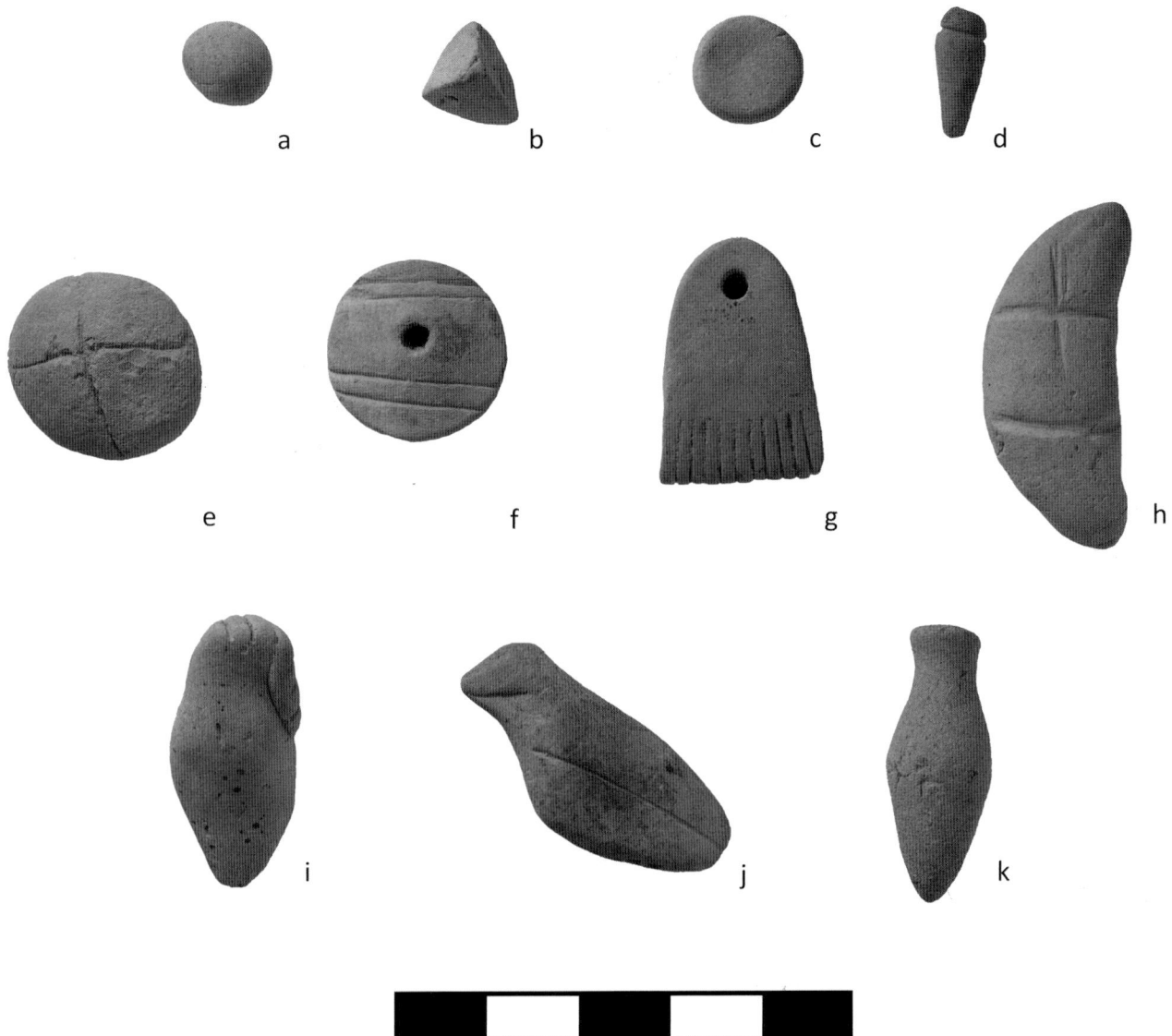

Figure 3.3: Compendium of different tokens: (a–d) undifferentiated tokens; (e–h) differentiated tokens; (i–k) figurative tokens (K. Sauer © Uruk-Warka-Sammlung des DAI, Universität Heidelberg).

with an accompanying inscription validating the transaction (*cf.* Oppenheim 1959 for a detailed description of the find).

Like tokens, hollow clay balls were made of clay. Their preparation required the forming by hand of a manageable, roundish to oval clay ball 5 to 7 cm in diameter, which was then hollowed out using the fingers (as evidenced by the finger impressions preserved in the interiors, *cf.* Schmandt-Besserat 1980, 364). The surface was not subjected to any special treatment except for some slight smoothing and sometimes the application of cylinder seals. The flattening of one side can also be observed; this was probably for ease of storage.

In contrast to strings of tokens or simple bags, these clay envelopes represent a certain guarantee for the respective transaction, since the counters themselves are enclosed (Amiet 1966, 70; Nissen *et al.* 2004, 48). In addition to

the enclosed tokens, there may be markings on the surface, though we cannot say much about what was counted or the quantities involved (Nissen *et al.* 2004, 48). This information can only be obtained at a later stage from the early texts. It is interesting to note that hollow clay balls went out of use once the archaic texts started appearing.

The notations on the surfaces are impressed either with a stylus or the fingertips; the tokens themselves could also be impressed (see Fig. 3.5b; see also for instance Sb 5340 from Susa: *cf.* Schmandt-Besserat 1992, fig. 71). Accordingly, it is fair to say that hollow clay balls represent a link with the later textual notations on tablets. However, the content of the hollow clay balls can seldom be determined, since in most cases the balls were either found intact making the content undeterminable, or they were broken and the content lost.[2] Of the more than 200 specimens of hollow clay balls we know of, only 18 specimens carry notations on their surface – mainly small round pinches and larger circles, plus elongated grooves (*cf.* Sauer and Sürenhagen 2016) resembling the arrangement of later numerical notations (*cf.* Damerow 2012, 160). Of these 18 hollow clay balls with markings on the exterior, the content of only 14 is known, and not all the examples display an equivalence between notation and content.

NUMERICAL TABLETS

Numerical tablets (*cf.* Fig. 3.6a) are flat clay lumps impressed with similar numerical notations to those found on hollow clay balls and sometimes sealed. In the archaeological record they appear simultaneously with hollow clay balls. The oldest known example of a numerical tablet stems from Middle Uruk levels at Hacınebi Tepesi, where it was found together with a hollow clay ball (Stein 2001, 289–291).

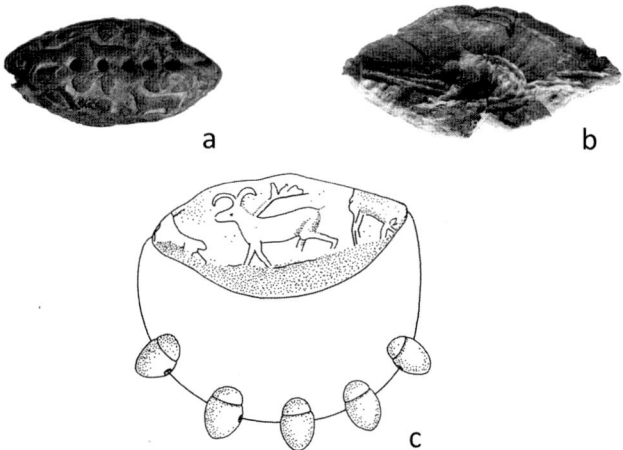

Figure 3.4: Archaic Bullae: (a) Bulla with Notations from Ḥabūba Kabīra-South, M II:125 (Rittig 2014, pl. 205.2); (b) interior view of a broken bulla from Ḥabūba Kabīra-South, M II:140 (Rittig 2014, pl. 208.1); (c) Reconstruction of a bulla with tokens secured on a string (Schmandt-Besserat 1992, fig. 53).

Another numerical tablet from the same period comes from Tall Brak, area CH (Oates 2002, 116).

These tablets are considered the immediate precursors of early writing, since the next logical step was not only to note down the quantities but also the commodities counted. But they also represent a logical step in optimising workflow. Instead of modelling a clay ball, enclosing tokens, closing and sealing the ball, and eventually marking it, one only needed a lump of clay that could be used as a script-bearer. Compared to later clay tablets their size is small, fitting comfortably into the palm of the hand (*cf.* Nissen 1986 and Reade 1992 for a few specimens made of gypsum). The clay was carefully prepared and slurred – a process comparable to the handling of later cuneiform tablets (*cf.* Taylor and Cartwright 2011 for the manufacture of tablets). However, the actual modelling of the tablets was anything but consistent. The surfaces of some specimens were carefully smoothed, the edges rounded, and the shape carefully hand-modelled, whereas others were less carefully processed (*cf.* Schmandt-Besserat 1992, 133). Generally, a diversity of shapes can be observed, from roughly oval, rounded specimens to small rectangular cushion-shaped tablets in the Late Uruk Period (*cf.* Englund 1998, 56–64 for an overview of the different tablet formats). This is a good illustration of the gradual standardisation one would expect at the formative stage of a novel technology.

The surfaces of numerical tablets were only rarely sealed, but they all carry numerical notations (*cf.* Englund 1998; Sauer and Sürenhagen 2016). As already stated, these notations not only resemble the notations on hollow clay balls but also those on the later accounts. The diversity in shape of the numerical tablets is mirrored by a corresponding variety in the forms and arrangements of the notations themselves. Specimens from Ğabal Arūda (*cf.* van Driel 1982, fig. 1a–b), Ḥabūba Kabīra-South (Rittig 2014, pl. 200.1–4) and Susa (*e.g.* Sb 6291: Schmandt-Besserat 1992, fig. 86), for example, carry elongated marks and small circular pinches. Others, from Godin Tepe (*cf.* Hallo 2011,

Figure 3.5: Hollow Clay Balls: (a) W 20987, 10 from Uruk (K. Sauer © Uruk-Warka-Sammlung des DAI, Universität Heidelberg); (b) M II:134 from Ḥabūba Kabīra-South (Schmandt-Besserat 1992, fig. 74); (c) Sb 6350 from Susa (Schmandt-Besserat 1992, fig. 11); (d) MII:133 Ḥabūba Kabīra-South (Rittig 2014, pl. 204.1).

a   b   c

Figure 3.6: Early administrative devices from Uruk: (a) Numerical tablet W 20239; (b) Tag W 20883 (c) Numero-ideographic Tablet W 20498.

fig. 4.43b–c), Susa (*e.g.* Sb 2312: Schmandt-Besserat 1992, fig. 83) and Uruk (*e.g.* W 20239, Englund and Nissen 2001, pl. 101), show wedge-shaped impressions and circles of different sizes. It is only with the appearance of the archaic texts that the numerical notations become standardised and organised (Damerow 2012, 162).

The distribution map (Fig. 3.7) shows clearly that while these accounting devices have been found all over the Near East, the find sites are exclusively Urukean, including Susa (*cf.* Le Brun and Vallat 1978) Ḫabūba Kabīra-South (*cf.* Rittig 2014), Ǧabal Aruda (*cf.* van Driel 1982), and Tall Šeḫ Hassan (*cf.* Boese 1995), or Urukean enclaves such as Hacınebi Tepesi (*cf.* Stein 2001) and Godin Tepe (*cf.* Hallo 2011). Most recently, a sealed numerical tablet has been found in Kani Shaie, a site in the Kurdistan Rgion of Iraq, emonstrating the presence of urukeans in this area as well. (The nature of the urukean presence at the site can not yet be determined; *cf.* Tomé *et al.* 2016). Unlike tokens and sealings, numerical tablets do not figure in local contexts.

The nature and quantity of the commodities thus controlled for cannot be determined. Robert Englund (1998, 49) conjectures that the devices found in Late Uruk sites in Syria and Anatolia contained representations of small numbers of animals and of grain measures, thus serving the bureaucratic needs of a local administration. He bases his assumption on the content of the slightly younger archaic texts, which also deal with intra-site activities. However, not all such devices were produced on the site. According to analysis of the clay, the numerical tablet from Hacınebi Tepesi originated in the Susiana region of south-western Iran (*cf.* Stein 2001, 290), so these devices may have served long-distance trade control purposes as well.

The archaic texts

As noted above, the earliest known examples of cuneiform texts come from the city of Uruk, and their precise dating is difficult. Nevertheless, the corpus is characterised by its strikingly conventionalised form with regard both to the writing medium (clay tablets) and script (proto-cuneiform), which in terms of graphic styles can be subdivided into two stages, Uruk IV and III, roughly corresponding to the building levels in Eanna (*cf.* Marzahn 2013 for the latest review of archaic writing techniques).

The archaic text corpus numbers some 6692 specimens (*cf.* Sallaberger and Schrakamp 2015), and the majority – roughly 90% (Woods 2010, 37) – are exclusively bureaucratic in nature. These written records deal mainly with herding, beer brewing, management of crops, and division of labour. Besides these archaic accounts there is a small group of slightly larger tablets that are organised differently, the so-called lexical lists. Literary texts, such as religious texts and historical accounts or letters, do not figure at all in the archaic corpus (Woods 2010, 34).

NUMERO-IDEOGRAPHIC TABLETS

Numero-ideographic tablets (Fig. 3.6c) carry information in the form of numerical notation and one or two graphemes, thus not only documenting the amount of a certain commodity but also the commodity itself (Englund 1998, 51; Damerow 2012, 160).

The tablets themselves do not show any special surface treatment. They are left plain if not sealed (Englund 1998, 51–52; Nissen *et al.* 2004, 56). According to Robert Englund (1998, 53), it is this kind of tablet that represents the 'missing link between numerical notations […] and the inception of proto-cuneiform'. This combination of numerical notation, grapheme, and sealing 'offered virtually unlimited opportunities for representing structured information' (Damerow 2012, 160). And these opportunities were used extensively, continually spawning new signs for use in the archaic texts.

Figure 3.7: Distribution of Urukean preliterate administrative devices (© K.Sauer).

Tags

Tags are small, perforated, cushion-shaped tablets that carry only a few signs and no numerical marks (see Fig. 3.6b). They were probably designed to hang on a string attached to a container (Nissen *et al.* 2004, 56). The signs used on the tags do not belong to the well-known repertoire of commodities. Some stand for personal or official titles, others for beverages and dried fruits (Englund 1998, 57).

Accounts

Amongst the large group of administrative documents there are different kinds of textual evidence that can be distinguished, ranging from single-entry accounts and tags to a diverse number of ever more complex accounts (*cf.* Englund 1998 for a detailed survey of the corpus). Increasing complexity is observable over time, both in layout and content (Green 1981, 352–356).

The more complex accounts stand out for the special treatment of the tablet surface, namely the division into registers with each chart containing single entries. Large accounts may display a summation on the tablet's reverse, merging the information given on the obverse (for a compilation of the archaic texts from Uruk see Englund 1994; 1998). Accounts of herding activities (*e.g.* Fig. 3.8a), for example, provide information on the kinds of animal (*e.g.* sheep or cattle), their sex and age (differentiating between adults and young animals), their total number, and also the volume of dairy products obtained in a certain period of time (for a detailed analysis of herding accounts, see Green 1980; Englund 1995a; 1995b).

Lexical Lists

The so-called 'lexical lists' are early sign and word lists comparable to modern dictionaries and encyclopaedias, though the cuneiform lists hardly provide any kind of subsidiary information (*cf.* Veldhuis 2014, 6). Lexical lists specify the words relating to a generic term such as 'vessels'.

At present, *c.* 700 copies of a total of 14 different lists dating to level IV and III styles are known (Veldhuis 2014, 32). These comprise semantic compilations of words and

a   b

Figure 3.8: Archaic tablets: (a) Account W 20274, 15; (b) Fragment of the list 'Vessels' (W 20258, 1 + 3) (K. Sauer © Uruk-Warka-Sammlung des DAI, Universität Heidelberg).

terms in a consistent, stereotypic form that can be divided into five categories: geography, fauna, flora, commodities, persons and literature (Englund 1998, 90). Amongst the oldest known lists from the Archaic IV writing phase are the 'Proto-Lu' (also known as 'Standard Professions List'), the 'Vessels', and the 'Metals' lists. From the Archaic III writing phase, the number of known lists increases considerably, with compilations on 'Titles and Professions', 'Vessels and Textiles', 'Wordlist C/Tribute', 'Metal', 'Wood', 'Cattle', 'Pigs', 'Officials', 'Fish', 'Cities', 'Geography', 'Grain', 'Birds' and 'Plants' (Englund and Nissen 1993, 13–37; Veldhuis 2014, 33).

The lexical lists can be easily distinguished from the other texts by their appearance. Their dimensions are usually larger than administrative tablets, and their surface is subdivided by a grid into fields and columns. Each field comprises a single entry that starts with a marker, the sign ⌐ (N$_1$) denoting the entity '1' in the sexagesimal system (*cf.* Englund and Nissen 1993, 9; for further information on the numerical sign systems, *cf.* Englund 1998, 111–120). The marker is followed by one or more ideograms, for example ▷ (DUG$_b$, 'Vessel'), followed by various ideograms denoting different qualities (size, form, *etc.*) as a description of the object in question (*cf.* Green and Nissen 1987 for the archaic sign list).

The consecutive entries contain very similar signs, compiling terms of certain categories without further elucidating them, as demonstrated here by the list 'Vessels': ▷ (DUG$_b$), ⊙ (DUG$_b$xŠE$_a$, 'vessel for grain/certain quantity of grain'), ⊙ (DUG$_b$xGAa/b, 'vessel for milk/certain quantity of milk'), and so on. Accordingly, the serial principles in

the archaic lists vary from text to text (Ross 2012, 302). The majority seem to follow a visual scheme, arranged primarily according to sign forms (*i.e.* scheme A, A1, A2 or AA, AB, AC, etc.; *cf.* Edzard 1999 for more detailed treatment). Later, the lists are arranged in different ways, *e.g.* in accordance with a logical/categorical principle (*i.e.* conceptually associated) or acoustically (*i.e.* words that sound alike).

The archaic lists could also be organised in a hierarchical sequence. The so-called 'Proto-Lu' list appears to be a case in point. With its 163 copies, the Proto-Lu is the best attested archaic list known and one of the most intriguing documents to be found among the early Mesopotamian texts (Englund and Nissen 1993, 17). From level Uruk III onwards, the list is canonised, *i.e.* copied in a largely consistent form still found in the later cuneiform corpora.

The Proto-Lu list registers titles and officials' names, for example the ⊟⊐ (NAM$_2$ URU$_{a1}$ 'the head of the city', *cf.* Englund and Nissen 1993, 69; Nissen *et al.* 2014, 153–157). In addition, the list refers to various cultic offices and job titles, such as the gardener, the cook, and so on. These terms appear to be sorted by rank and are thought to reflect actual social hierarchies, thus inviting a hypothetical reconstruction of the make-up of society during the Late Uruk Period (Englund and Nissen 1993, 19; Nissen *et al.* 2004, 55; Veldhuis 2014, 35).

When it comes to deciphering proto-archaic cuneiform, the lexical lists play a crucial role. The archaic lists were copied over and over again for several centuries in a more or less standardised form. Memorising and copying lists was one of the mainstays of scribal education, thus facilitating comparison between younger versions of lists and their archaic counterparts (Veldhuis 2014, 31).

Strikingly, the lexical lists contain numerous signs not found in contemporary administrative accounts (Nissen 1986, 329; Englund and Nissen 1993, 19; Veldhuis 2014, 52). Terms frequently used in the archaic accounts, such as the various designations for sheep, *e.g.* ⊶ (UDUNITA) 'ram' (see Fig. 3.8a), are not listed amongst the lexically arranged terms for animals or in a list of their own (as are cattle and swine). Instead, some terms are found scattered in 'Word List C' (compare to Fig. 3.8b) and 'Foods' (*cf.* Veldhuis 2014, 52).

The lists seem to assemble a whole range of different terms, regardless of whether they are needed or used otherwise. Also, as Niek Veldhuis (2014, 37) points out, there do not always appear to be strict compilation standards operating in the earliest records. The 'Vessels' list is usually supplemented with different terms for garments, but there are also examples ending with terms for metals instead (Veldhuis 2014, 37). This illustrates quite clearly an experimental stage subsequent to the introduction of a new technology. On the one hand, early scribes found themselves confronted with the need to memorise new information and deal with the new medium. On the other, they played around and tried out the possibilities offered by the innovation. In the case of archaic cuneiform, the novel technology is used in two different ways, not only for noting down accountings and administrative processes but also for literally listing things and collating words (Edzard 1999, 249; for a detailed discussion on the interpretation of the lists, see also Hruška 2005 and Veldhuis 2014).

Aliquid stat pro aliquo: Abstraction and innovation in early accounting devices

Niels Johannsen (2010, 61) reminds us that 'technological change also depends on concrete processes of creative thought, which are based on the cognitive ability to reconfigure certain aspects of the world'. In the case of the preliterate bureaucratic devices and proto-cuneiform, we are dealing with 'systems of symbols that represent cognitive constructions' (Damerow 2012, 153); more precisely with 'external representations of mental constructions' (Damerow 2012, 155). We are in other words faced with different levels of representational abstraction.

Following linguistic theory, tokens are representations of things in the classic sense of '*aliquid stat pro aliquo*' (*cf.* Ehlich 2007, 625; Assmann 2011, 35–38).

Object A → is represented by A' (token)

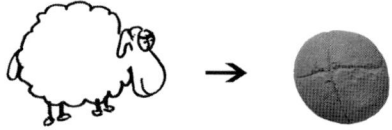

With the emergence of more complex administrative devices and early writing, a further stage of abstraction supervenes: the depiction of the representation, *i.e.* the sign. The representation loses its immediate connection to the object depicted.

Object A → represented by Token A' → is transformed into A'', the depiction of the representation (sign)

The sign A'' standing for the representation A' of an object can be seen in the marks on hollow clay balls. A'' represents the tokens stored inside the clay balls, not the immediate objects counted (Ehlich 2007, 625). A'' is also represented by the numerical notations on the numerical tablets and later by proto-cuneiform ideographs.

The early documents were still context-dependent (Kehnel and Panagiatopoulos 2015, 2), meaning that to understand numerical tablets and hollow clay balls, the reader had to have a certain degree of background knowledge about regarding the commodity in question, the person or institution responsible, and whether the function of the document was a receipt or a statement of output (*cf.* Damerow 2012, 161). This changed with the introduction of ideographic signs adding 'a greater variety of semantic coding' (Damerow 2012, 161) to the notational system in order to retain this kind of information as well. In this context, the term 'ideograph' actually reflects the abstraction very well, since it denotes the 'graphic symbolization of an idea' (Powell 2009, 3).

Regrettably, what sounds so neat and easy in theory is not so easy to identify in the actual findings. As we have seen, interpreting the correlations between tokens and archaic cuneiform signs has been a matter of intense debate. Nevertheless, it is possible to draw parallels between certain tokens and archaic signs (see Fig. 3.9), especially in the denominations for animals, textiles, and vessels (*cf.* Schmandt-Besserat 1992, 142–153), so there appears to be a degree of semantic continuity (*cf.* Sauer and Sürenhagen 2016, 32–35).

An even more abstract concept found in these early devices is the concept of numbers. Numbers and their depictions are more than mere representations, they '*enthalten in sich eine mentale Verarbeitung, die über die* […] *Gestalten der Dinge hinausgeht*' (Ehrlich 2007, 626). Accordingly, beyond representing a concrete object, numbers stand for an idea of the number of things, for example 'one sheep' or 'ten jugs of oil'. However, according to Peter Damerow (2012, 158), the early signs

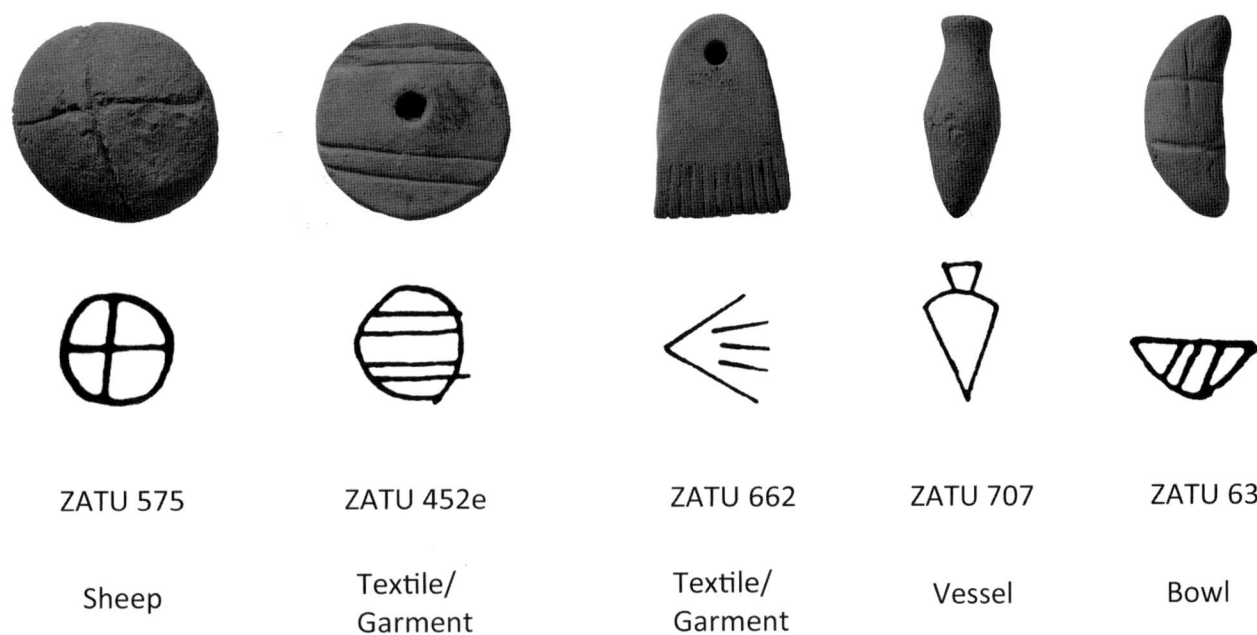

ZATU 575	ZATU 452e	ZATU 662	ZATU 707	ZATU 63
Sheep	Textile/Garment	Textile/Garment	Vessel	Bowl

Figure 3.9: Correlation of selected tokens and archaic signs (after D. Schmandt Besserat 1992).

for numbers do not represent 'context independent numbers […]; they can be used […] even without any cognitive numerical contrast'. Therefore, 'one sheep' or 'ten jugs of oil' could also be read as 'one unit of sheep' and 'ten units of containers of oil,' without necessarily specifying the quantity. Damerow (1996, 192) calls this the 'relative representation of numbers'. Moreover, what we see in the archaic record is a substitution of values (Nissen *et al.* 2014, 47), *i.e.* a certain number of units represented by certain tokens is replaced by a symbol standing for the summation of the tokens (*e.g.* 30 units of sheep could either be represented by 30 single counters or by 3 tokens each standing for the unit 10). This made it possible to depict larger quantities that would have been too laborious to represent via one-to-one-correspondence.

It would go beyond the scope of this paper to illustrate the diversity of numerical systems applied in the archaic texts. But since, with very few exceptions, the numerical notations on the archaic texts correspond to notations in later texts from the third millennium, their different values and systems can be reconstructed, indicating that there were different systems in use depending on the commodities accounted for (*e.g.* grain, liquids, timekeeping, and so on; *cf.* Englund 1998, 111–127).

Thus, tokens probably initially represented the unification of both object and quantity rolled into one. It is only with the use of numerical notations on hollow clay balls, numerical tablets, and later the emergence of archaic writing that this ambiguity was relinquished. Numbers were dissociated from the signs for commodities, thereby creating two types

of signs: numerals and pictographs (*cf.* Schmandt-Besserat 1992, 192; Damerow 1996, 219).

Dealing with innovations in Uruk-Period Mesopotamia

In the analysis of innovative technologies, various points require careful consideration. Most importantly, innovations are more than novel 'ideas, practices or objects' (Rogers 2003, 12); innovations are bundles of otherwise invisible knowledge condensed into a visible technology, or, as Everett Rogers (2003, 136) puts it, technologies consist of 'hardware' (the tool embodying the technology) and 'software' (the knowledge on which the tool is based). One might also add a third component, the 'bios' – to stay with Rogers' metaphor – encompassing social background, meaning, and symbolism. As Brian Pfaffenberger (1992, 500) points out, new technologies 'show the imprint of the context from which it arose, since system builders must draw on existing social and cultural resources'. It is only through appropriation that a new technology or an invention becomes a genuine innovation (*cf.* Burmeister and Müller-Scheeßel 2013, 2).

Thus, in conducting innovation studies one needs to consider several aspects. Such studies involve not only the analysis of actual innovations themselves but also of their preconditions (social and technological), and the extent to which they asserted themselves, meaning how they were appropriated (or not). Why do innovations occur, and why are they appropriated? Is it simply necessity being the mother of invention? In dealing with historical events that date back several millennia, it is fundamentally impossible to recover the actual origins of innovations (the moment that an innovation was invented), but it is worth trying to get as far back as we can.

As Rogers (2003, 137) points out, the innovation-development process often 'begins with the recognition of a problem or need'. In the case of the Urukean novelties, preliterate accounting devices and the emergence of writing, we can assume the need (rapidly evolving urban society) and we can trace the precursors (*inter alia* accounting with tokens, securing with seals). Johannsen (2010, 63) suggested that 'technological experience may, in other words, play a decisive role in creative cognition'. In the case of early Mesopotamian administrative tools, we see that they mainly consist of raw clay. Clay as a carrier for writing was employed for the first time in the Uruk Period and remained in use over thousands of years in the cuneiform world. Clay is abundant in the Middle East, it is readily accessible and easy to work with. And more importantly, clay as a working material was in use long before its suitability as a writing surface was recognised (*cf.* Wengrow 1998; for clay as script-bearer, *cf.* Balke *et al.* 2015). It had been used as a building material, in ceramic production, and in the manufacture of various commodities. Also, clay is highly plastic, malleable, and durable after drying. Its surface is easy to handle and can be painted, carved, or stamped. These properties were certainly not the least important assets encouraging its use for writing and drawing on.

But there is more to the appropriation of innovations. Avinoam Meir (1988, 234) inquires whether 'resources and the availability of the innovation [could] be considered sufficient conditions' for an appropriation. According to his concept of the 'adoption environment' (1988), multiple factors must be considered, for example the 'public and institutional frameworks' representing the 'cultural, societal, political, and geographical context pertaining to the particular innovation' (1988, 236). It is not only access to an innovation and knowledge about how to produce and apply it that leads to its appropriation. This model enables us to explain not only why innovations work, but also why they are sometimes rejected.

The rejection of innovations: Never change a winning system?

The rejection of innovations, in our case the rejection of novel means of supervision and accounting in the Uruk Period, can be elucidated very briefly by looking at the example of Arslantepe, a major centre in southern Anatolia and part of the wider exchange and communication networks of the time (*cf.* Frangipane 1997; 2011). Whereas the precursors of writing and administrative control are found distributed across Mesopotamia at Urukean sites – except for sealings – they are absent at Arslantepe (*cf.* Frangipane 2010, 27).

Ongoing excavations on this site have uncovered a centre of complex administration with evidence of large-scale organisation of storage and redistribution and the extensive

and sophisticated use of sealings (called 'cretulae' by the excavators; *cf.* Frangipane 2007 for an extensive study of these artefacts), which can be regarded as documents pertaining to economic transactions (*cf.* Frangipane 2007; 2012). They represent a temporary recording system (Frangipane 2012a, 118) consisting solely of sealings. Some were found stacked in the corner of a small room, ready to be put in order and probably stowed away temporarily (Frangipane 2012a, 118). Others had been dumped and used for fillings. Furthermore, sealings found in the collapsed debris indicate the existence of an upper storey where they may have been stored (Frangipane 2012a, 119).

As Marcella Frangipane (2012a, 119) points out, recording and managing storage with sealings was 'widely consolidated by millennia-old practice' and thus did not require 'the use of scribes for their basic operations'. Furthermore, in the case of Arslantepe, there is strong and visible (*i.e.* iconographical) continuity in the Late Chalcolithic sealing assemblages (Frangipane 2010, 37). The evidence from the site suggests an established, sophisticated system with a hierarchically organised class of officials (Frangipane 2011, 978). Moreover, what distinguishes Arslantepe from other major sites such as Uruk, is the absence of urbanisation and a heightened degree of interaction with the rural territory and the basically autonomous population (Frangipane 2010, 37; 2012b, 35). In contrast to major Urukean centres, the centralised administration at Arslantepe did not have to deal with high population densities and major interdependencies between the main cities and the surrounding countryside that would have required more refined control methods.

Thus, not only the on-site preconditions differ from those leading to the emergence of writing in southern Mesopotamia, but maybe also the bios – to return to Rogers' modified metaphor. It is entirely conceivable that the class of officials at Arslantepe had no interest in changing a traditional and well-established administrative system that not only fully satisfied their needs but probably also underlined their status.

Conclusions: The innovative potential of urukean bureaucracy

Regarding early administrative devices and archaic texts as innovative technologies, it has become clear that this innovative bundle was not the result of a straightforward *creatio ex nihilo* delivering a ready-made instrument. As Julian Reade (1992, 177) puts it, the 'range of types of early record that have been found at Uruk point to fertile experimentation'. What we see is the outcome of the dynamic transformative processes underlying constant change and sophistication, beginning with groups of tokens and different modes of storing and accounting them, continuing with the application of different symbolic notations probably denoting quantities, and culminating in the development of a flexible symbolic system facilitating and standardising the efficient control of accounting and communication. The artefacts discussed chime in perfectly with what V. Gordon Childe proclaimed 60 years ago: they are 'concrete expressions and embodiments of human thought and ideas' (1956, 1).

Once writing had asserted itself, it did not remain untouched but was constantly re-shaped. The first semi-pictographic symbols were transformed in to more and more abstract forms – the characteristic cuneiform wedges. And even these were subject to constant change (*cf.* Krebernik and Nissen 1994, 283–286). At the same time, tokens – the first administrative accounting devices – were not discarded. Interestingly, they remained in use parallel to writing, though not playing a very prominent role (*e.g.* in neo-Assyrian times in the 1st millennium BC, *cf.* MacGinnis *et al.* 2014). Hence, the employment of innovative, more complex technologies does not necessarily mean that old technologies are entirely superseded. Innovations may offer better solutions for some people (writing facilitates memorisation and bureaucracy) but not necessarily for everyone (the shepherd still using the 'old-fashioned' tokens to count his sheep).

Revisiting the quote by Nissen *et al.* at the head of this article, we can say that studies on the dynamic processes surrounding the emergence of writing are impossible without considering the social background and the implications for the handling of complex information. In the case of Uruk-Period Mesopotamia, the emergence of script and the deployment of complex bureaucratic mechanisms were a response to the needs of a complex society – a society with a growing population that had to be catered for. The introduction of new technologies in agriculture and production had led to an increase in the need for administration, control, and differentiation on the social level as well. But as Johannsen (2010, 61) points out, 'the acceptance or rejection of a new technology, whatever its origin, is always a social decision'. Furthermore, as noted by Jennifer Ross (2012, 314), in this specific context communication and administration needed to be formalised and regularised to deal with the imminent challenges.

It is striking that after Uruk presence in Syria, Anatolia and Iran had come to an end, the administrative novelties introduced fell into desuetude. Why? The knowledge of these technologies was given, as was the material needed to produce them – the ever-abundant clay. The simplest answer to that question would be that there was no need for these bureaucratic instruments. Only some hundreds of years later, in a period where we see the emergence of regional princedoms, was writing re-introduced to Syria (*c.f.* the archives of Ebla: Archi 2015).

Notes

1 The data presented here is part of the author's ongoing doctoral research conducted in the framework of Mini-Cluster 8.1 'Appropriating Innovations: Entangled Knowledge in Late Neolithic and Early Bronze Age Eurasia' of the University of Heidelberg's Cluster of Excellence 'Asia and Europe in a Global Context.'

2 There has been a series of analyses done on closed envelopes in order to determine their content: Drillon *et al.* 1987; Damerow and Meinzer 1995 and most recently Woods 2012.

References

Adams, R. McC. (1981) *Heartland of Cities.* Chicago, IL, University of Chicago Press.

Alagaze, G. (1989) The Uruk Expansion. Cross-Cultural Exchange in Early Mesopotamian Civilization. *Current Anthropology* 30, 571–608.

Algaze, G. (1993) *The Uruk World System. The Dynamics of Expansion of Early Mesopotamian Civilization.* Chicago, IL and London, University of Chicago Press.

Algaze, G. (2001) Initial Social Complexity in Southwestern Asia: The Mesopotamian Advantage. *Current Anthropology* 42, 199–233.

Algaze, G. (2013) The End of Prehistory and the Uruk Period. In H. Crawford (ed.), *The Sumerian World,* 68–94. London, Routledge.

Amiet, P. (1966) *Elam.* Auvers-Sur-Oise, Archée.

Amiet, P. (1972) *Glyptique Susienne.* Paris, Geuthner.

Archi, A. (2015) *Ebla and Its Archives: Texts, History, and Society.* Studies in Ancient Near Eastern Records Vol. 7. Boston and Berlin, de Gruyter.

Assmann, A. (2011) *Einführung in die Kulturwissenschaft. Grundbegriffe, Themen, Fragestellungen.* Berlin, Erich Schmidt.

Assmann, A. and Assmann, J. (2003) Schrift. *Reallexikon der Literaturwissenschaft* 3, 393–399.

Bachmann, F. (2011) Vierösengefäße der Urukzeit. Untersuchungen zu Typologie und Chronologie einer frühgeschichtlichen Gefäßgattung auf Grundlage des Fundmaterials aus Tell Sheikh Hassan/Syrien. Unpublished thesis, Freie Universität Berlin.

Balke, T., Panagiatopoulos, D., Sarri, A. and Tsouparopoulou, C. (2015) Ton. In T. Meier, M. R. Ott and R. Sauer (eds), *Materiale Textkulturen, Konzepte – Materialien – Praktiken,* 277–292. Berlin and Munich, de Gruyter.

Boese, J. (1995) *Ausgrabungen in Tell Sheikh Hassan* 1: *Vorläufige Berichte* über *die Grabungskampagnen 1984–1990.* Saarbrücken, SDV.

Burmeister, S. and Müller-Scheeßel, N. (2013) Innovation as Multi-Faceted Social Process: An Outline. In S. Burmeister, S. Hansen, M. Kunst and N. Müller-Scheeßel (eds), *Metal Matters. Innovative Technologies and Social Change in Prehistory and Antiquity,* 1–11. Rahden/Westf., Marie Leidorf.

Butterlin, P. (2014–2016) Uruk-Kultur. In *Reallexikon der Assyriologie* 14, 487–494. Berlin and Boston, MA, de Gruyter.

Childe, V. G. (1956) *Society and Knowledge.* London, Allen and Unwin.

Collon, D. (1987) *First Impressions: Cylinder Seals in the Ancient Near East.* London, British Museum Publications.

Damerow, P. (1996) *Abstraction and Representation: Essays on the Cultural Evolution of Thinking.* Dordrecht, Kluwer Academic Publishers.

Damerow, P. (2012) The Origins of Writing and Arithmetic. In J. Renn (ed.), *The Globalization of Knowledge in History: Based on the 97th Dahlem Workshop*. Max Planck Research Library for the History and Development of Knowledge. Berlin, <http://edition-open-access.de/studies/1/10/index.html> (28.07.2017), 153–173.

Damerow, P. and Meinzer, H.-P. (1995) Computertomographische Untersuchung ungeöffneter archaischer Tonkugeln aus Uruk, W 20987,2.11 und 12. *Baghdader Mitteilungen* 26, 7–11.

Drillon, F., Laval-Jeantet, M. and Lahmi, A. (1987) Étude en laboratoire de seize bulles mesopotamiennes appartenant au départment des antiquités orientales. In J.-L. Huot (ed.), *Préhistoire de la Mésopotamie. La Mésopotamie préhistorique et l'exploration récente du Djebel Hamrin*, 335–344. Paris, CNRS.

Edzard, D. O. (1999) Sumerisch-akkadische Listenwissenschaft und andere Aspekte altmesopotamischer Rationalität, in K. Gloy (ed.), *Rationalitätstypen* 246–267. Freiburg and Munich, Alber.

Ehlich, K. (2007) *Sprache und sprachliches Handeln. Band 3: Diskurs – Narration – Text – Schrift*. Berlin, de Gruyter.

Englund, R. K. (1994) *Archaic Administrative Texts from Uruk. The Early Campaigns*. Archaische Texte aus Uruk 5. Berlin, Gebrüder Mann.

Englund, R. K. (1995a) Late Uruk Period Cattle and Dairy Products: Evidence from Proto-Cuneiform Sources. *Bulletin of Sumerian Agriculture* 8/2, 33–48.

Englund, R. K. (1995b) Late Uruk Pigs and Other Herded Animals. In U. Finkbeiner, R. Dittmann and H. Hauptmann (eds), *Beiträge zur Kulturgeschichte Vorderasiens. Festschrift für Rainer Michael Boehmer*, 121–133. Mainz, Philipp von Zabern.

Englund, R. K. (1998) Texts from the Late Uruk Period. In P. Attinger and M. Wäfler (eds), *Mesopotamien. Späturuk-Zeit und Frühdynastische Zeit*. Orbis Biblicus et Orientalis 160/1, 15–215. Freiburg and Göttingen, Vandenhoeck & Ruprecht.

Englund, R. K. and Nissen, H. J. (1993) *Die lexikalischen Listen der archaischen Texte aus Uruk*. Archaische Texte aus Uruk 3. Berlin, Gebrüder Mann.

Englund, R. K. and Nissen, H. J. (2001) *Archaische Verwaltungstexte aus Uruk: Die Heidelberger Sammlung*. Archaische Texte aus Uruk 7. Berlin, Gebrüder Mann.

Frangipane, M. (1997) A 4th-Millennium Temple/Palace Complex at Arslantepe-Malatya. North-South Relations and the Formation of Early State Societies in the Northern Regions of Greater Mesopotamia. *Paléorient* 23/1, 45–73.

Frangipane, M. (2002) 'Non-Uruk' Developments and Uruk-linked Features on the Northern Borders of Greater Mesopotamia. In J. N. Postgate (ed.), *Artefacts of Complexity – Tracking the Uruk in the Near East*. Iraq Archaeological Reports 4, 123–148. Warminster, Aris and Philipps.

Frangipane, M. (ed.) (2007) *Arslantepe Cretulae. An Early Centralised Administrative System before Writing. Rome,* CIRAAS.

Frangipane, M. (2010) Arslantepe. Growth and Collapse of an Early Centralised System: The Archaeological Evidence. In M. Frangipane (ed.), *Economic Centralisation in Formative States. The Archaeological Reconstruction of the Economic System in 4th Millennium Arslantepe,* 23–42. Studia die Preistoria Orientale 3. Rome, Sapienza Universià di Roma, Dipartimento di Scienze Storiche Archeologiche e Antropologiche dell'Antichità.

Frangipane, M. (2011) Arslantepe-Malatya: A Prehistoric and Early Historic Center in Eastern Anatolia. G. McMahon and S. Steadman (eds), *The Oxford Handbook of Ancient Anatolia (10,000–323 BCE)*, 968–992. Oxford, Oxford University Press.

Frangipane, M. (2012a) The Evolution and Role of Administration in Anatolia: A Mirror of Different Degrees and Models of Centralisation. In M. E. Balza, M. Giorgieri and C. Mora (eds), *Archivi, depositi, magazzini presso gli i ittiti: nuovi materiali e nuove ricerche*, 111–126. Studia Mediterranea 23. Pavia, Italian University Press.

Frangipane, M. (2012b) Fourth Millennium Arslantepe: The Development of a Centralised Society Without Urbanisation. *Origini* 34, 19–40.

Glassner, J.-J. (2003) *The Invention of Cuneiform. Writing in Sumer*. Baltimore, MD, Johns Hopkins University Press.

Green, M. W. (1980) Animal Husbandry at Uruk in the Archaic Period. *Journal of Near Eastern Studies* 39/1, 1–35.

Green, M. W. (1981) The Construction and Implementation of the Cuneiform Writing System. *Visible Language* 15/4, 345–372.

Green, M. W. and Nissen, H. J. (1987) *Zeichenliste der archaischen Texte aus Uruk*. Archaische Texte aus Uruk 2. Berlin, Gebrüder Mann.

Hallo, W. (2011) The Godin Period VI Tablets. In H. Gopnik and M. S. Rothman (eds), *On the High Road. The History of Godin Tepe, Iran*, 116–117. Costa Mesa, CA, Mazda Publications.

Helwing, B. (2002) *Hassek Höyük II. Die spätchalkolithische Keramik*. Tübingen, Wasmuth.

Hilgert, M. (2010) 'Text-Anthropologie'. Die Erforschung von Materialität und Präsenz des Geschriebenen als hermeneutische Strategie. *Mitteilungen der Deutschen Orient-Gesellschaft* 142, 87–126.

Hruška, B. (2005) Prolegomena zur ältesten mesopotamischen Listenwissenschaft (Uruk, Fāra, Abū Ṣalabīḫ). *Archiv Orientální* 73, 273–289.

Johannsen, N. (2010) Technological Conceptualization: Cognition on the Shoulders of History. In L. Malafouris and C. Renfrew (eds), *The Cognitive Life of Things: Recasting the Boundaries of the Mind*, 59–69. Cambridge, McDonald Institute for Archaeological Research.

Kehnel, A. and Panagiatopoulos, D. (2015) Schriftträger – Textträger: Ein Kurzportrait (statt Einleitung). In A. Kehnel and D. Panagiatopoulos (eds), *Zur materiellen Präsenz des Geschriebenen in frühen Gesellschaften*, 1–14. Berlin, de Gruyter.

Krebernik, M. and Nissen, H. J. (1994) Die sumerisch-akkadische Keilschrift. In H. Günther (ed.), *Schrift und Schriftlichkeit. Ein interdisziplinäres Handbuch internationaler Forschung*, 274–288. Berlin, de Gruyter.

Le Brun, A. (1999) Hacinebi et Suse. *Paléorient* 25/1, 139–140.

Le Brun, A. and Vallat, F. (1978) L'origine de l'écriture à Suse. *Cahiers de la Délégation Archéologique Française en Iran* 8, 11–59.

MacGinnis, J., Monroe, M. W., Wicke, D. and Matney, T. (2014) Artefacts of Cognition: The Use of Clay Tokens in a Neo-Assyrian Provincial Administration. *Cambridge Archaeological Journal* 24, 289–306.

Mallowan, M. E. L. and Rose, J. C. (1935) Excavations at Tall Arpachiyah 1933. *Iraq* 2, 1–178.

Marzahn, J. (2013) Keilschrift schreiben. In M. van Ess, N. Crüsemann and M. Hilgert (eds), *Uruk – 5000 Jahre Megacity,*

Begleitband zur Ausstellung 'URUK – 5000 Jahre Megacity im Pergamonmuseum', Staatliche Museen zu Berlin, 25. April – 8. September 2013, 177–183. Petersberg, Imhof.

McMahon, A. (2009) The Lion, the King and the Cage: Late Chalcolithic Iconography and Ideology in Northern Mesopotamia. *Iraq* 71, 115–124.

Meir, A. (1988) Adoption Environment and Environmental Diffusion Processes: Merging Positivistic and Humanistic Approaches. In P. J. Hugill and B. D. Dickson (eds), *The Transfer and Transformation of Ideas and Material Culture,* 233–247. Austin, TX, Texas A&M University Press.

Michalowski, P. (1993) Literacy in Early States: A Mesopotamianist Perspective. In D. Keller-Cohen (ed.), *Literacy: Interdisciplinary Conversations,* 49–70. Cresskill, NJ, Hampton Press.

Nissen, H. J. (1986) The Archaic Texts from Uruk. *World Archaeology* 17/3, 317–334.

Nissen, H. J., Damerow, P. and Englund, R. K. (2004) *Frühe Schrift und Techniken der Wirtschaftsverwaltung im alten Vorderen Orient – Informationsspeicherung und -verarbeitung vor 5000 Jahren,* Berlin, Franzbecker.

Oates, J. (2002) Tell Brak: The 4th Millennium Sequence and Its Implications. In J. N. Postgate (ed.), *Artefacts of Complexity – Tracking the Uruk in the Near East.* Iraq Archaeological Reports 4, 111–122. Warminster, Aris and Philipps.

Oates, J. and Jasim, S. A. (1986) Early Tokens and Tablets from Mesopotamia: New Information from Tell Abada and Tell Brak. *World Archaeology* 17/3, 348–362.

Oates, D. and Oates, J. (2002) The Reattribution of Middle Uruk Materials at Brak. In E. Ehrenberg (ed.), *Leaving No Stones Unturned: Essays on the Ancient Near East and Egypt in Honour of Donald P. Hansen,* 145–154. Winona Lake, IN, Eisenbrauns.

Oates, J., McMahon, A., Karsgaard, P., Al Quntar, S. and Ur, J. (2007) Early Mesopotamian Urbanism: A New View from the North. *Antiquity* 81, 585–600.

Oates, D. and Oates, J. (1993) Excavations at Tell Brak 1992–93. *Iraq* 55, 155–199.

Oppenheim, L. A. (1959) On an Operational Device in Mesopotamian Bureaucracy. *Journal of Near Eastern Studies* 18, 121–128.

Ott, M. R. and Kiyanrad, S. (2015) Geschriebenes. In T. Meier, M. R. Ott and R. Sauer (eds), *Materiale Textkulturen, Konzepte – Materialien – Praktiken,* 157–168. Berlin and Munich, de Gruyter.

Otto, A. (2009–2011) Siegelpraxis. B. Archäologisch. *Reallexikon der Assyriologie* 12, 469–474.

Pfaffenberger, B. (1992) The Social Anthropology of Technology. *Annual Review of Anthropology* 21, 491–516.

Piquette, K. and Whitehouse, R. D. (2013) Introduction: Developing an Approach to Writing as Material Practice, in K. Piquette and R. D. Whitehouse (eds), *Writing as Material Practice. Substance, Surface and Medium,* 1–13. London, Ubiquity Press.

Pittmann, H. (2013) Seals and Sealings in the Sumerian World. In H. Crawford (ed.), *The Sumerian World,* 319–344. London, Routledge.

Postgate, N. J. (ed.) (2002) *Artefacts of Complexity: Tracking the Uruk in the Near East.* Iraq Archaeological Reports 4. Oxford, British Institute for the Study of Iraq.

Postgate, N. J., Wang, T. and Wilkinson, T. (1995) The Evidence for Early Writing: Utilitarian or Ceremonial? *Antiquity* 69, 459–480.

Powell, B. B. (2009) *Writing. Theory and History of the Technology of Civilization.* Malden, MA, Wiley-Blackwell.

Reade, J. E. (1992) An Early Warka Tablet. In B. Hrouda, S. Kroll and P. Z. Spanos (eds), *Von Uruk nach Tuttul: Eine Festschrift für Eva Strommenger. Studien und Aufsätze von Kollegen und Freunden,* 177–179. Munich and Vienna, Profil-Verlag.

Renfrew, C. (2001) Symbol before Concept. Die Macht des Symbols und die frühe Gesellschaftsentwicklung. In J. Fried and J. Süßmann (eds), *Revolution des Wissens, Von der Steinzeit bis zur Moderne,* 21–39. Munich, Beck.

Rittig, D. (2014) Siegel, Siegelbilder und ihre Träger. In D. Rittig, E. Strommenger und D. Sürenhagen (eds), *Die Kleinfunde von Habuba Kabira-Süd.* Ausgrabungen in Habuba Kabira II. Wissenschaftliche Veröffentlichungen der Deutschen Orient Gesellschaft 141, 327–366. Wiesbaden, Harrassowitz.

Rogers, E. M. (2003) *Diffusion of Innovations.* 5th ed. New York, Free Press.

Ross, J. C. (2014) Art's Role in the Origins of Writing: The Seal-Carver, the Scribe, and the Earliest Lexical Texts. In B. A. Brown and M. H. Feldman (eds), *Critical Approaches to Ancient Near Eastern Art,* 295–317. Boston, MA and Berlin, de Gruyter.

Rothman, M. S. (ed.) (2001) *Uruk Mesopotamia and Its Neighbors. Cross-Cultural Interactions in the Era of State Formation.* Santa Fe, NM and Oxford, School of American Research Press.

Sallaberger, W. and Schrakamp, I. (2015) *Associated Regional Chronologies for the Ancient Near East and the Eastern Mediterranean* III: History & Philology. Turnhout, Brepols.

Sauer, K. and Sürenhagen, D. (2016) Zählmarken, Zeichenträger und Siegelpraxis. Einige Bemerkungen zu vor- und frühschriftlichen Verwaltungshilfen in frühsumerischer Zeit. In C. Tsouparopoulou and T. Balke (eds), *Materiality of Writing in Early Mesopotamia.* Materiale Textkulturen 13, 209–241. Berlin, de Gruyter.

Schmandt-Besserat, D. (1980) The Envelopes that Bear the First Writing. *Technology and Culture* 21, 357–385.

Schmandt-Besserat, D. (1992) *Before Writing. From Counting to Cuneiform.* Austin, TX, University of Texas Press.

Stein, G. J. (1999) *Rethinking World Systems. Diasporas, Colonies, and Interaction in Uruk Mesopotamia.* Tucson, AZ, University of Arizona Press.

Stein, G. J. (2001) Indigenous Social Complexity at Hacınebi (Turkey) and the Organization of Uruk Colonial Contact. In M. S. Rothman (ed.), *Uruk Mesopotamia and Its Neighbors. Cross-Cultural Interactions in the Era of State Formation,* 265–305. Santa Fe, NM and Oxford, School of American Research Press.

Stein, G. and Özbal, R. (2007) A Tale of Two Oikumenai: Variation in the Expansionary Dynamics of ‹Ubaid and Uruk Mesopotamia. In E. C. Stone (ed.), *Settlement and Society. Essays Dedicated to Robert McCormick Adams,* 329–342. Los Angeles, CA and Chicago, IL, Cotsen Institute of Archaeology University of California.

Sürenhagen, D. (1986) The Dry-Farming Belt: The Uruk Period and Subsequent Developments. In H. Weiss (ed.), *The Origins of Cities in Dry-Farming Syria and Mesopotamia in the Third Millennium B.C.,* 7–43. Guilford, CT, Four Quarters Publisher.

Sürenhagen, D. (2014) Die Keramik. In D. Rittig, E. Strommenger und D. Sürenhagen (eds), *Die Kleinfunde von Habuba Kabira-Süd.* Ausgrabungen in Habuba Kabira II. Wissenschaftliche

Veröffentlichungen der Deutschen Orient Gesellschaft 141, 3–197. Wiesbaden, Harrassowitz.

Taylor, J. and Cartwright, C. (2011) The Making and Re-Making of Clay Tablets. *Scienze dell'Antichità* 17, 297–324.

Tomé, A., Cabral, R. and Renette, S. (2016) The Kani Shaie Archaeological Project. In K. Kopanias and J. MacGinnis (eds), *The Archaeology of the Kurdistan Region of Iraq and Adjacent Regions*, 427–434. Oxford, Archaeopress.

Van Driel, G. (1982) Tablets from Jebel Aruda. In G. van Driel, T. J. H. Krispijn, M. Stol and K. R. Veenhof (eds), *ZIKIR ŠUMIM. Assyriological Studies Presented to F. R. Kraus on the Occasion of his 70th Birthday*, 12–25. Leiden, Brill.

Veldhuis, N. C. (2014) *History of the Cuneiform Lexical Tradition.* Guides to the Mesopotamian Textual Record 6. Münster, Ugarit Verlag.

Wengrow, D. (1998) 'The Changing Face of Clay': Continuity and Change in the Transition from Village to Urban Life in the Near East. *Antiquity* 72, 783–795.

Wickede, A. v. (1990) *Prähistorische Stempelglyptik in Vorderasien.* Munich, Profil-Verlag.

White, R. (1989) Visual Thinking in the Ice Age. *Scientific American* 261/1, 74–81.

Woods, C. (2010) The Earliest Mesopotamian Writing. In C. Woods (ed.), *Visible Language, Inventions of Writing in the Ancient Middle East and Beyond*, 33–50. Chicago, IL, Oriental Institute of the University of Chicago.

Woods, C. (2012) Early Writing and Administrative Practice in the Ancient Near East. New Technology and the Study of Clay Envelopes from Choga Mish. *The Oriental Institute News & Notes* 215, 3–8.

Wright, H. T. and Rupley, E. S. A. (2001) Calibrated Radiocarbon Age Determinations of Uruk-Related Assemblages. In M. S. Rothman (ed.), *Uruk Mesopotamia and Its Neighbors. Cross-Cultural Interactions in the Era of State Formation*, 85–148. Santa Fe, NM and Oxford, School of American Research Press.

Wright, H. T., Miller, N. and Redding, R. (1981) Time and Process in an Uruk Rural Centre. M. T. Barrelet (ed.), *L'Archéologie de l'Iraq. du début de l'époque néolithique à 333 avant notre ère: perspectives et limites de l'interprétation anthropologique des documents*, 265–284. Paris, CNRS.

Chapter 4

Uruk, Pastoralism and Secondary Products: Was it a Revolution? A View from the Anatolian Highlands

Maria Bianca D'Anna and Giulio Palumbi

Introductory remarks

Innovation is a multidimensional concept since it entails the complex interactions of materiality, techniques, knowledge and culture at both an individual and a collective level. The processes that lead to innovations consist not only of technological novelties. Rather they are shaped by, and themselves shape, social practices. In the relations between humans, animals, objects and places, new things or ideas may also be invented or borrowed. Tradition and change are the main subject of our investigations of past societies, and innovations can readily be correlated to social change in more or less complex concatenations of cause and effect. When the concept of change is enriched by that of technological innovation, we may easily end up constructing evolutionary narratives. Whatever temporal depth is attributed to them, innovations are often conceptualised as events. They are events in the Žižekian sense of something that 'changes the rules of what is possible' and 'retroactively creates its own past' (Žižek, 2013). While this may reinforce teleological accounts of the past, we believe that innovations do not necessarily imply unidirectional and linear change from the simple to the complex, from egalitarian to hierarchical, from low-tech to high-tech. As Van der Leeuw has put it (1989), innovation is a complex process consisting of a number of components that may also encompass adoption, diffusion and rejection.

In this paper, we aim to investigate changes in socio-cultural traditions and innovations that involved human-animal interaction. Our aim is to stress that these relationships did not always follow cause-effect and evolutionary trajectories. Instead, they reflected non-linear and rhizomatic patterns depending on the diversity of the social, economic and cultural structures inventing, adopting, re-appropriating or rejecting them. In particular, we would like to identify the different roles of specialised animal husbandry strategies in different social and chronological contexts and understand how and where a secondary product revolution (SPR) did or did not accompany the emergence of political hierarchies. Our site of reference is Arslantepe in eastern Anatolia, which, during the late 4th and early 3rd millennium BC, provides the backdrop to the development of a complex relationship between cultural change and innovation, linked to trans-regional interaction networks in which the communities living on the site were engaged.

The paper takes two thoroughly debated topics as its starting point: the Uruk phenomenon and SPR. What the archaeologists of south-western Asia mean by the Uruk phenomenon is the development of urbanisation and proto-state societies in southern Mesopotamia and Khuzestan and the spread of southern material culture into northern Mesopotamia, Syrian Jezira, western Iran and south-eastern Anatolia. The SPR is a powerful model for problematising the emergence of social and political complexity in connection with specialised forms of subsistence economy and labour organisation in 4th-millennium Mesopotamia. This model, elaborated by Andrew Sherratt in a famous article published in 1981, conceptualised a radical change in terms of human-animal interaction (Sherratt 1981; Greenfield 2005). Alongside the invention of the plough, which implied an intensification of agricultural production and the exploitation of a broader range of soils, there were several other factors operative in 4th-millennium south-western Asia that favoured or represented a 'revolution', including the cart, which would have permitted bulk transport and faster movement between distant territories; large-scale use of milk and its transformation into different kinds of dairy product, which changed human eating habits and involved an intensification of protein intake because produce could be stored for longer; and the development of a large-scale wool industry that boosted long-distance trade with artefacts that were comparatively easy to transport. According to Sherratt, this process of innovation was not only a revolution in economic and productive terms, it also triggered the development of complex societies and the emergence of a centralised economy.

One of the markers of the Uruk economy was certainly the radical change in animal husbandry strategies. In this period, a specialisation in sheep and goat husbandry is recorded for the entire Mesopotamian region and represents an innovative change in comparison to the more balanced animal breeding strategies of the previous Ubaid period in which cattle and pigs featured prominently (Vila 1998; Berthon 2015; Dahl 2015). In Late Uruk southern Mesopotamia, there appear to have been links of a fairly coherent nature between urbanisation, early state societies and a centralised economy on the one hand (in short, a new social and political order and ideology) and technological achievements and innovations on the other.

There is general unanimity that the emergence of complex societies in southern Mesopotamia was accompanied by a shift from village-based flax manufacture to widespread household and 'industrial' wool-textile production (McCorriston 1997). Textual, archaeological and iconographic evidence from the alluvium testify that in the temple-based tributary economy of the earliest cities of southern Mesopotamia (Pollock 1999), not only primary but also secondary products such as wool-textile and cheese were exchanged against labour and constituted the financial basis of the powerful elites. As a consequence, certain types of work performed by some people, such as the activities of women associated with weaving, were progressively marked by repetitiveness, efficiency and segmentation. This increased dependency and reduced the freedom of large sectors of population. The Uruk iconographic repertoire of administrative, ceremonial and luxury material (*i.e.* glyptic and stone vessels) stresses the importance of wool and textile production. In particular, women seem to have been largely engaged in weaving activities. Also, archaic texts from Uruk refer to the new role of sheep in wool production (Green 1980) and tell of textile production and different kinds of dairy produce (see, for instance, Wagensonner 2015). According to Nissen (1986), it is possible to infer that religious and political institutions owned flocks that were entrusted to shepherds responsible for dealing with births, deaths and changes in the composition of the herd.

Archaeozoological, textual and iconographic evidence indicates a widespread adoption of husbandry strategies focused on caprines and an emphasis on wool and textile production during the late 4th millennium BC in southern Mesopotamia (McCorriston 1997). The introduction of these specialised husbandry strategies may have been associated with a new organisation of the territory surrounding the Mesopotamian cities, with a shift of the flocks from agricultural land to more marginal lands. McCorriston has defined this as a process of extensification of productive land, leaving space for the intensification of cereal cultivation in more productive and easily accessible stretches of arable land. In conjunction with an increasing demand for wool production, this process may have been one of the factors that according to McCorriston triggered off Uruk expansion outside the alluvium, if it was not indeed the main factor involved, as suggested by Porter (2014).

If this new focus on sheep and goats was a sign of the growing importance of these animals and their secondary products in the economies of the Uruk period, we would not expect to find much local differentiation in the archaeozoological data from the sites that were part of the Uruk phenomenon. However, within the overall specialisation in this type of livestock, the available data on the use of sheep and goats show a high degree of variability. Concerning the identification of species, the data militate against a higher ratio of sheep to goats and also against an exclusive relationship between specialised sheep- and goat-raising practices and secondary products such as milk, wool or hair. There is little indication that the production of wool or dairy foods was of central importance for these specialised husbandry strategies. And as Vila (1998, 127–128) contended some years ago, the data seem to suggest that sheep and goats were raised in order to exploit both primary (meat) and secondary products (milk and wool).

Late 4th-millennium Arslantepe (period VIA)

We now move to the extreme north-west of the Uruk world in eastern Anatolia, where Arslantepe, located 10 km south of the right bank of the Euphrates, is the largest site on the Malatya plain. Here a team from Rome's 'Sapienza' university has unearthed an uninterrupted archaeological sequence spanning from the end of the 5th millennium BC to the Middle Ages. In this paper, we focus on the 4th and early 3rd millennia BC, namely the Late Chalcolithic 3–5 and Early Bronze Age 1a periods in Anatolian chronology (Tab. 4.1).

During the first half of the 4th millennium BC (period VII in the site sequence; 3900–3350 BC), the settled area

Table 4.1: 4th and early 3rd mill. BCE occupation at Arslantepe: absolute and regional chronologies

Arslantepe sequence	Anatolian upper Euphrates periods	Absolute dates BC	Southern Mesopotamia periodisation
VI B2	Early Bronze Age Ib	2900–2750	Early Dynastic I
VI B1	Early Bronze Age Ia	3100–2900	Jemdet Nasr
VI A	Late Chalcolithic 5	3400–3100	Late Uruk
VII	Late Chalcolithic 4	3800–3400	Middle Uruk

at Arslantepe expanded and was organised differently in terms of its functions (Frangipane 2012a, 20–27). Several layers of small dwellings have been excavated on the northeastern periphery of the mound, while a substantial building (possibly an elite residence) has also been uncovered, extending across the western slope of the mound. At the very end of phase VII, the construction of a monumental tripartite ceremonial building, the so-called Temple C, clearly represents a further sign of social differentiation. The central hall of Temple C is larger than 120 square meters and could accommodate a large number of people assembled around a platform with an open fireplace. The discovery of more than 1000 mass-produced bowls inside this building provides strong evidence of meal distribution and consumption on the spot as a ritualised and communal practice (D'Anna and Guarino 2012). The presence of a hundred clay sealings (*cretulae*) found in one of its side rooms indicates bureaucratised control over the distribution of foodstuffs, confirming the emergence of elites whose power was probably based on some form of control over primary resources.

Throughout period VII animal husbandry was balanced, with a significant presence of cattle and pigs too, but at the end of the period, for the first time, we have a prevalence of sheep and goats in Temple C. This hints at the new economic role played by caprines in food distribution practices, a role presumably bound up with the emergence of a new kind of political organisation. As is commonly agreed, this change is consistent with what was happening in the Late Chalcolithic world of northern Mesopotamia, for example at Tell Brak, before there was any 'direct contact' with the Uruk world (Vila 1998).

At the end of the 4th millennium BC, during phase VIA, the size of the settlement shrinks and occupation at Arslantepe consists of what has been defined as a proto-palatial complex (Frangipane 2012a, 29) and a few outstanding elite residences. Recent excavations have proved that these two areas are closely interrelated. The public area was a multi-functional complex consisting of a series of buildings, each of them with a specific function: representative, bureaucratic, economic, ceremonial (Fig. 4.1).

Though the social and political complexity of this period at Arslantepe and the food provisioning activities undertaken there do echo structural features of the coeval Uruk world, Arslantepe VIA never became a real city, neither in dimensional nor in qualitative terms. Moreover, the local material culture only partially recalls the southern models.

In this very singular context, is there in fact any clear evidence for an SPR encouraging the centralisation of resources and ultimately the formation of political and social complexity similar to that recorded in the Mesopotamian alluvium? Data on phase VIA fauna clearly confirms the existence of new, specialised and heavily caprine-oriented

animal herding, which is in line with the trend recorded in the final stages of phase VII in Temple C. The kill-off patterns identified on the basis of the dental remains of sheep and goat prove that animals were exploited for meat consumption rather than for wool or milk (Siracusano and Bartosiewicz 2012). Different kinds of meat were however preferred in different commensal contexts and on different occasions (Bartosiewicz 2010). For example, more than 1,500 animal bones, mainly caprine from medium- and low-quality cuts, were found in redistribution unit A340, where meals were probably disbursed as provisions in exchange for labour. Conversely in Temple B, a relatively small structure difficult to get at, where restricted events including the preparation and consumption of lavish amounts of food were performed, the incidence of caprine bones is less pronounced (D'Anna 2012).

A functional analysis of the period VIA pottery assemblage based on the formal characteristics of the vessels, their capacity and surface alterations (D'Anna 2010) complicates the picture. There are certainly no strainers, filters or lids that would have been useful for processing milk into soft dairy products or cheeses. Because of their restricted opening, the so-called semi-fine necked jars of various dimensions (Fig. 4.2, left), a class of pots in which the influence of Uruk models is strong and which occur in all period VIA contexts, seem to be more suitable for keeping liquids or semi-liquid foods rather than dry goods. Residue analyses have not been carried out yet and it would be foolhardy to speculate on any likely scenarios. However, the few spouted or unspouted bottles (Fig. 4.2, right) found concentrated in specific areas (a stocking area and Temple B) in period VIA contexts could have been used for beer and wine, while the jars could have been used to store semi-liquid products such as yoghurt, soft cheese, porridge or animal fats.

The more frequent occurrence of the necked jars in comparison to the few, and generally small, bottles would suggest that storing semi-liquid foodstuffs was of greater economic importance. In the 'ration-meal' redistribution circuits, different kinds of food preparation may have played a major role in comparison to cooked foodstuffs, along with meat that could be roasted or smoked elsewhere and then brought to redistribution unit A340. By contrast, the preparation and consumption of cooked food (including 'special' *i.e.* non-caprine meat) may have been a feature of religious/ceremonial events with greater restriction of access (D'Anna and Jauss 2015).

With regard to the production and use of wool and goat hair as opposed to other fibres, we only have indirect evidence to go on, for instance, the imprints of cloth and ropes from the back of *cretulae*. Analyses carried out by Romina Laurito (2007) on more than 300 clay sealings have revealed that flax was more widely employed than goat hair. Laurito also identified a large variety of different cloths and possibly weaving techniques impressed on the reverse of

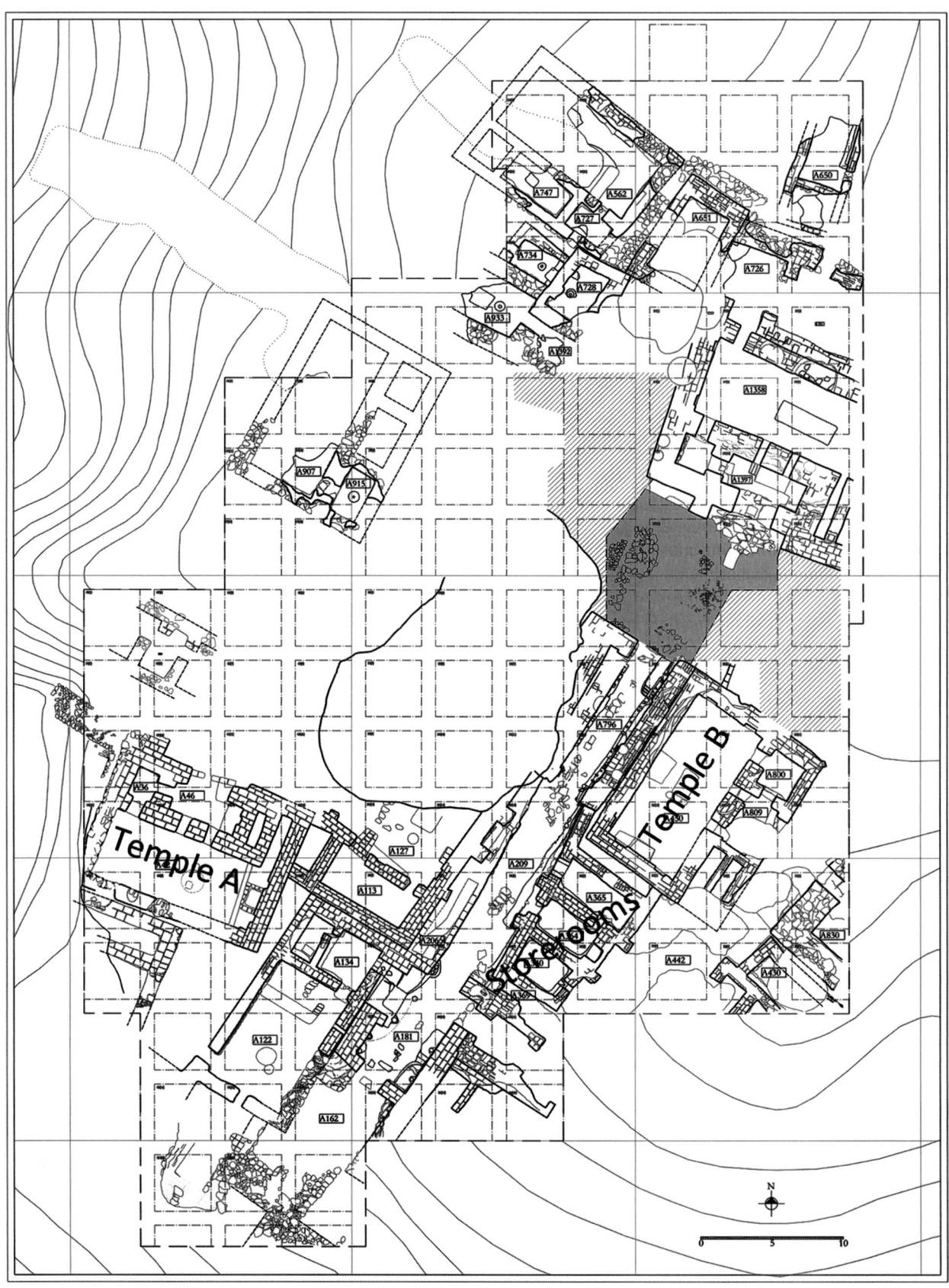

Figure 4.1: The period VI A buildings (courtesy Missione Archeologica Italiana nell'Anataolia Orientale).

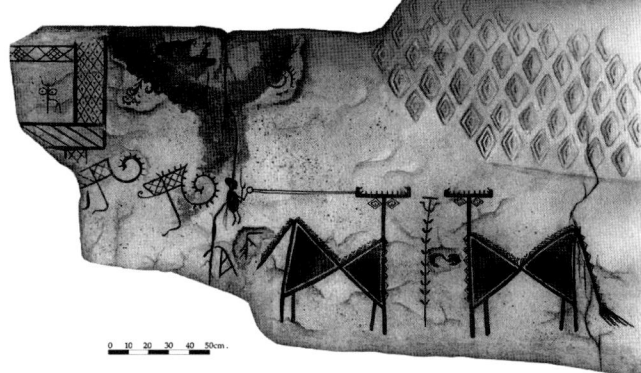

Figure 4.3: Arslantepe VI A. The corridor's wall painting (courtesy MAIAO).

Figure 4.2: Arslantepe VI A. One Light-coloured Necked jar and two bottles found in Temple B (courtesy MAIAO).

more than 160 *cretulae*, but unfortunately no conclusive statements can be made on the kinds of fibre used. Only a few spindle whorls and loom weights have been found in period VIA public contexts, and the majority of weaving tools were found in the residential sector.

In the period VIA palatial complex we have no evidence of large-scale transformation activities on milk and wool. These activities were probably independent of the direct economic interests of the Arslantepe elites. In particular, textile production could still have been in the hands of local households in the rural villages of the plain, carried out on a non-industrial scale and less mercantile in scope than in the southern Mesopotamian cities. It is possible that dairy products were also prepared outside the Arslantepe public building and then brought to the political centre to be distributed to different economic circuits, notably the circuit connected with food provisioning. However, in the public *milieu* of period VIA, in spite of a general overwhelming presence of sheep and goat bones, and in particular those of sheep, the mortality curves dovetail perfectly with the kill-off patterns of meat consumption. Caprine bones are even more heavily present in contexts connected with food redistribution practices, thus suggesting that meat was

basically exploited to finance the centralised system and subsidise the mobilisation of labour.

As for the organisation of pastoral activities, we do not know who owned the animals consumed inside the public buildings. Was it the centralising political elites, who directly managed these activities in accordance with strategies serving their own economic interests? Or did the households living in small villages around the settlement not only produce textiles but also manage livestock? As a matter of fact, the existence of specialised pastoralist communities in the Malatya region during the 4th millennium BC is still hypothetical. The absence of settlements contemporary to the VIA phase is puzzling. The convenient answer would be to take this as negative evidence of a pastoral mobile community, but this is not without its problems. At all events, while pastoral groups and secondary products have remained invisible in the archaeological record of Arslantepe in phase VIA so far, the ideological apparatus of the local elites foregrounded agricultural activities in the glyptic and wall paintings. Both testify to the use of cattle for traction, for threshing with the *tribulum* (on a clay sealing) and possibly for ploughing (*cf.* corridor wall painting: Fig. 4.3).

Moreover, at the very end of period VII we find a few pieces of ceramics that was totally new. This is the so-called Red-Black Burnished Ware that makes up 10% of the ceramic assemblage during period VIA. Period VIA RBBW comprises a small number of vessel types and invariably features a strictly alternating chromatic pattern: the black surface is outside in the closed vessels (jugs and small- to medium-sized jars) and inside in the open ones (bowls, mugs, and high-stemmed bowls). These ceramics could be the expression of an identity that linked this hand-made bichrome ceramic tradition developing in the Anatolian highlands during the 4th millennium BC with the new specialised husbandry practices, as epitomised by this red-black jar bearing the representation of a caprid (Fig. 4.4).

Figure 4.4: Arslantepe VI A. The capride applied on a Red-Black Burnished jar, particular during the excavation (courtesy MAIAO).

Early third-millennium Arslantepe (Phase VIB1)

A devastating fire completely destroyed the period VIA buildings and the way of life associated with it. It has been suggested that the destruction of the public buildings could have been the result of a systemic crisis befalling the centralised institution as the Anatolian highland environment may not have been conducive to the large-scale surplus production of needed to fuel the redistributive system controlled by the local elites (Frangipane 2012b).

After this destruction, a number of radical and long-term cultural changes characterised the history of Arslantepe in the early stages of the 3rd millennium. To understand these changes, we need to shift the focus of our attention to another phenomenon that was radically different in nature from Uruk: the so-called Kura-Araxes (KA) or Early Transcaucasian culture. As of the second half of the 4th millennium BC, the KA culture started developing among the communities of the southern Caucasus, north-western Iran and the eastern Anatolian highlands as a well-codified package of material and symbolic traditions (Sagona 1984; 2014; Greenberg and Palumbi 2014).

Among these traditions, ceramics seem to have functioned as an identity marker for the Kura-Araxes communities. Kura-Araxes ceramics are often characterised by a special attention to technical and aesthetic features, ranging from the burnishing of surfaces to a recurring set of decorative motifs, and finally a standard red-and-black chromatic effect between the black exterior and red interior surfaces of the same vessel. A rather standardised morphological repertoire is also characteristic of Kura-Araxes ceramics: truncated-conical necked jars, large S-shaped bowls and circular lids invariably fitted with handles represent clearly recognisable traits of the Kura-Araxes ceramic tradition.

As for metallurgical traditions, some typical body ornaments are also associated with this culture: double spiral

headed pins, hair spirals, and diadems, the latter often found in funerary contexts. The Kura-Araxes funerary customs display a large variety burial practices, with stone-lined cists among the most distinctive funerary structures.

Kura-Araxes communities seem to have lived in small villages composed of mono- or bi-cellular residential units. Here, three-leaved fireplaces or horseshoe-shaped andirons decorated with anthropomorphic or zoomorphic motifs were central features of what appears to have been a pronounced ritualisation of the everyday life (Sagona 1998; Smogorzewska 2004).

As for the primary economy, archaeozoological and paleobotanical data indicate that Kura-Araxes communities maintained a mixed agro-pastoral economy based on cereal agriculture and on non-specialised husbandry strategies mainly focusing both on cattle and caprines (Sagona and Zimansky 2009).

At some point in its history, the Kura-Araxes culture, from being a specific tradition of the South Caucasian highlands, became one the most geographically extended cultural horizons in western Asia (Smith 2005). This so-called 'Kura-Araxes expansion' took the form of a large-scale circulation of people that spread traits derived from their cultural model across a vast area extending from the Zagros Mountains in Iran to the Anatolian Upper Euphrates and from this latter region to the Amuq plain as far as southern Levant (Greenberg and Palumbi 2014). What is striking is that the first regions to be caught up in this development at the beginning of the 3rd millennium BC – the Upper Euphrates Valley in Anatolia and the Kangavar Valley in Iran – were the very same highland regions that had been previously involved in the so-called Uruk 'expansion', albeit in a very special and unique way. We believe that in the Upper Euphrates region the Late Chalcolithic innovations pertaining to animal production may have played a major role in the expansion of the Kura-Araxes culture taking place in this region at the very beginning of the 3rd millennium BC.

This is clearly exemplified by Arslantepe phase VIB1, following the destruction of the public building from the Uruk period.

With the exception of an enigmatic mud-brick construction (Palmieri 1981; Frangipane and Palmieri 1983), most of the architectural evidence from period VIB1 that was excavated at Arslantepe during the 1970s–1990s consisted of a series of levels of wattle-and-daub huts and large open areas (Frangipane 2012b, 239). However, the excavations carried out during the last few years have revealed that this phase was actually a more complex and substantial matter than has been assumed so far (Fig. 4.5).

In terms of the early VIB1 phase, the finds were exclusively light architecture in the form of wooden or wattle-and-daub huts, suggesting short-lived periods of temporary occupation (Frangipane 2014). Subsequent levels

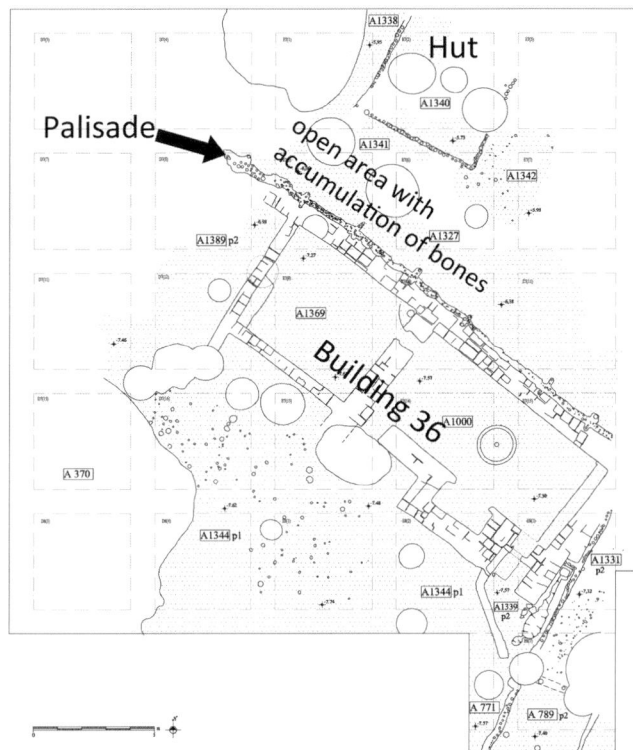

Figure 4.5: Arslantepe VI B1. Plan of the architectures in the Northern part of the settlement (courtesy MAIAO).

Figure 4.6: Arslantepe VI B1. The hearth in the main room (A1000) of Building 36 (courtesy MAIAO).

Figure 4.7: Arslantepe VI B1. The nine spearheads found in the 'Royal Tomb'.

testify to a progressive re-occupation of the higher part of the ancient mound, with more substantial structures built inside and outside a massive rectangular palisade.

Without going into the finer details of the various architectural phases recorded in phase VIB1, a large mud-brick building (Building 36) on the southern side of the palisade repays closer attention. It consisted of a long rectangular room with a large circular fireplace and a smaller adjoining storage room with a large number of *pithoi* (Frangipane 2014). The large dimensions of this building, the presence of a monumental fireplace and the adjoining store-room, where two metal spear-heads were also found, indicate that Building 36 was not an ordinary dwelling but a structure with special functions, possibly ceremonial and commensal. The main room was large enough for an assembly of people to sit on a bench and around the big hearth, which was more than 1m in diameter (Fig. 4.6).

As Frangipane (2014) has pointed out, the Arslantepe settlement maintained its central political and symbolic significance in the region even after the collapse of the monumental building from the Uruk period. This significance is a plausible assumption given the presence of the so-called Royal Tomb probably dating back to the final years of phase VIB1, an elite burial featuring an amazing number of metal grave goods (Frangipane *et al.* 2001; Palumbi 2008) (Fig. 4.7).

Thus the beginning of the 3rd millennium BC saw the emergence of new forms of power based on strength, warfare and the accumulation of prestige goods. This species of power was radically different from the bureaucratic power of the Uruk period and also from that of the coeval Jemdet Nasr period in southern Mesopotamia. Available evidence suggests that at the very beginning of the 3rd millennium in southern Mesopotamia institutions of power were still founded on the administered distribution of staple products. These continued to be disbursed in exchange for labour in the form of food rations given out in wheel-thrown conical bowls, the ceramic 'type fossil' of the Jemdet Nasr period and the equivalent of the bevelled-rim bowls from the Uruk period (Pollock 1990; 1992).

At Arslantepe, the lasting central symbolic significance of the site despite ongoing changes in the nature, structure and representation of power is testified to by ceremonial events that were however different in nature from those of the Uruk period. They include large-scale feasting taking place outside a large wattle-and-daub hut, where about 6000 animal bone fragments, mostly stemming from the best meat cuts, were found (Siracusano and Bartosiewicz 2012, 119–120). According to Giovanni Siracusano, both the quantitative and qualitative data point to the extraordinary nature of this dump. The bones are probably leftovers from large-scale commensal events involving large numbers of people and taking place periodically on the highest part of the mound (Siracusano and Palumbi 2014). The animal species identified on this site show a marked prevalence of caprines, representing as much as 93% of all the finds. This preference for caprines is further confirmed by data from the rest of the VIB1 settlement (Siracusano and Bartosiewicz 2012, fig. 3).

The comparison of faunal data from phase VIB1 at Arslantepe with those of the earlier phase VIA indicate a strong continuity in specialised husbandry strategies focusing on sheep and goats, suggesting a long-term impact of the 4th-millennium husbandry strategies rooted in the entrenchment of a centralised economy and exploited in the framework of an elite redistributive economy.

Despite this basic continuity, some changes in animal strategies between the 4th and the 3rd millennium BC can however be noted, such as the decrease in the sheep-goat ratio from 3:1 in phase VIA to 1.7:1 in phase VIB1 (Siracusano and Bartosiewicz 2012, fig. 1) and a slight change towards greater exploitation of milk highlighted by the mortality profiles (Siracusano and Bartosiewicz 2012, 119).[1] However, as in period VIA, neither strainers nor specific ceramic shapes clearly associated with milk-processing or the production of dairy products have yet been found in the ceramic repertoire from phase VIB1.

Though sheep meat was used in the 4th millennium BC as the foodstuff of choice for redistributive activities, the model of husbandry practised in the 3rd millennium BC did not specialise in one single product, either primary or secondary.

For wool, the data available for period VIB1 is inconclusive. On the one hand, mortality profiles show no clear evidence in favour of wool exploitation, and textile or weaving tools are very rarely found in phase VIB1 (Laurito *et al.* 2014, 159). On the other hand, the analyses carried out on some textile fragments preserved in the Royal Tomb have shown that they were made of goat hair and characterised by extremely fine, high-quality fabrics, indicating the use of thin yarns, an increasing variety of wool fibres produced during this period, and highly skilled weaving technologies (Frangipane *et al.* 2009, 16–20; Laurito *et al.* 2014, 160).

On the whole, specialised herding strategies focusing on caprines and the widespread use of light architecture are strongly characteristic of VIB1 occupation of Arslantepe by a pastoral society. However, the economic activities of these pastoral groups do not seem to have focused on the production of secondary products, while the consumption of mutton was a feature of both everyday meals and exceptional commensal events.

The basic continuity of husbandry strategies between phases VIA and VIB1 contrasts with the fact that pastoral occupation in phase VIB1 at Arslantepe marks a radical cultural change over and against the previous phase. This change has to be seen in the framework of an active process of cultural innovation resulting from new interactions with the Kura-Araxes world that were integrated into a continuity framework featuring some characteristics preserved from past local traditions. The majority of phase VIB1 ceramics was hand-made, mostly red and black (Fig. 4.8). In formal terms, however, the VIB1 ceramic repertoire combines typical Kura-Araxes phenomena, such as cylindrical necked jars, circular lids and the widespread use of handles with a firing technique that replicates the same shifting chromatic pattern as we find in the 4th millennium Red-Black Burnished Ware and that differs from the fixed alternation between red and black surfaces in the KA vessels.

There are also clear signs of coexistence between new (Kura-Araxes) and past traditions recorded in the Royal Tomb. The construction of a stone lined cist, the morphologies of the red-and-black jars and the presence of body ornaments such as diadems, hair spirals and double

Figure 4.8: Arslantepe VI B1. Ceramics found in situ in Building 36 (courtesy of MAIAO).

headed spiral-pins all clearly recall the Kura-Araxes world, whereas the metal spearheads with their silver inlaid decorations and the wheel-made light-coloured jars appear to stand for the heritage deriving from the metallurgical and potting traditions of the Uruk period (Frangipane *et al.* 2001; Palumbi 2008). At the same time, continuities with 4th millennium domestic traditions are also visible in the persistent use of the omphalos-shaped circular fireplaces found not only in Building 36 but also in other huts (Balossi Restelli 2015). Finally, the specialised husbandry strategy focusing on sheep and goat represents the last and by no means unimportant element of structural continuity between the late 4th and early 3rd millennium BC at Arslantepe.

Putting all this data together, the mixture of old (Uruk and Anatolia-related) and new (Kura-Araxes-related) elements strongly suggests that the communities living at Arslantepe at the beginning of the 3rd millennium BC were local shepherds, possibly the direct descendants, in both social and productive terms, of those specialised pastoral communities that had been generated and integrated (or, in less euphemistic terms, exploited) by the centralising elites emerging in the Upper Euphrates during the 4th millennium BC.

Data from phase VIB1 at Arslantepe highlight the fact that these pastoral communities did not simply survive the collapse of the centralised economic institutions that generated and possibly exploited them. In the early 3rd millennium BC they did become the protagonists of a long-term process of cultural innovation.

Conclusion

The data on the SPR at our disposal does not indicate a unitary and linear process of innovation throughout the entire 4th millennium BC in Mesopotamia, but rather processes of adoption and selection according to different economic, social and cultural contexts.

In southern Mesopotamia, the SPR has to be seen in the context of an urban way of life and a temple-based tributary economy. The intensification of the production of wool, textiles, milk and cheese was probably designed both to satisfy the staple needs of larger concentrations of people and to create economic surplus. In this socio-economic context, specialised pastoralism was a primary activity creating secondary structural products to feed a complex economic and social system based on productive specialisation and interdependence between producers and consumers.

As Greenfield (2005) has pointed out in his work on the SPR, one of the problems with Sherratt's model is that the appearance of the secondary products may have varied in time from region to region and that not all products may have arrived or been adopted as a unitary package.

Arslantepe seems to bear this out. The specialisation and possibly the intensification of husbandry strategies recorded

at the site from the mid-4th millennium BC onwards were not directly or exclusively connected to the exploitation of milk and wool, but rather to meat consumption and the accumulation of animal capital for exchange, storage and eventual consumption. Furthermore, this trend does not seem to have been adopted or emulated in a straightforward way from Uruk communities, since it preceded the Uruk influence at the site. In the non-urbanised context of Arslantepe, the exploitation of meat was a way of financing the economy of a centralised institution and building up wealth in the framework of relations of inequality. And this was ultimately the innovative side of the phenomenon. The central role of meat seems to have continued to figure in the dietary habits of the early 3rd millennium specialist pastoralists just as extensively as in the feasting activities performed by those same shepherds on the site.

However, Andrew Sherratt was not wrong in hypothesising that the development of different subsistence specialisations, such as those of agriculturalists and mobile pastoralists, may have led to the existence of polyethnic social systems (Frangipane 2015). With respect to Arslantepe, matters seem to have developed in this direction, and this may in fact be one of the keys to understanding the dynamics of long-term cultural change in this region.

This long-term cultural change started in the early 3rd millennium BC[2] as a result of a process of cultural appropriation headed by pastoral groups selecting and re-adapting Kura-Araxes cultural elements for the construction of their new cultural identity. This new identity was built on a strong pastoral infrastructure requiring mobility practices that may have strengthened a sense of collective identity different from that of the contemporary agricultural communities of the region. At Arslantepe this pastoral infrastructure was already very strong at the end of the Late Chalcolithic and was integrated into the centralised economic system as a 'structuring' component, however submitted or exploited to some degree. With the collapse of this system at the beginning of the Bronze Age, specialised pastoralism acted as a vector of cultural and eventually political innovation. The adoption of new traits linked to the Kura-Araxes model – traits that were the expression of small-scale communities, household production and kinship-based social organisation – may have been an attempt to reconstruct a new social order and collective social identity as an alternative, possibly even subversive, model to the centralised systems of the Chalcolithic.

We began our remarks by questioning the assumption of linear correlations between innovations and cultural change and more generally of unidirectional changes from simple to complex, from egalitarian to hierarchical, from low-tech to high-tech. The case of Arslantepe shows that similar innovations can occur in different social and political settings and can accordingly pervade, shape and be shaped differently by diverse *milieus*. More generally speaking, innovations can be the results of non-linear re-combinations of existing elements in which situational and relational dimensions become crucial factors in processes of appropriation.

Acknowledgements

We would like to thank Romina Laurito for valuable information on textile topics and Marcella Frangipane for her insightful suggestions for improving this paper. The paper has greatly benefited from the comments of an anonymous reviewer.

Notes

1   Haskel Greenfield notes that, as published at present, the mortality profiles of both phase VIA and VIB1 animals lump together kill-off patterns for both sheep and goats, thus representing a levelling form of 'average' information that may not reflect the specific use made of the products of each of these species.

2   The complexity of the Mesopotamian case, and especially the interdependences between socio-political complexity and SPR, is also attested to by what happened in other regions of the Uruk world. This is a complex topic that goes beyond the scope of the present article. However, it is worth briefly recalling that the end of the Uruk period marks important changes not only in Anatolia but also in northern Mesopotamia, though the adoption of the Uruk lifestyle was here much more pervasive and 'global' than north of the Taurus range. Though in northern Mesopotamia continuity in material culture between the Late Uruk and the 'early' Early Bronze Age is strong, some regions underwent a process of profound ruralisation, for example the Charchemish-Birecik sector along the mid Euphrates River valley (Ricci 2013). Along the whole 'fragile crescent' it was during the 'late' Early Bronze Age that large-scale wool production and the emergence of specialised herders boosted urbanisation. This also accounts for the occupation of more marginal landscapes during a period that was marked by the consolidation of strong city-based elites (Lawrence 2012; Lawrence and Wilkinson 2015, 337).

References

Balossi Restelli, F. (2015) Hearth and Home. Interpreting Fire Installations at Arslantepe, Eastern Turkey, from the Fourth to the Beginning of the Second Millennium BCE. *Paléorient* 41/1, 129–154.

Bartosiewicz, L. (2010) Herding in Period VIA. Development and Transformation from Period VII. In M. Frangipane (ed.), *Economic Centralisation in Formative States. The Archaeological Reconstruction of the Economic System in 4th Millennium Arslantepe*. Studi di Preistoria Orientale 3, 119–148. Rome, Sapienza University.

Berthon, R. (2015) Animal Resources in the Late Uruk Period Food Practices. In M. B. D´Anna, C. Jauss and J. C.

Johnson (eds), Food and Urbanization. Material and Textual Perspectives on Alimentary Practice in Early Mesopotamia. *Origini* 37, 43–45.

Dahl, J. (2015) The Production and Storage of Food in Early Iran. In M. B. D´Anna, C. Jauss and J. C. Johnson (eds), Food and Urbanization. Material and Textual Perspectives on Alimentary Practice in Early Mesopotamia. *Origini* 37, 67–72.

D'Anna, M. B. (2010) The Ceramic Containers of Period VIA. Food Control at the Time of Centralisation. In M. Frangipane (ed.), *Economic Centralisation in Formative States. The Archaeological Reconstruction of the Economic System in 4th Millennium Arslantepe*. Studi di Preistoria Orientale 3, 167–191. Rome, Sapienza University.

D'Anna, M. B. (2012) Between Inclusion and Exclusion. Feasting and Redistribution of Meals at Late Chalcolithic Arslantepe (Malatya, Turkey). In S. Pollock (ed.), *Commensality, Social Relations and Ritual: Between Feasts and Daily Meals*. eTopoi Journal for Ancient Studies, Special Volume 2, 97–123.

D'Anna, M. B. and Guarino, P. (2012) Pottery Production and Use at Arslantepe between Periods VII and VIA. Evidence for Social and Economic Change. *Origini* 34, 59–77.

D'Anna, M. B. and Jauss, C. (2015) Cooking at 4th Millennium BCE Chogha Mish (Iran) and Arslantepe (Turkey). Investigating the Social via the Material. In S. Kerner, C. Chou and M. Warmind (eds), *Commensality and Social Organisation, Food and Identity*, 65–85. London, Bloomsbury.

Frangipane, M. (2012a) Fourth Millennium Arslantepe: The Development of a Centralised Society without Urbanisation. *Origini* 34, 19–40.

Frangipane, M. (2012b) The Collapse of the 4th Millennium Centralised System at Arslantepe and the Far-Reaching Changes in 3rd Millennium Societies. *Origini* 34, 237–260.

Frangipane, M. (2014) After Collapse: Continuity and Disruption in the Settlement by Kura-Araxes-Linked Pastoral Groups at Arslantepe-Malatya (Turkey). New Data. *Paléorient* 40/2, 169–182.

Frangipane, M (2015) Different Types of Multiethnic Societies and Different Patterns of Development and Change in the Prehistoric Near East. *Procedings of the National Academy of Sciences* 112/30. DOI: 10.1073/pnas.1419883112.

Frangipane, M. and Palmieri, A. (1983) Cultural Developments at Arslantepe at the Beginning of the Third Millennium. *Origini* 12, 523–574.

Frangipane, M., Di Nocera, G. M., Hauptmann, A., Morbidelli, P., Palmieri, A., Sadori, L., Schultz, M. and Schmidt-Schultz, T. (2001) New Symbols of a New Power in a 'Royal' Tomb from 3000 BC Arslantepe, Malatya (Turkey). *Paléorient* 27/2, 105–139.

Frangipane, M., Andersson Strand, E., Laurito, R., Möllerwiering, S., Nosch, M.-L., Rast-Eicher, A. and Wisti Lassen, A. (2009) Arslantepe, Malatya (Turkey): Textiles, Tools and Imprints of Fabrics from the 4th to the 2nd Millennium BCE. *Paléorient* 35/1, 5–29.

Green, M. W. (1980) Animal Husbandry at Uruk in the Archaic Period. *Journal of Near Eastern Studies* 39/1, 1–35.

Greenberg, R. and Palumbi, G. (2014) Corridors and Colonies: Comparing Fourth – Third Millennia BC Interactions in Southeast Anatolia and the Levant. In B. Knapp and P. van

Dommelen (eds), *The Cambridge Prehistory of the Bronze and Iron Age Mediterranean*, 111–138. Cambridge, Cambridge University Press.

Greenfield, H. J. (2005) A Reconsideration of the Secondary Products Revolution: 20 Years of Research in the Central Balkans. In J. Mulville and A. Outram (eds), *The Zooarchaeology of Milk and Fats. Proceedings of the 9th ICAZ Conference, Durham 2002*, 14–31. Oxford, Oxbow Books.

Greenfield, H. J. (2010) The Secondary Products Revolution: The Past, the Present and the Future. *World Archaeology* 42/1, 29–54.

Laurito, R. (2007) Ropes and Textiles. In M. Frangipane (ed.), *Arslantepe Cretulae. An Early Centralised Administrative System Before Writing*, 381–396. Rome, Sapienza University.

Laurito R., Lemorini, C. and Perilli, A. (2014) Making Textiles at Arslantepe, Turkey, in the 4th and 3rd Millennia. Archaeological Data and Experimental Archaeology. In C. Breniquet and C. Michel (eds), *Wool Economy in the Ancient Near East and the Aegean*, 151–168. Oxford, Oxbow Books.

Lawrence, D. (2012) *Early Urbanism in the Northern Fertile Crescent: A Comparison of Regional Settlement Trajectories and Millennial Landscape Change.* PhD dissertation, Durham, Durham University. Available at Durham E-Theses Online: http://etheses.dur.ac.uk/5921/.

Lawrence, D. and Wilkinson, T. J. (2015) Hubs and Upstarts: Pathways to Urbanism in the Northern Fertile Crescent. *Antiquity* 89, 328–344.

McCorriston, J. (1997) The Fiber Revolution: Textile Extensification, Alienation, and Social Stratification in Ancient Mesopotamia. *Current Anthropology* 38/4, 517–535.

Nissen, H. (1986) The Archaic Texts from Uruk. *World Archaeology* 17/3, 317–334.

Palmieri, A. (1981) Excavations at Arslantepe (Malatya). *Anatolian Studies* 31, 101–119.

Palumbi, G. (2008) *The Red and Black. Social and Cultural Interaction between the Upper Euphrates and Southern Caucasus Communities in the Fourth and Third Millennium BC.* Rome, Sapienza University.

Pollock, S. (1990) Political Economy as Viewed from the Garbage Dump: Jemdet Nasr Occupation at the Uruk Mound, Abu Salabikh. *Paléorient* 16/1, 57–75.

Pollock, S. (1992) Bureaucrats and Managers, Peasants and Pastoralists, Imperialists and Traders: Research on the Uruk and Jemdet Nasr Periods in Mesopotamia. *Journal of World Prehistory* 6/3, 297–336.

Pollock, S. (1999) *Ancient Mesopotamia. The Eden that Never Was.* Cambridge, Cambridge University Press.

Porter, A. (2014) *Mobile Pastoralism and the Formation of Near Eastern Civilisations.* Cambridge, Cambridge University Press.

Ricci, A. (2013) An archaeological landscape study of the Birecik-Carchemish Region (Middle Euphrates River Valley) between the 5th and the 3rd millennium BC. PhD thesis, Kiel, University of Kiel.

Sagona, A. (1984) *The Caucasian Region in the Early Bronze Age.* British Archaeological Reports, International Series 214. Oxford, BAR.

Sagona, A. (1998) Social Identity and Religious Ritual in the Kura-Araxes Cultural Complex: Some Observations from Sos Höyük. *Mediterranean Archaeology* 11, 13–25.

Sagona, A. (2014) Rethinking the Kura-Araxes Genesis. *Paléorient* 40/2, 23–46.

Sagona, A. and Zimansky, P. (2009) *Ancient Turkey.* London, Routledge.

Sherratt, A. (1981) Plough and Pastoralism: Aspects of the Secondary Products Revolution. In I. Hodder, G. Isaac and N. Hammond (eds), *Pattern of the Past: Studies in Honour of David Clarke*, 261–305. Cambridge, Cambridge University Press.

Siracusano, G. and Bartosiewicz, L. (2012) Meat Consumption and Sheep/Goat Exploitation in Centralised and Non-Centralised Economies at Arslantepe, Anatolia. *Origini* 34, 111–123.

Siracusano, G. and Palumbi, G. (2014) 'Who'd be Happy, Let Him Be So: Nothing's Sure about Tomorrow.' Discarded Bones in an Early Bronze I Elite Area at Arslantepe (Malatya, Turkey): Remains of Banquets? In P. Bieliński, M. Gawlikowski, R. Koliński, D. Ławecka, A. Sołtysiak and Z. Wygnańska (eds), *Proceedings of the 8th International Congress on the Archaeology of the Ancient Near East, 30 April – 4 May 2012.* University of Warsaw 3, 349–365. Wiesbaden, Harrassowitz.

Smith, A. (2005) Prometheus Unbound: Southern Caucasia in Prehistory. *Journal of World Prehistory* 19, 229–279.

Smogorzewska, A. (2004) Andirons and their Role in Early Transcaucasian Culture. *Anatolica* 30, 151–177.

Van der Leeuw, S. (1989) Risk, Perception, Innovation. In S. E.van der Leeuw and R. Torrence (eds), *What's New? A Closer Look at the Process of Innovation*, 300–329. London, Unwin Hyman.

Vila, E. (1998) *L'exploitation des animaux en Mésopotamie aux IVe et IIIe millénaires avant J.-C.* Paris, CNRS.

Wagensonner, K. (2015) Vessels and other Containers for the Storage of Food According to the Early Lexical Record. In M. B. D'Anna, C. Jauss and J. C. Johnson (eds), Food and Urbanization. Material and Textual Perspectives on Alimentary Practice in Early Mesopotamia. *Origini* 37, 15–27.

Žižek, S. (2013) *Events. Philosophy in Transit.* London, Penguin.

Chapter 5

The 'Green Revolution' in Prehistory: Late Neolithic Agricultural Innovations as a Technological System

Maria Ivanova

Introduction

In the early 1960s, large areas of Asia and Latin America experienced a severe food crisis. China suffered the consequences of the Great Famine of 1958–1961, and mass starvation was threatening India. The breeding of improved crop varieties combined with the expanded use of fertilisers, other chemical inputs, mechanisation, and irrigation, led to a dramatic increase of agricultural productivity in the developing countries. Cereal production in Asia doubled between 1970 and 1995 and, instead of a widespread famine, calorie availability per person increased by nearly 30% (Thapa and Gaiha 2014, 74). The new 'package' of crops and techniques spread quickly. The area planted to High Yield Varieties (HYV) in the developing countries increased from about 30% in 1970 to 70% in 1990 (Hazell 2003). It needs to be stressed that it was only the adoption of the entire package of crops and technological inputs at the same time that guaranteed a satisfactory increase in production. The dramatic rise in agricultural productivity worldwide is known as the 'Green Revolution'. Importantly, this 'revolution' involved a series of agricultural techniques that had previously been well-established in the temperate zone. New was merely their adaptation to the agriculture of the developing nations in the tropical and subtropical area (Jones 1977, 55).

Such dramatic expansions in productivity have been rare in history. Archaeologists have been aware for a long time that the evidence for animal traction in tillage and transportation clusters in a narrow horizon around 3500 BCE (Sherratt 1981; 2004). This horizon has been documented in a vast area between Southwest Asia and Western Europe, possibly indicating a very fast (spread and) adoption of the technology. While the geographic origin of animal traction and other associated technical innovations has been a major topic of archaeological debate, the mechanisms of their introduction have rarely been questioned. The inherent benefits of animal traction, notably its application in tillage, seem to be a sufficiently convincing explanation for its wide adoption. Plow cultivation, it has been argued, enables farmers to increase the productivity of their land. Over and against manual cultivation, plough farming is assumed to be associated with higher output. It saves labour, enables farmers to increase their cultivated areas, increases the capacity for surplus, and promotes extensive cultivation regimes (Sherratt 1981, 287; Bogucki 1993, 498; Halstead 1995; Johannsen 2005, 47).

The present paper casts doubt on these apparently reasonable assumptions. We argue that animal traction was only one element in a system of interdependent agricultural techniques that developed around the middle of the 4th millennium BCE and facilitated an increase in agricultural production comparable to the 'Green Revolution'. As in the example of the 'Green Revolution', the techniques involved were not new, nor were they particularly beneficial in themselves. It was their integration into a coherent system that ensured their success.

ANTRAC (animal traction) in agricultural research

Empirical studies comparing the efficiency of manual versus animal traction (ANTRAC) farming are hard to find. The available data originate mainly from sub-Saharan Africa, where numerous attempts have been made to introduce animal traction in farming through development programs. In francophone west Africa, for example, over 125 projects with this aim were carried out between the 1930s and 1980s, but long-term adoption was limited to a few geographic areas (Delgado and McIntire 1982, 189). Adoption rates in other parts of the semi-arid tropics were similarly low. Thus, the rejection of ANTRAC developed into an important topic of debate in the 1970s and 1980s, with a body of literature evaluating unsuccessful programs and exploring the constraints of adoption. We use the available literature as a source of empirical data helpful in understanding the benefits and drawbacks of the two farming regimes.

There are several important caveats to consider when using these data. As in other present-day contexts, research on animal traction in sub-Saharan Africa deals with natural environments, crops, and technical implements that differ significantly from the ones in prehistoric Europe. Moreover, the latter were embedded in a substantially different socio-historical context. We do not intend to simply project the findings of research on recent farming practice back into prehistory. Rather, our aim is to better understand the relation between two regimes of manual and animal traction farming in the same environment. In a second step, a comparison of the regional findings can help recognise patterns that cut across regions and thus achieve an informed understanding of the constraints and potentialities of the two systems that transcends mere common sense.

The limitations of ANTRAC farming

The firm trust in the benefits of draft animals that has driven the majority of agricultural development programs makes the non-adoption of animal traction appear puzzling. Why do farmers reject a labour-saving technology that promises to increase land productivity? The relevant research has come up with some unexpected findings in connection with the profitability of ANTRAC and its effects on agricultural production.

Area under cultivation

The assumption has been that animal traction enables farmers to cultivate significantly larger areas through a reduction in the labour time required per hectare (Kjaerby 1983, 29; Starkey and Mutagubya 1992, 15). The dataset of Barrett *et al.* from Burkina Faso shows that farms employing draft animals do indeed cultivate larger areas (Barrett *et al.* 1982, tab. 4.15). However, the authors note that the number of workers in traction households was also higher. When calculated per active worker, the areas under cultivation were only marginally (by 5%) larger in ANTRAC farms compared to manual farms (Fig. 5.1). In this specific case, the employment of animal power did not bring about any actual land extensification. Similar observations have been made in Mali, where ANTRAC was found to be associated with a rise in total acreage, but also with higher household sizes (Jolly and Gadbois 1996, 460). A likely explanation is that, while speeding up tillage, the

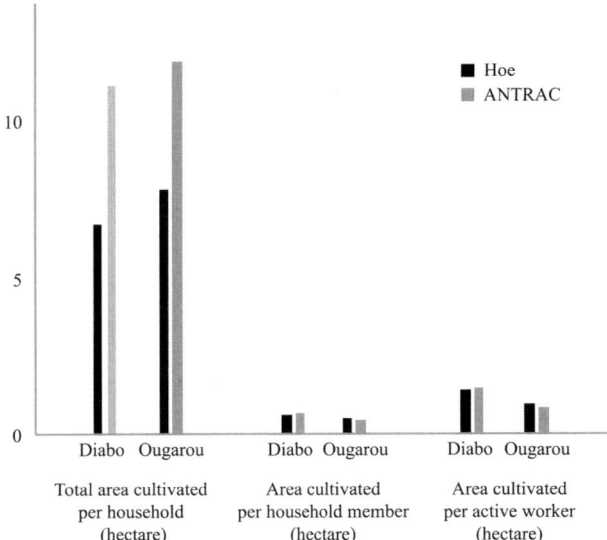

Figure 5.1: Area cultivated by hoe and ANTRAC households in two areas in Eastern Burkina Faso according to survey data from 1978 and 1979 (data from Barrett et al. 1982, tab. 4.15).

use of draft animals also exacerbates weeding and harvesting bottlenecks. In practice, then, the number of active workers on a farm remains the limiting factor for land extensification, regardless of tillage technology.

Labour input per hectare

The expectation that draft animals will save human labour in agriculture is supported to some extent by the available data. Compared to hoe farms, active workers on ANTRAC farms spent fewer hours on cropping per hectare of land (Barrett *et al.* 1982, tab. 4.19). However, this applies only to actual work in the fields. When maintenance of the draft animals was included, the total labour input of ANTRAC households per cropped unit of land was in fact higher (Delgado 1989, tab. 1) (Fig. 5.2). Accordingly, labour-intensive off-season work significantly offset the labour-saving effects of animal draft. In fact, taking account of the labour allocated to the draft team, input during the peak season was only insignificantly lower on ANTRAC farms. The assumption that animal traction would reduce labour bottlenecks and facilitate timeliness was therefore not fulfilled in this case.

Land productivity (output per hectare)

Lower yields under ANTRAC have been documented in several cases, and most authors seem to agree that draft animal power does not increase land productivity. Jolly and Gadbois (1996, 460, tab. 2), for example, state that in Mali cereal yields per hectare decreased with the introduction of draft animals. In this case, pure hoe farms obtained the highest yields per active labour unit. For Tanzania, Starkey and Mutagubya (1992, 15) concede that 'yield increases per unit area do not necessarily take place with animal traction, and higher yields

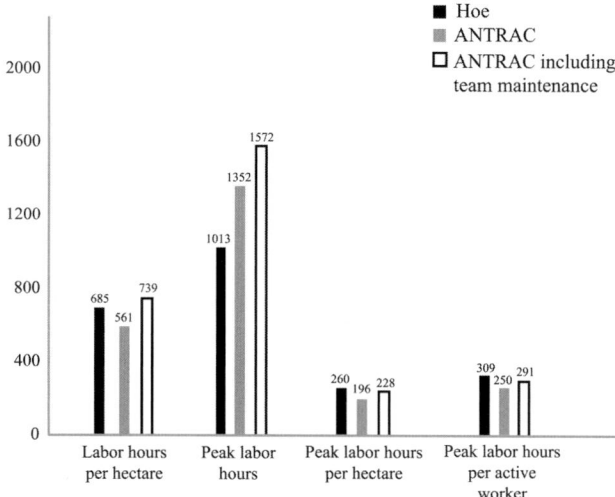

Figure 5.2: Labour hours in hoe and ANTRAC households according to survey data from 1978 and 1979 in Eastern Burkina Faso (data from Delgado 1989, tab. 1, compiled from Barrett et al. 1982, tabs. 4.15, 4.19, 5.2 and team maintenance data on page 66).

may actually be obtained from intensive hand hoe cultivation'. Even tractors do not always raise yields, as demonstrated by the numerous tractorisation projects in sub-Saharan Africa (Binswanger and Pingali 1988, 91). Decreasing yields under animal traction regimes are often related to weed infestation. Weed control was, for example, the limiting factor when ox-ploughs were introduced in some areas of Tanzania. The drop in production per farmer could not even be counterbalanced by planting larger areas, and, in comparison, hoe cultivation was both more productive and less demanding in terms of land requirements (Kjaerby 1983).

Labour productivity (output per active person)

Given the increasing labour demands and decreasing yields in ANTRAC farming described above, the finding that draft animal power does not necessarily boost labour productivity is not surprising. Even if crop income per hour of cropping labour and per hour of peak cropping labour are clearly higher on oxen farms, the actual product of labour is not higher when the maintenance of draft animals is added. Delgado and McIntire (1982, 193) estimate that profitability may be even lower for farmers who maintain their own draft team, unless they can rent the animals out.

The benefits of ANTRAC farming

The studies discussed above show that ownership and employment of draft animals in farming are not invariably superior to manual cultivation. The high costs of animal maintenance combined with the decrease in land productivity under ANTRAC represent the major constraints of the technology. For many smallholders, acquiring draft animals was clearly not profitable. However, for some

farmers the traction system yielded acceptable returns. The crucial question is therefore under what conditions ANTRAC becomes superior to manual cultivation. If off-field investment in labour for tending the animals is the most important factor, then farmers stepping up the utilisation of animals should potentially make ANTRAC sustainable.

This expectation is indeed supported both by survey data and modelling. The analysis of empirical data in Jaeger and Malton (1993), for example, makes it clear that high levels of utilisation were crucial for the profitable adoption of animal traction, while Jolly and Gadbois (1996) convincingly argue that if pure hoe farms were better off than 'semi-equipped', the 'fully equipped' group was clearly privileged in terms of productivity. A simulation model based on data from Niger and Nigeria provides a similar picture. ANTRAC only yields satisfactory returns when the package includes a cart. For millet-sorghum farmers in Nigeria, in particular, animal traction would not be a plausible option under any circumstances because of the prohibitively high market prices for ox-carts (Jansen 1993).

The most immediate effect of draft animals in transportation is an increase in the training of the animals. Well-trained draft teams are faster in tillage work, and training enhances work quality. Moreover, nutrition for work animals can be significantly improved by the use of animal traction in transportation (Starkey and Mutagubya 1992, 23). Stocking crop residues and fodder generally only becomes feasible when farmers start using carts, as collecting and storing is difficult without transport. Experience from West Africa demonstrates that animal-drawn carts are one of the best means of improving nutrition for work animals, and the improved condition of the draft animals has a substantially positive effect on operation times (Starkey and Mutagubya 1992, 23).

Animal transport also facilitates the recycling of nutrients and thus enhances the productivity of land. Manure is very heavy and, depending on the distance from the fields, large loads of it are usually only transported by means of animal traction. As pointed out by Starkey and Mutagubya (1992, 16), animal-drawn transport makes the manuring of distant plots a practicable proposition, and farmers are likely to travel further to their fields. Access to effective transport can thus increase both the range of travel and manuring by several kilometres, so that more land becomes profitable for farming. While simple animal-drawn vehicles such as travois and sledges are far superior to human carriage and pack animals, the use of wheeled vehicles dramatically increases the capacity to move heavy loads. Compared to a typical sledge load of about 250 kg (Dennis 1999; Isaakidou 2011, 104), loads of up to 900 kg can be transported by a two-oxen cart. When average speed is taken into account, the transport capacity (ton km/h) of wheeled vehicles is nearly five times higher than that of sledges (Dennis 1999, tab. 1).

In conclusion, empirical research on agricultural development suggests that better integration of plant and

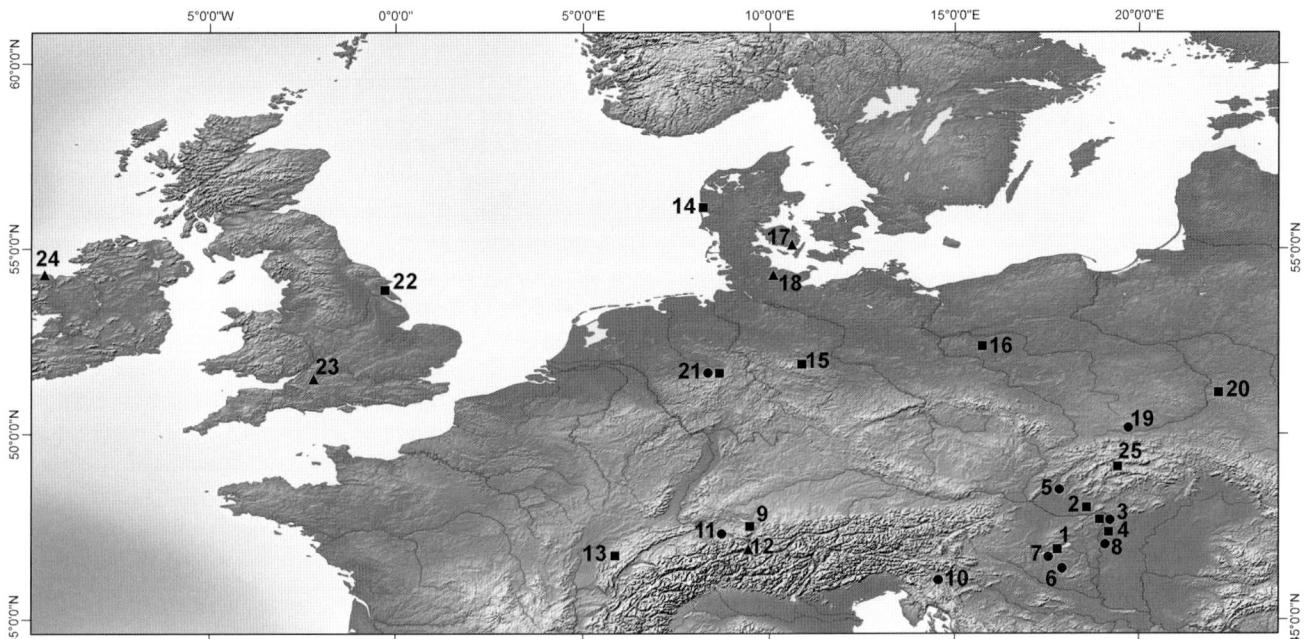

Figure 5.3: Map showing the distribution of ANTRAC in Europe (c. 3600/3500–3000 BCE). (1) Balatonőszöd-Temetői dűlő; (2) Svodin; (3) Budakalász; (4) Alsonémedi; (5) Radošina; (6) Kaposvár; (7) Boglárlelle; (8) Szigetszentmarton; (9) Arbon Bleiche 3; (10) Stare gmajne; (11) Zürich AKAD; (12) Chur; (13) Chalain; (14) Troldebjerg; (15) Derenburg; (16) Bytyń; (17) Højensvej 7; (18) Flintbek; (19) Bronocice; (20) Krężnica Jara; (21) Lohne-Züschen; (22) Etton; (23) South Street Long Barrow; (24) Céide Fields. square – paired draft; circle – vehicles; triangle – plough.

animal production through utilisation of draft animals in both tillage and transportation is one of the keys to ANTRAC's success. Isolated indications of ANTRAC use from the 6th and 5th millennia BCE in Europe (Isaakidou 2006; Tarrús *et al.* 2006) contrast with a striking upsurge of evidence around the middle of the 4th millennium. A comparison of the three regions that feature prominently in the research done on the introduction of animal traction in prehistoric Europe so far – the Carpathian Basin, the North Alpine area, and the North European Plain/Denmark - illustrates the emergence of a profitable ANTRAC farming system based on tighter crop-animal integration (Fig. 5.3).

The Carpathian Basin

The Carpathian Basin is regarded as one of the earliest centres of animal traction in Europe. However, we have almost no osteological evidence for the use of work animals from this region. A deformation of the horn core possibly caused by a yoke has been identified on the individual from Pit 1612 form Balatonőszöd, a feature with a radiocarbon date of 3140–2990 BCE (Horváth 2010, 3, fig. 13). A remarkable find was reported from Svodin in Slovakia. Two cattle skeletons uncovered next to each other in a pit (Grave 328/74) were identified as belonging to castrated adult males. Pathological changes of the older individual, including spondylitic lesions on the spine and broadening of the metatarsus, attested to its use as a draft animal, while

the anomalous development of the lumbar vertebrae of the younger individual indicated abnormal stress in its early years. The asymmetries in these pathological changes strongly suggest the use of these two individuals for traction as a pair (Fabis 2005). Although cattle burials have been frequently excavated from sites of this period, no further pathologies have been reported so far. Striking are the 'double burials' of adult cows accompanied by 10–18-month-old calves, such as those from Budakalász and Alsónémedi (Bökönyi 1951; Gál 2009). It has been suggested that they may represent draft cattle, an interpretation that appears likely since, as noted by Johannsen, breaking in a young animal by pairing it with an experienced adult is widespread practice (Johannsen 2011). A draft team consisting of a six-year-old cow and her calf aged less than one year has been observed in eastern Romania by Bartosiewicz *et al.* (1997, 17).

Depictions of cattle pairs, such as the small copper statuette from the Lisková cave in northern Slovakia and the box-shaped ceramic artefacts with applications of cattle protomes from Radosina, Kaposvár, and Boglárlelle, indicate that paired traction was practiced around the middle of the 4th millennium (Němejcová-Pavúková and Bárta 1977; Ecsedy 1982, fig. 8; Struhár *et al.* 2010; Bondár 2012). The models from Budakalász and Szigetszentmarton unambiguously representing wagons with wheels stem from the end of the same millennium (Maran 2004; Siklósi 2009). Employment in tillage has not been demonstrated via direct evidence. The farming system in which draft cattle teams were employed is poorly known. There are only a few studies of animal bone material from the Boleráz/Baden sites showing that in terms of numbers cattle predominated (Hoekman-Sites and Giblin 2012, tab. 2). Glume cereals, in particular emmer, were the most important crops (Gyulai 2010, 88–90).

The North Alpine area

In the later 4th millennium, cattle herding in the North Alpine area underwent a succession of changes. While in the faunal assemblages of the Pfyn period (*c.* 4000–3500 BCE) from Lake Zurich cattle were represented by up to 60% of NISP, their proportion decreased to 20% in the following Horgen period (*c.* 3400–3000 BCE). At the same time, there was a rise in the number of animals that reached an advanced age, with cattle aged 10–14 years outnumbering their 3–10-year-old counterparts. Remarkable is the small body size of the females in comparison to earlier periods, which may be an indication for selective breeding and/ or isolated populations. The number of adult males significantly increased from the Pfyn to the Horgen period (Hüster-Plogmann 2002, 105). Body-part distributions show that cattle were now slaughtered inside the villages, a practice that had not been customary before. Very similar developments were observed in other parts of the North Alpine region, for example at Arbon-Bleiche 3 and Chalain 3 (Arbogast 1997, 663–666; Deschler-Erb and Marti-Grädel 2004). A change in the use of cattle is also discernible in the tendency for the front limbs of cattle to widen from 3400 BCE onwards, a phenomenon interpreted by Deschler-Erb and Marti-Grädel (2004, 172) as indicating that they were increasingly pressed into service for traction purposes.

Traction in pairs is supported by the discovery of what may be a wooden yoke from Arbon Bleiche 3 (Leuzinger 2002). Criss-cross plough marks from Chur/ Areal Ackermann and Castaneda in Graubünden, the former probably dating back to the transition from Pfyn to Horgen, attest to the use of draft animals in tillage (Rageth 1992; Pétrequin *et al.* 2006). A travois and a yoke found at Chalain 19 (*c.* 3100 BCE) and the wooden wheels from Zürich AKAD (*c.* 3300 BCE) and Stare Gmajne in the Ljubljana marshes (*c.* 3160 BCE) testify to the use of draft animals in transportation (Ruoff and Jacomet 2002; Pétrequin *et al.* 2006a; Pétrequin *et al.* 2006b; Velušček 2009). While manuring was practiced at least from the beginning of the 4th millennium onwards, as indicated by stable cereal isotopes from Hornstaad Hörnle qualifying for inclusion in 'medium' and 'high' manuring categories, direct evidence for an intensification of manuring practices coeval with the introduction of wheeled transport is not yet available (Bogaard *et al.* 2013, fig. 4C). The analysis of plant remains in cattle dung from Arbon Bleiche 3 indicated winter stalling in the settlement and a diet of leafy fodder and cereals (Kühn and Hadorn 2004).

The archaeobotanical assemblages from the Horgen period are indicative of long-term and intensive cultivation in stable plots (Brombacher and Jacomet 1997) and show increasing agricultural intensity (Hosch and Jacomet 2004, 134). A remarkable shift in cereal crops took place at this time. Tetraploid free threshing wheat, the predominant crop of the earliest farming communities in the northern Alpine foreland (Maier 1996; 1998; Jacomet 2007), was replaced in the late 4th millennium by glume cereals, notably emmer and barley. This trend is documented in several parts of the region, for example at Chalain and Clairvaux, and at a series of sites on Lake Zurich and Lake Constance (Lundström-Baudais 1986; Hosch and Jacomet 2004, 124, fig. CD77; Pétrequin *et al.* 2006a, 115). It was accompanied by conspicuous changes in harvesting techniques. Petrequin *et al.* (2006a) observe that a variety of harvesting tools existed during the Pfyn/Cortaillod period (4000–3500 BCE), though only one of them remained in use in the later 4th millennium (Fig. 5.4). Experiments demonstrated that this 'Horgen-type' harvesting knife was most efficient for quickly plucking the ears in densely sown cereal fields, avoiding weeds and leaving straw in place. In correspondence analysis, the Horgen-type knife clearly stood out among all types of harvesting tools tested. Pétrequin *et al.* (2006a, 118) relate this marked shift in harvesting strategy to the introduction of the ard plus increasing cultivation areas and decreasing weed control. This hypothesis finds support in the composition of weed assemblages associated with cereals at Arbon Bleiche, which predominantly contained medium- and high-growing taxa (Hosch and Jacomet 2004, 137).

In summary, the later 4th millennium was a period of increasing agricultural intensity and a re-structuring of the farming system in the North Alpine area, during which the role of animal traction was modified. Changes included more regular use of traction teams for tillage with the ard and transportation with travois and carts, as well as related developments in cereal crops, new harvesting techniques and weeding strategies, and possibly intensive manuring (which had already been practiced in the preceding period). These changes ensued after a 'crisis' in food production around 3700–3600 BCE in many lakeshore settlements, during which declining crop yields were offset by higher levels of hunting and foraging, followed by a settlement hiatus lasting several hundred years (Schibler *et al.* 1997).

The North European Plain and Denmark

Evidence for cattle traction turns up in this area around the middle of the 4th millennium (the ENII period according to the chronology of southern Scandinavia). Draft-related morphological changes to cattle phalanges, notably pronounced and extreme extensions of articular surfaces (lipping and broadening) have been documented at several sites in Denmark (Johannsen 2006). Burials of pairs of

French Jura   Trois-Lacs   C. Switzerland   Zurich   Constance   Federsee

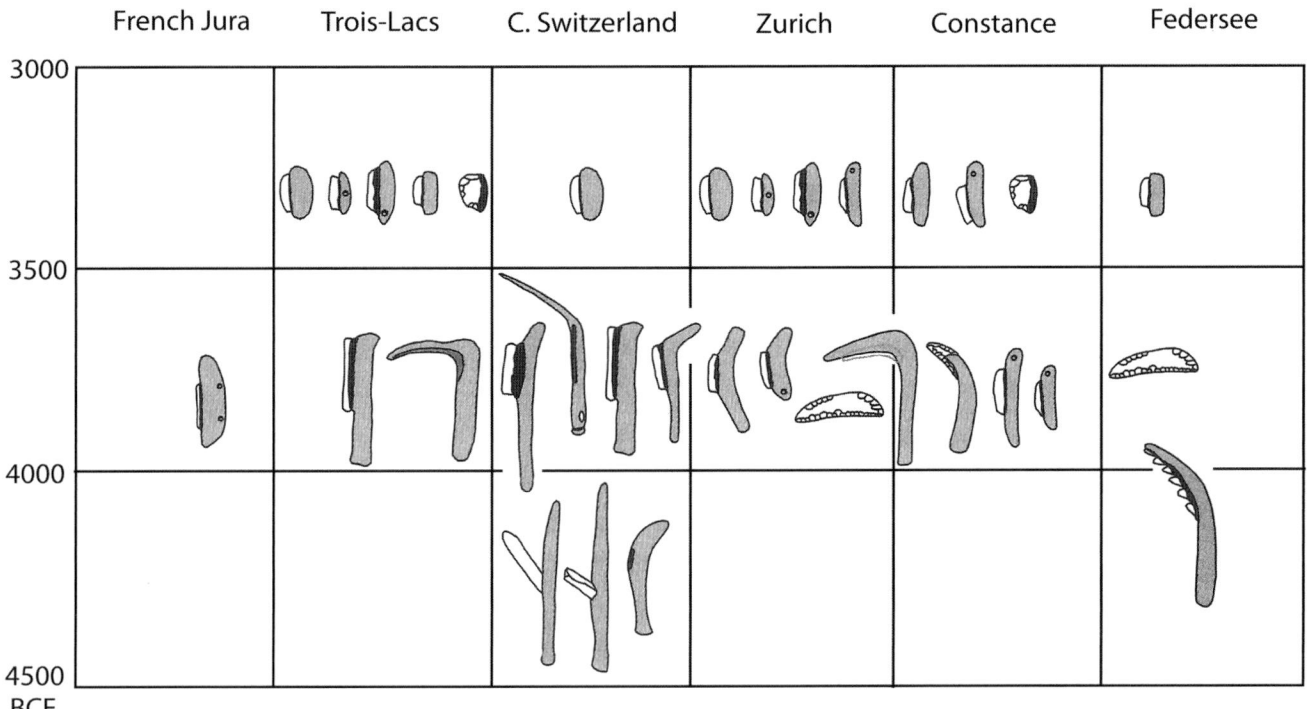

Figure 5.4: Harvesting tools in the North Alpine area (after Pétrequin et al. 2006).

adult cattle, such as the find from Derenburg (Dohle and Stahlhofen 1985), may be associated with paired traction, while depictions of draft teams like the famous copper pair from the hoard of Bytyń and the handle of a ceramic vessel from Krężnica Jara (Dinu 1981; Matuschik 2002) emphasise the importance of animal traction in this region.

Plow marks show that cattle were put to work in tillage from the ENII period onwards. Sixteen instances of plough marks from the 4th millennium have been documented under barrows, four from the EN II (3500–3300 BCE), ten from MN I and II (3300–3000 BCE), and two further examples attributed to either EN II or MN I/II (Thrane 1989). A *terminus ante quem* for the plough marks below the long barrow of Højensvej 7, Fyn provides a ^{14}C date of 3770–3637 BCE on charred hazelnut from a pit cutting the marks (Sørensen and Karg 2014, 104). Two sets of plough marks under the barrow at Flintbek LA 3 assigned to Phases 4 and 5 precede the construction of megalithic tombs at this site and can therefore be dated before 3480–3450 BCE (Mischka 2011, 756). Unambiguous evidence for wheeled vehicles also materialises around this time, with the cart tracks from Flintbek having been recently dated to 3420–3385 BCE (Mischka 2011, 755). Another well-known find, the depiction of a two-axle wagon on the vessel from Bronocice, comes from Horizon III at this site and dates to 3637–3373 BCE (Bakker *et al.* 1999, tab. 1).

Recent results from the stable isotope analysis of charred grains indicate that in the fourth millennium BCE manuring was an established farming practice in northern Europe

(Bogaard *et al.* 2013; Kanstrup *et al.* 2014). Ceremonial aggregation sites from the second half of the 4th millennium, such as Sarup, yielded samples with δ^{15}N values consistent with little or no manuring, while samples from contemporary settlement sites showed high levels of manuring and apparently originated from intensively cultivated plots (Bogaard *et al.* 2013, fig. 4). More data are needed not only to verify this difference between ceremonial and habitation sites, but also to compare cultivation regimes before and after the introduction of the ard. At present, lack of data from the early 4th millennium makes this impossible.

The evidence for animal use in tillage, transportation, and manuring coincides with the increasing human impact on the landscape of northern Europe. Several proxies for anthropogenic influence show substantial changes between 3750 and 3500 BCE, such as a decline in forests and an extension of arable land (Feeser *et al.* 2012; Kirleis and Fischer 2014). Wood charcoal and charred seeds of wild plants at sites from this period suggest shifts in forest composition to light-demanding species (Whitehouse and Kirleis 2014). Moreover, the intensification in animal traction coincided with a notable shift in cropping practices. In ENII, free-threshing wheat, the predominant cereal crop since the introduction of farming, was replaced rather abruptly by glume cereals such as emmer and barley (Kirleis and Fischer 2014, fig. 6). This shift to glume cereals is very similar to the situation in the North Alpine area.

In short, a new farming regime was adopted in northern Europe around 3500 BCE. Traction is well documented from the middle of the 4th millennium onwards, osteopathologies indicate the intensive use of cattle as work animals, and finds show that draft teams were employed both in tillage and in transportation with wheeled vehicles. The new system involved a shift in cereal crops from naked wheat to glume cereals. The use of large quantities of manure was part of this regime, but the available evidence does not permit any conclusions about the intensification of manuring in line with the expansion of draft animal tillage and transportation in the second half of the 4th millennium. The increasing impact of human activity on the landscape of northern Europe with the introduction of ard cultivation and animal traction is striking. These developments are closely paralleled in the North Alpine area. In both regions, the relative abruptness of change is highly important and suggests the adoption of a package of interrelated techniques rather than a gradual development.

Demography in Late Neolithic Europe

It is difficult to establish how the changes in farming regimes that took place around the middle of the 4th millennium BCE correlated with population changes. Studies on a continent-wide scale based on radiocarbon dates as a proxy for human activity and population size show considerable regional variation and significant fluctuations, but the picture is too coarse-grained for the aims of the present paper (Shennan *et al.* 2013). The analysis by Hinz *et al.*, using pollen records and radiocarbon dates as proxies for human populations during the 4th millennium BCE in northern Europe, provides a more detailed account (Hinz *et al.* 2012). The curves of pollen and ^{14}C-dates indicate a rise in activity either from 4000 or from 3800/3700 cal BCE peaking around 3500 and declining after 3350 BCE, with a notable depression around 3100 BCE (exceptions are the Danish Isles, eastern central Sweden, and the Swedish Baltic Isles with no evidence of decline after 3350 BCE). Strikingly, the same patterns of population expansion and contraction, with an upsurge around 3800/3700 and a decline after 3350 BCE were also observed by Hinz *et al.* in regions where farming started considerably earlier than in northern Europe and southern Scandinavia, such as central Germany and Kujawy in Poland. Therefore, it appears likely that in northern Europe these fluctuations were not simply a consequence of the initial introduction of farming but were related rather to the above-mentioned changes in farming regimes.

Discussion and conclusions

The fact that agricultural systems changed in the second half of the 4th millennium BCE is by no means a novel archaeological insight. It has also been widely recognised that animal traction played an important role in this process. The question why these changes happened, however, has received

relatively little attention, and the advantages of ANTRAC farming in terms of increasing land productivity and saving labour have often been taken for granted. In this paper, we have shown that the ownership and use of draft animals in farming are not invariably superior to manual cultivation. To understand the use of animal power in prehistoric Europe farming, we need to interpret the evidence for traction in the context of the entire farming system.

There is a growing body of evidence supporting of a model of early farming in Europe as small-plot, intensive, 'garden' cultivation (Bogaard 2004; Hosch and Jacomet 2004; Isaakidou 2011; Bogaard *et al.* 2013; Saqalli *et al.* 2014). In recent pre-mechanised cultivation in Greece and Asturias discussed by Isaakidou (2011), animal draft and intensive garden cultivation were not mutually exclusive practices. Cows were employed for ploughing, but their main contribution was in transport rather than tillage – they pulled sledges with manure, moved harvests from the fields and fodder to the byres. Apart from draft power, cows provided other valuable products such as stall manure, milk and calves. Isaakidou's suggestion that a similar regime may have been implemented by the first farmers in Europe is plausible, since practices such as traction, manuring and foddering seem to have been intrinsic to European farming from the very beginning (Isaakidou 2006; Bogaard *et al.* 2013). However, Isaakidou's study does not offer any explanation for the sudden upsurge of evidence for animal traction around 3500 BCE, coinciding with population growth and increasing human impact on the environment.

In our opinion, the essence of the major transformation in farming regimes taking place in some regions of Europe during the later 4th millennium was not extensification but tight integration of plant and animal production. At this time, ANTRAC tillage was combined with animal traction in transportation and threshing, cultivation of hulled cereals, manuring, stabling, and foddering to produce an efficient crop-livestock system. While the separate techniques were often cost- and labour-intensive in isolation, they became affordable and efficient when applied as a package. Wheeled transport, which apparently appeared around this time, played a pivotal role, permitting a dramatic increase in the capacity for moving bulky loads in farming, notably fodder and manure. This new farming regime based on ANTRAC had one critical advantage over the previous systems. Against established opinion, empirical studies comparing the efficiency of manual versus animal traction farming demonstrate that this advantage did not lie in larger cultivation areas, higher yields or lower labour input. The point was that better crop-animal integration enabled ANTRAC farmers to improve land productivity and ensure self-reliance. There are unmistakable parallels between these developments and the 'Green Revolution' of the 1960s. Both involved previously existing crops and techniques combined in a new 'package' that, if introduced at the same time, brought about an unprecedented increase in productivity. From a technological perspective, the ANTRAC complex of the later 4th millennium is not a revolutionary innovation requiring top-down introduction coordinated by an elite but an adjustment and extension of traditional farming skills in land and livestock management.

References

Arbogast, R.-M. (1997) La grande faune de Chalain 3. In P. Pétrequin (ed.), *Les sites littoraux néolithiques de Clairvaux-les-lacs et de Chalain (Jura) III: Chalain station 3, 3200–2900 av. J.-C.* 2, 641–692. Paris, Éditions de la maison des sciences de l'homme.

Bakker, J. A., Kruk, J., Lanting, A. E. and Milisauskas, S. (1999) The Earliest Evidence of Wheeled Vehicles in Europe and the Near East. *Antiquity* 73, 778–790.

Barrett, V., Lassiter, G., Wilcock, D., Bakerand, D. and Crawford E. (1982) *Animal Traction in Eastern Upper Volta: A Technical, Economic and Institutional Analysis.* MSU International Development Paper 4. East Lansing, Department of Agricultural Economics, Michigan State University.

Bartosiewicz, L., Neer, W. van and Lentacker, A. (1997) *Draught Cattle: Their Osteological Identification and History.* Tervuren, Musée Royal de l'Afrique Centrale.

Binswanger, H. and Pingali, P. (1988) Technological Priorities for Farming in Sub–Saharan Africa. *The World Bank Research Observer* 3/1, 81–98.

Bogaard, A. (2004) *Neolithic Farming in Central Europe. An Archaeobotanical Study of Crop Husbandry Practices.* London, Routledge.

Bogaard, A., Fraser, R., Heaton, T. H. E., Wallace, M., Vaiglova, P., Charles, M., Jones, G., Evershed, R. P., Styring, A. K., Andersen, N. H., Arbogast, R.-M., Bartosiewicz, L. S., Gardeisen, A., Kanstrup, M., Maier, U., Marinova, E., Ninov, L., Schäfer, M. and Stephan, E. (2013) Crop Manuring and Intensive Land Management by Europe's First Farmers. *Proceedings of the National Academy of Sciences* 110/51, 12589–12594.

Bogucki, P. (1993) Animal Traction and Household Economies in Neolithic Europe. *Antiquity* 67, 492–503.

Bökönyi, S. (1951) Untersuchung der Haustierfunde aus dem Gräberfeld von Alsónémedi. *Acta Archaeologica Academiae Scientiarum Hungaricae* 1, 72–79.

Bondár, M. (2012) A New Late Copper Age Wagon Model from the Carpathian Basin. In P. Anreiter, E. Bánffy, L. Bartosiewicz, W. Meid and C. Metzner-Nebelsick (eds), *Archaeological, Cultural and Linguistic Heritage. Festschrift for Erzsébet Jerem in Honour of Her 70th Birthday*, 79–91. Budapest, Archaeolingua.

Brombacher, C. and Jacomet, S. (1997) Ackerbau, Sammelwirtschaft und Umwelt: Ergebnisse archäobotanischer Untersuchungen. In J. Schibler, H. Hüster-Plogmann, S. Jacomet, C. Brombacher, E. Gross-Klee and A. Rast-Eicher (eds), *Ökonomie und Ökologie Neolithischer und Bronzezeitlicher Ufersiedlungen am Zürichsee. Ergebnisse der Ausgrabungen Mozartstrasse, Kanalisationssanierung Seefeld, AKAD/Pressehaus und Mythenschloss in Zürich*, 220–229. Zürich, Zürich and Egg.

Delgado, C. and McIntire, J. (1982) Constraints on Oxen Cultivation in the Sahel. *American Journal of Agricultural Economics* 64, 188–196.

Delgado, C. L. (1989) The Changing Economic Context of Mixed Farming in Savanna West Africa: A Conceptual Framework Applied to Burkina Faso. *Quarterly Journal of International Agriculture* 28, 351–364.

Dennis, R. (1999) Meeting the Challenge of Animal-Based Transport. In P. Starkey and P. Kaumbutho (eds), *Meeting the Challenges of Animal Traction. A Resource Book of the Animal Traction Network for Eastern and Southern Africa*, 150–169. London, Intermediate Technology Publications.

Deschler-Erb, S. and Marti-Grädel, E. (2004) Viehhaltung und Jagd. Ergebnisse der Untersuchung der handaufgelesenen Tierknochen. In S. Jacomet, U. Leuzinger and J. Schibler (eds), *Die Jungsteinzeitliche Seeufersiedlung Arbon Bleiche 3. Umwelt und Wirtschaft*. Archäologie im Thurgau 12, 158–251. Frauenfeld, Amt für Archäologie.

Dinu, M. (1981) Clay Models of Wheels Discovered in Copper Age Cultures of Old Europe Mid-Fifth Millenium B.C. *The Journal of Indo-European Studies* 9, 1–14.

Dohle, H.-J. and Stahlhofen, H. (1985) Die neolithischen Rindergraber auf dem 'Lowenberg' bei Derenburg, Kr. Wernigerode. *Jahresschrift für Mitteldeutsche Vorgeschichte* 68, 157–177.

Ecsedy, I. (1982) Későrézkori leletek Boglárlelléről. Late Copper Age Finds from Boglárlelle. *Communicationes Archaeologicae Hungariae*, 15–29.

Fabis, M. (2005) Pathological Alteration of Cattle Skeletons – Evidence for the Draught Exploitation of Animals? In J. Davies, M. Fabis, I. Mainland, M. Richards and R. Thomas (eds), *Diet and Health in Past Animal Populations. Current Research and Future Directions. Proceedings of the 9th Conference of ICAZ, Durham, August 2002*, 58–62. Oxford, Oxbow Books.

Feeser, I., Dörfler, W., Averdieck, F.-R. and Wiethold, J. (2012) New Insight into Regional and Local Land-Use and Vegetation Patterns in Eastern Schleswig-Holstein during the Neolithic. In M. Hinz and J. Müller (eds), *Siedlung, Grabenwerk, Großsteingrab. Studien zu Gesellschaft, Wirtschaft und Umwelt der Trichterbechergruppen im nördlichen Mitteleuropa. Frühe Monumentalität und soziale Differenzierung* 2, 159–190. Bonn, Habelt.

Gál, E. (2009) Animal Bone Offerings from the Baden Cemetery at Budakalász. In M. Bondár and P. Raczky (eds), *The Copper Cemetery of Budakalász*, 372–378. Budapest, Pytheas.

Gyulai, F. (2010) *Archaeobotany in Hungary. Seed, Fruit, Food and Baverage Remains in the Carpathian Basin from the Neolithic to the Late Middle Ages.* Budapest, Archeolingua.

Halstead, P. (1995) Plough and Power: The Economic and Social Significance of Cultivation with the Ox-Drawn Ard in the Mediterranean. *Bulletin on Sumerian Agriculture* 8, 11–22.

Hazell, P. B. R. (2003) The Green Revolution. In J. Mokyr (ed.), *Oxford Encyclopedia of Economic History*, 478–480. Oxford, Oxford University Press.

Hinz, M., Feeser, I., Sjögren, K.-G. and Müller, J. (2012) Demography and the Intensity of Cultural Activities: An Evaluation of Funnel Beaker Societies (4200–2800 cal BC). *Journal of Archaeological Science* 39/10, 3331–3340.

Hoekman-Sites, H. A. and Giblin, J. I. (2012) Prehistoric Animal Use on the Great Hungarian Plain: A Synthesis of Isotope and

Residue Analyses from the Neolithic and Copper Age. *Journal of Anthropological Archaeology* 31, 515–527.

Horváth, T. (2010) Transcendent Phenomena in the Late Copper Age Boleráz/Baden Settlement Uncovered at Balatonőszöd-Temetői dűlő: Human and Animal 'Depositions'. *Journal of Neolithic Archaeology* 12. DOI: http://dx.doi.org/10.12766/jna.2010.54.

Hosch, S. and Jacomet, S. (2004) Ackerbau und Sammelwirtschaft. Ergebnisse der Untersuchung von Samen und Früchten. In S. Jacomet, U. Leuzinger and J. Schibler (eds), *Die Jungsteinzeitliche Seeufersiedlung Arbon Bleiche 3. Umwelt und Wirtschaft*. Archäologie im Thurgau 12, 112–157. Frauenfeld, Amt für Archäologie.

Hüster-Plogmann, H. 2002. Früheste archäozoologische Hinweise zur Nutzung von Rindern als Zugtiere in neolithischen Siedlungen der Schweiz. In J. Köninger, M. Mainberger, H. Schlichtherle and M. Vosteen (eds), *Schleife, Schlitten, Rad und Wagen. Zur Frage früher Transportmittel Nördlich der Alpen* (Hemmenhofener Skripte 3), 103–106. Freiburg, Janus.

Isaakidou, V. (2006) Ploughing with Cows: Knossos and the 'Secondary Products Revolution'. In D. Serjeantson and D. Field (eds), *Animals in the Neolithic of Britain and Europe*, 95–112. Oxford, Oxbow Books.

Isaakidou, V. (2011) Farming Regimes in Neolithic Europe: Gardening with Cows and Other Models. In A. Hadjikoumis, E. Robinson and S. Viner (eds), *Dynamics of Neolithisation in Europe: Studies in Honour of Andrew Sherratt*, 90–112. Oxford, Oxbow Books.

Jacomet, S. (2007) Neolithic Plant Economies in the Northern Alpine Foreland from 5500–3500 cal BC. In S. Colledge and J. Conolly (eds), *The Origins and Spread of Domestic Plants in Southwest Asia and Europe*, 221–258. Walnut Creek, CA, Left Coast Press.

Jaeger, W. K. and Matlon, P. J. (1993) Utilization, Profitability, and the Adoption of Animal Draft Power in West Africa. *American Journal of Agricultural Economics* 72, 35–48.

Jansen, H. G. P. (1993) Ex-Ante Profitability of Animal Traction Investments in Semi-Arid Sub-Saharan Africa: Evidence from Niger and Nigeria. *Agricultural Systems* 43/3, 323–349.

Johannsen, N. N. (2005) Palaeopathology and Neolithic Cattle Traction: Methodological Issues and Archaeological Perspectives. In J. Davies, M. Fabis, I. Mainland, M. Richards and R. Thomas (eds), *Diet and Health in Past Animal Populations: Current Research and Future Directions,* 39–51. Oxford, Oxbow Books.

Johannsen, N. N. (2006) Draught Cattle and the South Scandinavian Economies of the 4th Millennium BC. *Environmental Archaeology* 11, 35–48.

Johannsen, N. N. (2011) Past and Present Strategies for Draught Exploitation of Cattle. In A. Arbarella and A. Rentakoste (eds), *Ethno-Zoo-Archaeology: The Present and Past of Human-Animal Relationships*, 13–19. Oxford, Oxbow Books.

Jolly, C. M. and Gadbois, M. (1996) The Effect of Animal Traction on Labour Productivity and Food Self-Sufficiency: The Case of Mali. *Agricultural Systems* 51/4, 453–467.

Jones, D. M. (1977) The Green Revolution in Latin America: Success or Failure? In G. S. Elbow (ed.), *International Aspects of Development in Latin America: Geographic Aspects.*

Conference of Latin Americanist Geographers 6, 55–63. Muncie, CLAG Publications.

Kanstrup, M., Holst, M. K., Jensen, P. M., Thomsen, I. K. and Christensen, B. T. (2014) Searching for Long-Term Trends in Prehistoric Manuring Practice. $\delta^{15}N$ Analyses of Charred Cereal Grains from the 4th to the 1st Millennium BC. *Journal of Archaeological Science* 51, 115–125.

Kirleis, W. and Fischer, E. (2014) Neolithic Cultivation of Tetraploid Free Threshing Wheat in Denmark and Northern Germany: Implications for Crop Diversity and Societal Dynamics of the Funnel Beaker Culture. *Vegetation History and Archaeobotany* 23/1, 81–96.

Kjaerby, F. (1983) *Problems and Contradictions in the Development of Ox-Cultivation in Tanzania. Centre for Development Research (Copenhagen, Denmark).* Research Report No. 66. Uppsala, Scandinavian Institute of African Studies.

Kühn, M. and Hadorn, P. (2004) Pflanzliche Makro- und Mikroreste aus Dung von Wiederkäuern. In S. Jacomet, U. Leuzinger and J. Schibler (eds), *Die Jungsteinzeitliche Seeufersiedlung Arbon Bleiche 3. Umwelt und Wirtschaft*. Archäologie im Thurgau 12, 327–348. Frauenfeld, Amt für Archäologie.

Leuzinger, U. (2002) Holzartefakte. In A. De Capitani, S. Deschler-Erb, U. Leuzinger, E. Marti-Grädel and J. Schibler (eds), *Die Jungsteinzeitliche Seeufersiedlung Arbon-Bleiche 3. Funde.* Archäologie im Thurgau 11, 76–114. Frauenfeld, Amt für Archäologie.

Lundström-Baudais, K. (1986) Etude paléethnobotanique de la station III de Clairvaux. In P. Pétrequin (ed.), *Les sites littoraux néolithiques de Clairvaux-les-Lacs (Jura) I. Problématique générale. L'exemple de la station III,* 311–392. Paris, Editions de la Maison des Sciences de l'Homme.

Maier, U. (1996) Morphological Studies of Free-Threshing Wheat Ears from a Neolithic Site in Southwest Germany, and the History of the Naked Wheats. *Vegetation History and Archaeobotany* 5, 39–55.

Maier, U. (1998) Der Nacktweizen aus den neolithischen Ufersiedlungen des nördlichen Alpenvorlandes und seine Bedeutung für unser Bild von der Neolithisierung Mitteleuropas. *Archäologisches Korrespondezblatt* 28, 205–218.

Maran, J. (2004) Die Badener Kultur und ihre Räderfahrzeuge. In M. Fansa and S. Burmeister (eds), *Rad und Wagen – Der Ursprung einer Innovation. Wagen im Vorderen Orient und Europa.* Archäologische Mitteilungen aus Nordwestdeutschland Beiheft 40, 265–282. Mainz, Philipp von Zabern.

Matuschik, I. (2002) Kupferne Rindergespann-Darstellungen der mitteleuropäischen Kupferzeit. In J. Köninger, M. Mainberger, H. Schlichtherle and M. Vosteen (eds), *Schleife, Schlitten, Rad und Wagen. Zur Frage früher Transportmittel nördlich der Alpen.* Hemmenhofener Skripte 3, 111–122. Freiburg, Janus.

Mischka, D. (2011) The Neolithic Burial Sequence at Flintbek LA 3, North Germany, and Its Cart Tracks: A Precise Chronology. *Antiquity* 85, 742–758.

Němejcová-Pavúková, V. and Bárta, J. (1977) Äneolitische Siedlung der Boleráz-Gruppe in Radošina. *Slovenská archeológia* 25/2, 433–448.

Pétrequin, P., Lobert, G., Maitre, A. and Monnier, J.-L. (2006a) Les outils à moissonner et la question de l'introduction de l'araire dans le Jura (France). In P. Pétrequin, R.-M. Arbogast,

A.-M. Pétrequin, S. van Willigen and M. Bailly (eds), *Premiers chariots, premiers araires. La diffusion de la traction animale en Europe pendant les IVe et IIIe millénaires avant notre ère.* CRA Monographies 29, 107–120. Paris, CNRS.

Pétrequin, P., Pétrequin, A.-M., Arbogast, R.-M., Maréchal, D. and Viellet, A. (2006b) Travois et jougs néolithiques du lac de Chalain à Fontenu (Jura France). In P. Pétrequin, R.-M. Arbogast, A.-M. Pétrequin, S. van Willigen and M. Bailly (eds), *Premiers chariots, premiers araires. La diffusion de la traction animale en Europe pendant les IVe et IIIe millénaires avant notre ère.* CRA Monographies 29, 87–105. Paris, CNRS.

Rageth, J. (1992) Chur-Areal Ackermann, Jungsteinzeitliche Siedlungsreste und Spuren eines Pflugackerbaus. *Archäologie in Graubünden. Funde und Baufunde,* 31–42. Chur, Bündner Monatsblatt.

Ruoff, U. and Jacomet, S. (2002) Die Datierung des Rades von Zürich-Akad und die stratigraphische Beziehung zu den Rädern von Zürich-Pressehaus. In J. Köninger, M. Mainberger, H. Schlichtherle and M. Vosteen (eds), *Schleife, Schlitten, Rad und Wagen. Zur Frage früher Transportmittel nördlich der Alpen.* Hemmenhofener Skripte 3, 35–37. Freiburg, Janus.

Saqalli, M., Salavert, A., Brehard, S., Bendrey, R., Vigne, J.-D. and Tresset, A. (2014) Revisiting and Modelling the Woodland Farming System of the Early Neolithic Linear Pottery Culture (LBK), 5600–4900 B.C. *Vegetation History and Archaeobotany* 23/1, 37–50.

Schibler, J., Jacomet, S., Hüster-Plogmann, H. and Brombacher, C. (1997) Economic Crash in the 37th and 36th Century BC cal in Neolithic Lake Shore Sites in Switzerland. *Anthropozoologica* 25/16, 553–570.

Shennan, S., Downey, S. S., Timpson, A., Edinborough, K., Colledge, S., Kerig, T., Manning, K. and Thomas, M. G. (2013) Regional Population Collapse Followed Initial Agriculture Booms in Mid-Holocene Europe. *Nature Communications* 4/2486. DOI: 10.1038/ncomms3486.

Sherratt, A. (1981) Plough and Pastoralism: Aspects of the Secondary Products Revolution. In I. Hodder, G. Isaac and N. Hammond (eds), *Pattern of the Past. Studies in Honour of David Clarke.* Cambridge, Cambridge University Press.

Sherratt, A. (2004) Wagen, Pflug, Rind: Ihre Ausbreitung und Nutzung – Probleme der Quelleninterpretation. In M. Fansa and S. Burmeister (eds), *Rad und Wagen – Der Ursprung einer Innovation. Wagen im Vorderen Orient und Europa.* Archäologische Mitteilungen aus Nordwestdeutschland Beiheft 40, 409–428. Mainz, Philipp von Zabern.

Siklósi, Z. (2009) Absolute and Internal Chronology of the Late Copper Age Cemetery at Budakalász. In M. Bondár and P. Raczky (eds), *The Copper Cemetery of Budakalász,* 457–774. Budapest, Pytheas.

Sørensen, L. and Karg, S. (2014) The Expansion of Agrarian Societies towards the North – New Evidence for Agriculture during the Mesolithic/Neolithic Transition in Southern Scandinavia. *Journal of Archaeological Science* 51, 98–114.

Starkey, P. and Mutagubya, W. (1992) *Animal Traction in Tanzania: Experience, Trends and Priorities.* Chatham, Ministry of Agriculture.

Struhár, V., Soják, M. and Kučerová, M. (2010) An Aeneolithic Copper Yoked-Ox Statuette from the Lisková Cave (Northern Slovakia). In J. Šuteková, P. Pavúk, P. Kalábková and B. Kovár (eds), *PANTA RHEI. Studies in Chronology and Cultural Development of South-Eastern and Central Europe in Earlier Prehistory Presented to Juraj Pavúk on the Occasion of His 75th Birthday,* 449–467. Bratislava, Comenius University.

Tarrús, J., Saña, M., Chinchilla, J. and Bosch, A. (2006) La Draga (Banyoles, Catalogne): traction animale à la fin du VIe millénaire. In P. Pétrequin, R.-M. Arbogast, A.-M. Pétrequin, S. van Willigen and M. Bailly (eds), *Premiers chariots, premiers araires. La diffusion de la traction animale en Europe pendant les IVe et IIIe millénaires avant notre ère.* CRA Monographies 29, 25–31. Paris, CNRS.

Thapa, G. and Gaiha, R. (2014) Smallholder Farming in Asia and the Pacific: Challenges and Opportunities. In P. B. R. Hazell and A. Rahman (eds), *New Directions for Smallholder Agriculture,* 69–114. Oxford, Oxford University Press.

Thrane, H. (1989) Danish Plough-Marks from the Neolithic and Bronze Age. *Journal of Danish Archaeology* 8, 111–125.

Velušček, A. (2009) *Stare Gmajne Pile-Dwelling Settlement and Its Era. The Ljubljansko barje in the 2nd Half of the 4th Millennium BC.* Ljubljana, Založba ZRC.

Whitehouse, N. J. and Kirleis, W. (2014) The World Reshaped: Practices and Impacts of Early Agrarian Societies. *Journal of Archaeological Science* 51, 1–11.

Chapter 6

The Spread of Productive and Technological Innovations in Europe and the Near East: An Integrated Zooarchaeological Perspective on Secondary Animal Products and Bronze Utilitarian Metallurgy

Haskel J. Greenfield

Introduction

In this paper, I wish to attempt something new (for me at least). Rather than focusing on a single region (or site) and one individual innovation, I would like to bring together various strands that have characterised my research work over the past 35 years. These are the study of the origins and spread of (a) animal secondary products and (b) utilitarian bronze metallurgy for processing animal foods. Few people (who are not archaeologists) realise how the history of these two innovations has transformed the way we live our lives today. In some respects, they were more important than the so-called 'Neolithic Revolution' because they led to, or were part of, the productive specialisation and intensification that were essential underpinnings in the development of complex and urban societies (*i.e.* states, empires and whatever comes next). This was recognised by scholars long ago (*e.g.* Childe 1951a; 1951b; White 1959) but is often forgotten by more recent generations with their focus on 'initial origins' (Lull and Mico 2011). These innovations ultimately affected all levels of society and enabled social hierarchies and social complexity to ramify over time and in space.

Before I go on, I have to acknowledge my indebtedness to a number of people, particularly to Andrew Sherratt and Tony Legge, both of whom passed away recently. They both pointed a way out of that intellectual morass typical of Anthropological Archaeology that many of us found ourselves bogged down in during the late 1970s and early 1980s. Like various others, I discovered that my interest in the later prehistory and early historical periods of the Old World was not (and is still not) fashionable in North American academia, with the exception of a few major urban institutions (but that is a different subject). This of course puts interested students at a disadvantage in the job market. The value of much of what happens in the Old World is minimised in many North American anthropology departments with their overwhelming focus on their own backyards.

But let us return to the subject at hand. While over the past 35 years Andrew Sherratt has provided the theoretical background for much of the work conducted on the rise of complex Bronze Age societies in Europe (Sherratt 1972; 1973; 1976; 1980; 1981; 1982; 1983; 1997), Sebastian Payne and Tony Legge have supplied some of the methodological rigour required to test his theoretical model with the aid of what was then a relatively new sub-discipline called zooarchaeology *(e.g.* Payne 1972; 1973; Legge 1981; Legge *et al.* 1991) . However, since at that time there were not many systematically compiled or analysed data sets from the regions of interest, I had to go out and collect my own data (as indeed any young or old scholar should to test their ideas. Since then, I have collected data from several countries in western Eurasia and will draw upon the results here with a view to foregrounding some common themes.

What I am setting out to do is to summarise and eventually weave together the two strands of my work on domestic animal exploitation strategies and the spread of butchering technology with a view to increasing our understanding of the evolution of productive intensification. Both impose changes on lifestyles, not only among the elite but also in the lower strata of Near Eastern and European society (and beyond).

Secondary products and the origins of productive intensification

After V.G. Childe's seminal works on the evolution of human society, the first scholar to genuinely home in on the relationship between domestic animal secondary products, metallurgy and productive intensification in a global sense was Andrew Sherratt with his Secondary Products Revolution (2PR) model. Before we go on, let me quickly try to summarise it, at least for the purposes of this article.

In 1981, Sherratt first published his model of the Secondary Products Revolution to explain the dramatic changes in economic organisation (subsistence, settlement and trade) in the Near East and Europe between the end of the Neolithic and the beginning of the Bronze Age (Sherratt 1981; 1983; 1986; 1997; 2006). He hypothesised that the rise of early complex societies was the result of innovations in domestic animal production and related technologies. The 2PR model also tried to answer the essential question of whether early domestic animals were domesticated for their primary or secondary products.

The core of his proposal was that there was a shift from an emphasis on the primary products of domestic livestock to encompass both their primary and secondary products. Primary products are those that one has to kill the animal to obtain. They can be taken from animals only once in their lifetime (*e.g.* blood, meat and fat, bone, skin/fur) (*e.g.* Århem 1989). Secondary products, by contrast, are those products that can be taken from a domestic animal at recurring intervals throughout its lifetime. Here one does not have to kill the animal to extract the resource or product in question (milk, wool, traction, dung, *etc.*). For dogs, such products/resources would be guard duty, companionship and hunting functions. Secondary product exploitation requires an appreciation of the animal for what it can produce while it is still alive. This requires a change in the way one looks at animals – immediate versus long-term delays in one's investment.

Sherratt proposed that sheep, goat and cattle were originally domesticated for primary product exploitation and were only later exploited for their secondary products. For example, they could not have been immediately exploited for their secondary products since primitive breeds of cattle, sheep and goats would not yield large quantities of milk, wild sheep do not have woolly coats and there is no evidence of ploughs or wagons in the archaeological record until the Chalcolithic period. He argued that it would take several millennia of genetic manipulation to breed milking cows and woolly sheep. As a result, the origins of the large-scale and intensive use of

domestic livestock for their secondary products in Europe and the Near East did not begin with the earliest Neolithic cultures but appeared much later, during the Chalcolithic and the Bronze Age.

In a separate (albeit related) stream of research, Sherratt investigated the spread of metallurgical innovations and their effects upon European society (Sherratt 1976; 1997). He foresaw how the introduction of metallurgy fed into the social systems, ultimately leading to the rise of social complexity in certain regions. At the time, he was trying to shift the discussion on the origins of metallurgy away from technical and technological matters to the social realm. As with most archaeologists (of the time and even today), there is an overwhelming focus on the beginnings of metallurgy and the 'fancy' display items. Yet, there is also a need to understand the technological innovations that affected the 'masses', or the lower strata of society, who would not have had access to the 'goodies' that are often at the heart of research. They would have had to wait until metal became cheap and strong enough to be affordable at all levels of society.

Investigating the Secondary Products Revolution

The 2PR model was investigated intensively almost from the outset. Throughout the 1980s and 1990s, a number of researchers, primarily zooarchaeologists (myself, L. Horwitz, S. Davis, I. Davidson, *etc.*) and archaeologists (T. Levy, J. Chapman, *etc.*) were intensively involved in its investigation (see citations in Greenfield 2010; 2014). Its relevance for researchers far beyond the original geographic limits of the model have and continue to investigate this shift in exploitation strategies as far afield as China and Peru (Hesse 1986; Li *et al.* 2014).

By the early 2000s, most of us believed that Sherratt's 2PR model had been validated. True, there were always expressions of disbelief (Chapman 1982; Bogucki 1984; 1986; Vosteen 1996) but the weight of the evidence still lay with the proponents of the model (Sherratt 1996; Greenfield 2005b). In recent years, the issue has been unexpectedly readdressed from an entirely new vantage. Residues of milk lipids were found to exist in Neolithic ceramics from the Near East and Europe by Richard Evershed and his colleagues (*e.g.* Copley *et al.* 2003; Evershed 2008; Evershed *et al.* 2008; Outram *et al.* 2009; Salque *et al.* 2012). Immediately, this complicated the issue, forcing faunal specialists to go rushing back to their data to see if they had got it all wrong. This prompted many of us to begin to reinterpret faunal data in ways that we had not envisioned when they were collected, which incidentally highlights the importance of not discarding old collections.

The lipid data disproved one essential aspect of the 2PR model, which originally proposed that milking began with the Chalcolithic. Instead, the lipid data confirmed that milking was associated with very early ruminant animal domestication, at least far back as the origins of ceramics (or containers for holding and processing milk). It is difficult to investigate whether milking made its appearance any earlier because the ceramics we have only date back to *c.* 6500 BCE. This is a few thousand years after the earliest domestication of sheep and goats and about a thousand years after cattle domestication. Equids and camels were domesticated much later.

With the appearance of Neolithic ceramics bearing preserved milk lipids, it dawned on me that most analyses had framed the question inappropriately and in fact cited Sherratt's original thesis incorrectly. It was not about the 'origins' of milking (*e.g.* Vigne and Helmer 2007) but about 'origins' of productive intensification . This was the original intent and focus of the 2PR model: to try to explain how these changes had major repercussions in society and the economy (Greenfield 2005b; 2010). Furthermore, it is not possible to determine the first time an innovation appears by looking at the faunal record. A change of this nature, particularly in subsistence, can only be monitored in the faunal record when it permeates society and transforms behaviour on a large scale.

In fact, as I intend to demonstrate, the zooarchaeological evidence (particularly from Europe) shows that intensive milking appeared much later – in the Chalcolithic of the Near East and Europe. Studies on human dental calculus suggest a similar time frame (Warinner *et al.* 2014). This time frame coincides with the appearance of ploughs, wagons and woolly sheep.

So, were animals intensively milked during the Neolithic? The easy answer is 'Yes – but probably not on a large scale'. The lipid analysis of large samples of sherds have demonstrated that milk products were of minor quantitative importance for human diet, with low frequencies of ceramic sherds bearing traces of milk lipids (8–25% of the Neolithic assemblages tested) (Copley *et al.* 2003; Craig *et al.* 2005; Evershed *et al.* 2008; Smyth and Evershed 2015), which is hardly surprising given the absence of specialised milk-vessel types in the Neolithic (Sherratt 1981; 1996).

Zooarchaeology, faunal remains and origins of milking

The analysis of faunal remains is one way of measuring productive intensification and is especially relevant since the 2PR model is all about how humans changed the way they exploited domestic livestock. Even though they use the same data set, the study of faunal remains differs from zooarchaeology. Zooarchaeology takes an analytical perspective. It is in place when animal remains from archaeological contexts are analysed from a behavioural or human vantage (Olsen and Olsen 1982; Hesse and Wapnish

1985; Reitz and Wing 2008). It is interested in how faunal remains reflect human exploitation, the local economy and other aspects of human behaviour. My research combines archaeology, zooarchaeology and investigative techniques of a scientific nature to answer behavioural questions. Bones are not the only way of doing this but I have found them very productive.

Harvest profiles and origins of milking in the Near East

What does zooarchaeology say? What can it tell us? One of the most fruitful pathways is reconstructing age at death, also known as the 'harvest profiles' of domestic livestock (Legge 1981; Cribb 1984; Hesse 1984; Redding 1984; Greenfield 1988). This can show us when animals lose their usefulness to humans for a particular product, when they are weeded out of the herd. It is not the same thing as reconstructing a live herd, it focuses rather on the age distribution of animals when they are withdrawn from the herd.

Changes in exploitation strategy would be reflected in changes in harvest profile. A good example is provided by Payne's (1973) theoretical models for the harvest profiles of specialised herds – *e.g.* milk, meat and wool production. He uses sheep as an example but we can substitute cattle and other livestock instead. Essentially, a sheep-wool production harvest profile would probably resemble a cattle harvest profile when the animals are used for traction. The ultimate goal is to keep as many of them alive through adulthood until they are no longer useful.

One of the major problems with the use of caprine (sheep and goat) data in investigating the origins of milking has been that most analysts simply dump all the data from both sheep and goats into a generalised sheep/goat age distribution curve. This poses any number of analytical issues, especially because these are two different species involved with different kinds of products that can be produced from them (*i.e.* sheep wool versus goat hair). Mix up the data like this and all you end up with is an average for the combined distribution, which has little interpretive utility. A good example from the Near East that has been lauded as demonstrating the advent of milking is the work of Helmer *et al.* (2007). In this figure, I have taken Helmer *et al.*'s published data and reconfigured it to fit onto a commonly used cumulative frequency graph (Fig. 6.1; Tab. 6.1). I have only included data from the Neolithic and the Bronze Age (and kept out the Chalcolithic) in order to highlight the differences in patterns between the two periods instead of obscuring them. It is very clear that there is no obvious or significant difference in the harvest profiles between the two periods. The data suggests continuity over time in domestic ovicaprine exploitation strategy – in other words, that milk was present. The harvest profiles (age

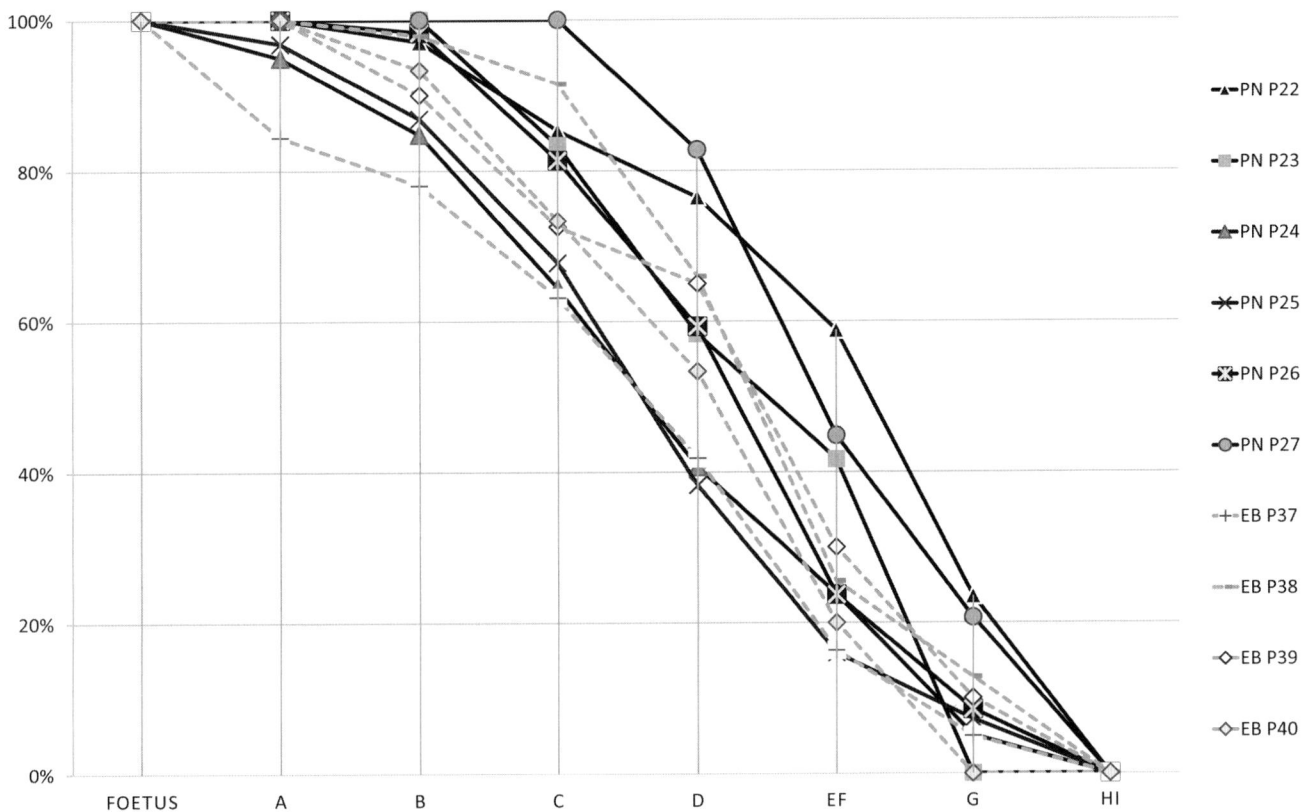

Figure 6.1: Harvest profile of domestic Ovis *and* Capra *age classes generated from data in Table 6.1 (Helmer* et al. *2007).*

distributions) slither back and forth across each other. Given this lack of change over time, it is not surprising that Helmer *et al.* (2007) argue that milking was important right from the beginning of the Neolithic. The lack of difference between the two periods fits in very nicely with the ceramic lipid data suggesting that milk was present from the beginning of the Pottery Neolithic. But with this type of analysis one cannot determine if either of the taxa were actually milked or whether the data simply represent a spurious relationship because the two taxa have been combined.

My proposal is that their harvest profiles are meaningless. The apparent similarity between the Neolithic and Bronze Age simply derives from mixing sheep and goat data together. In general, sheep and goats usually have very different culling patterns, which is particularly apparent in the periods after the Neolithic. This is the topic I should now like to turn to.

Separating the sheep from the goats

When I attempted to compare Helmer *et al.*'s data with those produced by of other scholars in the region, I encountered a fundamental problem. Most other studies publish their data in age groups that are as basic as possible (A, B, C *etc.*). These groups are relatively narrow in scope. However, I

could not fully deconstruct Helmer *et al.*'s data down to the basic age groups for a comparison with most of the research done in the region because they conflate four of the age groups into two larger age groups (E–F and H–I). So not only is there a conflation of taxa (sheep and goats) but also a conflation of age classes.

I believe there is a better way of presenting data for the reconstruction of exploitation practices and for any comparative discussions. For illustration purposes, I shall draw upon some of my own data from the EBA site of Titriş Höyük (south-east Turkey, not far from Helmer *et al.*'s Syrian-focused data sets) (Greenfield 2002a; Allentuck and Greenfield 2010). Here I have tried to separate the sheep from the goats as far as possible. And if you do that, very different culling patterns emerge. The culling evidence reveals very different strategies (Fig. 6.2; Tab. 6.2). Sheep are culled at a much younger age and in larger quantities. The goat distribution shows a very clear shift towards the culling of much older animals. The implication that can be drawn from this is that goats were being kept alive longer and were probably exploited more for their secondary product (*i.e.*, milk). Very interestingly, the sheep and cattle distributions are very similar. While both can yield secondary products (*e.g.* milk or wool), the harvest profile seems to point in the direction of primary product exploitation – *i.e.*, meat.

Table 6.1: Data used to generate Figure 6.1 (Helmer et al. 2007)

Provenance Site	Horizon	Period	Sub-period	Code	Age classes (TNF/NISP)								Foetus	Age class cumulative frequency %							Source
					A	B	C	D	EF	G	HI	Total		A	B	C	D	EF	G	HI	
El Kowm	2	PN		P22	0	1	4	3	6	12	8	34	100	100	97	85	76	59	24	0	Helmer 2000a
Sotto		PN		P23	0	0	2	3	2	5	0	12	100	100	100	83	58	42	0	0	Helmer unpublished
Halula	25	PN		P24	3	6	12	14	10	11	3	59	100	95	85	64	41	24	5	0	Saña Segui, pers. comm.
Halula	26	PN		P25	8	25	48	74	56	22	18	251	100	97	87	68	38	16	7	0	Saña Segui, pers. comm.
Aswad		PN		P26	0	1	10	13	21	9	5	59	100	100	98	81	59	24	8	0	Helmer & Gourichon in press
Seker		PN		P27	0	0	0	5	11	7	6	29	100	100	100	100	83	45	21	0	Gourichon unpublished
Khirbet Derak		Halaf		P28	1	1	6	2	10	9	12	41	100	98	95	80	76	51	29	0	Helmer unpublished
Kosak Shamali		Ubaid		P29	0	1	3	2	7	0	1	14	100	100	93	71	57	7	7	0	Gourichon & Helmer 2003
Kosak Shamali		Ubaid	Late	P30	0	0	0	2	9	6	2	19	100	100	100	100	89	42	11	0	Gourichon & Helmer 2003
Kosak Shamali		Ubaid	Post	P31	0	0	1	4	11	7	3	26	100	100	100	96	81	38	12	0	Gourichon & Helmer 2003
Kosak Shamali		Uruk		P32	0	1	2	5	6	8	5	27	100	100	96	89	70	48	19	0	Gourichon & Helmer 2003
El Kowm	2	Uruk		P33	0	7	14	49	36	11	23	140	100	100	95	85	50	24	16	0	Vila 1998
Sheikh Hassan		Uruk	Middle	P34	4	13	71	233	173	88	62	644	100	99	97	86	50	23	10	0	Vila 1998
Sheikh Hassan		Uruk	Late	P35	2	2	38	63	33	19	9	166	100	99	98	75	37	17	5	0	Vila 1998
Mashnaqa		Uruk		P36	0	3	23	49	58	21	6	160	100	100	98	84	53	17	4	0	Vila unpublished
Rawda		EBA		P37	22	9	21	30	36	16	7	141	100	84	78	63	42	16	5	0	Vila & Al Basso 2005
Sidon		EBA		P38	0	1	3	12	19	6	6	47	100	100	98	91	66	26	13	0	Vila 2006
Mishrife		EBA		P39	0	4	7	3	14	8	4	40	100	100	90	73	65	30	10	0	Vila unpublished
Byblos		EBA		P40	0	1	3	3	5	3	0	15	100	100	93	73	53	20	0	0	Vila 1998

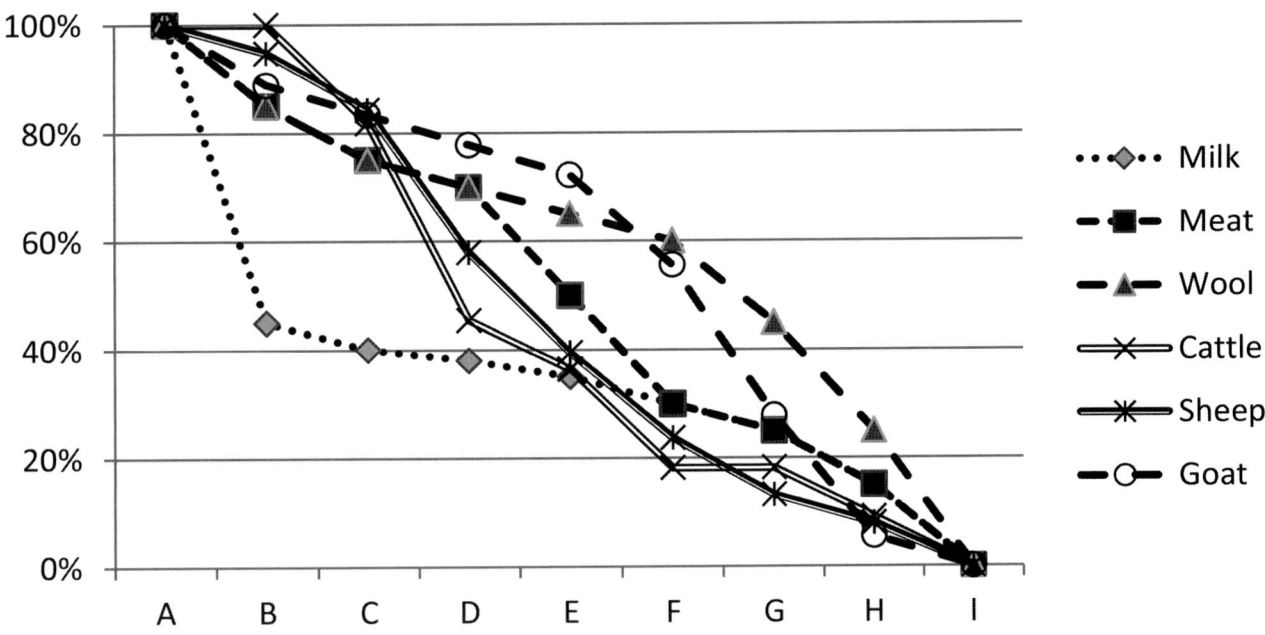

Figure 6.2: Harvest profile of separate domestic sheep, goat and cattle from Early Bronze Age Titriş Höyük generated from data in Table 6.2.

Table 6.2: Data used to generate Figure 6.2 (Allentuck and Greenfield 2010; Greenfield and Allentuck 2011)

Age class	Bos taurus			Ovis aries			Capra hircus		
	Raw count (NISP)	% mortality	% survivorship	Raw count (NISP)	% mortality	% survivorship	Raw count (NISP)	% mortality	% survivorship
A	0	0.00	100.00	0	0.00	100.00	0	0.00	100.00
B	0	0.00	100.00	2	5.26	94.74	2	11.11	88.89
C	2	18.18	81.82	4	10.53	84.21	1	5.56	83.33
D	4	36.36	45.45	10	26.32	57.89	1	5.56	77.78
E	1	9.09	36.36	7	18.42	39.47	1	5.56	72.22
F	2	18.18	18.18	6	15.79	23.68	3	16.67	55.56
G	0	0.00	18.18	4	10.53	13.16	5	27.78	27.78
H	1	9.09	9.09	2	5.26	7.89	4	22.22	5.56
I	1	9.09	0.00	3	7.89	0.00	1	5.56	0.00
Total	11	100.00%		38	100.00%		18	100.00%	

This raises the question why they were not exploited for their secondary products as well. The answer is that the data derive from the end of their lifeline-we are seeing the end product of their productive existence. The urban data pattern represents where they are being consumed (*i.e.*, as meat). The Titriş Höyük data represent an urban, not a village subsistence pattern.

Clearly, younger sheep and cattle were sent to the cities for consumption. But why did the same thing not happen with goats? The best answer I can suggest is that maybe goats were kept locally by families, either individually or in small herds, and their culling profile reflects this pattern.

When I reconfigure my own data from the EBA site of Titriş Höyük (which separates the sheep from the goats, even when using Helmer's age-grouping system) to match Helmer

et al.'s data (Fig. 6.3), I find that Helmer *et al.*'s data plot somewhere in between the distribution of my separate sheep and goat data. Clearly, Helmer *et al.*'s data are simply an average of the two species and are analytically meaningless when presented in this manner.

As the reader can see, separating sheep from goat mandibles is essential. In fact, it is one of the easiest osteological elements to deal with in the two taxa. The technique has been around since the 1960s, when Boessneck (and colleagues) first began to systematically publish their findings on the nature of the differences between the two taxa (Boessneck *et al.* 1964; Boessneck 1969). I am just as guilty as others of having combined the two taxa in certain of my early zooarchaeological studies. With hindsight, such an approach is misleading and should be avoided.

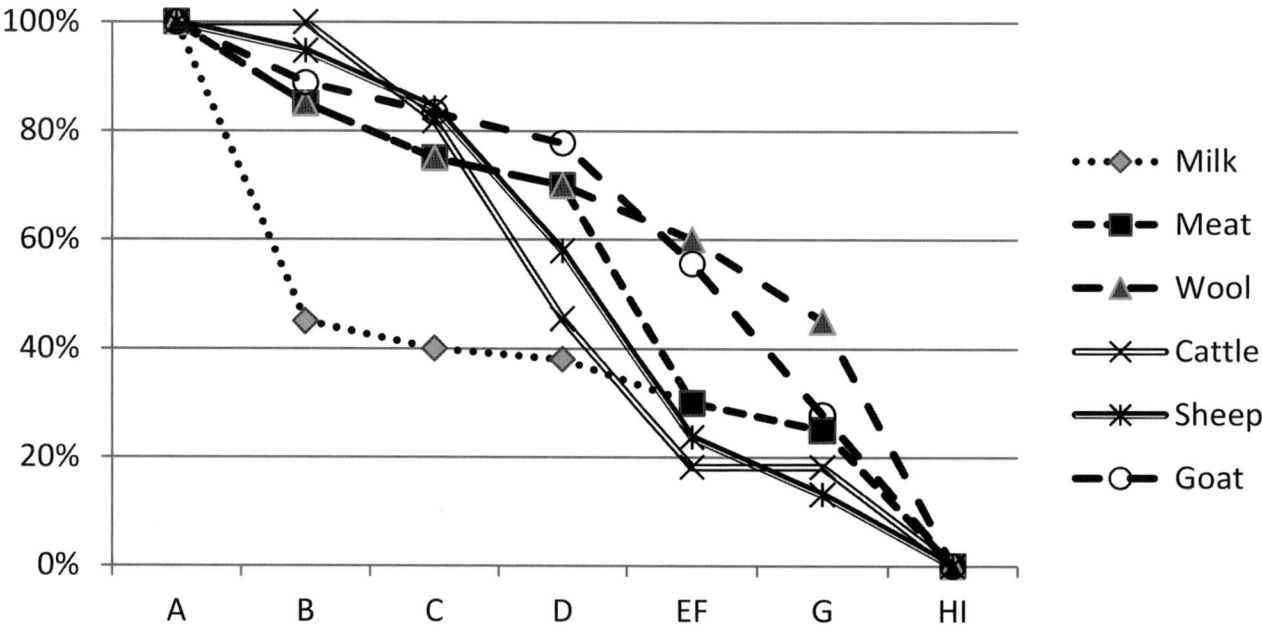

Figure 6.3: Figure 6.2 data reconfigured to match Figure 6.1 (Helmer et al. 2007).

It is also important to note that none of the harvest profile patterns that I or others have generated can match up to Payne's theoretical models. They are theoretical models based on modern industrial-style herds and were never expected to reflect ancient subsistence or early urban-oriented production needs. This is something that I and others have long maintained (Payne 1973; Greenfield 1988).

Central Balkans

To return to the subject at hand, how does one investigate productive intensification? First of all, one must define what one means by intensification – from my perspective, it is any dramatic change in the nature of exploitation strategies that results in a much higher or a steeply increasing level in the production of food, goods and/or services in return for the labour expenditure involved (see *e.g.* http://dictionary.cambridge.org/dictionary/british/intensify). This would be where the return for any 'investment of the means of production and labour per unit of ground area or, in animal husbandry, per head of livestock' would be greater (http://encyclopedia2.thefreedictionary.com/Intensification+of+Agriculture). In any zooarchaeological analysis, this would be identifiable from a change in the nature of domestic livestock exploitation strategy that suggests that one taxon (*e.g.* sheep) was being more intensively exploited for a new product (*e.g.* milk) or an increased range of products (*e.g.* milk and wool).

I would like to use the data from my long-term research on the central Balkans of SE Europe as an example of how one identifies intensification in the archaeological record. In these data, one can see how productive strategies changed and intensified over time as these new breeds and technologies spread to Europe. But in order to do this, one needs long-term data from many sites spread out over the region in each period. Only under such conditions can one examine the long-term changes required for any investigation of changes in exploitation strategies possibly reflecting general patterns across the region rather than being unique to a single site or valley.

In my study, I collected data that ranged from the Early Neolithic to the Early Iron Age. There were over a dozen sites with a range of deposits from lowland, midland and highland environments, and from riverine to mountain pasture. It was therefore possible to monitor changes in harvest profiles over a long period of time and in a variety of different environments within a single period of time.

The harvest profiles of pigs display no major shift in exploitation strategies over time (Fig. 6.4). This is not to be expected since pigs do not have any secondary products and their exploitation will accordingly remain constant from the beginning of their domestication (or from their introduction into Europe from the Near East with the advent of the Early Neolithic).

By contrast, the cattle harvest profile shows a demonstrable shift from the Neolithic to the post-Neolithic toward the culling of much older animals (Fig. 6.5). This would imply the incorporation of their exploitation for both primary and secondary products (*e.g.* traction and/or milking) during the post-Neolithic, since many more adult animals were now being kept alive.

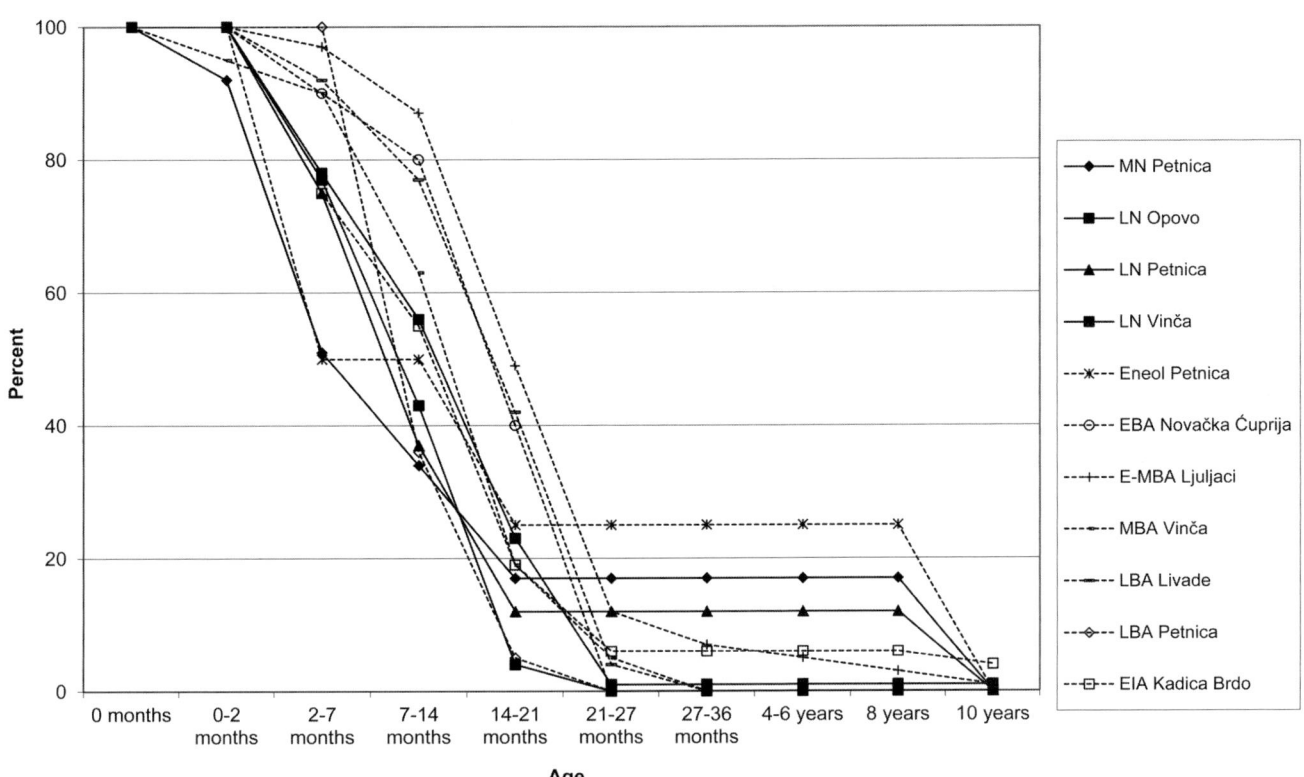

Figure 6.4: Harvest profile based on domestic pig mandibles from central Balkans from Neolithic through Bronze Age sites (Greenfield and Arnold 2014).

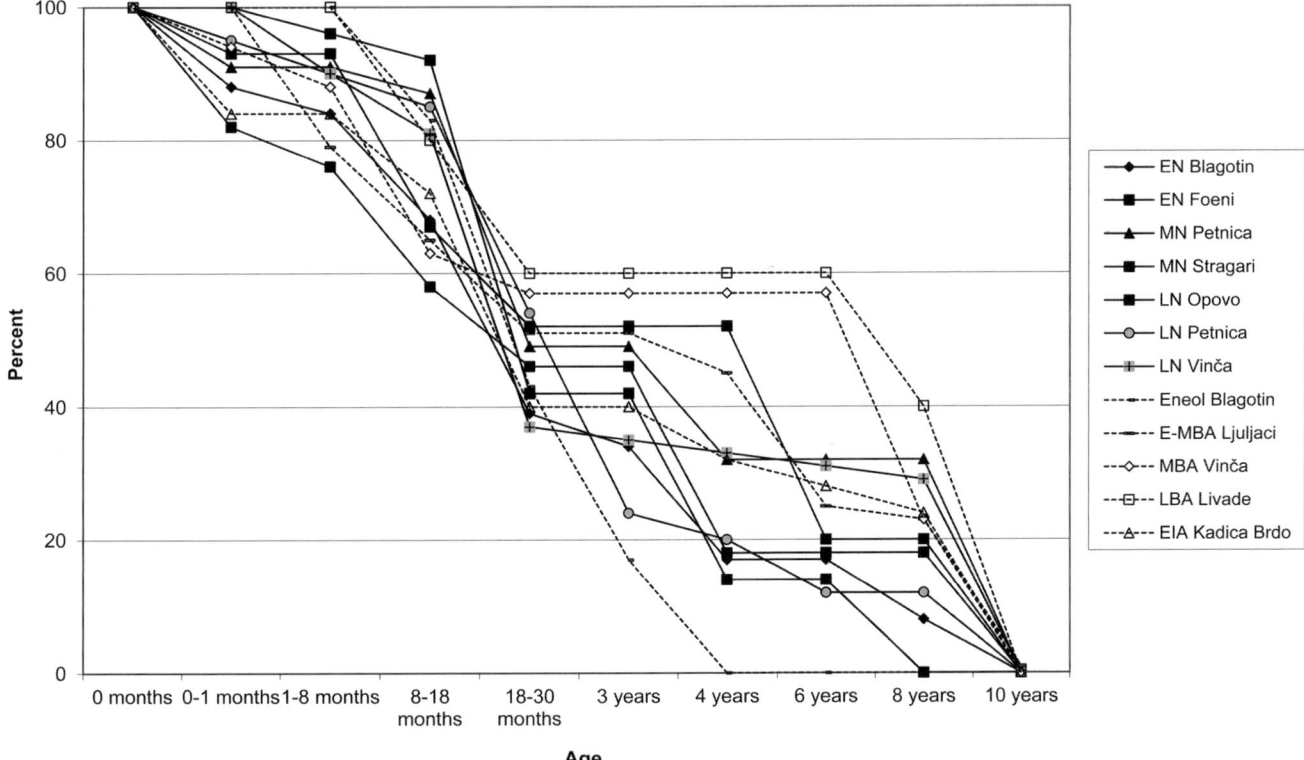

Figure 6.5: Harvest profile based on domestic cattle mandibles from central Balkans from Neolithic through Bronze Age sites (Greenfield and Arnold 2014).

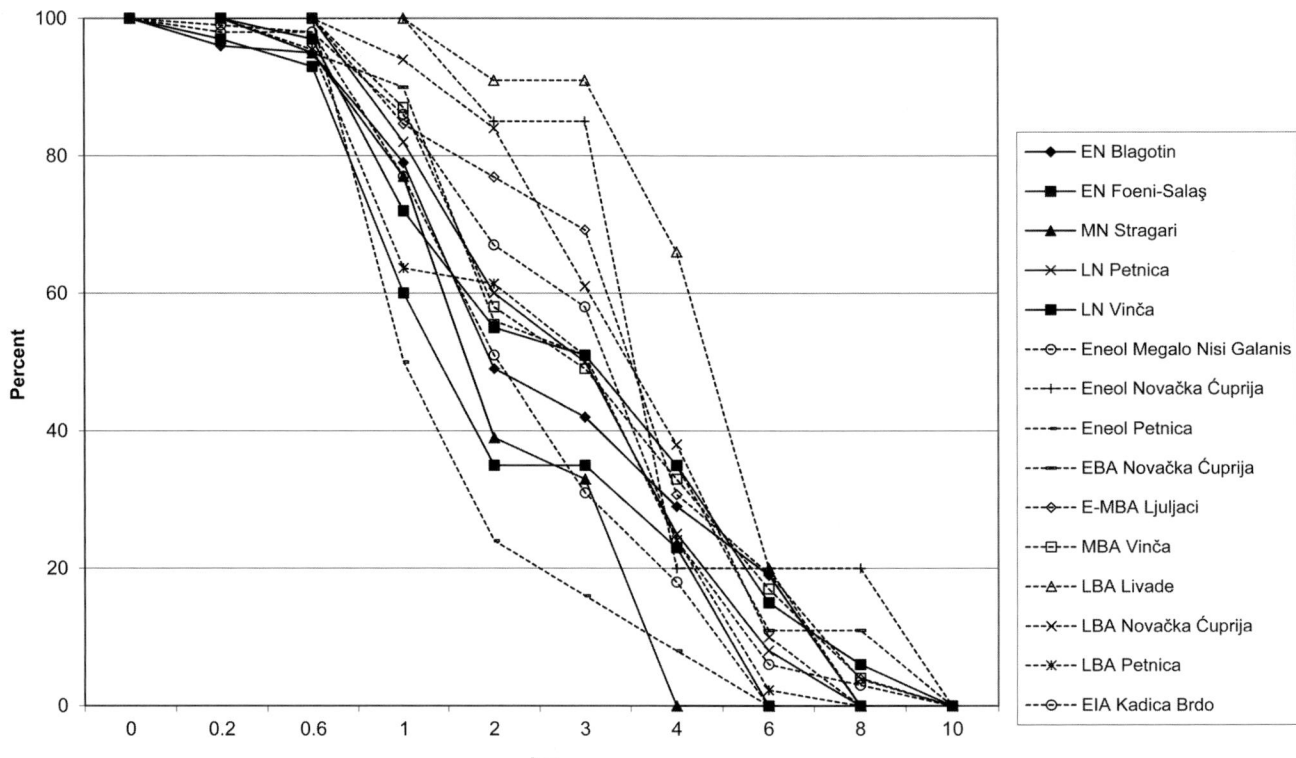

Figure 6.6: Harvest profile based on domestic sheep and goat (combined) mandibles from central Balkans from Neolithic through Bronze Age sites (Greenfield 2005a).

The combined caprine (also known as ovicaprine) harvest profiles from the Neolithic and post-Neolithic similarly show a demonstrable shift towards the culling of more mature animals (Fig. 6.6). This culling pattern appears highly typical of the exploitation of these taxa for both their primary and secondary products.

It could also be argued that the trend in both cattle and ovicaprines simply reflects more conservative herd management practices. However, two strands appear to militate against this kind of explanation. First, various artefacts appearing during this time indicate that we are not dealing with simple conservatism. In the Near East and Europe, both the plough and wagon start to be widely disseminated during the Chalcolithic. Second, if conservative herd management began to prevail, one would expect it to win out equally in pig farming. However, there is no evidence for this.

As we have seen above, one of the major issues in the analysis of ovicaprine harvest profiles is the heuristic problem that exists when sheep and goat remains are combined to form the amorphous caprine (or ovicaprine). When sheep and goats are separated in these assemblages, there is an average of goat:sheep ratio of 1:5 (Greenfield and Arnold 2014). The goat pattern would be swamped by the overwhelming number of sheep remains. Hence, it can be readily argued that sheep plus goat is not evident

of any particular exploitation strategy or that it simply represents what happened with the sheep since theirs are the overwhelming majority of remains. The only effective way of responding to this objection is to separate the sheep from the goats. To do this, we must move away from the use of cumulative frequency graphs simply because there are not enough goat remains in many of the assemblages analysed to build harvest profiles based on mandibular tooth eruption and wear data.

I have also successfully used another graph technique to illustrate differences in age-at-death data under such conditions – insufficient frequencies in many age classes. Ternary (or tripole) graphs use only three variables, which enables us to use all of the age-at-death information available, including both the post-cranial and the cranial remains of the taxa under consideration. In this analysis, only three (and not 8–10) variables (or age classes) are employed as a reflection of culling strategies: very immature (usually <1 year and infants and juveniles), sub-adult (usually 1–3 years in sheep and goats or up to 4 years in larger taxa such as cattle and horses), and adult (Fig. 6.7). These age classes conflate the data from different stages of growth. When sheep and goats are separately plotted on the ternary graph, a very interesting pattern immediately emerges. They are at opposite ends of the graph. The sheep pattern emphasises the culling of young animals (mostly 'very immature'), while the goat

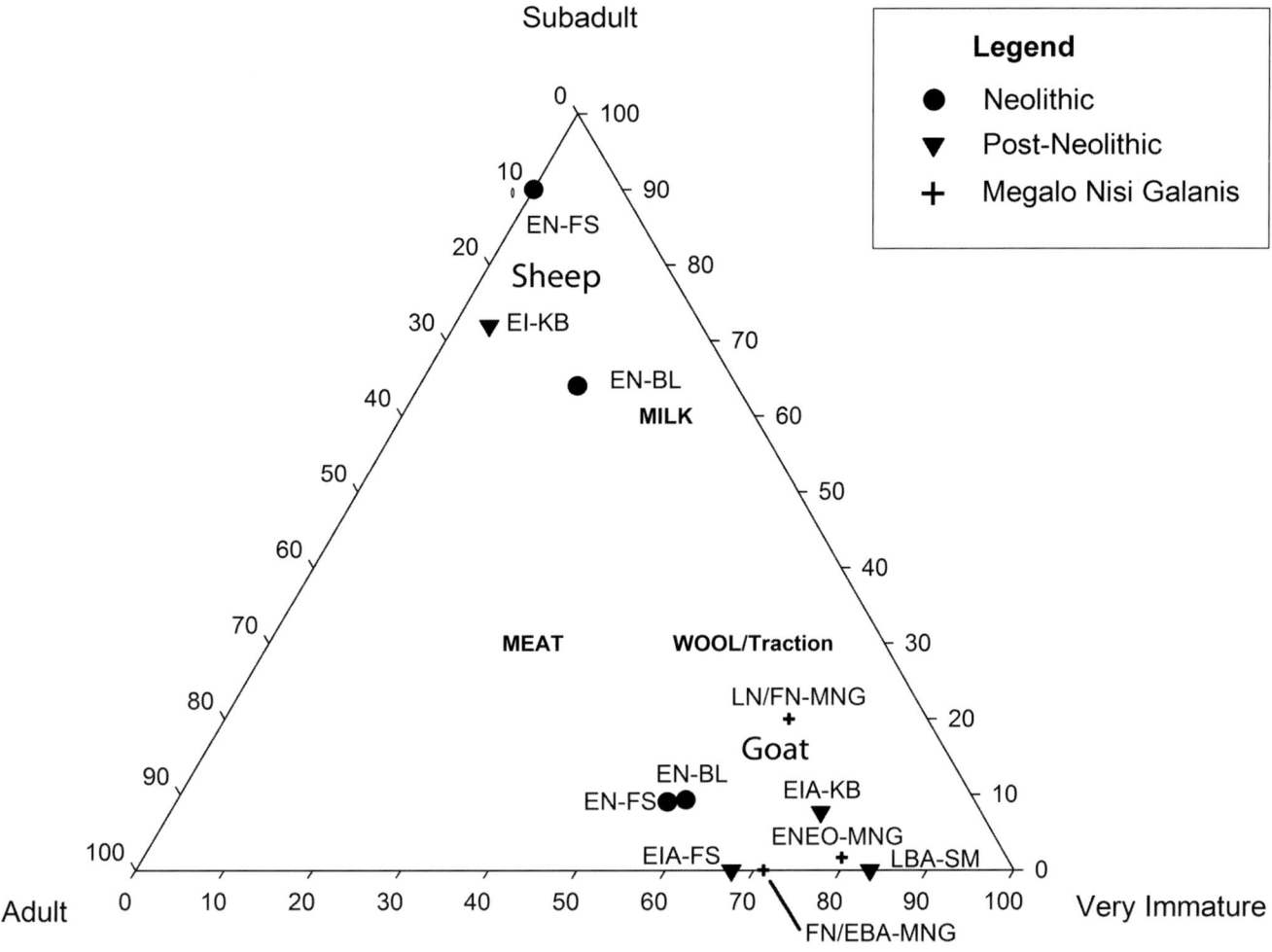

Figure 6.7: Ternary graph of sheep and goat mandibular remains that are separated from each other (Greenfield and Arnold 2014).

pattern emphases the culling of adults. For goats, this pattern exists from the beginning of Neolithic, which implies that in their case exploitation strategies did not change over time. However, the sheep pattern shifts from the Neolithic to the post-Neolithic, with slightly greater culling emphasis on older animals in the post-Neolithic. It is thus clear that sheep were exploited for their primary products in the Neolithic and more broadly exploited for their secondary products in the post-Neolithic.

Also, the data clearly show that any attempt to match these archaeological assemblages with modern industrial herd patterns does not really work. There are many variables that intervene to modify the assemblages, both cultural and natural. Hence, when we talk about exploitation for primary products in a theoretical sense, it does not necessarily mean that all the relevant assemblages will fall on or near the theoretical, industrially-based modern meat point on the graph. Otherwise, one would say that goats were used for wool/traction, while sheep were used for milk. In a real-world subsistence exploitation strategy, this clearly makes no sense.

So what was being milked in the European Neolithic? The overwhelming evidence suggests that goats were the candidates. The milking of goats began in the early Neolithic and has continued until today.

Mapping the origins and adoption of utilitarian metallurgy

The second of my long-term research interests has focused upon the spread of metallurgy, albeit not from a traditional metal-object perspective. I was not interested in treading the beaten path common to most metallurgical studies and emphasising the analysis of the few metal objects that have come down to us from antiquity. My goal has been to identify and understand the shift from stone- to metal-based technology for the 'common man', in everyday life, as it took place in households across the region from the Near East to Europe. In other words, I wanted to investigate when metal became such a commonly used item as to have an appreciable impact on daily lives in all domains of society. As I will show, this (like the intensification of secondary product exploitation) occurred later in time than is usually acknowledged.

One of the major issues plaguing traditional approaches to the origins and spread of metallurgy is how to find evidence of its early manifestations. And is that evidence representative of what was originally out there?

After 200 years of archaeological research on the subject (if we begin with Thomsen's Three Age system proposal that was intimately based on the appearance of metal objects in graves), we know a great deal about where metallurgy began, the types of artefacts that appeared at various periods in time, the processes of production and consumption, *etc.* But we know relatively little about the rate and nature of its spread, the frequency of use in daily life, particularly at the various levels of what was fast becoming an increasingly hierarchically differentiated social (including political) and economic world.

Why do we know so little about its rate of spread and the nature of its adoption? One reason is that there are a host of forces, both cultural and natural, reducing the number and variety of metal tools preserved in the archaeological record. These include recycling (most broken metal objects were probably recycled since they were a precious commodity and we find relatively few tools that were discarded because they were broken or worn), decomposition (due to soil and weather conditions or fires), differential preservation (different tool shapes with different functions and composition will preserve differentially – *e.g.* larger tools survive better than smaller ones), raw material type (*e.g.* iron tools decompose much faster than copper-based tools), the production process (types of smelting and smithing) and social factors (*e.g.* feasting, ritual, funerary and other activities that may destroy items or take them out of circulation through burials and/or destruction).

As a consequence, our understanding of the nature and use of early metals in society is quite biased. The number and types of metal tools recovered from archaeological contexts are quite small. They do not reflect the full range of types, functions or uses. There are spatial biases (no equal preservation within and between sites, cultures and regions), hierarchical biases (within a single society) and temporal biases (some periods have more than others).

Therefore, to make progress, I felt that we needed new means of documenting the changes in technology. The easy part would have been to focus on the other common raw material on the sites, *i.e.* stone, and its changing distribution. This was successfully accomplished by Steve Rosen, beginning in the 1980s (Rosen 1997). One of the problems that he and others encountered is that stone is haphazardly collected in later periods, particularly in assemblages collected before sieving became widespread (Ford *et al.* 1984; Eriksen 2010; Piličiauskas and Osipowicz 2010). Therefore, its frequency would be compromised. Combined with the poor preservation or extensive recycling of metal,

this created a dilemma that was only resolved with the advent of zooarchaeology.

Through experimentation, I discovered that it was possible to distinguish the presence of stone and metal tools (particularly for slicing activities) through the often microscopically visible grooves on animal bones resulting from butchering activities (Greenfield 1999; 2000). Animal bones, particularly those with evidence of ancient butchering, are far more common on sites than metal artefacts. There are anything from thousands to millions of them on sites, thus (unlike the odd metal object) yielding statistically high frequencies. As we shift our focus from the relatively few metal tools to the large collections of faunal remains, our data set increases substantially. With these kinds of data, one can begin to compare different time periods, regions, sites, parts of sites, *etc.* One can quantitatively compare stone and metal butchering-mark frequencies over time as a proxy measure for the frequency of their importance in this everyday activity that would have been present in every household and every community across regions small or vast.

The focus of my interest has been on slicing marks that leave distinctive grooves on bones as a result of butchering activities. Different stages of the butchering process have long been identified by identification and location of slicing marks, including slaughtering, disarticulation, skinning and filleting (Binford 1981). Dismemberment, a term often used synonymously with disarticulation, is distinguished here as the function of a chopping action that cuts through elements. Up to the advent of iron axes, chop marks are often difficult to identify. Even bronze axes do not consistently leave the kind of impressions on bone that enables the grooves to be identified as a result of the use of metal tools (Mathieu and Meyer 1997).

By focusing on the numerous slicing marks, it is possible to create an independent measure of relative importance for different raw materials, at least for this type of activity. This quantification enables us to monitor the nature and rate of the spread/adoption of a particular utilitarian technology, the shift over time to a new raw-material base (from stone to metal) and a shift in associated activity (butchering) over time and space (Greenfield 2013).

Method

After the initial insight that it might be possible to distinguish metal from chipped stone knife blade marks on bone, my research shifted toward the conduct of systematic experiments. Tools of a known shape and raw material were drawn across wood and bone surfaces in order to observe and identify any consistencies in morphology. Essentially, wood was a better surface since it was of consistent thickness, density and angle (which was not the case with bone, making experimentation much more difficult). It quickly became apparent that it was possible to determine whether slice marks had been made by metal or stone tools. There was a direct relationship between the type of raw material and the morphology of the groove left as a result of slicing activities by butchers. The criteria for identification have been described in a number of publications (Greenfield 1999; 2002b; 2006).

Attempts to do this type of analysis without any kind of magnification have proven to be largely fruitless, particularly in the transition periods where both technologies may have been employed. Analysis is dependent on microscopy. Two levels of microscope technology are employed:

1. Light Optical Microscopes (LOM) for initial and bulk analyses where all specimens are examined. This level of analysis identifies most non-slicing marks and removes them from the analysis. It also supplies an initial identification of the type of material (metal or stone). However, these need to be checked at a high level of magnification to ensure consistency of identification and to make any measurements.
2. Scanning Electron Microscopes (SEM), which use electrons instead of light to produce an image.

I started by using an environmental version of SEM, but I was not satisfied with the quality of the image since most of the electrons were absorbed by the bone. Also, the chamber of the SEM is too small to hold most bones. The solution is to make high-quality moulds of the grooves (from silicone dental material), coat them with gold or gold palladium and subject them to SEM analysis. However, this is a costly and time-consuming process. Accordingly, I find it best to use the LOM to select for samples of higher-quality grooves for SEM analysis (1 stroke only in a groove). If there is more than a single stroke in a groove (and this can be most easily identified by the number of mini-grooves at either end), then it is usually possible to identify the raw material. However, it is best when the grooves come from a relatively flat area of the bone. When they are taken from a sharply angled or curved surface, the slicing activity often removes essential diagnostic characteristics as the pressure of the blade shifts in the middle. Generally speaking, then, only a small percentage of any assemblage is subjected to SEM analysis, which is used to test the consistency of LOM observations before subjecting them to further analysis. It is important to record both types of identification in any database because the data may need to be re-examined at the SEM level before it can be deemed accurate.

One of the advantages of SEM is that it provides resolution (strikingly clear images), magnification (from 10–500 microns) and depth of field (more area can be viewed in focus at the same time than with most other forms of microscopy) on a three-dimensional surface.

The initial analysis should always be conducted with an LOM on the original bone specimen. On the other hand, working with silicone moulds (dental moulding compound) has several advantages too. It is easy to transport (lightweight, soft and pliable), sensitive (can filter into microscopic spaces and shows fine striations) and complies with widespread legislation on antiquities, so that it can be transported from one country to another. Furthermore, if hard cortical bone surfaces are chosen for moulding, it does not damage the original specimen and is highly favoured by most museum curators for analyses of this kind.

Some of the characteristics of grooves created by metal knife blades are sharp V- shaped grooves, a distinct apex at the bottom of the groove, symmetrically angled sides, an absence of lateral striations and the production of straight and clean-cut slicing grooves that are more even in depth.

By contrast, slicing grooves made by chipped stone blades will differ depending on whether they were unifacially or bifacially produced and whether they were retouched or not. Nonetheless, some general characteristics for stone tools still emerge. They are asymmetrical in shape, the ventral side rises steeply and smoothly, the dorsal side rises more gradually and the gradually rising side will have one or more striations parallel to the apex of the cut. The number of lateral striations depends on whether it has been retouched or not.

Data sets and regions

To investigate the spread of the adoption of utilitarian metallurgy with my slicing-mark studies, I have collected and processed data from the length and breadth of three geographical regions. These are the southern Levant (Israel and Palestinian Authority), Serbia and Poland. The analysis of the data from these three 'hot spots' gives us an idea of the rate of spread across this vast geographic continuum.

Southern Levant

The data from the southern Levant clearly demonstrate that there is no evidence of utilitarian metallurgy in the periods preceding the Bronze Age (Tab. 6.3). All the data from PPNB Jericho were recently re-examined in 2014 and they have either been reassigned to later periods or recognised as inaccurate identifications. The ability to return and check problematic identifications is one of many advantages of working with well-curated assemblages.

The situation begins to change during the latter part of the EBA with the appearance in assemblages of small numbers of grooves that seem to have been made by metal knives. But these low frequencies invariably come from a single bone whose temporal origin is suspect. For example, at Be'erotayim the bone could derive from either EB I or MB I since it was impossible to distinguish the two deposits stratigraphically. A slightly different issue plagues the high frequencies of metal slice marks in the original EB data from Jericho (Greenfield 2005a). The apparently high

Table 6.3: Frequency distribution of metal and stone slicing marks from the southern Levantine sites (Greenfield 2013)

Period	Sub-period	Site	Stone		Metal	
			No.	%	No.	%
Protoneolithic		Jericho	4	100.00	0	0.00
PrePottery Neolithic	A	Jericho	17	100.00	0	0.00
PrePottery Neolithic	B	Jericho	15	88.24	2	11.76
PrePottery Neolithic	B	Yiftahel	4	100.00	0	0.00
PrePottery Neolithic	C	Atlit yam	305	100.00	0	0.00
Pottery Neolithic		Jericho	1	100.00	0	0.00
Pottery Neolithic		Lod	23	100.00	0	0.00
Pottery Neolithic		Neve Yam	10	100.00	0	0.00
Pottery Neolithic		Tel Dan	10	100.00	0	0.00
Pottery Neolithic		Tel Hereiz	12	100.00	0	0.00
Chalcolithic	Early	Gilat	50	100.00	0	0.00
Chalcolithic		Halif	1	100.00	0	0.00
Chalcolithic		Palmahim Quarry	1	100.00	0	0.00
Chalcolithic		Shiqmim	67	100.00	0	0.00
Chalcolithic		Tsaf	97	100.00	0	0.00
Early Bronze Age	I	Afridar	8	88.89	1	11.11
Early Bronze Age	I	Afridar (1963)	10	100.00	0	0.00
Early Bronze Age	I	Ashkelon Marina	6	85.71	1	14.29
Early Bronze Age	I/IV	Rujm Be'erotayim	6	100.00	0	0.00
Early Bronze Age	IA	Azor	3	100.00	0	0.00
Early Bronze Age	IA	Halif	12	100.00	1	0.00
Early Bronze Age	IB	Afek	17	89.47	2	10.53
Early Bronze Age	IB	Halif	4	100.00	0	0.00
Early Bronze Age	IC	Halif	1	100.00	1	0.00
Early Bronze Age	II	Afek	7	77.78	2	22.22
Early Bronze Age	II	Qashish	2	100.00	0	0.00
Early Bronze Age	II/III	Arad	355	94.41	23	6.12
Early Bronze Age	II/III	Yarmouth	330	91.92	29	8.08
Early Bronze Age	IIIA1	Halif	50	90.91	5	9.09
Early Bronze Age	IIIA2	Halif	123	96.85	4	3.15
Early Bronze Age	IIIB1	Halif	48	90.57	5	9.43
Early Bronze Age	IIIB2	Halif	190	95.48	9	4.52
Early Bronze Age		Dalit	4	100.00	0	0.00
Early Bronze Age		Jericho	10	58.82	7	41.18
Early Bronze Age		Tel Kinrot	1	100.00	0	0.00
Middle Bronze Age	IIA	Afek	6	5.71	99	94.29
Middle Bronze Age	IIA	Ashkelon tell	1	2.94	33	97.06
Middle Bronze Age	IIA/IIB	Qashish	0	0.00	4	100.00
Middle Bronze Age	IIB	Afek	4	15.38	22	84.62
Middle Bronze Age	IIB	Qashish	0	0.00	5	100.00
Middle Bronze Age	IIC (end of MB IIB)	Qashish	0	0.00	1	100.00
Middle Bronze Age	Tell and Tomb	Jericho	13	6.07	201	93.93
Late Bronze Age	I	Afek	0	0.00	1	100.00
Late Bronze Age	I	Qashish	0	0.00	8	100.00
Late Bronze Age	I	Halif	12	13.64	76	86.36
Late Bronze Age	I/II	Afek	1	33.33	2	66.67
Late Bronze Age	IA	Halif	32	23.53	104	76.47
Late Bronze Age	IB	Halif	0	0.00	14	100.00
Late Bronze Age	IC	Halif	2	18.18	9	81.82
Late Bronze Age	ID	Halif	0	0.00	2	100.00
Late Bronze Age	II	Afek	5	5.56	85	94.44
Late Bronze Age	IIA	Halif	8	12.50	56	87.50
Late Bronze Age	IIB	Halif	35	12.11	254	87.89
Late Bronze Age		Afek	0	0.00	2	100.00

frequency of metal marks in the EB III deposits appears to coincide with the initial appearance of metal knives (Philip 1989) and the transition to tin-bronze technology in the EB III (Genz 2000; Greenfield 2013) across the region. However, since they derive from tombs that were used over several generations and/or centuries, recent temporal reanalysis of these remains has moved most of them out of the EB III and into the more amorphous EB IV/MB I/ IB or MB II range.

As a result, most of the data that I collected early on in my studies and that seemed to suggest a degree of utilitarian metallurgy in EB I-III have disappeared after closer examination of their context. Unfortunately, there are very few EB IV/MB I assemblages from the southern Levant since most sites were abandoned at the end of the EB III for 400–500 years. It is with MB II that we begin to see significant changes in butchering technology. This change looks almost instantaneous because of the missing transitions between the end of EB III and beginning of MB II.

Accordingly, it is in the MB II that we have the first very clear indications that metal knife blades were much more frequent than stone tools. This significant change in raw material from stone to metal blades probably begins at the end of EB III, continues slowly through EB IV/MB I and picks up speed in MB II (75% average) where metal then dominates the assemblages. The frequency of metal knife grooves continues to increase throughout the MB until it reaches 90% or more in the LB and almost 100% in the Iron Age.

One might ask why it took so long for metal knives to be developed for butchering activities. Metal daggers had been around since early in the EB. However, they do not appear to have been used for activities in daily life. Part of the problem probably lies in the shape of the instrument. Daggers are designed for stabbing, not for slicing. They tend to be thicker than knives and have sharp edges on both sides. It is very difficult to effectively fillet meat off a bone with a dagger because one has to constantly press down on one of the sharp edges to maintain pressure. Knives solve this dilemma by extending the metal from the blade through the centre of the handle (which is often sheathed in bone, wood, *etc.*) and by having one sharp side and one blunt

side. Such tools only appear late in the EBA (Philip 1989; Moorey 1994). Even after the appearance of the earliest bronze knives, it clearly took time for the new technology to spread and become widely available to the lower classes of society in the region.

Serbia

Data from two sites in Serbia – Ljuljaci and Petnica – have been systematically analysed in a similar fashion. Petnica has a long sequence from the Middle Neolithic to the Early Iron Age, while Ljuljaci fills in the gap with MBA remains (Bogdanović 1986; Greenfield 1986; Starović 1993; Orton 2008).

When the data from the two sites are combined, it is quite interesting to observe the almost total lack of metal slicing marks from the periods preceding the MBA (Tab. 6.4).

While the traditional typochronology for the Vatin culture of Serbia extends into the EB, most of the radiocarbon dates from Ljuljaci seem to imply that its occupation occurred in the early-to-middle second millennium BCE. The implication is that the vast majority of deposits and remains are from the MBA. When this is taken into consideration, the EB disappears for the most part in the Serbian database and the analysis of more sites proves necessary before we can say anything definite about the pace of spread for utilitarian metal objects.

At this point in my analyses of the data from this region, it would seem that metal butchering marks only become common in the MBA. But the situation appears to change with respect to the LB-EIA deposits from Petnica. In these deposits, the majority of remains demonstrate metal slicing marks (58%), but the frequency is far lower than in the MBA data from Ljuljaci. Is it possible that Ljuljaci was a more important site than Petnica and that there were disparities caused by differential access to such goods depending on one's position in the settlement status hierarchy? This is another issue that needs to be explored in the future.

A small quantity of remains with metal knife marks appears at Petnica throughout the Neolithic-Chalcolithic periods. But most of them can probably be attributed to two variables. 1) These data were collected during the earliest

Table 4. Frequency distribution of metal and stone slicing marks from the central Balkan sites of Petnica and Ljuljaci (Greenfield 2000)

Period (Culture)	Stone		Metal	
	#	%	#	%
Middle Neolithic (Vinča B)	16	94.12%	1	5.88%
Late Neolithic (Vinča C)	20	90.91%	2	9.09%
Late Neolithic (Vinča D)	36	83.72%	7	16.28%
Eneolithic (Baden-Kostolac)	19	86.36%	3	13.64%
Early-Middle Bronze (Vatin)	2	15.38%	11	84.62%
Late Bronze Age-Early Iron Age (Hallstatt A-B)	24	58.54%	17	41.46%
Vinča: Roman** pits	33	91.67%	3	8.33%

**Roman pits, filled with animal bones, intrusive into Vinča C horizon (94% Vinča ceramics)

stages of my study and may not have been as accurate as current research. 2) The data from Petnica derive from early excavations when some contamination between layers occurred. Hence, I would certainly not insist that the presence of a few metal cut marks in these earlier periods has any implications for what we know about the behaviour of the ancients.

As noted above, the major problem with the Serbian data is the absence of significant EBA samples for analysis. This is similar to the problem in the southern Levant for EB IV/MB I.

The Serbian data suggest that utilitarian metallurgy for animal butchering only became widespread in the early second millennium BC. This is more or less contemporary with the picture from the southern Levant.

Poland

I applied a similar method in analysing data from central Poland (Tab. 6.5). In this study, the remains from 17 sites ranging from the Middle Neolithic to the Late Bronze Age (LBA) were analysed to find out more about when utilitarian metallurgy may have appeared for the first time (Marciniak and Greenfield 2013). Contrary to the other regions, all the intervening periods are well represented in the study. The data are more than suggestive of major differences between central Europe and the Balkans.

Metallurgy begins to appear in the MBA. There is one fragment from the EBA that is thought to bear a metal slicing mark. But it is the only one of its kind, and that makes it suspicious. In the MBA the frequency of metal slicing marks increases (30%). It is in the LBA that utilitarian metallurgy begins to spread throughout the region (90% of remains). This was the key period when utilitarian metallurgy established itself widely in central Europe.

Synopsis

When the three regions are compared, a number of conclusions can be drawn about the temporal distribution of metal slicing marks. First, it is apparent that although the overall pattern is very similar, there are differences between the three regions. In all of them, the use of bronze metallurgy for butchering makes its appearance and then grows rapidly. However, the speed at which this happened is difficult to

judge because crucial intermediate periods are missing in both the southern Levant and in Serbia.

Second, metal butchering marks begin towards the end of EBA and then grow quite suddenly in popularity in the eastern Mediterranean regions during their respective MBA. This more or less follows the appearance of metal knives (*e.g.* Tylecote 1987; Philip 1989; McIntosh 2009; Kienlin 2010). The first metal slicing mark also appears in the Polish EBA, but it is only a single fragment. It is in the next period (the MBA) that we begin to see changes occurring in Poland. But only in the following period (LBA) does metallurgy become the dominant technology.

Third, in general the transition to utilitarian tool metallurgy is not part of the 'Copper Age' world with its emphasis on prestige and decorative items (Lichardus 1991; Levy 2007; Anthony 2010). Bronze only becomes the dominant technology much later after tin bronze appears. In general, this process occurs later in central Europe than in the eastern Mediterranean, whether measured by relative periodicity or by absolute dating (Kienlin and Roberts 2009; Kienlin 2010). Even the timing of the MBA varies between regions. There is a temporal disparity in the various periods of the Bronze Age between central Europe, SE Europe and the southern Levant. In the southern Levant, the MBA begins at the end of the third millennium BCE. In Serbia, it begins early in the second millennium BCE, although the data from LBA Petnica suggest that it may have occurred even later in some sites. In Poland, it begins in the latter part of the second millennium BCE. Polish LBA is later than Serbian LBA. In a sense, Poland was a backwater, lagging far behind south-east Europe and the Levant with this kind of technological progress. The fact that the periods all have the same name implies the rapid spread of this new technology across a broad swath of the Old World from the Near East to central Europe. But this is simply not accurate. There is at least half a millennium between the two extreme ends of this geographic continuum. It appears that large-scale metallurgy arrives much later in central Europe (*c.* 1000 years) in comparison to the eastern Mediterranean regions (Levant).

Fourth, a major consequence of the spread and adoption of bronze metallurgy over time and space is that we see a change in the function and the relevant domains of metal items from the passive (ornament/display) to the active (utilitarian/use). In the southern Levant and Serbia this began in the MBA, but only much later in Poland. These transformations profoundly affected all levels of society. Since food-processing tools are one of the basic building blocks of production, metallurgy moved down from the elite to pervade the domains of everyday life. Approaching the issues from this perspective enables us to look beyond the elite of society and focus on what was happening in the 'lower strata'. The widespread adoption of utilitarian metallurgy changes the nature of Near Eastern

Table 6.5: Frequency distribution of metal and stone slicing marks from the central Poland sites (Marciniak and Greenfield 2013)

Period	Metal #	Metal %	Stone #	Stone %
Early Neolithic		0.00	41	100.00
Middle Neolithic		0.00	4	100.00
Late Neolithic		0.00	29	100.00
Early Bronze Age	1	6.25	15	93.75
Middle Bronze Age	3	30.00	7	70.00
Late Bronze Age	11	91.67	1	8.33

and European cultures in many dimensions by taking the role of metallurgy beyond the social dimension (as a new way of displaying status differences, ritual and warfare) to the economic sphere (where it becomes one of the basic building blocks of production). This in its turn led to the creation of new types of employment and encouraged the circulation of goods and services. As new specialised tools are created and production intensifies, the nature of society becomes more complex and hierarchically differentiated. This is a pattern that we find wherever this transition occurs.

Conclusions

In this paper, I have reviewed my efforts to investigate two different production modes that transformed Old World societies and led to an intensification of production. With the exception of goats (probably milked from the incipient or early Neolithic onwards), domestic livestock (*i.e.* sheep and cattle) was increasingly exploited as of the Chalcolithic for their secondary animal products. Some species did not have certain secondary products until after that time (*i.e.* woolly sheep appear for the first time or the relevant technology was not yet available (*i.e.* wagons and ploughs). Metallurgy expands into the functional or utilitarian domain to encompass the processing of basic foods (*i.e.* meat) over a thousand years later. Nevertheless, when these innovations are adopted by local societies, they institute a synergistic effect on the economic and social systems in which they are embedded, leading to increased levels of productive intensification. Secondary animal products show us how production can devolve from the community to households in connection with developments like the spread of settlement across regions, the colonisation of new frontiers (leading to vertical transhumance) in the search for more grazing space for what were likely larger flocks, the development of wool industries and the increasing movement of both basic food-stuffs and fine trading goods (*e.g.* fine fabrics, metal objects, *etc.*) across short and long distances. In the case of metallurgy, there would have been increased demand on metallurgists to make more tools for basic food production and processing. All of this would have entailed increased specialisation in production and increased intensification of production as more types and quantities of goods came into circulation (*cf.* Sherratt 1981). When production intensifies appreciably, productive specialisation increasingly becomes the norm. With the growth of societal complexity, there is greater specialisation in all domains of life (Wailes 1996). Specialised metallurgists and herders were two essential features of productive systems beginning in the Mesopotamian Chalcolithic and the Levantine EBA simply because it was in those periods that the societies in question evolved into early state and urban societies. In the Balkans and central Europe there is less evidence for productive specialisation in food production during the prehistoric periods. This is probably a result of the lower degree of social complexity there. Yet specialisation becomes apparent in the production of metals in all these regions. Whether this is a cause or a result of the increased scale of production and circulation of goods is difficult to determine (it is much like a 'chicken and egg' question).

This poses the question of what productive intensification actually is. Is it the same as specialisation? The answer is simply 'No'. Specialisation and intensification are different processes inherent in the organization of production. But they may also be part of the same larger phenomenon involving the growth of hierarchical societies. Specialisation involves more people doing the same thing over and over again. As one gains more experience and knowledge, the quality of the goods may get finer, and usually does, (Costin 1991; Blackman *et al.* 1993). But this is not the same as intensification. Intensification involves significant changes in both the quality and quantity of production (Morrison 1994; Stark 1995).

The kinds of change that I have identified in the archaeological record as signals of productive intensification are changes in the nature of domestic livestock exploitation from the harvest profiles and changes in the way food was processed through analysis of the frequencies of metal slicing marks on bones. Society is transformed by cheap metallurgy (enabling it to be incorporated into food production) and changes in food processing (including new forms of animal exploitation). These changes in the scale and nature of domestic animal exploitation and metallurgy occur in the post-Neolithic or later in different areas and increasingly later in time as we move farther away from the Near East.

In conclusion, we have to ask (particularly for the next generation) what we should be investigating. Should we be focusing on initial origins or intensification? What can we see archaeologically? My thoughts are essentially this: Any search for initial origins is elusive. It is akin to the search for the 'missing link'. It is hard to find a first occurrence of something, particularly in the archaeological record, which is a palimpsest of activities often occurring over vast periods of time.

Long ago, archaeologists recognized that it was impossible to find the earliest (also known as 'first') agricultural settlement or domestic animal. This was part of the shift away from cultural history that was part of the processual revolution in archaeology (Binford 1965). Yet this lesson appears to be increasingly forgotten in the post-modern/processual archaeological world. It is almost impossible, except maybe via aDNA, to find the 'first' person with the lactose-tolerant gene or the first cow with the capacity to give milk without offspring (*e.g.* Salque *et al.* 2012). In such cases, it could never be fully 'proven' that this was the 'first' such individual because just as with the search for the 'first humans' there could always potentially be another or an earlier one. While the search for 'origins' provides an emotional fillip for scientists and is great for public consumption, it is really a fruitless goal.

Furthermore, such issues cannot be determined analytically with traditional archaeological or zooarchaeological methods. What those methods can do is to provide data, via such things as the through excavation of early sites or the taxonomic identification of early domestic animals. These can then be subjected to aDNA or other future methods of analysis.

Those of my colleagues who are still looking for the 'first milking animal' or the first use of metal should perhaps begin to think differently. It is much more profitable to increase our understanding of the process by which productive intensification and specialisation became such a significant part of our lives. It is only when the scale of change becomes very great that we see changes in the scale of production leading to intensification in the archaeological record. Origins are elusive, but processes are definable and operational. This is the focus the next generation should aim at!

Acknowledgements

My research was conducted and reported on under the auspices of the following institutions: University of Manitoba (Dept. of Anthropology and St. Paul's College), University of Cambridge (McDonald Institute for Archaeological Research and St. John's College), W.F. Albright Institute for Archaeological Research, University of Belgrade and Indiana University. Funding for this research came from a number of sources including IREX, Fulbright-Hayes, University of Manitoba (UM/SSHRC) and the Social Sciences and Humanities Research Council of Canada. A special vote of thanks goes to Arkadiusz Marciniak, Željko Jež, Tina Greenfield, Steve Rosen, Liora Horwitz and the many student assistants, excavators and curators of the assemblages used in the analysis. Finally, I should like to express my gratitude to Joseph Maran and Philipp W. Stockhammer, who organised this conference and encouraged me to think in broader terms about the theoretical basis underlying much of my research. Without them I would continue to feel like a dilettante gathering data from disparate regions without really knowing what it was for. Now I understand what I was doing all along! Without their unstinting support, none of this could have been possible.

References

Allentuck, A. and Greenfield, H. J. (2010) The Organization of Animal Production in an Early Urban Center: The Zooarchaeological Evidence from Early Bronze Age Titriş Höyük, Southeast Turkey. In D. V. Campana, P. J. Crabtree, S. D. DeFrance, J. Lev-Tov and A. M. Choyke (eds), *Anthropological Approaches to Zooarchaeology: Colonialism, Complexity and Animal Transformations*, 12–29. Oxford, Oxbow Books.
Anthony, D. (ed.) (2010) *The Lost World of Old Europe: The Danube Valley, 5000–3500 BC.* New York and Princeton, NJ, The Institute for the Study of the Ancient World.
Århem, K. (1989) Maasai Food Symbolism – The Cultural Connotations of Milk, Meat, and Blood in the Pastoral Maasai Diet. *Anthropos* 84, 1–23.
Binford, L. R. (1965) Archaeological Systematics and the Study of Culture Process. *American Antiquity* 31, 203–210.
Binford, L. R. (1981) *Bones: Ancient Men and Modern Myths.* New York, Academic Press.
Blackman, M. J., Stein, G. J. and Vandiver, P. B. (1993) The Standardization Hypothesis and Ceramic Mass Production: Technological, Compositional, and Metric Indexes of Craft Specialization at Tell Leilan, Syria. *American Antiquity* 58/1, 60–80.
Boessneck, J. (1969) Osteological Differences between Sheep (Ovis aries Linne) and Goats (Ovis capra Linne). In D. Brothwell and E. S. Higgs (eds), *Science in Archaeology*, 2nd ed., 331–359. London, Thames and Hudson.
Boessneck, J., Mueller, H. H. and Teichert, M. (1964) Osteologische Unterscheidungsmerkmale zwischen Schaf (Ovis aries Linne) und Ziege (Capra hircus Linne). *Kuhn-Archiv* 78, 1–129.
Bogdanović, M. (1986) *Ljuljaci. Naselje Protovatinske I Vatinske Kulture.* Kragujevac, Narodni Muzej.
Bogucki, P. (1984) Linear Pottery Ceramic Sieves and Their Economic Implications. *Oxford Journal of Archaeology* 3/1, 15–30.
Bogucki, P. (1986) The Antiquity of Dairying in Temperate Europe. *Expedition* 28/2, 51–58.
Chapman, J. C. (1982) The Secondary Products Revolution and the Limitations of the Neolithic. *Bulletin of the Institute of Archaeology* (University of London) 19, 107–122.
Childe, V. G. (1951a) *Man Makes Himself.* New York, Mentor.
Childe, V. G. (1951b) *Social Evolution.* London, C. A. Watts & Co.
Copley, M. S., Berstan, R., Dudd, S. N., Docherty, G., Mukherjee, A. J., Straker, V., Payne, S. and Evershed, R. P. (2003) Direct Chemical Evidence for Widespread Dairying in Prehistoric Britain. *Proceedings of the National Academy of Sciences of the USA* 100, 1524–1529.
Costin, C. L. (1991) Craft Specialization: Issues in Defining, Documenting, and Explaining the Organization of Production. *Archaeological Method and Theory* 3, 1–56.
Craig, O. E., Chapman, J. C., Heron, C., Willis, L. H., Bartosiewicz, L., Taylor, G., Whittle, A., and Collins, M. (2005) Did the First Farmers of Central and Eastern Europe Produce Dairy Food? *Antiquity* 79, 882–894.
Cribb, R. L. D. (1984) Computer Simulation of Herding Systems as an Interpretive Heuristic Device in the Study of Kill-Off Strategies. In J. Clutton-Brock and C. Grigson (eds), *Animals and Archaeology 3. Early Herders and Their Flocks.* British Archaeological Reports, International Series 227, 161–170. Oxford, BAR.
Eriksen, B. V. (ed.) (2010) *Lithic Technology in Metal Using Societies. Proceedings of a UISPP Workshop, Lisbon, September 2006, Moesgaard.* Jutland Archaeological Society 67. Højbjerg, Jutland Archaeological Society.
Evershed, R. P. (2008) Experimental Approaches to the Interpretation of Absorbed Organic Residues in Archaeological Ceramics. *World Archaeology* 40/1, 26–47.
Evershed, R. P., Payne, S., Sherratt, A. G., Copley, M. S., Coolidge, J., Urem-Kotsu, D., Kotsakis, K., Özdogan, M., Özdogan, A.,

Nieuwenhuysen, O., Akkermans, P. M. M., Bailey, D., Andeescu, R.-R., Campbell, S., Farid, S., Hodder, I., Yalman, N., Özbaşaran, M., Bycaky, E., Garfinkel, Y., Levy, T. and Burton, M. M. (2008) Earliest Date for Milk Use in the Near East and Southeastern Europe Linked to Cattle Herding. *Nature* 455, 528–531.

Ford, S., Bradley, R., Hawkes, J. and Fisher, P. (1984) Flint-Working in the Metal Age. *Oxford Journal of Archaeology* 3/2, 157–173.

Genz, H. (2000) The Organisation of Early Bronze Age Metalworking in the Southern Levant. *Paléorient* 26/1, 55–65.

Greenfield, H. J. (1986) *The Paleoeconomy of the Central Balkans (Serbia): A Zooarchaeological Perspective on the Late Neolithic and Bronze Age (4500–1000 BC).* British Archaeological Reports, International Series 304. Oxford, BAR.

Greenfield, H. J. (1988) The Origins of Milk and Wool Production in the Old World: A Zooarchaeological Perspective from the Central Balkans. *Current Anthropology* 29/4, 573–593.

Greenfield, H. J. (1999) The Origins of Metallurgy: Distinguishing Stone from Metal Cut Marks on Bones from Archaeological Sites. *Journal of Archaeological Science* 26/7, 797–808.

Greenfield, H. J. (2000) Monitoring the Origins of Metallurgy: An Application of Cut Mark Analysis on Animals Bones from the Central Balkans. *Environmental Archaeology* 5, 119–132.

Greenfield, H. J. (2002a) Faunal Remains from the Early Bronze Age Site of Titriş Höyük, Turkey. In H. Buitenhuis, A. M. Choyke, M. Mashkour and A. H. Al-Shiyab (eds), *Archaeozoology of the Near East V. Proceedings of the ICAZ-SW Conference.* Archaeological Research and Consultancy 62, 252–261. Groningen, Rijksuniversitit Groningen.

Greenfield, H. J. (2002b) Distinguishing Metal (Steel and Low-Tin Bronze) from Stone (Flint and Obsidian) Tool Cut Marks on Bone: An Experimental Approach. In J. R. Mathieu (ed.), *Experimental Archaeology: Replicating Past Objects, Behaviors, and Processes.* British Archaeological Reports, International Series 1035, 35–54. Oxford, BAR.

Greenfield, H. J. (2005a) The Origins of Metallurgy at Jericho (Tel es-Sultan): A Preliminary Report on Distinguishing Stone from Metal Cut Marks on Mammalian Remains. In H. Buitenhuis, A. Choyke, L. Martin, L. Bartosiewicz, and M. Mashkour (eds), *Archaeozoology of the Near East VI. Proceedings of the 6th International Symposium on the Archaeozoology of Southwestern Asia and Adjacent Areas.* ARC Publications 123, 183–191. Groningen, Archeological Research and Consultancy.

Greenfield, H. J. (2005b) A Reconsideration of the Secondary Products Revolution: 20 Years of Research in the Central Balkans. In J. Mulville and A. Outram (eds), *The Zooarachaeology of Milk and Fats. Proceedings of the 9th ICAZ Conference, Durham 2002,* 14–31. Oxford, Oxbow Books.

Greenfield, H. J. (2006) Slicing Cut Marks on Animal Bones: Diagnostics for Identifying Stone Tool Type and Raw Material. *Journal of Field Archaeology* 31, 147–163.

Greenfield, H. J. (2010) The Secondary Products Revolution: The Past, the Present and the Future. *World Archaeology* 42/1, 29–54.

Greenfield, H. J. (2013) 'The Fall of the House of Flint': A Zooarchaeological Perspective on the Decline of Chipped Stone Tools for Butchering Animals in the Bronze and Iron Ages of the southern Levant. *Lithic Technology* 38/3, 161–178.

Greenfield, H. J. (ed.) (2014) *Animal Secondary Products: Domestic Animal Exploitation in Prehistoric Europe, the Near East and the Far East.* Oxford, Oxbow Books.

Greenfield, H. J. and Allentuck, A. (2011) Neighbourhood Differences in Animal Exploitation and Consumption Patterns in an Early (Early Bronze Age) Urban Center: The Zooarchaeology of Titriş Höyük, SE Turkey. In S. Morton, D. Butler and K. Reese-Taylor (eds), *It's Good to Be King: The Archaeology of Power and Authority. Proceedings of the 41st Chacmool Conference, November 7–10, 2008,* 171–180. Calgary, University of Calgary and Chacmool Archaeological Association.

Greenfield, H. J. and Arnold, E. R. (2014) 'Crying over spilt milk': An Evaluation of Recent Models, Methods, and Techniques on the Origins of Milking during the Neolithic of the Old World. In H. J. Greenfield (ed.), *Animal Secondary Products: Domestic Animal Exploitation in Prehistoric Europe, the Near East and the Far East,* 130–185. Oxford, Oxbow Books.

Helmer, D., Gourichon, L. and Vila, E. (2007) The Development of the Exploitation of Products from Capra and Ovis (Meat, Milk and Fleeces) from the PPNB to the Early Bronze in the northern Near East (8700 to 2000 BC cal.). *Anthropozoologica* 42/2, 41–69.

Hesse, B. (1984) These are Our Goats: The Origins of Herding in West-Central Iran. In J. Clutton-Brock and C. Grigson (eds), *Animals and Archaeology 3. Early Herders and Their Flocks.* British Archaeological Reports, International Series 202, 243–264. Oxford, BAR.

Hesse, B. (1986) Buffer Resources and Animal Domestication in Prehistoric Northern Chile. *ArchéoZoologie* 1, 73–85.

Hesse, B. and Wapnish, P. (1985) *Animal Bone Archaeology: From Objectives to Analysis.* Washington, D.C., Taraxacum Press.

Kienlin, T. L. (2010) *Traditions and Transformations: Approaches to Eneolithic (Copper Age) and Bronze Age Metalworking and Society in Eastern Central Europe and the Carpathian Basin.* British Archaeological Reports, International Series 2184. Oxford, Archaeopress.

Kienlin, T. L. and Roberts, B. (eds.) (2009) *Metals and Societies: Studies in Honour of Barbara S. Ottaway.* Universitätsforschungen zur prähistorischen Archäologie 169. Bonn, Habelt.

Legge, A. J. (1981) Aspects of Cattle Husbandry. In R. Mercer (ed.), *Grimes Graves, Norfolk Excavation I: 1971–72.* Department of the Environment Research Reports 11, 79–103. London, HMSO.

Legge, A. J., Williams, J. and Williams, P. (1991) The Determination of Season of Death from the Mandibles and Bones of the Domestic Sheep (Ovis aries). *Rivista di Studi Liguri* 157/1–4, 49–65.

Levy, T. E. (2007) *Journey to the Copper Age: Archaeology in the Holy Land.* San Diego, CA, San Diego Museum of Man.

Li, Z., Campbell, R. B., Brunson, K. R., Yang, J. and Tao, Y. (2014) The Exploitation of Domestic Animal Products from the Late Neolithic Age to the Early Bronze Age in the Heartland of Ancient China. In H. J. Greenfield (ed.), *Animal Secondary Products: Domestic Animal Exploitation in Prehistoric Europe, the Near East and the Far East,* 56–79. Oxford, Oxbow Books.

Lichardus, J. (ed.) (1991) *Die Kupferzeit als historische Epoche (The Copper Age as a Historical Epoch).* Saarbrücker Beiträge zur Altertumskunde 55. Bonn, Habelt.

Lull, V. and Mico, R. (2011) *Archaeology of the Origin of the State: The Theories.* Oxford, Oxford University Press.

Marciniak, A. and Greenfield, H. J. (2013) A Zooarchaeological Perspective on the Origins of Metallurgy in the North European Plain: Butchering Marks on Bones from Central Poland. In S. Bergerbrant and S. Sabatini (eds), *Counterpoint: Essays in Archaeology and Heritage Studies in Honour of Professor Kristian Kristiansen.* British Archaeological Reports, International Series 2508, 457–468. Oxford, Archaeopress.

Mathieu, J. and Meyer, D. A. (1997) Comparing Axe Heads of Stone, Bronze, and Steel: Studies in Experimental Archaeology. *Journal of Field Archaeology* 24, 333–351.

McIntosh, J. (2009) *Handbook to Life in Prehistoric Europe.* Oxford, Oxford University Press.

Moorey, P. R. S. (1994) *Ancient Mesopotamian Materials and Industries.* Oxford, Oxford University Press.

Morrison, K. D. (1994) The Intensification of Production: Archaeological Approaches. *Journal of Archaeological Method and Theory* 1/2, 111–159.

Olsen, S. L. and Olsen, J. W. (1982) A Comment on Nomenclature in Faunal Studies. *American Antiquity* 46/1, 192–194.

Orton, D. (2008) Beyond Hunting and Herding: Humans, Animals, and the Political Economy of the Vinča Period. Unpublished thesis, University of Cambridge.

Outram, A., Stear, N. A., Bendrey, R., Olsen, S. L., Kasparov, A., Zaibert, V., Thorpe, N. and Evershed, R. P. (2009) The Earliest Horse Harnessing and Milking. *Science* 323/5919, 1332–1335.

Payne, S. (1972) Partial Recovery and Sample Bias: The Results of Some Sieving Experiments. In E. S. Higgs (ed.), *Papers in Economic Prehistory,* 49–64. Cambridge, Cambridge University Press.

Payne, S. (1973) Kill-Off Patterns in Sheep and Goats: The Mandibles from Aşvan Kale. *Anatolian Studies* 23, 281–303.

Philip, G. (1989) *Metal Weapons of the Early and Middle Bronze Ages in Syria-Palestine.* Oxford, BAR.

Piličiauskas, G. and Osipowicz, G. (2010) The Processing and Use of Flint in the Metal Ages: A Few Cases from the Kernavė and Naudvaris sites in Lithuania. *Archaeologia Baltica* 13, 110–124.

Redding, R. (1984) Theoretical Determinants of a Herder's Decisions Modeling Variation in the Sheep/Goat Ratio. In J. Clutton-Brock and C. Grigson (eds), *Animals and Archaeology 3: Early Herders and Their Flocks.* British Archaeological Reports, International Series 202, 223–242. Oxford, BAR.

Reitz, E. J. and Wing, E. S. (2008) *Zooarchaeology.* 2nd ed. Cambridge, Cambridge University Press.

Rosen, S. A. (1997) *Lithics after the Stone Age: A Handbook of Stone Tools from the Levant.* Walnut Creek, CA, Altamira Press.

Salque, M., Bogucki, P. I., Pyzel, J., Sobkowiak-Tabaka, I., Grygiel, R., Szmyt, M. and Evershed, R. P. (2012) Earliest Evidence for Cheese Making in the Sixth Millennium BC in Northern Europe. *Nature* 493, 522–525.

Sherratt, A. G. (1972) Socio-Economic and Demographic Models for the Neolithic and Bronze Ages of Europe. In D. L. Clark (ed.), *Models in Archaeology,* 477–542. London, Methuen.

Sherratt, A. G. (1973) The Interpretation of Change in European Prehistory. In C. Renfrew (ed.), *The Explanation of Culture Change,* 419–428. London, Duckworth.

Sherratt, A. G. (1976) Resources, Technology and Trade in Early European Metallurgy. In G. G. Sieveking, I. H. Longworth and K. E. Wilson (eds), *Problems in Economic and Social Archaeology,* 557–582. London, Duckworth.

Sherratt, A. G. (1980) Water, Soil and Seasonality in Early Cereal Cultivation. *World Archaeology* 11/3, 311–330.

Sherratt, A. G. (1981) Plough and Pastoralism: Aspects of the Secondary Products Revolution. In I. Hodder, G. Isaac and N. Hammond (eds), *Pattern of the Past: Studies in the Honour of David Clarke,* 261–306. Cambridge, Cambridge University Press.

Sherratt, A. G. (1982) Mobile Resources: Settlement and Exchange in Early Agricultural Europe. In C. Renfrew and S. Shennan (eds), *Ranking, Resource and Exchange: Aspects of the Archaeology of Early European Society,* 13–26. Cambridge, Cambridge University Press.

Sherratt, A. G. (1983) The Secondary Products Revolution of Animals in the Old World. *World Archaeology* 15/1, 90–104.

Sherratt, A. G. (1986) Wool, Wheels, and Ploughmarks: Local Developments or Outside Introductions in Neolithic Europe? *Bulletin of the London University Institute of Archaeology* 23, 1–15.

Sherratt, A. G. (1996) 'Das sehen wir auch den Rädern ab': Some Thoughts on M. Vosteen's 'Unter die Räder gekommen'. *Archäologische Informationen* 19/1–2, 155–172.

Sherratt, A. G. (1997) *Economy and Society in Prehistoric Europe: Changing Perspectives.* Princeton, NJ, Princeton University Press.

Sherratt, A. G. (2006) La traction animale et la transformation de l'Europe néolithique. In P. Pétrequin, R.-M. Arbogast, A.-M. Pétrequin, S. van Willigen, and M. Bailly (eds), *Premiers chariots, premiers araires: La diffusion de la traction animale en Europe pendant les IVe et IIIe millénaires avant notre ère,* 329–360. Paris, CNRS.

Smyth, J. and Evershed, R. P. (2016) Milking the Megafauna: The Role of Organic Residue Analysis in Understanding Early Farming Practice. *Journal of Environmental Archaeology* 21, 214–229.

Stark, M. T. (1995) Economic Intensification and Ceramic Specialization in the Philippines: A View from Kalinga. *Research in Economic Anthropology* 16, 179–226.

Starović, A. (1993) Petnica. Unpublished thesis, University of Belgrade.

Tylecote, R. F. (1987) *The Early History of Metallurgy in Europe.* London, Longman.

Vigne, J.-D. and Helmer, D. (2007) Was Milk a 'Secondary Product' in the Old World Neolithisation Process? Its Role in the Domestication of Cattle, Sheep and Goats. *Anthropozoologica* 42/2, 9–40.

Vosteen, M. (1996) Unter die Räder gekommen. Untersuchungen zu Sherratts 'Secondary Products Revolution'. *Archäologische Berichte* 7. Bonn, Habelt.

Wailes, B. (1996) *Craft Specialization and Social Evolution: In Memory of V. Gordon Childe.* Philadelphia, PA, The University Museum of Archaeology and Anthropology and University of Pennsylvania.

Warinner, C., Hendy, J., Speller, C., Cappellini, E., Fischer, R., Trachsel, C., Arneborg, J., Lynnerup, N., Craig, O. E., Swallow, D. M., Fotakis, A., Christensen, R. J., Olsen, J. V., Liebert, A., Montalva, N., Fiddyment, S., Charlton, S., Mackie, M., Canci, A., Bouwman, A., Rühli, F., Gilbert, M. T. P. and Collins, M. J. (2014) Direct Evidence of Milk Consumption from Ancient Human Dental Calculus. *Scientific Reports* 4/7104. DOI: 10.1038/srep07104.

White, L. (1959) *The Evolution of Culture.* New York, McGraw-Hill.

Chapter 7

Early Wagons in Eurasia: Disentangling an Enigmatic Innovation

Stefan Burmeister

Earliest adoption of wheeled vehicles

The wagon performed a triumphant feat in the 4th millennium BCE. Hardly any other technology spread in prehistoric times so rapidly and sustainably. The wagon of the 4th millennium was a success story that in its essential technical features displayed impressive continuity for over two thousand years and was only superseded by the railway in the 19th century CE and the motor car in the 20th. Today, we recognise the importance of wagons for mobility and transport. In retrospect, their significance for pre-modern societies also seems evident to us. The advantages they brought are obvious.

Nevertheless, much is still puzzling about this innovation. The oldest archaeological evidence of the use of wheeled vehicles dates from the mid-4th millennium BCE and the following centuries. The distribution map of these early finds confronts us with the first enigma posed by the innovation of wheeled vehicles (Fig. 7.1). At present, a small number of clay tablets from Uruk with pictograms featuring wagons and tracks beneath a burial mound in Flintbek in northern Germany are the oldest evidence we have. Via Bayesian modeling of ^{14}C-data, the tracks can be dated to the years 3420–3385 cal BCE (Mischka 2011); and if we follow the approach of Dietrich Sürenhagen (1999, 117) at least some of the clay tablets derive from settlement layers of Uruk IVc and thus also stem from the beginning of the second half of the 4th millennium.[1] The earliest distribution of wheeled vehicles as reflected in these two find sites already marks the extreme poles of their expansion. The map does not indicate any point of departure or any diffusion in space. The wagon was – hey presto! – just there. And this in an area extending from the Alps to the Caucasus, from Northern Germany to Mesopotamia – and perhaps even to Pakistan.[2]

Mapping the archaeological evidence of wagons for the 3rd millennium BCE, we see little spread beyond the early horizon, only an increase in density. With its first appearance, this technology already reached its full extension for the next two millennia.

True, this map reflects only an advanced stage in the use of wagons. An innovation has to be widely established before it makes its mark on the archaeological records. If we are interested in seeking the origins of wheeled vehicles and the earliest horizon of their diffusion we have to go back several generations before their first appearance in the archaeological record around the middle of the 4th millennium. The beginnings of this technology certainly lie in the first half of the 4th, if not in the 5th millennium. And looking at the early evidence, we see significant regional differences: in northern Europe, primarily pictorial representations; in the Circum-Alpine area, finds of wheels and axles in the vicinity of lakeside settlements; in the middle Danube, wagon models made of clay; in the North Caucasus, wagon models and burials with interred wagons; and in the Near East, pictographic characters and wagon models. In most cases, these respective sources surfaced around 3500 BCE, *e.g.* pictorial representations in northern Europe, kurgans in the North Caucasus, writing in Mesopotamia. So, strictly speaking, the early evidence of wheeled vehicles confronts us with a change in various cultural forms of representation. But we still need an explanation why in their different ways wagons suddenly and simultaneously show up in the cultural repertoire of these highly diverse societies.

Another point is remarkable. On the one hand, technologies, like any kind of social production, are based on a variety of conditions. But they are equally embedded in a variety of cultural practices. The adoption of a new

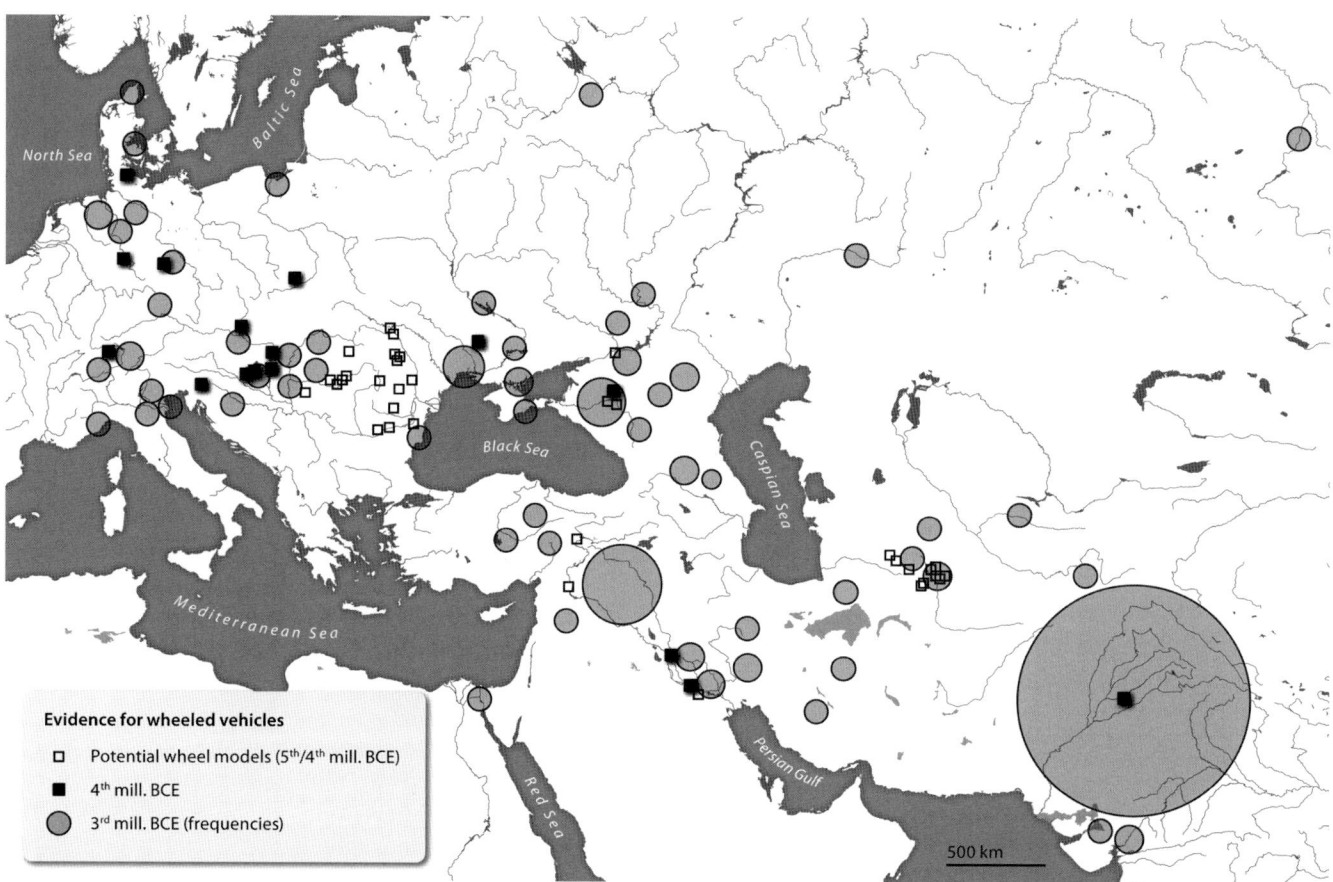

Figure 7.1: Evidence for wheeled vehicles in the 4th and 3rd mill. BCE (mapping 5th and 4th mill. Stefan Burmeister/mapping 3rd mill. Atlas of Innovations, TOPOI Research group D-6/map design Dirk Fabian, ingraphis). For reference literature see: Burmeister 2011; Dinu 1981; Kirtcho 2009; Woolley 1955, 28.

technology will only work if it fits in existing knowledge systems and activity patterns (Burmeister and Müller-Scheeßel 2013). A prerequisite for successful innovation is that the adopting society must be ready for the new technology; André Leroi-Gourhan (1973, 375) refers to a *milieu favorable* as a necessary precondition for successful innovation. The societies adopting the wagon in the 4th millennium could hardly be more different. In the North Pontic steppes and the North Caucasus, we have nomadic and semi-nomadic pastoralists. In the Northern European Plain, we are dealing with the farmers of the Funnel Beaker culture, who only a few generations before had taken up a Neolithic way of life. In Mesopotamia, we see the geographical and social counter-pole: early city-states (at the time Uruk already had up to 50,000 inhabitants) with a centralised temple economy, administration and the advent of script. In all these societies, the wagon was obviously readily adaptable, suggesting diverse functionality and minor technological complexity or a degree of technical sophistication that could be mastered by all social systems.

Different construction traditions already apparent in the early horizon of wagon use and the lack of evidence for contacts spanning the overall distribution area have sparked off a discussion on whether wheeled vehicles may have developed in parallel and independently of each other (one indication might be geographical distances and independent construction methods, *cf. e.g.* Vosteen 2002; Kenoyer 2004 and others) or whether a specific region can claim to be the epicenter of wagon development (Maran 2004; Sherratt 2004; Matuschik 2006). The proponents of monocentric origins postulate different regions as the starting point for the spread of this technology. The records themselves provide little or no clarification. Assuming different parallel origins does nothing to facilitate the interpretation of this amazingly rapid spread, it only creates a new problem. What possible explanation can there be for the fact that wagons not only developed independently in several places, but that they did so more or less at the same time?[3]

Earlier finds of wheel-shaped ceramic discs in the North Pontic steppes and on the lower Danube dating back to *c.* 4000 BC (*e.g.* Dinu 1981; Gusev 1998, 23–25) have been interpreted as model wheels, which may indicate a knowledge of wagons. But the information we have is too vague to support such far-reaching conclusions. And suggestions that the origins are to be found in the urban civilisations of Mesopotamia (*e.g.* Sherratt 2004) are ultimately based on assumptions rooted in historical narratives that are demonstrably both hypothetical and ideological in character.

Road capability

The functionality of the wagon is another puzzle. For the assessment of its practical use manoeuvrability is of crucial importance. Single-axle vehicles – *i.e.* carts – are easy to steer. Carts can be turned around the pivot of the axle without further ado. But without a steerable front axle, the two-axle vehicle (the wagon in the strict sense of the word) can only be turned with considerable effort by lateral pulling or pushing. The earliest evidence of a steerable front axle dates from the first half of the 1st millennium BCE. A centre pivot plate found on a bog trackway in Lower Saxony, northern Germany has a ^{14}C-date of 769–408 cal BCE (Hayen 1983, 452).[4] A wagon burial at Ca'Morta, Italy from the first half of the 7th century BCE and other wagons from the slightly younger Late Hallstatt princely graves bear clear evidence of lateral pivoted poles (Pare 1987a; 1987b, 210).[5] In other words, for the first two to three thousand years the wagons we are dealing with here had no steerable front axle and could only move straight ahead.

Under these conditions, roads are especially important for the efficiency of overland transport with wagons. Considering that in the 4th millennium large areas of northern and central Europe were forests and the other regions also had little to offer in the way of suitable terrain, the question arises how the wagon was used here. The conditions for overland road transport were only in place in some areas from the 1st millennium BCE onwards. Overland transport with wagons requires regular road maintenance, and this in its turn calls for the presence of superordinate central authorities. The only examples we have of a general commitment to providing and maintaining suitable roads are the Roman Empire and, very much later, a number of instances in the late Middle Ages. Despite the well-developed infrastructure of the Roman Empire, land transport with wagons was a very unattractive business. For wagons, both the loading capacity and the daily carrying capacity measured in kilometres were greatly inferior to shipping transport. On the whole, supra-regional transport with wagons was relatively inefficient. Only in a local context did the picture change. Here wagons had a real significance as transport vehicles.[6]

The importance of the early wagons certainly had little to do with supra-regional transport or a notion of improved mobility. Numerous finds of broken wheels and axles impressively

demonstrate the technical vulnerability of wagons and their limited mobility (*e.g.* Burmeister 2002; 2004).

Technological knowledge

The fact that early wagons were apparently of little practical use has prompted the assumption that these vehicles were primarily used in cultic activities rather than as vehicles enhancing people's daily economic activities (Vosteen 1999). I will come back to the issue of wagon use. Our present idea of mobility, which centres on faster coverage of spatial distances, is hardly applicable to the prehistoric wagon. This potential has only been fully exploited from the 19th century CE onward. But wheeled vehicles must have held a great fascination for prehistoric people. There appears to be no other explanation for the rapid spread of wagons in the 4th millennium BCE. Andrew Sherratt (2004, 422) refers to this spread as a 'conflagration', Bakker *et al.* (2005–2006, 22) as a chain reaction. These images both suggest uncontrolled spread. It is a diffusion process that we can only wonder at.

For a long time, archaeological innovation research focused on surface phenomena such as the development of purpose-bound actions and devices generally subsumed under the term 'technology'. There was little true appreciation of the fact that we are dealing here with specific knowledge systems. Knowledge is a critical resource. It is the crucial element in any innovation process, and it is the prerequisite for awareness of new features and their adoption and adaptation. In the following, I shall be considering the spread of wagons from this vantage.

It should be clear by now that the fixed front axle was a drawback for the spread of these vehicles. The wagon itself was not the medium of diffusion, so it will not have served as a model and template for the societies adopting it. The diffusion must have taken place in the form of knowledge transfer – as, for example, with metallurgy. In the latter case, it was not enough to pass on some metal objects, the production of which could be inferred by a process of reverse engineering. Metallurgy is a complex knowledge system made up of various complexes of knowledge on procedures and recipes plus experience in dealing with such procedures (Kuipers 2013).

By comparison, knowledge of wagon technology is relatively simple. In essence, it consists of four components:

- the principle of rotation;
- a wagon box or a similar construction fixed on the axles enabling the transport of persons and goods;
- power transmission from the draught animals to the chassis and wheels, including techniques of harnessing and bridling and the use of the draught-pole;
- use of animal traction.

All these functionally important elements can be registered via observation and do not require intensive learning processes.

On the user side, the following requirements must be satisfied:

- the intellectual capacity to transfer the principle of rotation to a horizontal axis;
- building material suitable for the construction of wagons, suitable tools, and technical skills;
- handling and management of appropriate draught animals, their training (this takes generally one year), and enough economic potential to be able to divert the animals from other existential purposes, like their use as food.[7]

The intellectual ability to understand the principles of the wheel can generally be assumed; craft skills can also be seen as given in the societies in question. The critical point was probably the use of draught animals. Animal husbandry could not be learned simply by observing wagons. It was a cultural system in its own right that needed to be known – or had to be developed.

Given that appropriate draught animals were available in the Early Bronze Age, Copper Age, in the young Neolithic or the early Neolithic – depending on the region we are looking at – knowledge transfer was not likely to have posed any major problems.

I borrow my considerations here from a new approach proposed by Jochen Büttner,[8] who in an attempt to understand the modes of knowledge transfer distinguishes between production knowledge and functional knowledge. In this endeavour, he goes in search of regularities in technical objects that do not derive from their function(s) but can be attributed to specific manufacturing knowledge.

If we try to track down technical regularities in early wagons, the archaeological evidence (or lack of it) quickly puts a spoke in our wheel. Most of it comes in the guise of representations of wagons (depictions and models) that are so schematic that they reveal hardly any technical details. Real finds of vehicles or vehicle parts are very much the exception. They are also in such a poor state of preservation that technical details are usually barely detectable.

Neolithic axles from Lower Saxony

To exclude differing local traditions within the same sample, investigations should be carried out on a small scale, even if this means operating with numbers that are very small for a statistically significant analysis. My concern here is with a group of axle finds from a North German bog which all were found along the 2.5-km Neolithic trackway XV (Le) dating back to the 24th century BCE (Fansa and Schneider 1994). Along this trackway, 13 axles or axle fragments were found among various other wagon parts. They were all deposited in the bog at more or less the same time, but they certainly do not come from one deposition. Seven axles are preserved well enough for metric analysis (Tab. 7.1). So far, this is the largest group of wagon parts from the 4th to the 3rd millennium BCE to be found in one spot.

The axles can be subdivided according to the different sections of the construction (Fig. 7.2): in the middle the axle-bed, *i.e.* the surface bearing the superstructure; on each side a stop ring preventing the rotating wheel from coming into contact with the wagon box; the surface of revolution, where the wheel turns around the axle; and on the outer ends a projection holding the linchpin securing the wheel on the axle. In their total length, the axles range from 184 to 207 cm.[9] As overall length differs, so do the gauges, *i.e.* the distance between the wheels on one axle. They range from 141 to 156 cm. The only indications of standardisation

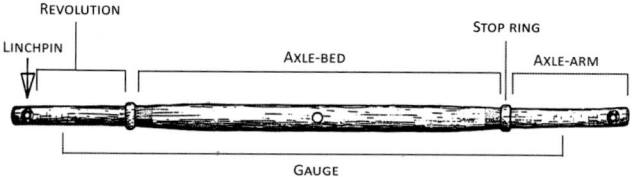

Figure 7.2: Structural parts of a Neolithic axle, northern Germany, 24th century BCE.

Table 7.1: *Dimensions of constructive axle parts from the Neolithic trackway XV (Le) in the Meerhusener Moor, Lower Saxony; 24th century BCE. The range in the diameter of axle-arms is caused by abrasion through the rotating wheel. The severest abrasion is on the upper side of the axle-arm; the higher measure can be seen as close to the original diameter. The objects are housed at the Landesmuseum für Natur und Mensch Oldenburg (LNMO). All measurements in cm.*

Find no. (excavation)	Inv. no. (LNMO)	Wood	Total length	Axle-bed (length)	Axle-bed (diam)	Stop ring (width)	Axle-arm (length)	Revolution (length)	Projecting end (length)	Axle-arm (diam)	Gauge
F 3	C 824	Oak			5.7	3	39.8	26.5		5.2–5.7	
F 4	C 825	Oak	184	115		3	31.5	25.2	6.3	5.7–6.1	146.2
F 8	C 826	Oak					> 31	> 23	8	5.2–5.8	
F 11	C 829	Oak		114		–	*c.* 35	*c.* 27.5	7.5	4.7–5.2	*c.* 141.5
F 13	C 831	Oak					> 32	> 26	6		
F 14	C 832	Oak	207.4	116		4.4	41.3	32	9.3		155.8
F 27	C 840	Oak?	198	114	8–9	4.5	37.5	30	7.5	5–6	153

that are discernible are found in the axle-bed, the section supporting the wagon box. This measures 115 ± 1 cm. This is surprising, as one would have expected norms, especially in the gauge sector. Here uniformity would be desirable for the construction of pathways and the formation of ruts.

The axles display no other regularities. There are no concerted dimensional ratios, for example, (the ratio in length between axle-bed and axle-arm). The variability here sometimes exceeds 10%. Even in parts where correlations would be expected for purely technical reasons, for example, the ratio between gauge and axle diameter, there is no consistency. Production principles as defined by Jochen Büttner are absent. The axles are more or less one-offs, some are round in the cross section, others rectangular. Only the size of the wagon box seems to have been standardised.

Mental patterns

Switching from local evidence to overall findings, we quickly come across a wide range of technological variability. As we have seen, two different design principles formed in the 4th millennium, (a) two-wheeled carts with a rotating axle and fixed wheels and (b) four-wheeled wagons with two fixed axles and rotating wheels. The two principles make differing technical demands on axle and wheel production.

The early disc wheels of the four-wheeled wagons follow a uniform construction plan. They are made of one wooden board, and the hub was a thickening of several centimetres around the axle hole Fig. 7.3(1). This type of wheel is found in the bogs of the Netherlands, northwest Germany, and Denmark (Burmeister 2004) as well as in the Eurasian steppes (Gej 2004; Tureckij 2004). Technically, such hubs have a weak spot. When the wheels start sloping – which is only a matter of time – due to pressure the hub will break sooner or later. In both regions, technical solutions were found to offset this vulnerability. In north-western Europe, the loose hub (Hayen 1972), in the steppes (at the same time) the loose hub (Shishlina *et al.* 2014) and the hemispherical hub (Belinskij and Kalmykov 2004) were improvements to minimise the risk of hub breakage (Fig. 7.4).

Let us now look briefly at the early disc wheels on carts in the Circum-Alpine region. These wheels were tightly fixed to the axle and usually consist of two or three planks held together by wooden strips Fig. 7.3(2). The construction principle is identical in all cases, the only variables are the number and arrangement of the strips. The sole exception to this general construction principle is one of the oldest wheels of this kind from Lake Zurich that was made using only one plank (Schlichtherle 2004).

This brief excursus on the details of wagon construction should not be misunderstood as going off at a tangent for an unmotivated foray into the history of technology. My focus is still on the knowledge aspect of the wagon innovation. This focus is inevitably a little fuzzy. Due to the paucity of archaeological records, lack of evidence blinds us to the other construction features of wagons. We usually only see axles and wheels. At all events, this brief digression has brought it home to us that the general idea of the 'wheeled vehicle' principle was already subject to specific regional

Figure 7.3: (1) Wheel from Kideris, Ringkøbing amt, Denmark (Burmeister 2004, 322 fig. 1.3); 3rd mill. BCE. The disc wheel is broken into two halves, the attached segment at the bottom is a repair of ancient breakage. These wheels were rotating on the axle; (2) Wheel from Stare gmajne, Ljubljansko barje, Slovenia (Velušček et al. 2009, fig. 8.10, b); 2nd half 4th mill. BCE. These wheels were fixed to the rotating axle.

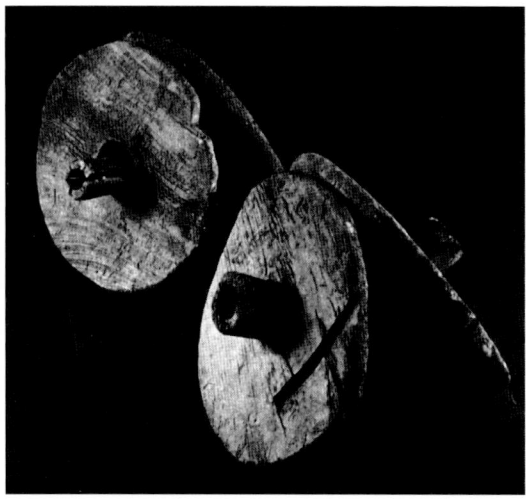

Figure 7.4: (1) Wheels from Glum, Oldenburg, Germany with loose hub; 1st half of 2nd mill. BCE (Hayen 1972, pl. 1.1); (2) Wheel with hemispherical hub from Great Ipatovskij Kurgan, burial 168, Stavropol, Russia; mid-3rd mill. BCE (Belinskij and Kalmykov 2004, 206 fig. 6).

variations in the 4th millennium BCE. Coming back to the distinction between production knowledge and functional knowledge, we can say that vehicle construction did not follow internal production rules. Of course, the highly incomplete records should prevent us from jumping to hasty conclusions, but it still seems fair to assume that it was not production knowledge but functional knowledge that guided the approach to engineering. There were regional solutions for specific requirements, and local traditions emerged accordingly.

Here I believe we see an initial answer to the question why the wagon spread so quickly. The wagon was passed on as a mental pattern. There was an idea of its possibilities – and an idea of its function. It did not need lengthy learning and mediation processes. The simplicity of the technology and the fact that the working principle behind wheeled vehicles can easily be comprehended by mere observation made for rapidity in the transfer of knowledge. Where the prerequisites of craftsmanship and suitable draught animals were given, there were no practical obstacles to the adoption of wagons. Via simple knowledge transfer, the wagon technology could spread quickly – and could be quickly integrated into the adopting societies. In this way, the wagon as an innovation covered not only large spatial, but also major cultural distances.

Auto-mobility

Suggesting how the wagon was able to spread so quickly is of course not the same as answering the question *why* this happened. Reconstructing functions and assets in retrospect is a treacherous business. In the past, a functionalistic perspective and a narrative of progress have biased our views on innovation processes. Frequently, the axiomatic assumption that successful innovations are a function of

the technical and economic assets they offer has been the ideological substructure of reasoning. And a focus on the instrumental rationality of new techniques can also hinder our understanding of innovation. Numerous recent studies clearly show the irrationality involved in the adoption of new technologies (*e.g.* Norman 1988; Suchman and Bishop 2000). Innovations can fail, and in many cases the reasons for the decision to accept or reject them lies outside the specific technology itself. If consumers fail to establish an affiliation with the new features of an innovation, then it too is doomed to fail. The potential reasons are legion and often have little to do with the new features themselves. Vice versa, innovations can enhance individual and social action and produce entirely new aims (Withagen *et al.* 2012). The social appropriation of an innovation – which I understand as the actual innovation process – requires innovators to become aware of new possibilities for action. So the crucial question is: Did the wagon offer new possibilities for action in the 4th millennium BCE, and if so, what were they?

In terms of mobility and transport wagons, certainly offered some potential for extending the options for action. But here we have to make regional distinctions. The question is whether wagon use was an asset in the overall perspective, independently of regional and cultural conditions.

Many scholars have argued that vehicles were an elite phenomenon. From the outset, the wagon was a high-ranking and – at least in Mesopotamia – a divine vehicle (Salonen 1951, 66–76; Civil 1968, 3). It stood for a completely new kind of movement: self-movement – auto-mobility – and the driver was physically elevated above the crowd. It was an experience of movement that was almost disembodied. The wagon driver did not arrive, he appeared. One can easily imagine that this engendered a new sense of self-esteem. It is here that we can intuit the intense fascination that vehicles exerted on people – as they still do today.

The innovation represented by the wagon is the product of successful interplay between technical potential and the ideological needs of human agents. Wagons are *social practice* resulting both from the interests and capabilities of the user and from the technical 'affordances' of the vehicles. This innovation cannot be properly understood – or properly investigated – from the perspective of human agency or technical rationality alone.

The overwhelming success of the wagon was most likely due to the new form of movement it provided: disembodied auto-mobility. In terms of perception, it brilliantly combined locomotion with social elevation. This made vehicles particularly interesting for elites – regardless of the kind of society they belonged to. The sheer simplicity of wagon technology and the knowledge transfer associated with it accelerated its dissemination to a degree unequalled by almost any other technology before or since. To use a term coined by Jürgen Renn (2012), knowledge of wagon technology was a fellow traveller. It was not the actual purpose of communication, it materialised like an appurtenance in the course of other exchange processes. Distribution networks may have been provided by copper metallurgy. The areas favouring the spread of prestigious copper tools from approximately 4600 BCE are also areas with evidence of early wagon use. It is here that we need to look, not for the origins of the wagon but for the mechanisms of its diffusion. These were the networks in which, as far as we can see, prestigious objects were distributed and prestigious knowledge communicated. It is in this context that the innovation represented by the wagon demands to be seen.

Notes

1   For further discussion, see Bakker *et al.* 1999; Burmeister 2011.

2   A small fragment of a cart model found in a house layer from the early Ravi phase in Harappa can also be dated to the years 3500–3300 BCE (Kenoyer 2004, 91). No other fragments were found in the following excavation campaigns, so we need to bear in mind that this solitary find may be an intrusion from higher layers; for critical comments, see also Albert Lanting (Bakker and Lanting 2005–2006, 29).

3   'Wheeled toys' in Mexico and El Salvador clearly demonstrate that the principle of the wheel was indeed invented independently several times. In the early post-classic period (900–1250 CE), we clearly see the wheel implemented as a means of locomotion (Ekholm 1946; Haberland 1965; Diehl and Mandeville 1987). Nevertheless, no wagons developed in ancient America. While the Mesoamerican civilizations certainly had the potential for this innovation, the main reason for the absence of this technology is generally held to be the lack of suitable draught animals (Diehl and Mandeville 1987, 244).

4   The ^{14}C-date (Hv 2455 ± 65 BP) is calibrated with Oxcal 4.2; the given range covers a 95.4% probability. The wagon part was found among the construction wood of the trackway V (Pr) in the Aschener Moor. Dendro-dates from the trackway support an early date at the end of the 7th century BCE (Bauerochse *et al.* 2014, 488 fig. 3).

5   Swedish rock art shows four-wheeled vehicles with a center pivot plate suggesting a steerable front axle; see *e.g.* Larsson 2004, 394 fig. 18. The petroglyphs are broadly dated to the Bronze Age; however, they may be contemporaneous with the Central European wagon burials from the early Iron Age (Larson 2004, 392).

6   For further discussion, see Burmeister 2010; 2012, 91–92.

7   Usually oxen were used as draught animals. They needed to be trained for a long time and accordingly were no longer available as a source of food. Their maintenance costs were considerable (see Ebersbach 2002, 153–157). Draught animals are a means of production with high investment costs. There is an echo of this in the Code of Ḥammurabi, which states that two-thirds of the rent for a four-wheeled wagon (*ereqqu*) were payable for the draught team (Salonen 1951, 30).

8   I would like to thank Jochen Büttner of the Max Planck Institute for the History of Science, Berlin for his suggestions and for discussion on this issue.

9   All the dimensions have been taken from the unpublished documentation by the excavator Hajo Hayen and have also been ascertained from the original finds in the Museum für Natur und Mensch, Oldenburg.

References

Bakker, J. A., Kruk, J., Lanting, A. E. and Milisauskas, S. (1999) The Earliest Evidence of Wheeled Vehicles in Europe and the Near East. *Antiquity* 73, 778–790.

Bakker, J. A., Kruk, J., Lanting, A. E. and Milisauskas, S. (2005–2006) The Unabridged Text of the 'Antiquity' 1999 Paper. *Palaeohistoria* 47/48, 10–28.

Bakker, J. A. and Lanting, A. E. (2005–2006) Relevant Sections of Two Letters by A. E. Lanting to J. A. Bakker on the Book 'Rad und Wagen' (2004) and Related Subjects. *Palaeohistoria* 47/48, 29–37.

Bauerochse, A., Leuschner, B., Frank, Th., Metzler, A., Höppel, G. and Leuschner, H. H. (2014), Dendrochronologische Datierungen an Bauhölzern von Moorwegen Nordwestdeutschlands – Ergänzung, Korrektur und Neubewertung. *Archäologisches Korrespondenzblatt* 44, 483–494.

Belinskij, A. B. and Kalmykov, A. A. (2004) Neue Wagenfunde aus Gräbern der Katakombengrab-Kultur im Steppengebiet des zentralen Vorkaukasus. In M. Fansa and S. Burmeister (eds), *Rad und Wagen – Der Ursprung einer Innovation. Wagen im Vorderen Orient und Europa*. Archäologische Mitteilungen aus Nordwestdeutschland Beiheft 40, 201–220. Mainz, Philipp von Zabern.

Burmeister, S. (2002) 'Don't Litter' – Müll am steinzeitlichen Wegesrand. In S. Wolfram and M. Fansa (eds), *Begleitschrift zur Sonderausstellung Müll – Facetten von der Steinzeit bis zum Gelben Sack vom 06. September bis 30. November in Oldenburg*. Schriftenreihe des Landesmuseums für Natur und Mensch 27, 47–54. Mainz, Philipp von Zabern.

Burmeister, S. (2004) Neolithische und bronzezeitliche Moorfunde aus den Niederlanden, Nordwestdeutschland und Dänemark.

In M. Fansa and S. Burmeister (eds), *Rad und Wagen – Der Ursprung einer Innovation. Wagen im Vorderen Orient und Europa*. Archäologische Mitteilungen aus Nordwestdeutschland Beiheft 40, 321–340. Mainz, Philipp von Zabern.

Burmeister, S. (2010) Transport im 3. Jahrtausend v. Chr. Waren die Wagen ein geeignetes Transportmittel im Überlandverkehr? In S. Hansen, A. Hauptmann, I. Motzenbäcker and E. Pernicka (eds), *Von Majkop bis Trialeti. Gewinnung und Verbreitung von Metallen und Obsidian in Kaukasien im 4.–2. Jt. v. Chr. Beiträge des Internationalen Symposiums in Berlin vom 1.–3. Juni 2006*. Kolloquien zur Vor- und Frühgeschichte 13, 223–235. Bonn, Habelt.

Burmeister, S. (2011) Innovationswege – Wege der Kommunikation. Erkenntnisprobleme am Beispiel des Wagens im 4. Jahrtausend v. Chr. In S. Hansen and J. Müller (eds), *Sozialarchäologische Perspektiven: Gesellschaftlicher Wandel 5000–1500 v. Chr. zwischen Atlantik und Kaukasus*. Archäologie in Eurasien 24, 211–240. Mainz, Philipp von Zabern.

Burmeister, S. (2012) Der Mensch lernt fahren – zur Frühgeschichte des Wagens. *Mitteilungen der Anthropologischen Gesellschaft in Wien* 142, 81–100.

Burmeister, S. and Müller-Scheeßel, N. (2013) Innovation as a Multi-Faceted Social Process: An Outline. In S. Burmeister, S. Hansen, M. Kunst and N. Müller-Scheeßel (eds), *Metal Matters. Innovative Technologies and Social Change in Prehistory and Antiquity*. Menschen – Kulturen – Traditionen. Studien aus den Forschungsclustern des Deutschen Archäologischen Instituts 12, 1–11. Rahden/Westf., Marie Leidorf.

Civil, M. (1968) Išme-Dagan and Enlil's Chariot. *Journal of the American Oriental Society* 88, 3–14.

Diehl, R. A. and Mandeville, M. D. (1987) Tula, and Wheeled Animal Effigies in Mesoamerica. *Antiquity* 61, 239–246.

Dinu, M. (1981) Clay Models of Wheels Discovered in Copper Age Cultures of Old Europe Mid-Fifth Millennium B. C. *Journal of Indo-European Studies* 9, 1–14.

Ebersbach, R. (2002) *Von Bauern und Rindern. Eine* Ökosystemanalyse *zur Bedeutung der Rinderhaltung in bäuerlichen Gesellschaften als Grundlage zur Modellbildung im Neolithikum*. Basler Beiträge zur Archäologie 15. Basel, Schwabe.

Ekholm, G. F. (1946), Wheeled Toys in Mexico. *American Antiquity* 11, 222–228.

Fansa, M. and Schneider, R. (1994) Steinzeitlicher Pfahlweg XV (Le) im Meerhusener Moor zwischen Aurich-Tannenhausen, Landkreis Aurich, im Südosten und dem Ewigen Meer, Landkreis Wittmund, im Nordwesten. *Archäologische Mitteilungen aus Nordwestdeutschland* 17, 15–37.

Gej, A. N. (2004) Die Wagen der Novotitarovskaja-Kultur. In M. Fansa and S. Burmeister (eds), *Rad und Wagen – Der Ursprung einer Innovation. Wagen im Vorderen Orient und Europa*. Archäologische Mitteilungen aus Nordwestdeutschland Beiheft 40, 177–190. Mainz, Philipp von Zabern.

Gusev, S. A. (1998) K voprosu o transportnych sredstvach tripol'skoj kul'tury [On the Issue of the Tripolye Culture Transportation Means]. *Rossijskaja Archeologija* 1998, 15–28.

Haberland, W. (1965) Tierfiguren mit Rädern aus El Salvador. *Baessler-Archiv Neue Folge* 13, 309–316.

Hayen, H. (1972) Vier Scheibenräder aus dem Vehnemoor bei Glum (Gemeinde Wardenburg, Landkreis Oldenburg). *Die Kunde Neue Folge* 23, 62–86.

Hayen, H. (1983) Handwerklich-technische Lösungen im vor- und frühgeschichtlichen Wagenbau. In: H. Jankuhn, W. Janssen, R. Schmidt-Wiegand and H. Tiefenbach (eds), *Das Handwerk in vor- und frühgeschichtlicher Zeit 2. Archäologische und philologische Beiträge*. Abhandlungen der Akadamie der Wissenschaften Göttingen, Philologisch-Historische Klasse 3, Folge 123, 415–470. Göttingen, Vandenhoeck & Ruprecht.

Kenoyer, J. M. (2004) Die Karren der Induskultur Pakistans und Indiens. In M. Fansa and S. Burmeister (eds), *Rad und Wagen – Der Ursprung einer Innovation. Wagen im Vorderen Orient und Europa*. Archäologische Mitteilungen aus Nordwestdeutschland Beiheft 40, 87–106. Mainz, Philipp von Zabern.

Kirtcho, L. B. (2009) The Earliest Wheeled Transport in Southwestern Central Asia: New Finds from Altyn-Depe. *Archaeology, Ethnology and Anthropology of Eurasia* 37, 1, 25–33.

Kuipers, M. H. G. (2013) The Sound of Fire, Taste of Copper, Feel of Bronze, and Colours of the Cast: Sensory Aspects of Metalworking Technology. In M. L. S. Sørensen and K. Rebay-Salisbury (eds), *Embodied Knowledge. Perspectives on Belief and Technology*, 137–150. Oxford, Oxbow Books.

Larsson, T. B. (2004) Streitwagen, Karren und Wagen in der bronzezeitlichen Felskunst Skandinaviens. In M. Fansa and S. Burmeister (eds), *Rad und Wagen – Der Ursprung einer Innovation. Wagen im Vorderen Orient und Europa*. Archäologische Mitteilungen aus Nordwestdeutschland Beiheft 40, 381–398. Mainz, Philipp von Zabern.

Leroi-Gourhan, A. (1973) *Milieu et techniques*. 3rd ed. Paris, Michel.

Maran, J. (2004) Kulturkontakte und Wege der Ausbreitung der Wagentechnologien im 4. Jahrtausend v. Chr. In M. Fansa and S. Burmeister (eds), *Rad und Wagen – der Ursprung einer Innovation. Wagen im Vorderen Orient und Europa*. Archäologische Mitteilungen aus Nordwestdeutschland Beiheft 40, 429–442. Mainz, Philipp von Zabern.

Matuschik, I. (2006) Invention et diffusion de la roue dans l'Ancien Monde: L'apport de l'iconographie. In P. Pétrequin, R.-M. Arbogast, A.-M. Pétrequin, S. van Willigen and M. Bailly (eds), *Premiers chariots, premier araires. La diffusion de la traction animale en Europe pendant les IVe et IIIe millénaires avant notre ère*. Centre National de la Recherche Scientifique Monographies 29, 279–297. Paris, CNRS Éditions.

Mischka, D. (2011) The Neolithic Burial Sequence at Flintbek LA 3, North Germany, and its Cart Tracks: A Precise Chronology. *Antiquity* 85, 442–458.

Norman, D. A. (1988) *The Psychology of Everyday Things*. New York, Basic Books.

Pare, C. F. E. (1987a) Bemerkungen zum Wagen von Hochdorf. In *Vierrädrige Wagen der Hallstattzeit. Untersuchungen zu Geschichte und Technik*. RGZM Monographien 12, 128–133. Mainz, Römisch-Germanisches Zentralmuseum.

Pare, C. F. E. (1987b) Der Zeremonialwagen der Hallstattzeit – Untersuchungen zu Konstruktion, Typologie und Kulturbeziehungen. In *Vierrädrige Wagen der Hallstattzeit. Untersuchungen zu Geschichte und Technik*. RGZM Monographien 12, 189–248. Mainz, Römisch-Germanisches Zentralmuseum.

Renn, J. (2012) Survey: Knowledge as a Fellow Traveller. In J. Renn (ed), *The Globalization of Knowledge in History*, 205–243. Berlin, Edition Open Access.

Salonen, A. (1951) *Die Landfahrzeuge des Alten Mesopotamien nach sumerisch-akkadischen Quellen (mit besonderer Berücksichtigung der 5. Tafel der Serie ḪAR-ra = ḫubullu).* Helsinki, Suomalainen Tiedeakatemia.

Schlichtherle, H. (2004) Wagenfunde aus den Seeufersiedlungen im zirkumalpinen Raum. In M. Fansa and S. Burmeister (eds), *Rad und Wagen – Der Ursprung einer Innovation. Wagen im Vorderen Orient und Europa.* Archäologische Mitteilungen aus Nordwestdeutschland Beiheft 40, 295–314. Mainz, Philipp von Zabern.

Sherratt, A. (2004) Wagen, Pflug, Rind: Ihre Ausbreitung und Nutzung – Probleme der Quelleninterpretation. In M. Fansa and S. Burmeister (eds), *Rad und Wagen – Der Ursprung einer Innovation. Wagen im Vorderen Orient und Europa.* Archäologische Mitteilungen aus Nordwestdeutschland Beiheft 40, 409–428. Mainz, Philipp von Zabern.

Shishlina, N. I., Kovalev, D. S. and Ibragimova, E. R. (2014) Catacomb Culture Wagons of the Eurasian Steppes. *Antiquity* 88, 378–394.

Suchman, L. and Bishop, L. (2000) Problematizing 'Innovation' as a Critical Project. *Technology Analysis & Strategic Management* 12/3, 327–347.

Sürenhagen, D. (1999) *Untersuchungen zur Relativen Chronologie Babyloniens und angrenzender Gebiete von der ausgehenden ᶜUbaidzeit bis zum Beginn der Frühdynastisch II-Zeit 1. Studien zur Chronostratigraphie der südbabylonischen Stadtruinen von Uruk und Ur.* Heidelberger Studien zum alten Orient 8. Heidelberg, Heidelberger Orientverlag.

Tureckij, M. A. (2004) Wagengräber der grubengrabzeitlichen Kulturen im Steppengebiet Osteuropas. In M. Fansa and S. Burmeister (eds), *Rad und Wagen – Der Ursprung*

einer Innovation. Wagen im Vorderen Orient und Europa. Archäologische Mitteilungen aus Nordwestdeutschland Beiheft 40, 191–200. Mainz, Philipp von Zabern.

Velušček, A., Čufar, K. and Zupančič, M. (2009) Prazgodovinsko leseno kolo z osjo s kolišča Stare gmajne na Ljubljanskem barju. Prehistoric wooden wheel with an axle from the pile-dwelling Stare gmajne at the Ljubljansko barje. In: A. Velušček (ed), *Koliščarska naselbina Stare gmajne in njen čas. Ljubljansko barje v 2. polovici 4. tisočletja pr. Kr. Stare gmajne pile-dwelling settlement and its era. The Ljubljansko barje in the 2nd half of the 4th millennium BC.* Opera Instituti Archaeologici Sloveniae 16, 197–222. Ljubljana, Inštitut za arheologijo ZRC SAZU.

Vosteen, M. U. (1999) *Urgeschichtliche Wagen in Mitteleuropa. Eine archäologische und religionswissenschaftliche Untersuchung neolithischer bis hallstattzeitlicher Befunde.* Freiburger Archäologische Studien 3. Rahden/Westf., Marie Leidorf.

Vosteen, M. U. (2002) Die fünffache Erfindung von Rad und Wagen. In J. Köninger, M. Mainberger, H. Schlichtherle and M. U. Vosteen (eds), *Schleife, Schlitten, Rad und Wagen. Zur Frage früher Transportmittel nördlich der Alpen.* Hemmenhofener Skripte 3, 143–148. Gaienhofen-Hemmenhofen, Janus.

Withagen, R., de Poel, H. J., Araújo, D. and Pepping, G.-J. (2012) Affordances can Invite Behavior: Reconsidering the Relationship between Affordances and Agency. *New Ideas in Psychology* 30, 250–258.

Woolley, L. (1955) *Ur Excavations IV. The Early Periods. A report on the sites and objects prior in date to the third dynasty of Ur discovered in the course of the excavations.* Philadelphia, PA, Johnson Fund of the American Philosphical Society.

Chapter 8

Contextualising Innovation: Cattle Owners and Wagon Drivers in the North Caucasus and Beyond

Sabine Reinhold, Julia Gresky, Natalia Berezina, Anatoly R. Kantorovich, Corina Knipper, Vladimir E. Maslov, Vladimira G. Petrenko, Kurt W. Alt and Andrey B. Belinsky[1]

Innovation is the social act of appropriating new practices or techniques into an existing life. During the late 4th millennium BC, the populations of the North Caucasus and the neighbouring steppe adopted animal traction and vehicles into their lifescapes. The representation of this innovation, however, suggests different intellectual discourses in the appropriation process. The Maikop communities selected the powerful driving force – cattle teams – for their burial representations, whereas the steppe communities chose to highlight the means of transportation – wagons. While the one emphasised a new form of extended labour and neglected the objects of traction, the other highlighted the new means of transportation and mobility, disregarding the 'engines' in the process. This article discusses the different ways the same innovation can be appropriated in different communities and draws upon bioarchaeological studies to question the practical relevance of these innovations for the everyday life of the adopting societies.

Introduction

During the mid-4th and early 3rd millennia BC, fundamental technological innovations and new knowledge spread across wide parts of western Eurasia. In the last few years, the process once referred to as the 'secondary products revolution' (Sherratt 1981; 1983) has become a major and multifaceted topic in archaeological research (Fansa and Burmeister 2004; Greenfield 2010; Hansen 2011). The social mechanisms behind the appropriation of new techniques, exploitation strategies and material objects represent extended, multidimensional processes with rapid leaps and inconsistent developments in different regions of Eurasia. Exploitation of secondary animal products, such as milk, is now discussed as part of early Neolithic praxis (Greenfield 2010; Halstead and Isaakidou 2011, 67), and the debate on the 'secondary products revolution' has shifted from the question of origins to inquiries into processes of intensification and diffusion (Beaujard 2011; Müller 2013).

V. Gordon Childe's idea of technological 'revolutions' as the driving forces in historical change still dominates the debate (Childe 1936; Sherratt 1981, 287–299 fig. 10.16; Greene 1999). At the same time, social transformations preceding the horizon of innovations in the mid-4th millennium BC are beginning to shed light on the societal foundations of the essential changes. They demonstrate an intensification of social relations involving growing co-habitation groups (Ur 2010; Müller 2015), an increasing focus on individuals, differentiation and competition (Govedarica 2004; Furholt and Müller 2011; Stein 2012), the emergence of conflict (Erdal and Erdal 2012), transformations in the production and use of material culture (Oates *et al.* 2007) and the transformation of landscapes (Steadman 2005; Furholt and Müller 2011). The growing social demands posed by coping with greater accumulations of people and increasing competition would be a reasonable basis for the adoption of new technological solutions and would promote the transfer of skills and knowledge (Müller 2013; Hansen 2014).

Two of the critical regions in the transfer and development of new technologies in western Eurasia are the Caucasus mountain system and the neighbouring steppe zone to the north (Fig. 8.1). Thanks to its geographical position,

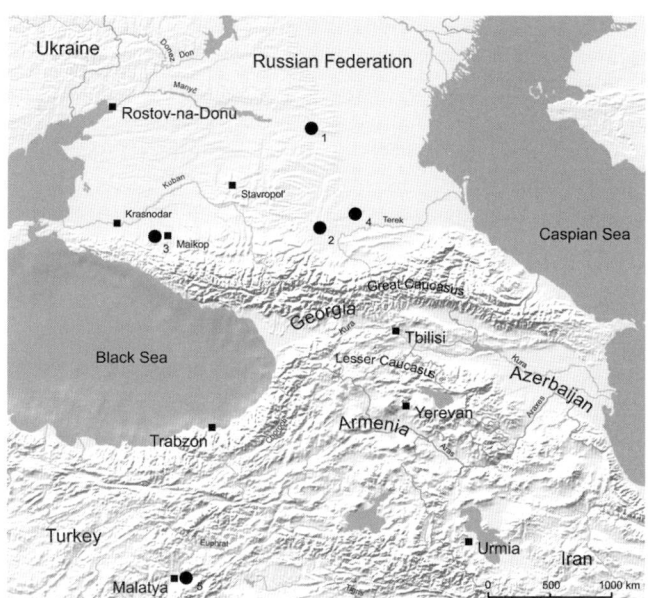

Figure 8.1: The topography of Caucasia and the neighbouring areas with the sites mentioned in the text: (1) Sharakhalsun 2; (2) Mar'inskaya 5; (3) Maikop; (4) Mozdok-Komarovo; (5) Arslantepe.

Caucasia is one of the most important conveyors of knowledge, new technologies and people in both directions. Also, a wide range of resources predestined this area as a centre for development in the 4th and 3rd millennia BC, notably in metal production and other crafts.

Its position in a geographical contact zone is one of the factors explaining why during all epochs Caucasia linked a large variety of cultural complexes with different traditions and geographical relations (Kohl and Trifonov 2014) (Fig. 8.2). As early as the 5th millennium BC, a cultural division ran along the northern piedmont-steppe borderline dividing the region into a northern province closely connected to the lower Volga, the Don and the northern Black Sea (Korenevsky 2012) and a southern territory equivalent to the northernmost periphery of North Mesopotamia (Marro 2007). The Eneolithic of the steppe zone is characterised by low intensity in the exploitation of steppe environments (Shishlina 2008, 15–21, 222–227). In the North Caucasian piedmont zone and south of the main watershed, communities of the Svobodnoe-Meshoko (north) or Sioni-Tsopi (south) traditions developed, practising a small-scale system of mixed agro-pastoralism (Nechaev

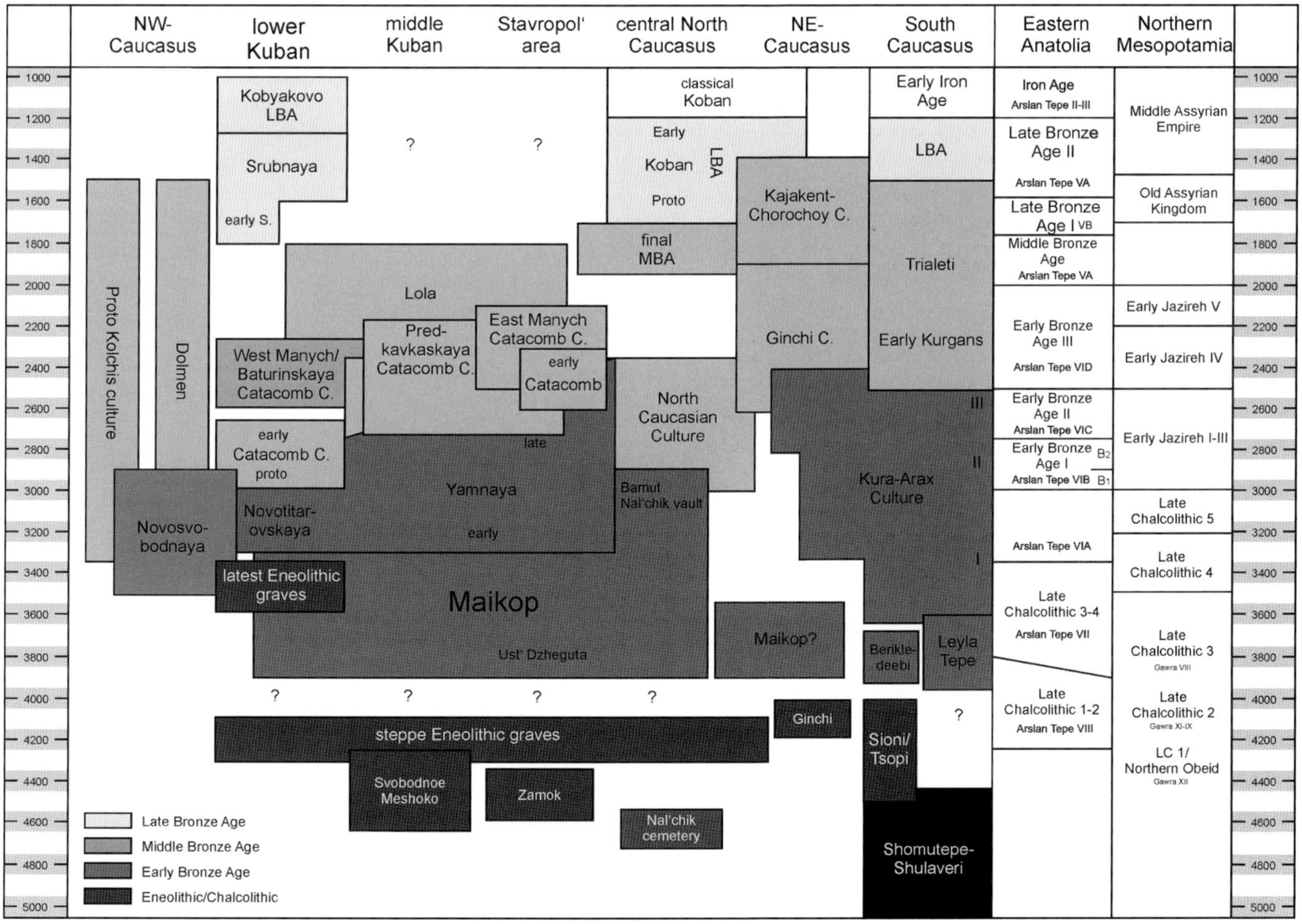

Figure 8.2: Chronology of Bronze Age cultures in the Caucasus and neighbouring areas.

1992; Nebieridze 2010). The following Early Bronze Age in the early 4th millennium BC saw the development of the Maikop phenomenon in the piedmonts of the North Caucasus, while the closely linked Leilatepe culture was established in the Southeast Caucasus (Kohl and Trifonov 2014). The communities in the steppe region persisted in a local Late Eneolithic tradition. During the late 4th and the early 3rd millennia BC, the Kura-Araxes phenomenon emerged south of the mountains, while in the steppe area north of the Caucasus Novotitarovskaya, a Caucasian variant of the Yamnaya culture, took shape. All three overlap chronologically but not geographically with the Late Maikop or Novosvobodnaya phase of the North Caucasian piedmonts. A similar divergence between steppe- and mountain-related cultures is identifiable in the Middle Bronze Age from the early 3rd millennium BC onwards. The local variants of the Catacomb culture in the steppe differ considerably from North Caucasian and Trialeti cultures in the piedmonts and the mountains north and south of the main Caucasus range.

The first pastoral groups inhabited the plateaus of southern Caucasia and the steppe zone north of the Caucasus as early as the late 5th millennium BC (Marro 2007), but only from the 3rd millennium BC onwards is an intensive exploitation of steppe and mountain pastures apparent. The 'opening up of the steppes' has long been associated with the development of a mobile, pastoral way of life (Merpert 1974; Shishlina 2008, 231–232). New technologies, in particular wheeled transport and most likely horsemanship, fostered this process with its increasing demands for new large pastures. But did these demands trigger the adoption of the new idea of exploiting animal labour?

The opening question: Why innovate?

Why innovate? What makes people change well-known practices for uncertain novelties (Bloch 1967)? Why did the communities in the Caucasus and beyond adopt new devices in the 4th and 3rd millennia BC that required new knowledge and most probably involved challenges never envisaged before? How did the different communities respond to the introduction of these new technologies? In our modern world, such questions would be answered with utilitarian criteria pertaining to an increase of productivity and economic efficiency or indirect benefits such as a higher connectivity, time-saving or greater personal convenience (Rogers 2003). Yet in a prehistoric world living in household economies and at a much slower tempo than today, these answers may perhaps be wrong (Bloch 1967).

Innovation, as opposed to invention, is in the first place an intellectual dialogue on the issue of what is 'new'. While invention is a creative act that produces radically new solutions for technical or social problems, innovation is the stage in which these solutions have to be approved by the community outside the inventor's inner circle. Thus, innovation is an entirely social act (Bloch 1967, 130–133). It rests upon the convincement potential of the groups inventing or first adopting the novelties in question and collective appropriation by a wider community (Hägerstrand 1967; Haggett 1991, 383–408).

Novelties are challenges, both physically and intellectually. Communication is the first precondition for the spatial diffusion of new technologies and the knowledge/know-how related to them. Local rumour, targeted information and convincing performance encourage the acceptance of technical innovations (Hägerstrand 1967, 149–263). Charismatic preachers can spread new ideas quickly, even against opposition (Cavalcanti 2005). Information flows travel by way of social networks, or larger migrant groups transfer some of their own technological apparatus to new areas (Burmeister 2000). Appropriation of new technologies by higher social classes encourages acceptance in subordinate groups (Haggett 1991, 386–387 fig. 13.3), while mastering the 'new' also opens up possibilities for ambitious individuals or groups to transcend the traditional scheme of things and enhance their social status.

A second precondition for innovation is a cultural disposition to accept new ideas, changing practices or exotic devices. True, we need to acknowledge that many of the western Eurasian innovations of the 4th and 3rd millennia BC did have utilitarian assets: larger loads and greater distances in transport on land and on water (Ward 2006; Burmeister 2011), enlargement of cultivated areas using ploughs (Sherratt 1981, 292–293), more efficient metal tools and weapons advancing both craftsmanship and warfare (*cf.* Hansen in this volume), the employment of iconographic seals or early script assisting in the administration of growing communal enterprises, and the use of wool, a fibre greatly superior to linen in cold, wet climates (Shishlina *et al.* 2003).

The first draught animals in the Caucasus

In this article, we compare different ways of appropriating the idea of animal labour, or more specifically, the use of cattle as draught animals in two rather different cultural spheres – the Caucasian mountain traditions and the steppe-related cultures of the Early and Middle Bronze Age. The survey proceeds from two complexes dating to the last third of the 4th millennium BC. They are nearly identical in absolute dating terms but have different cultural backgrounds involving different perspectives on the incorporation of animal energy into human societies.

Mar'inskaya 5

In 2009 a huge, single-standing burial mound was excavated near the village of Mar'inskaya (Kirov rayon, Stavropol region, Russia) in a joint expedition undertaken by the

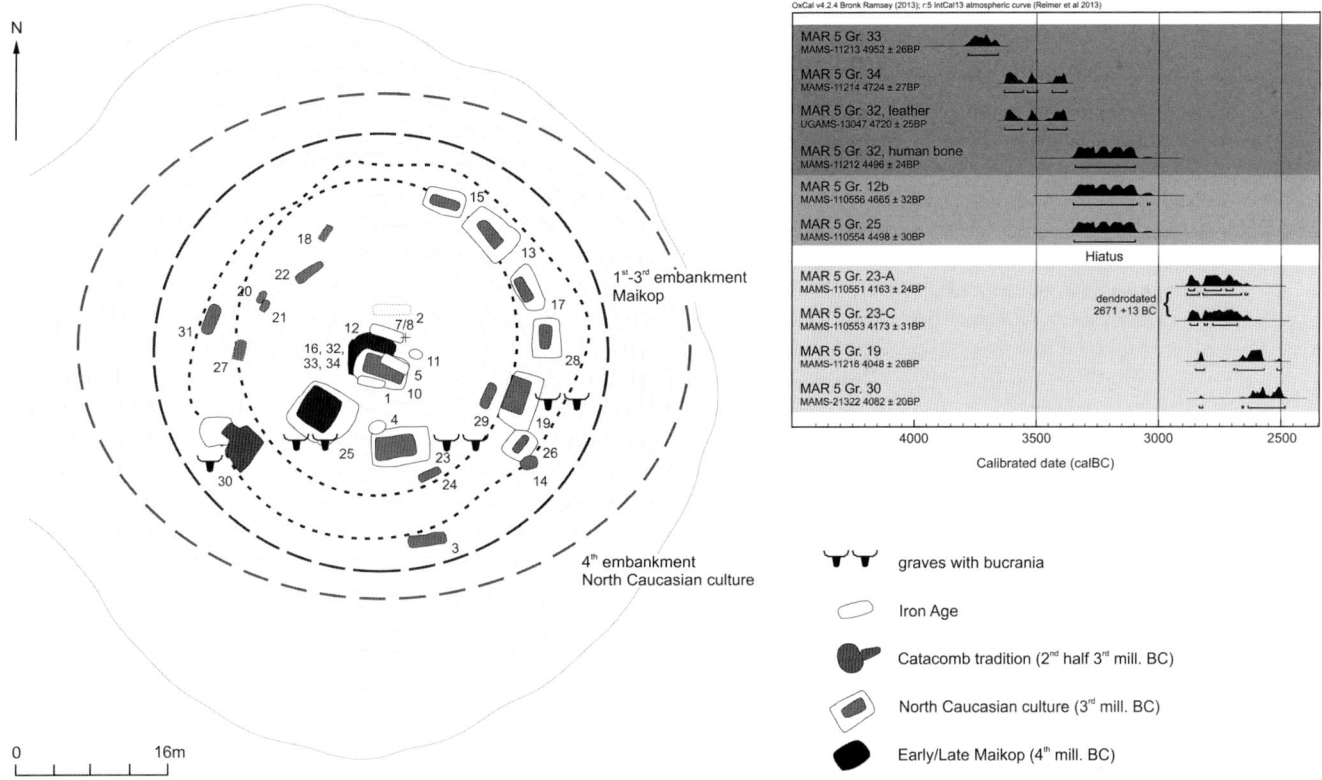

Figure 8.3: Mar'inskaya 5. General plan of the mound and radiocarbon dates of selected burials.

Lomonosov Moscow State University and the local heritage organisation 'Nasledie'. The mound (Mar'inskaya 5) was surrounded by a ditch, was slightly oval in form with a diameter of 34 to 40 metres, and was 4.3 metres high. Its Maikop stratigraphy has recently been published in Russian (Kantorovich *et al.* 2013) (Fig. 8.3).

In the central part of the mound, the excavations uncovered a sequence of six graves dating to the Maikop epoch and after a hiatus of 600 years another sequence of 16 graves dating to the North Caucasian culture, *i.e.* the local Middle Bronze Age (MBA). One catacomb grave, also MBA, intersected the ring of the North Caucasian graves. The entire sequence comprised four construction layers, three of them related to the Maikop mound. The stratigraphy, a series of radiocarbon dates and a first dendrochronological sequence, permits precise dating of practically all burials (Kantorovich *et al.* 2013, 92 tabs 1–2, figs 45–46).

The three later Maikop burials 16, 12 and 25 date back to the second half of the 4th millennium BC. Of interest is grave 25 (Kantorovich *et al.* 2013, 86–92 figs 34–44) (Tab. 8.1). This partly disturbed grave of a male individual was a square construction of wooden posts covered with a wooden ceiling (Fig. 8.4.3). It contained a partly preserved inventory (Fig. 8.5.1–7) with good parallels in the later part of the 4th millennium BC, *e.g.* in the west Caucasian site of Klady (Fig. 8.5.8–19). The most important discovery in this grave was a pair of bucrania deposited outside the wooden

chamber on its southern side *(*Kantorovich *et al.* 2013, 87–92, fig. 35, 41–42) (Fig. 8.4.1–2). They were found at the same level as the inhumation inside. Their faces were turned away from the grave, and both of them once had a looped bronze nose ring at the tip of the skull. One of these rings was found *in situ* on the nose of the western skull II (Fig. 8.4.3 no. 6). It is slightly longer and straighter than the second ring. This other ring had been dislocated when the chamber collapsed (Fig. 8.4.3 no. 5) but it left a green stain on the nose part of the second (eastern) skull I. It is shorter and slightly deformed. The better-preserved western skull had a horn span of about 0.8 m, the eastern skull was too badly preserved for measurement. Accordingly, sex determination on the basis of the very fragmented skulls was impossible, but the horn span and their massiveness suggest male individuals, probably oxen. In front of the skulls, traces of a wooden object of at least 0.5 m length were visible, possibly an object related to the cattle skulls.

Looped nose rings belong to a category of objects known from late Maikop sites in the North Caucasus (Munchaev 1975; 1994, 209–211 fig. 57; Korenevsky 2011, 86 figs 83, 84 4–5, 85 4–6.10) and from Alaça Höyük in Central Anatolia (Korenevsky 2011, 86 fig. 83, 9–10) (Fig. 8.7). One such Caucasian complex is grave 1 in mound 4 of the Klady site (Rezepkin 2012, fig. 7). (Fig. 8.5.8–19). The chronological position has now been confirmed by the radiocarbon dating of grave 25 in Mar'inskaya 5 to

Table 8.1: The Mari'nskaya 5/1 complexes with reliable remains of paired bucrania

Grave no.	Description	Dating
25	Disturbed inhumation burial in a square chamber surrounded by wooden posts & covered by wooden planks, outside the wooden chamber a stone filling & a stone packing on top of the grave & related mound embankment, the skeleton was placed in a crouched positon on organic material, bones had been dislocated, below the skeleton an area 2 m in diameter was plastered with clay, 2 dislocated bronze daggers, a gold ear–ring, a ceramic vessel & nearby a bone arrowhead, outside the chamber 2 cattle skulls with nose rings & a badly preserved wooden object (yoke?, fixing pole?), *Anthropology*: individual: male, 40–50 yrs, probable fracture of 3rd thoracic vertebra, entheseal changes particularly in the upper extremity, arthrosis of the cervical spine, ribs & left knee joint.	Late Maikop MAMS 110554: 4498 ± 30 BP 3334–3105 cal BC 1σ 3347–3095 cal BC 2σ charcoal
19	Inhumation burial in a deep, stepped burial pit covered with 8–9 wooden planks & a stone filling, stretched skeleton in supine position laid on longitudinal planks, vertical wooden constructions indicate a wooden chamber, on top of the feet a ceramic vessel at the southern end of the burial pit, a stone object near the hands, bronze objects (4 bracelets) & beads of faience, nacre & bronze deposited with the skeleton, outside the chamber: 2 cattle skulls, one bulky, one small from a juvenile individual & an upper & lower grinding stone at the northern end of the burial pit outside the wooden chamber. *Anthropology*: male, 25–30 yrs, 3 fractures of the hands.	North Caucasian Culture MAMS 11218: 4048 ± 26 BP 2619–2495 cal BC 1σ 2832–2482 cal BC 2σ human bone
23	Inhumation burial in a deep, stepped burial pit inside a wooden chamber surrounded by a layer of river pebbles, chamber roofing with four massive timber & wooden planks, at the sides of the chamber timber log constructions & in corners vertical posts, stretched skeleton in prone position laid diagonally into the chamber, no grave goods, outside the chamber on an extension of the pebble construction surrounding the chamber: 2 badly preserved cattle skulls at the eastern end of the burial pit outside the wooden chamber. *Anthropology*: male, 50–60 yrs, fractures of the hands.	North Caucasian Culture MAMS 110551: 4163 ± 24 BP MAMS 110552: 4158 ± 31 BP MAMS 110553: 4173 ± 31 BP radiocarbon dates for wiggle matching of dendrochronology, gaps 50 yrs tree rings from wooden objects dendro–dated to 2671 + 13 BC
30	Inhumation burial in a catacomb, the stretched skeleton in supine position on organic material (leather?), partly covered by red ochre, bronze beads & animal bones at the feet, a ceramic vessel beside the head in the western side of the chamber, outside the catacomb 1 cattle skull in entrance pit, the grave destroyed an older one of the North Caucasian Culture (Ind. II). *Anthropology*: male, 35–50 yrs (Ind. I)	Caucasian Piedmont (Suvorovo) Catacomb Culture MAMS 21322: 4082 ± 20 BP 2831–2577 cal BC 1σ 2848–2500 cal BC 1σ charcoal

4498 ± 30 BP, *i.e.* the last third of the 4th millennium BC (Fig. 8.3, Tab. 8.1). The stratigraphic position of the dendrochronologically investigated graves 12 and 25 suggests a dating around 3200 BC after calibration. The Mar'inskaya rings have internal diameters of 4.8–5.4 and 3.7–4.5 cm, overlapping straight ends and remains of leather fittings (Fig. 8.5.1–2, Fig. 8.6). Their ends are smooth and plain. Other similar examples have ornamented tips, *e.g.* an example from a destroyed mound near Maikop, or end in knot-like thickenings, *e.g.* rings from Chegem or Bamut (Korenevsky 2011, 86 figs 83.15, 83.1–3). Such rings are known from the 3rd millennium BC contexts in Mesopotamia, *e.g.* on the famous Ur standard or depictions on cylinder seals. Moreover, they were found as silver objects in the noses of oxen pairs in grave PG789 at the Ur royal cemetery (Woolley 1934, 64–65 pl. 35 a, b, pl. 92). Nose rings are among the most common implements for making cattle go in a particular direction.

The function of these objects has been under discussion for a long time. Despite the acknowledged lack of horse remains in Maikop contexts, Rauf M. Munchaev, the first

to summarise the arguments, suggested that they may have played a role in harnessing horses (Munchaev 1975, 390–391). Trifonov (1987, 21–23 fig. 1) argued for their use as cultic insignia of the Mesopotamian goddess Ishtar. The *in situ* find of the pair of rings in Mar'inskaya 5, however, points to a practical function for these objects and dates this type of harnessing known from Near Eastern art back to the 4th millennium BC (Balabina 2004, 205–210). It is legitimate to assume that the wooden remains in front of the skulls stem from a poorly preserved yoke or pole to which the animals were attached. So what we have here is the symbolic gift of a buried team of cattle (oxen?) that was involved in the funeral ceremony, most likely as working animals.

Remarkably, more graves at the same site yielded up offerings of cattle skulls (Kantorovich *et al.* 2013, 104). In burial 23 and in grave 19, both inhumations of the North Caucasian culture in radial positions (Fig. 8.8.1–2), such pairs were found at the feet of the buried males outside their actual chambers but still within the grave-pit. Both graves were intact. Grave 23 had no burial gifts but is one of the

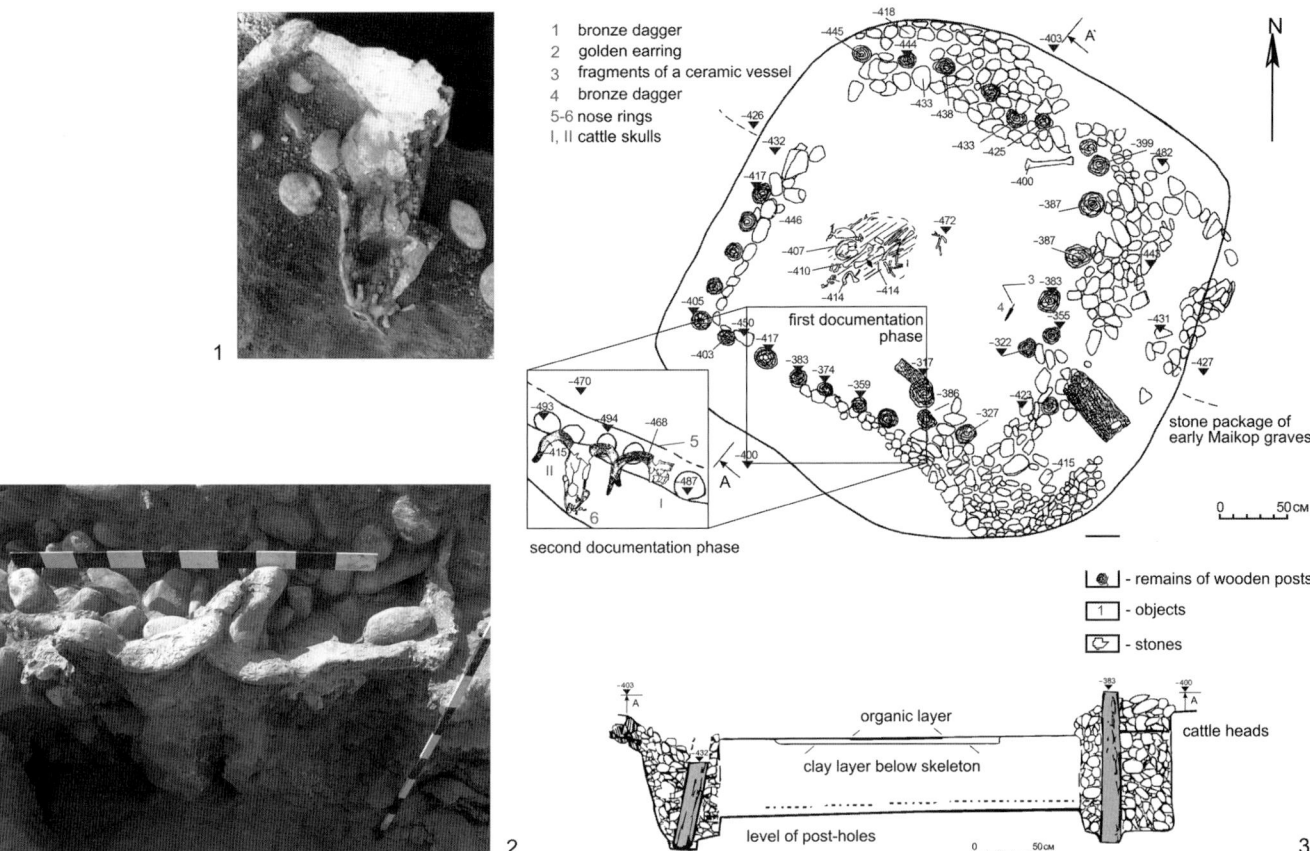

Figure 8.4: Mar'inskaya 5. Grave 25 with bucrania. (1) the western cattle skull II in situ; (2) both cattle skulls in situ; (3) plan and section (after Kantorovich et al. 2013, figs 34–35, 38, 41–42).

oldest complexes of the North Caucasian burial community on the site; grave 19 contained the wealthiest burial ensemble of all the contemporaneous graves. As in grave 25, the skulls were placed close to each other, however without nose rings. One skull in grave 19 belonged to a juvenile individual, and the horn spans of both individuals were comparatively small. The combination of adult and juvenile individuals in one pair is a practice frequently encountered in teams used to train labour animals (Rosenstock and Masson 2011). The fourth instance of a cattle skull offering was found in the entrance pit of Catacomb culture grave 30 (Fig. 8.8.3). This catacomb was empty except for the skeleton and some unspecific grave goods, so cultural assignment is difficult. Nevertheless, the radiocarbon dates for grave 19 and 30 confirm that they are more or less contemporaneous (Tab. 8.1).

Evidence for the practice of burying paired cattle skulls as *pars pro toto* for a pair of draught animals can be seen at Mar'inskaya 5 over a time span of more than 700 years in varying cultural contexts. This long sequence of cattle-head donations in one mound is unique in Caucasia. However, cattle as grave goods in, or on top of, burials are occasionally found in association with higher-status burials of the North Caucasian culture. In 2015, a grave

with two individuals dating to this epoch was discovered not far from Mar'inskaya at Lysogorskaya 6. It contained three pairs of cattle skulls, each with an adult and a juvenile individual. The cattle skulls were placed at the feet of the human inhumations outside their wooden (?) burial chamber. In a nearby grave, two poorly preserved cattle skulls were found on top of the stone burial cover of another wealthy inhumation (Berezin pers. comm.).

Sharakhalsun 2

Located about 300 km north-east of Mar'inskaya in the Kalmykian steppe, burial mound 6 in the cemetery of Sharakhalsun 2 revealed four complexes with remains of wooden wagons (Tab. 8.2). They are among the more than 20 wagon graves in the Stavropol' region that are in the pre-publication stage.[2] We focus here on grave 18, the radiocarbon date of which is almost identical with grave 25 at Mar'inskaya. The mound was uncovered during a rescue excavation by 'Nasledie' in 2001. It is part of a linear alignment of mounds situated on the right bank of the river Kalaus near the Manych water reserve. The mound was 50 m in diameter and 3 m high (Fig. 8.9). The stratigraphy is not entirely clear, but the first mound was

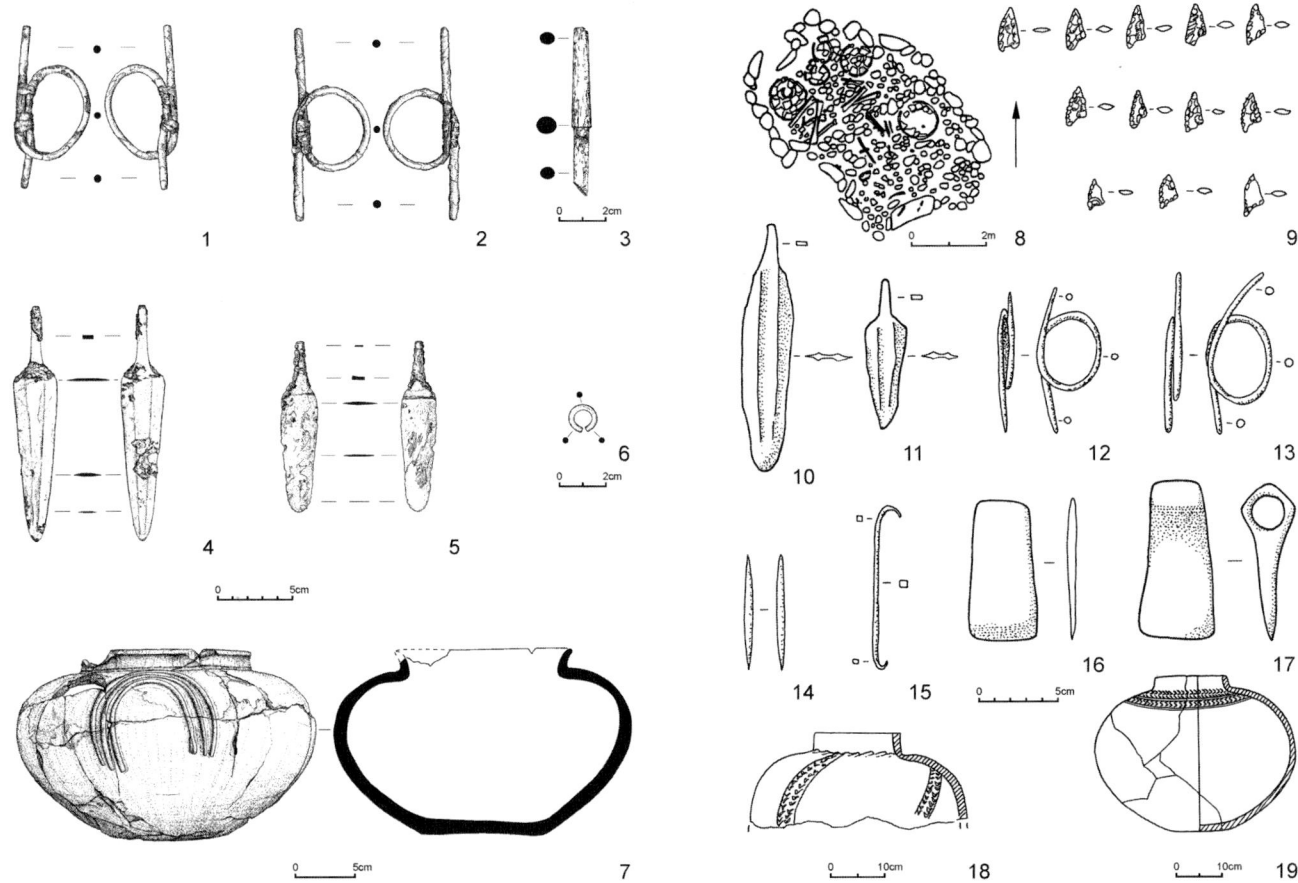

Figure 8.5: The remaining inventory from Mar'inskaya 5, grave 25 and the inventory Klady, mound 4, grave 1. (1–2, 12–13) nose rings; 3 bone projectiles; (4–5, 10–11) bronze daggers; (6) golden earring; (7, 18–19) ceramic vessels; (8) plan of Klady, mound 4, grave 1; (9) flint arrowheads; (14–15) bronze awls; (16–17) bronze axes (after Kantorovich et al. 2013, fig. 44; Rezepkin 2012, fig. 7).

Figure 8.6: Mar'inskaya 5. The looped nose rings from grave 25 (after Kantorovich et al. 2013, fig. 43). The longer one – on the left – was found in situ *on the head of the western bull II, the shorter ring on the right was found dislocated near the eastern bull head I.*

constructed by communities of the Steppe Maikop culture in the late 4th millennium BC (Yakovlev and Samoylenko 2003). During the 3rd millennium, Yamnaya groups used the Maikop mound and added several graves in central

positions (graves 4, 5, 16) and on the periphery (graves 3, 18). The third to fifth construction layers of the mound embankment can safely be attributed to these Yamnaya communities. The first and second are related to Maikop period internments.

Grave 18 (Fig. 8.10) was placed in a narrow, deep, catacomb-like shaft dug from the side into the existing mound. The stratigraphic relation to the central Yamnaya graves is unclear, but wooden parts of the wagon from grave 18 have been dated to 4500 ± 40 BP (Tab. 8.2) and link it to the early Yamnaya or steppe Majkop culture. At the bottom of the shaft, an exceptional burial of a middle-aged male was found, the skeleton crouched in a sitting position on a four-wheeled wooden wagon. The wooden parts of the wagon were poorly preserved, but it was still obvious that a standing vehicle had been squeezed into the burial chamber. The burial did not contain any other inventory apart from the wagon, and the unique burial position makes it difficult to assign the grave to a specific archaeological culture on the basis of burial customs. It belongs to a small group of intermediate burials in sitting position with both early Yamnaya and Maikop aspects.

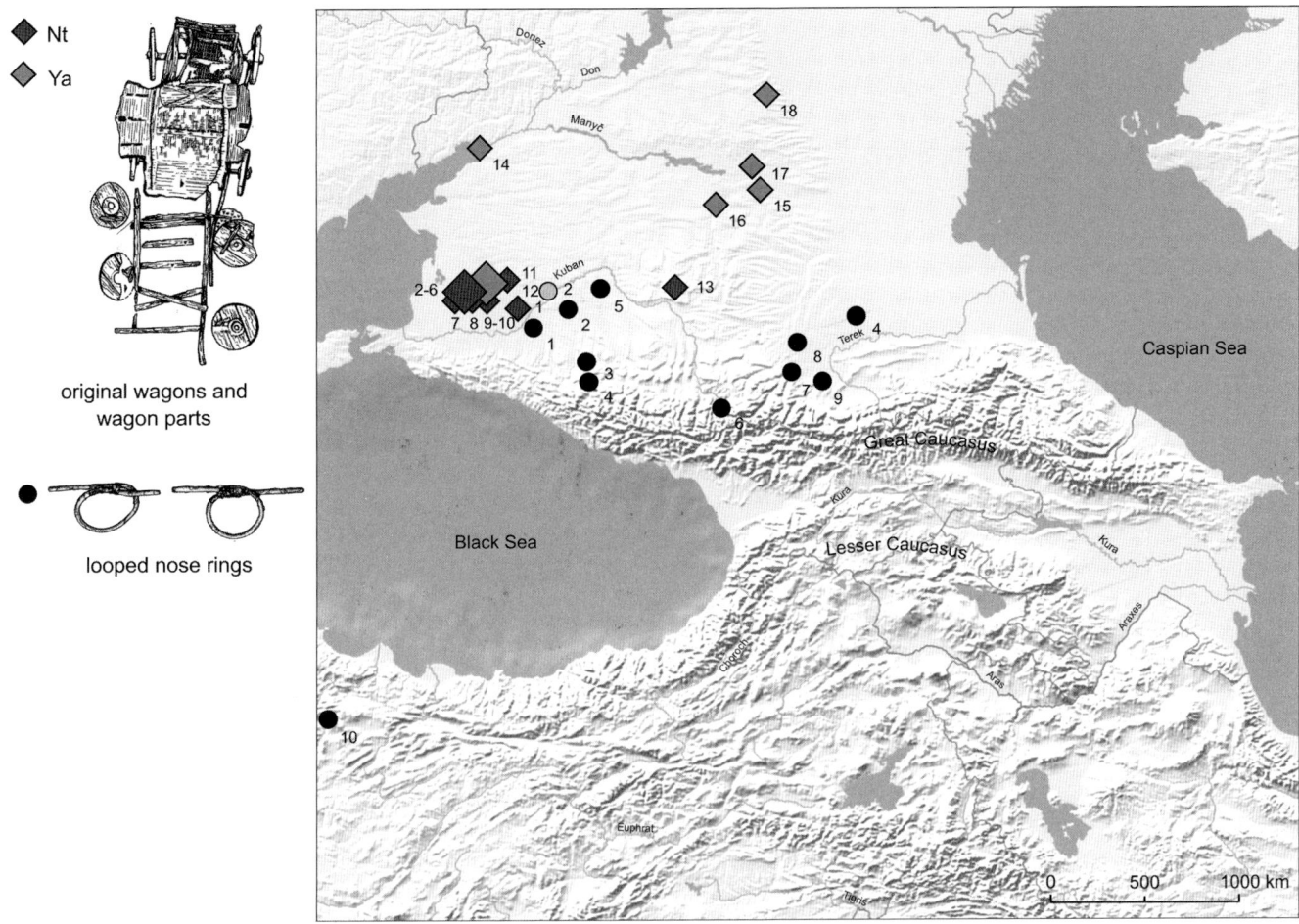

Figure 8.7: Distribution of early vehicles and looped nose rings in graves of the Late Maikop, Novotitarovskaya and Yamnaya communities (late 4th/early 3rd millennium BC). Early wagons: (1) Starokorsunskaja; (2) Ostannyj K1; (3) Lebedi I K2; (4) Lebedi II K2; (5) Malay I K9; (6) Oleny K2; (7) Novoveličkovskaja; (8) Novotitarovskaja; (9) Proletarskaja-86 K4; (10) Dinskaya K/11-12; (11) Baturinsk II K1; (12) Novokorsunskaja-85 K1/10; (13) Rasshevatsky 1 K/21; (14) Rostov-na-Donu; (15) Sharakhalsun 6 K2/18; (16) Zolotarëvka 1 K25; (17) Lola; (18) Tri Brata and 52 further burials with wagons from Novotitarovskaja sites + 33 unmapped Yamnaya wagon graves (based on Kaiser 2007, Mansfeld 2013, archive Nasledie). Looped nose rings: (1) Vency K3/4; (2) Chishcho; (3) Maikop, destroyed mound; (4) Klady K4/1, K1/25, K27/1; (5) U'lsky; (6) Kubina; (7) Chegem II; (8) Mar'inskaya 5; (9) Bamut K14; (10) Alaça Hüyük (based on Korenevsky 2013).

This wagon belongs to a well-known group of early wooden vehicles from the Northwest and North Caucasian steppe zone (Häusler 1982; Gei 2000; Kaiser 2007) (Fig. 8.7). Most of them are associated with the Northwest Caucasian Novotitarovskaya culture, but they have also been found in Yamnaya and Catacomb contexts. Elke Kaiser has collected the available data and discussed the difficulties posed by these grave goods (Kaiser 2007, 142–146 figs 12–13). Her interpretation, however, follows Stuart Piggott and others, who read wooden wagons as grave goods for members of higher social strata (Piggott 1992, 20–23; Kaiser 2007, 146–147).

The contexts with well-preserved wooden wagons have revealed a broad variety of ways in which these vehicles were handled. The variants include intact standing wagons in graves, wagons with dismantled wheels below the burials and wagon boxes used as grave ceilings. In some cases, wagons have also been found (intact or dismantled) outside the graves (Häusler 1982; Gei 2000, 119–123 figs 38, 128–131; Limberis and Marchenko 2002; Belinsky and Kalmykov 2004).

The wagon of Sharakhalsun 2/6, grave 18 is an intact vehicle placed in the burial catacomb. It was used to carry the dead. Two more Yamnaya burials with wagons from the same mound, grave 4 (Fig. 8.9; Fig. 8.11.2) in the centre and grave 3 in the periphery (Fig. 8.9; Fig. 8.11.1), featured dismantled wagons (Tab. 8.2). Their four wheels were placed flat in the corners of the burial pit, which in turn was covered with the wooden wagon box. Both graves held one male individual and are close in dates in the first quarter of the 3rd millennium BC. (Tab. 8.2). The complexes were AMS-dated in 2017. Grave 3 date to 4187 ±

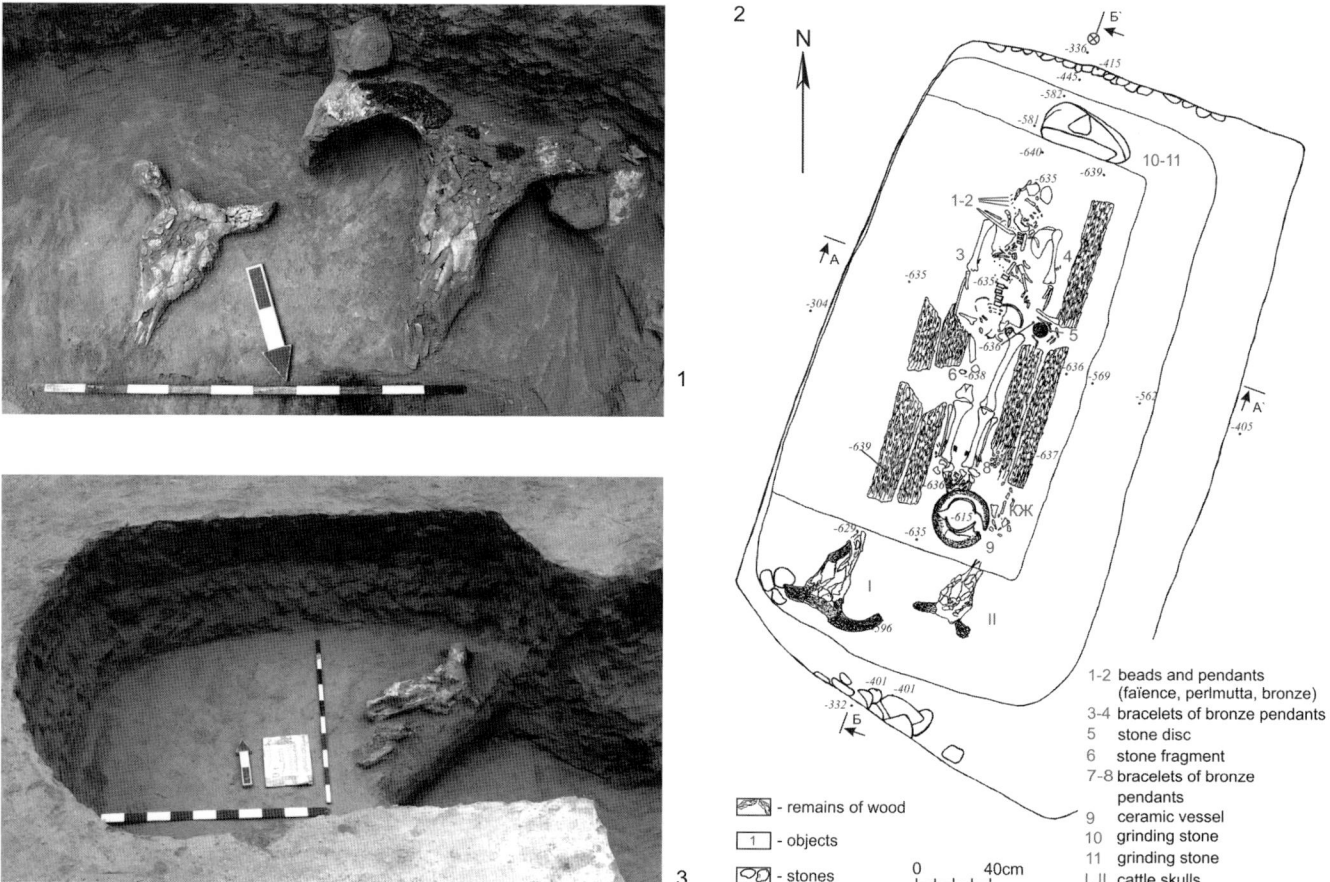

Figure 8.8: Mar'inskaya 5. Burials with cattle skulls from the North Caucasian and the Caucasian Catacomb cultures. (1–2) Grave 19; (3) grave 30.

22 BP and grave 4 to 4167 ± 22 BP. The latest wagon grave 9 most probably contained half of a four-wheeled vehicle squeezed into the entrance shaft of a catacomb. The remaining parts of the wheels, pole and wagon box were very well preserved (Fig. 8.10.3). Grave 9 is a product of the East Manych Catacomb culture and is radiocarbon dated to 3890 ± 40 BP (Tab. 8.2).

As in Mar'inskaya 5, the tradition of burying chosen individuals with symbols of traction was practised in Sharakhalsun 2 over a time span of at least 800 years. Unlike the first case study, this situation is not an exception. Successions of wagon burials have been repeatedly recorded in the area of the Novotitarovskaya variant of the Yamnaya phenomenon (Gei 2000). In Ostyanny, mound 1, two successive wagon graves 150 and 160 have been found (Gei 2000, 66–67 fig. 21). At Lebedy I, mound 2, four burials out of 15 contained a wagon. At Novovelichkovskaya three graves out of eleven were furnished with vehicles (Limberis and Marchenko 2002). Farther to the east, the large Ipatovo mound displayed three wagon burials of the Early and Late Catacomb cultures (Belinsky and Kalmykov 2004). Likewise, several mounds near Sharakhalsun contained one or several wagon graves, *e.g.* the Tri Brata kurgan, the Lola kurgan near Elista (Mansfeld 2013, 124–125 fig. 50) and several of the still unpublished mounds in the Stavropol archives (Belinsky and Kalmykov 2004, 217). However, of the entire Sharakhalsun 1–5 sequence of 48 excavated mounds, only two contained burials with remains of vehicles, four of them in mound 6 of Sharakhalsun 2.

In a wider perspective, these graves are part of a burial tradition found in the steppe zone from the lower Volga area across the entire northern Pontic region. The Yamnaya graves with wooden wagons are the most fervently discussed arguments in favour of the development of wheeled transportation in, or its transfer across, the Pontic area. Likewise, they are fundamental arguments for enhanced mobility and migrations in the steppe zone from the late 4th millennium BC onwards, even though most scholars have meanwhile acknowledged that it was only here that wooden vehicles were preserved in burials as real objects (Sherratt 2004; Burmeister 2011). The implementation of the idea of technical devices pulled by animals into local societies, however, differed considerably, as is demonstrated by the complexes we have described.

Table 8.2: The Sharakhalsun 2/6 complexes with remains of wooden wagons

Grave no.	Description	Dating
18	Inhumation burial, skeleton sitting on a wooden 4–wheeled wagon in a deep catacomb–like burial chamber, no inventory. *Anthropology:* male, 30–39 yrs, entheseal changes of the upper extremity & pelvis, 26 fractures, arthrosis of the whole spine, left ribs, upper right extremity & both lower extremities	Early Yamnaya Culture(?) GIN–12401: 4500 ± 40 BP 3356–3033 cal BC 1σ 3336–3105 cal BC 2σ wood
4	Inhumation burial in a stepped burial pit, the skeleton was placed in a contracted position in supine position on organic matter, on his knees a ceramic vessel, above the lower part of the burial pit on the step a 1.8 × 1.5 wide area organic mat (reed?) & the remains of a wooden wagon box, long wooden bars (pole?) & 2 dismantled wooden wheels preserved, 2 further wheels were traceable by disc & nave imprints, beside the wagon fragments of a ceramic vessel & animal bones. *Anthropology:* male, 30–39 yrs, entheseal changes of the patellar tendons, 2 fractures of the left hand & 1 of the left foot, arthrosis of the whole spine & ribs & both hip joints.	Yamnaya Culture by stratigraphy earlier than grave 3 MAMS 29245: 4167 ± 22 BP (2017) cal BC 2873–2698 1σ cal BC 2878–2667 2σ human bone GIN–12398: 3980 ± 100 BP 2829–2306 cal BC 1σ 2866–2205 cal BC 2σ human bone
3	Inhumation burial in a stepped burial pit, the skeleton was placed in a contracted position in supine position on organic matter, probably wrapped in & covered by mats with various weaving techniques, beside the feet remains of a wooden bowl & a stone pestle, at the skull antimony pendant, near the left hand bronze pricker, red ochre in the area of skull & pelvis, above the lower part of the burial pit on the step a 2.5 × 2.8 m wide area of organic mats (reed?), the remains of a wooden wagon box, a long wooden bar or pole & 4 dismantled wooden wheels, beside the wagon 1 ceramic vessel. *Anthropology:* male, 30–39 yrs, entheseal changes of the symphysis, left costoclavicular ligament & right 'tennis elbow', fracture of the right hand, arthrosis of cervical & lumbar spine & ribs.	Yamnaya Culture GIN–12398: 3980 ± 100 BP 2829–2306 cal BC 1σ 2866–2205 cal BC 2σ human bone by stratigraphy later than grave 4 MAMS 29247: 4187 ± 22 BP (2017) cal BC 2878–2708 1σ cal BC 2885–2680 2σ human bone
9	Inhumation burial in a catacomb, the skeleton was placed right sided in a contracted position on organic material (mats?), in front of the breast a wooden bowl, beside the feet a ceramic incense burner & remains of a wooden object, at the knees traces of ochre, & charcoal remains on top of the skeleton, near the feet a bronze pricker, the catacomb was closed with parts of a wagon, 2 wheels, their axles & parts of the wagon box are very well preserved. *Anthropology:* male, 50–69 yrs, entheseal changes of the upper extremities & pelvis, 9 fractures of the right hand, right foot, thoracic & lumbar spine, arthrosis of the whole spine, ribs, shoulder, elbow, hip, knee & feet joints.	Early Catacomb Culture GIN–12400 3890 ± 40 BP 2461–2310 cal BC 1σ 2474–2211 cal BC 2σ wood

Representation of power versus representation of mobility: cattle teams and wagons

If innovation is first of all the social act of incorporating new practices and devices into an existing social sphere, the different modes of presenting animal labour in the various Caucasian burial traditions suggest different intellectual discourses taking place in the appropriation process. While the Maikop and later North Caucasian communities selected representations of the powerful driving force (the cattle or oxen), the steppe communities of the Novotitarovskaya, Yamnaya and later the Catacomb culture chose to highlight the means of transportation – carts and wagons. The one emphasised the new forms of extended labour and neglected the objects of traction – sledges, carts, wagons or ploughs –

the others highlighted the new providers of transport and mobility and disregarded the 'engines'.

The presentation of looped nose rings with cattle skulls (as in Mar'inskaya 5) or without them accentuates the subjection of the animal. Yoking cattle to pull vehicles or ploughs represented a new level of domestication. Animals became increasingly objectified and turned into items of permanent availability and human control (Ingold 1994). Simultaneously, working with animals produced closer human–animal interaction and longer timespans of direct contact. The study by Eva Rosenstock and Astrid Masson (2011) lists the measures and resources involved in the training of cattle: castrating bulls to reduce animal aggression, yokes joining cattle pairs in such a way that

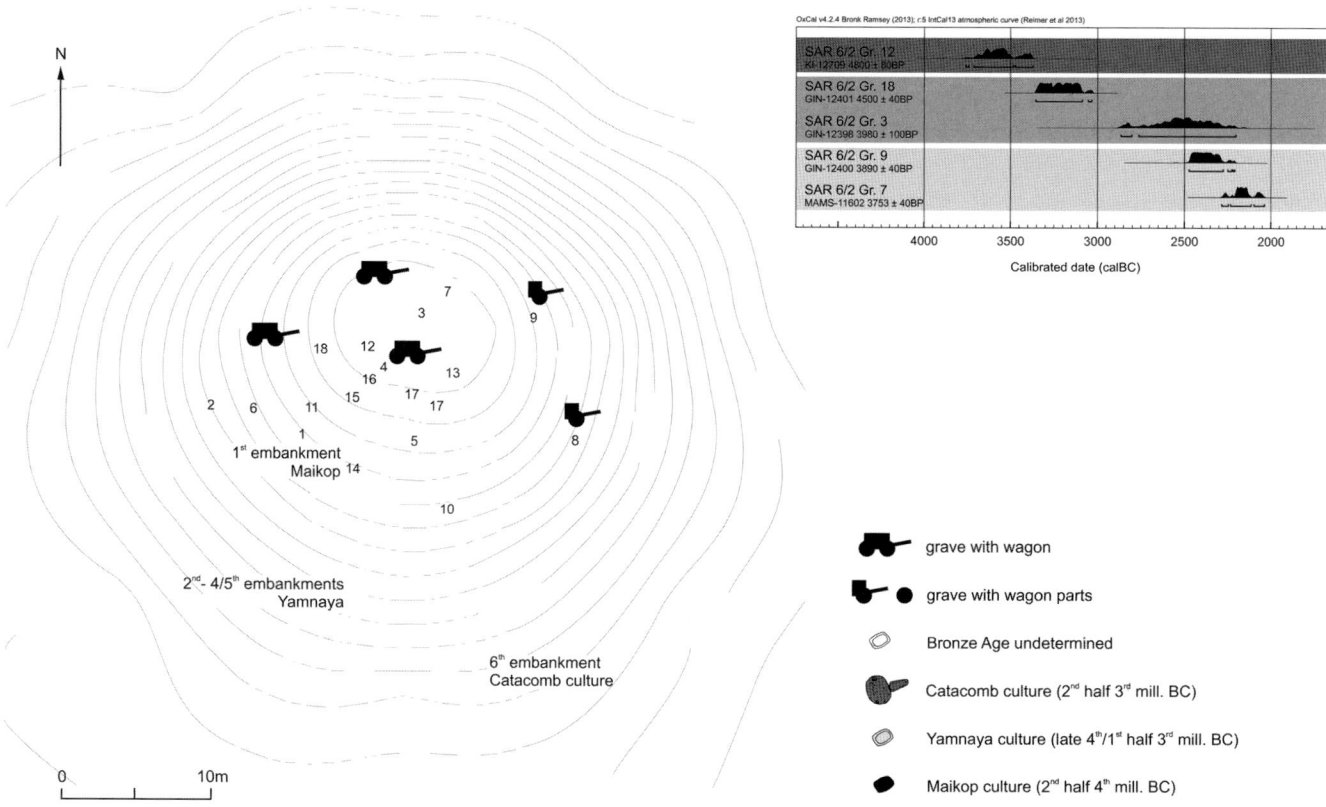

Figure 8.9: Sharakhalsun 2. General plan of mound 6 and radiocarbon dates of the burials.

one individual cannot move without the other, undercutting species-specific individual distances, heavy loads, and harnessing methods effectively controlling the animals. This goes far beyond earlier intrusions into animal lifestyles such as control of breeding or drawing off milk for the owners' purposes (Russell 2012, 209–211). Such training involves violence and the infliction of pain, the animals become lethargic, their spirits are broken. Even later forms of training are less violent using experienced and unexperienced pairs to teach and learn; draught animals are bound creatures.

The ubiquity of animals in human lifescapes is a factor for consideration in all the societies discussed above, societies that were starting to use cattle as working animals. At all events, the first training experiences must have been a very physical business and a highly asymmetrical power struggle. This raises questions about the first animal trainers and pinpoints the reasons for training working animals in the first place. Draught cattle increase the efficiency of transport or the working of the soil, but the husbandry costs are considerably higher since the animals need more calories and cannot graze when working (Halstead and Isaakidou 2011, 62).

The Maikop bronze nose rings must have been a highly effective tool to force the animals to bow to the will of their human masters. Is 'mastering the beast' thus the real narrative in the Maikop burial scenario with its pair of draught animals? A near-contemporaneous wall painting from Arslantepe VI A in corridor A796 dating to about 3350 BC presents a similar topic and identical actors (Fig. 8.12): a pair of over-sized cattle standing between a floral motif and controlled by a rather small human figure with the help of ropes fixed to their horns (Frangipane 1997, 64–67 fig. 15; 2012). The size of the actors involved in the scene accentuates the bravery of the physically small human in mastering the two huge bulls or oxen. The rhombic eyes of the cattle are repeated in the decoration of the wall with its similar rhombic motifs – the actions of the central figures were obviously designed to be observed by a larger audience.

Caucasian bulls have a long history in Bronze Age symbolism. Archaeological evidence and representations in the form of clay figurines tell us that cattle were introduced to the North Caucasus during the 5th millennium BC (Antipina and Lebedeva 2005). The placing of three cattle skulls in a separate pit in a mound near Mozdok-Komarovo (Fig. 8.13) in northern Ossetia is another early reflection of the symbolic importance of these animals in the Maikop context. The pit was located next to two other deep shafts with human burials that have been interpreted as sacrifices (Nagler 1996, 18, 51–52 tabs 54–58). These shafts are adjacent to a large empty burial(?) pit. The constellation is highly reminiscent of Maikop burial constructions, but the

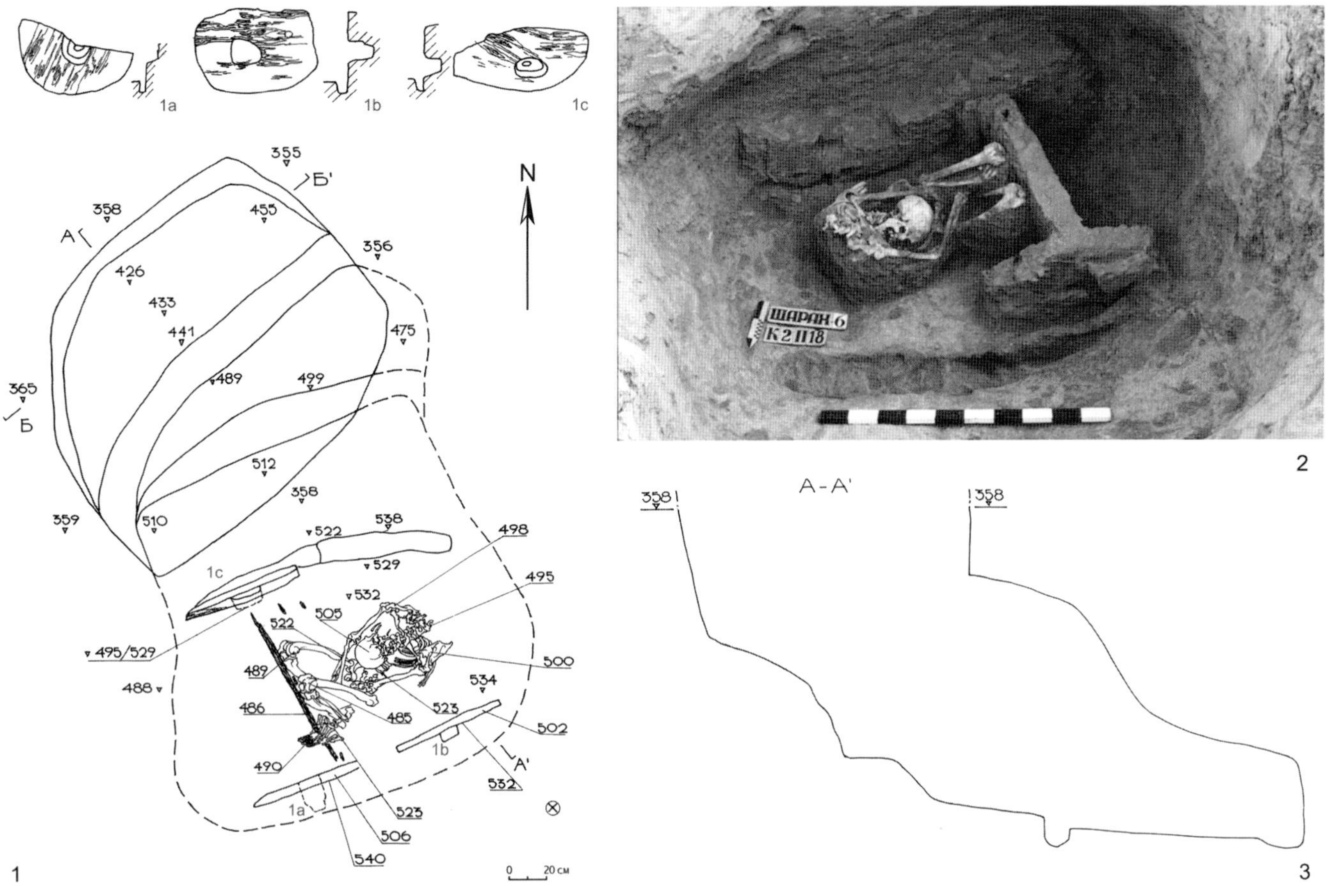

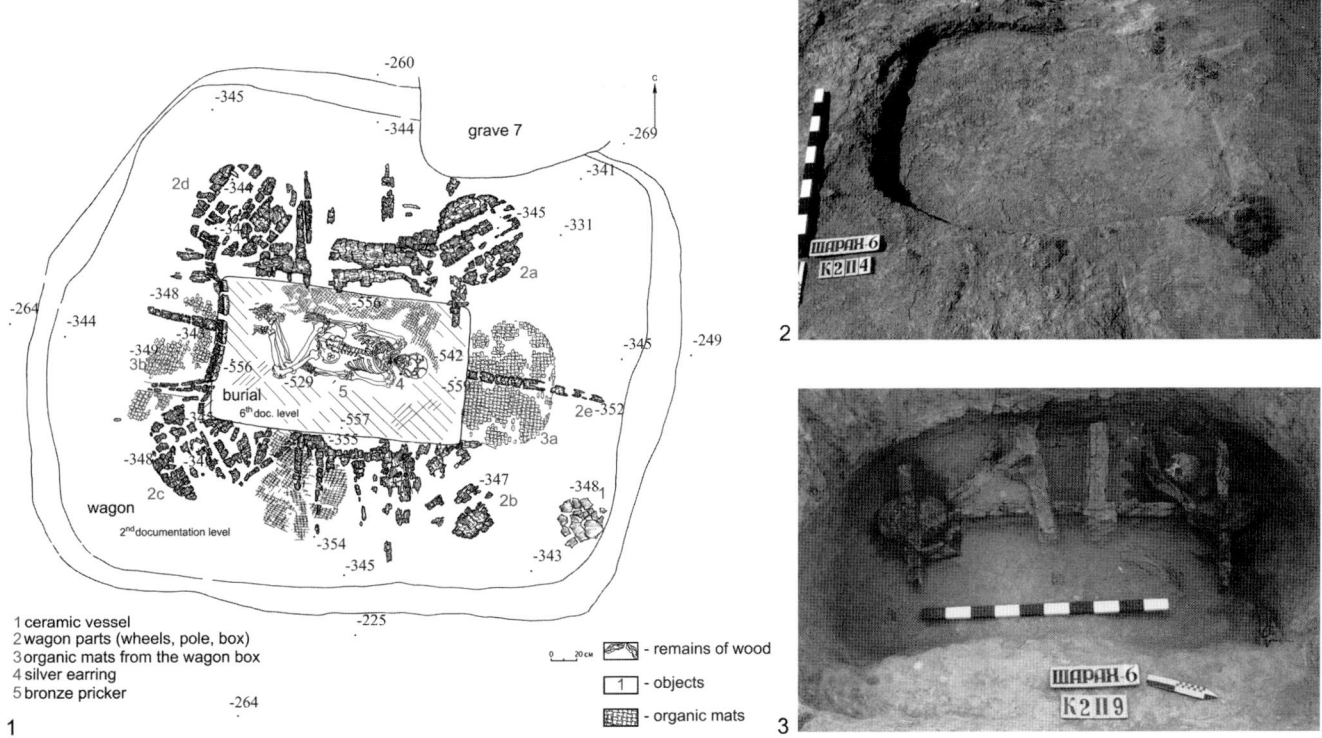

Figure 8.10: Sharakhalsun 2. Grave 18. (1) remains of wheel negatives in the chamber; (2) the buried driver; (3–4) plan and section of the grave shaft.

1 ceramic vessel
2 wagon parts (wheels, pole, box)
3 organic mats from the wagon box
4 silver earring
5 bronze pricker

- remains of wood
1 - objects
- organic mats

Figure 8.11: Sharakhalsun 2. Wagon graves. (1) Yamanya grave 3, 2nd (wagon) and 6th (burial).

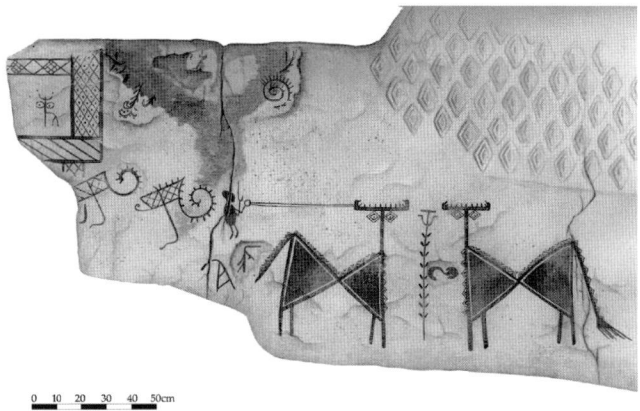

Figure 8.12: 'Mastering the beast'. Wall painting from Arslantepe VI corridor room A796 (courtesy MAIAO).

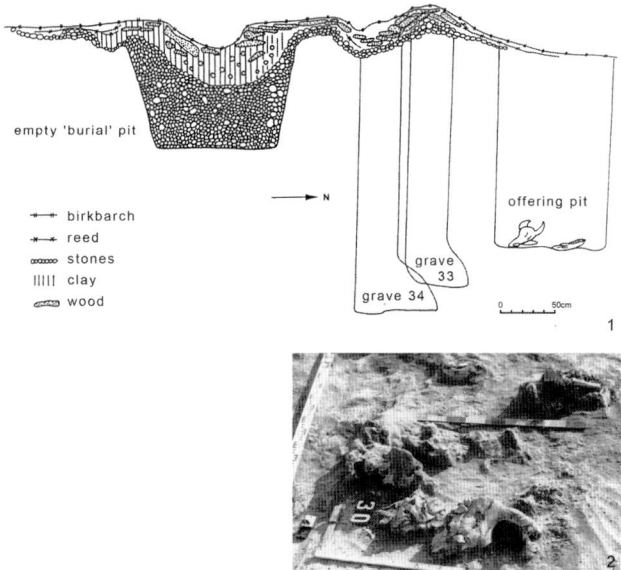

Figure 8.13: The 'burial' pit form Mozdok-Komarovo, grave 30, 33–34 and cattle skull offering. (1) Plan; (2) photograph of the bucrania (after Nagler 1996, tab. 54, tab. 58.1).

absence of objects make dating questionable (Korenevsky 2012, 99 fig. 92).

Spectacular representations of cattle have been found in the eponymous Maikop mound dating to the first half of the 4th millennium BC. Here four bovine figurines with large, sharp horns cast in silver and gold (Fig. 8.14.4–5) had been attached to poles by means of holes in their bodies (Fig. 8.14.1). However, as recent research by Juri Ju. Piotrovsky on the placement of the objects in this grave demonstrate, the famous baldachin mooted in one of the earliest publications obviously never existed (Munchaev 1975, 213–222, fig. 34; Piotrovsky and Bochkaryov 2013, 312–313, fig. 20.15–16). Other unpreserved objects near the skeleton were decorated with a large number of golden plaques depicting lions (Piotrovsky and Bochkaryov 2013,

Figure 8.14: The bull figurines from the Maikop mound. (1) poles with metal sheet-covering and figurines as part of a combined object; (2–3) golden lion and cattle plaques; (4–5) silver and gold figurines (1 after Tallgren 1926, fig. 55, 1, 2–4 after Piotrovsky and Bochkaryov 2013, figs 20.15, 22, 22.5 after Piotrovsky 1998, kat. 304).

314–315, fig. 20.20–21) and more cattle (Piotrovsky and Bochkaryov 2013, 315 fig. 20.22) (Fig. 8.14.2–3). At this early juncture already, the bulls were symbolically attached to a pole or a human figure and with their brute strength had to 'serve' their human 'masters' as representative grave goods.

Two silver vessels from the same complex bear depictions of wild cattle in the company of other wild animals (Uerpmann and Uerpmann 2011, 242–247 figs 5–8). We can understand these scenes adorning a precious grave good as an early form of the later well-known Mesopotamian narrative on 'mastering' wild and dangerous animals by a royal hero (Rice 1998, 94–102). Forcing 'wild' bulls to labour for humans is an act of civilisation, a defeat inflicted on nature by human will for which the Mesopotamian hero Gilgamesh stands as a symbol (Uerpmann and Uerpmann 2011, 89–93). The attached golden and silver bull figurines of Maikop in their supporting position and the *pars pro toto* representation of draught animals in graves can be interpreted as the symbolic translation of a similar story: dangerous animals bound and forced to work demonstrates human dominion over nature. Such a narrative would be fully in line with other transformations taking place in this epoch. Maikop is among the first cultures in Eurasia whose members began to transform the landscape by erecting artificial edifices. It is also the first epoch in which material culture was used prominently to mark out social hierarchies after death. Social hierarchies, *i.e.* the division into important and subordinate human individuals, are expressed in this way. Like other new forms animal exploitation – including wool, which is also found in Late Maikop contexts (Shishlina *et al.* 2003) – these are facets of a new conception of the interrelation between humanity and nature: a ritual domestication of landscape, a social domestication of subordinates, and an objectification of animals.

If the narrative of 'mastering the beast' is the subtext underlying the symbolic representation of draught animals in the Caucasian traditions, why did the steppe communities not highlight this topic in the same way? One reason may be a difference in the actual employment and the everyday importance of the new devices. The practical assets of the new vehicles are still the most plausible argument for the implementation of wheeled transport in steppe societies. The transport of infrastructure such as tents, containers and food supplies may have been the prime mover for steppe people to adopt carts and wagon. Providing a deceased individual with a wagon enabled him or her, in the afterlife, to participate in the mobile lifestyle of the community. Mobility is the pertinent narrative, not the subjugation of nature.

Giving the dead a wagon in the steppe zone was the equivalent of bestowing enormous wealth on them. While the piedmont and piedmont steppe zone inhabited by Maikop or North Caucasian Culture communities had the abundance of high-quality timber needed to build wooden wagons, steppe environments provide very little of the raw materials required to make functional vehicles. For the Early Catacomb Culture wagons in the large mound at Ipatovo, a sophisticated selection of suitable woods was used (Belinsky and Kalmykov 2004, 206–207). A similarly skilful composition of different woods is now known to have been used for the Late Catacomb wagon from Ulan IV (Shishlina *et al.* 2013, 121–122 fig. 2). In the steppe, the offering of a vehicle as an object of accumulated wealth was obviously more important than having a pair of trained cattle at one's disposal.

Who were the first individuals with draught animals?

The stratification of Maikop society is well attested by the ranked grave-good assemblages from the 4th millennium BC. Burials in Yamnaya communities, by contrast, did not involve material culture to an extent that would reliably indicate the social position of a buried individual (Kaiser 2011; Reinhold 2012). Wagons are, so far, the only indicators that have qualified as status markers in scholarly discussion.

The mere existence of burial mounds in steppe environments has long been considered an argument in favour of the mobility of these groups. More recently, new evidence like the seasons of the burial events or isotope analyses disclosing dietary habits have shored up this hypothesis (Shishlina 2008). In line with this research strategy, bioarchaeological investigations have been included in the present study. Who were the individuals involved in the process of animal subjugation? Were they part of an upper class able to afford better food avoid physical work (Knipper *et al.* 2015)? Are the wagons in

the steppe indeed indicators of advanced mobility on the part of the individuals buried with such an object, or of the community as a whole?

Anthropology and palaeopathology

Anthropological and especially palaeopathological methods provide information about people's health and their way of life. The various incidences of certain diseases such as caries frequency may indicate different diets, possibly based on different subsistence strategies. If the individuals from Mar'inskaya 5 and Sharakhalsun 6 buried with a wagon or with oxen nose rings actively drove wagons, then the traces of physical stress in their bones should differ over and against other members of the same group. Since we are possibly dealing with the earliest drivers of all, more physical stress might be conceivable due to difficulties in handling the draught animals or the new challenges involved in wagon functionality.

We cannot, however, answer these questions at present. Even though the number of wagon burials is not small – about 260–280 excavated complexes from the North Caucasian steppe zone (see Belinsky and Kalmykov 2004; Kaiser 2007) – most of the assemblages are still unpublished and only very few have been studied anthropologically. While this precludes statistically significant conclusions, interesting tendencies can still be observed.

The hypothesis that actual wagon drivers were both more accident-prone and more physically active would be borne out by more signs of physical stress on their skeletons. This hypothesis was evaluated with reference to the frequency and severity of (a) degenerative changes of the joints, (b) entheseal changes at muscle attachment sites and (c) fractures. Training animals is difficult when one is dealing with powerful, large and heavy beasts. Accidents during this process would lead to ruptured tendons or bone fractures. Early driving also has greater accident potential due to unpredictable movements by the animals and the velocity of aggravated falls. Fractures occurring most frequently in wagon drivers from Mar'inskaya 5 and Sharakhalsun 6 are fractures of the hands, followed by the spine and the feet. The fractures fully correspond to assumed occupational stress. Hands and feet are vulnerable because of close contact with the animals, while fractures of the vertebrae are frequently reported as resulting from falls from a certain height (Siebenga *et al.* 2006). However, the fracture pattern of the wagon drivers is not genuinely specific to them. It resembles the most frequently found fracture patterns for the deceased as a whole. The only exceptional case is the male individual in Sharakhalsun 2, burial mound 6, grave 18. He had a total of 26 fractures, which must have occurred at different times since at the time of his death some were fresh and others had healed well (Fig. 8.15). His fracture pattern is reminiscent of combat rather than everyday activities. Was he perhaps actively training cattle

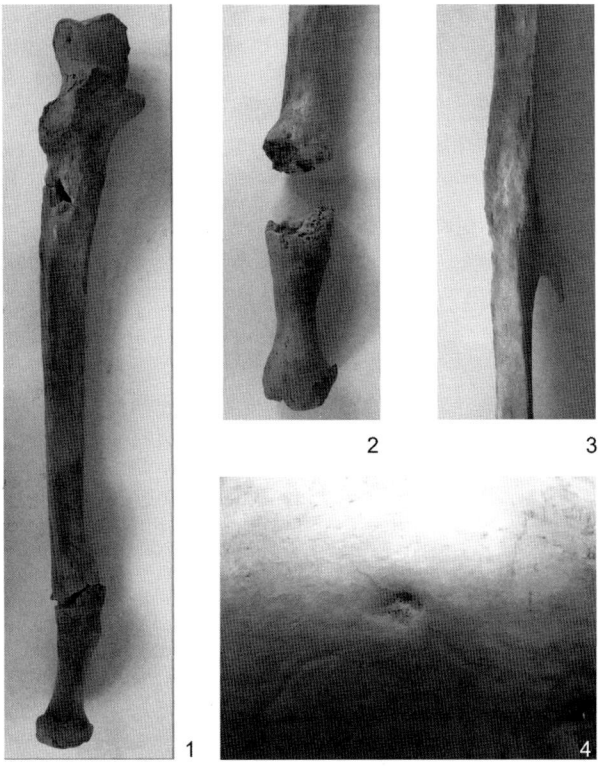

Figure 8.15: Sharakhalsun 2/6, Grave 18. Fractures of the earliest wagon driver in Eurasia. (1) right Ulna with unhealed Parry fracture; (2) detail of the distal third of fig. 1 showing the pseudoarthrosis; (3) right fibula with well-healed fracture and additional myositis ossificans; *(4) skull trauma.*

teams? The burial position undoubtedly presents him as a driver (Fig. 8.10.2).

The patterns of degenerative change in the bones of wagon drivers is, however, not distinct from that of other people (Tucker *et al.* 2017). All drivers demonstrate severe degeneration of the spine and ribs, followed by the joints in the lower extremities. However, degenerative changes could be age-related and are dependent, for instance, on metabolic factors as well as entheseal changes (Mariotti *et al.* 2007). Although their causes may vary widely, these changes can be used to estimate physical activity. In the case of the wagon drivers, the most severe changes occurred in the muscle attachment sites of the upper extremities and the pelvis.

Since the wagon drivers displayed no significant differences in their physical activity markers, anthropological indications for social status were investigated. In Mar'inskaya 5 and Sharakhalsun 6, only male individuals between 25 and 69 years were buried with a wagon. At other sites, however, infants and women were also supplied with wagons. This confirms the data we have from the northwest Caucasian complexes (Kaiser 2007, 142–144 figs 12–13). Age and sex were obviously not the key criteria for putting a wagon in someone's grave.

Belonging to a social elite may imply a better quality of food or other advantages that to some extent may be reflected in bone structures. A well-balanced or nutritious diet can be inferred from the state of the skeleton if it displays fewer traces of malnutrition or enamel hypoplasia. Furthermore, there have been suggestions that some skeletal changes are specific to elites (Temple 2010). However, the relevant Caucasian skeletons show neither dietary nor metabolic differences between wagon drivers and others. The case is similar with inflammatory diseases of the nose and ears, which are sometimes referred to as being typical of members of the lower class (Schultz 1982).

In view of the small number of individuals from Bronze Age contexts in the North Caucasus that have been studied in detail, it is hardly surprising that no specific anthropological or palaeopathological features for wagon drivers should have been observed. This applies to entheseal change, the number of fractures, the ratio of degenerative joint diseases and all other pathologies. Furthermore, the types of fracture are more or less the same as those found in individuals buried without a wagon. Wagon drivers appear to resemble the average Bronze Age Caucasian, and the signs of physical stress point to an engagement in animal husbandry and physically hard work that is similar to what we find in the majority of the population.

Stable isotope analyses

To gain more insight into the mobility patterns and dietary habits of the people who employed the innovations discussed above, a series of isotope analyses were applied to their skeletal remains. The primary aim was to collect data on certain aspects of their daily lifestyles that go beyond the information that can be drawn from archaeology or physical anthropological analysis.

Strontium isotope ratios ($^{87}Sr/^{86}Sr$) of tooth enamel reflect the geological conditions of the area in which an individual grew up (Bentley 2006). Because tooth crowns form at different ages in childhood (AlQahtani *et al.* 2010), Sr isotope ratios may vary among the teeth of the same individual if a person changed his/her place of residence early in life. Carbon ($\delta^{13}C$) and nitrogen ($\delta^{15}N$) isotope ratios of bone collagen reflect the ratios of major dietary components and may be an indication of increased consumption of meat and dairy products or freshwater and aquatic fish, infant nursing and the average proportion of C_3 and C_4 plants in the diet (Ambrose 1993). In an environmentally diverse landscape such as the Caucasus mountains and the steppe zone to the north, carbon and nitrogen isotope ratios also reflect environmental conditions such as humidity patterns and associated vegetation cover (Hollund *et al.* 2010). First of all, the combined analysis of strontium, carbon and nitrogen isotopes in different landscapes provides crucial information on the differentiation of the local baseline data, a precondition for the detection of human mobility. Also, it points up differences among the members of the burial communities that can be correlated with indications of social differentiation demonstrated in grave architecture and grave goods. Along with other venues, a preliminary study in the region of Stavropol' in the North Caucasus encompassed the sites Mar'inskaya 5 and Sharakhalsun 2 as referred to earlier. These represent different vegetation zones of the steppe depending largely on different humidity levels.

A precondition for any successful application of isotope analyses to indicate human mobility is the existence of adequate regional differences in the baseline values. If this condition is fulfilled, largely overlapping data ranges at variant sites and significant differences between strontium isotope ratios in early and late-forming teeth in the same individuals are indications of highly mobile lifestyles. Such patterns are to be expected in steppe communities, which after all practised mobile pastoralism. By contrast, the isotope ranges in representatives of more sedentary communities of the kind hypothesised for most North Caucasian piedmont cultures should be distinct and reveal little variation between early and late-forming molars in different individuals.

Preliminary findings from the burial assemblages referred to above document that $^{87}Sr/^{86}Sr$, $\delta^{13}C$ and $\delta^{15}N$ vary within certain ranges with some overlaps between sites. If strontium and carbon or nitrogen isotope data are combined, sufficiently variegated data clusters emerge for the sites in the steppe zone and the piedmont areas. This is an argument against highly mobile groups with home ranges covering a number of environmental and geological zones. However, each site also featured a few individuals whose stable isotope data were more similar to those of the majority of one of the other burial communities. Some burials, *e.g.* the oldest of the North Caucasian culture in mound Mar'inskaya 5 including grave 23 with a pair of bucrania, revealed noteworthy differences between early and late-developing teeth. These findings suggest deviant dietary habits and/or origin from another area and community. As we have seen earlier, such individuals may have been driving forces in the exchange of knowledge and a precondition for the spread of innovations. The home ranges of the local communities were spatially distinct, so frequent contacts did not occur naturally between the investigated groups. Accordingly, these individuals were certainly of special relevance.

All isotope data on the individuals buried with wagons or pairs of cattle skulls plot among those buried without. As indicated by the physical anthropological analyses, the stable isotope data displayed no evidence that inhumation with a wagon or with cattle offerings marks a distinct social group with access to certain kinds of higher-quality food (Knipper *et al.* 2015). These observations also apply to the Maikop individuals, who lived in a society that

used grave goods to emphasise social status. Moreover, the strontium isotope data from the wagon burials were indistinguishable from those of the other individuals and do not indicate any differences in mobility. Overall, the isotope data of the skeletal remains did not provide any evidence that the presence of wagons or cattle offerings as grave goods indicates membership of some kind of social elite characterised by regular access to certain types of food or by special mobility patterns.

Contextualising early traction in the Caucasus and beyond

The appropriation of animal labour and its integration into human lifescapes was a process that involved a considerable degree of violence and subjection. Transforming domestic cattle into working animals must have changed the attitudes of their owners and brought them into closer and more intensive contact with their livestock than ever before. This ground-breaking process took place during the early 4th millennium BC in various areas of Eurasia. It seems fair to suppose that the first stages in the subjugation and objectification of animals beyond previous forms of exploitation did not leave many archaeological traces. Animal representations, particularly of cattle, are found in Caucasian settlements as early as the 5th millennium BC, first miniature wheels can be traced back to the 4th millennium BC (Trifonov 2004; Mansfeld 2013).

In the heterogeneous, hilly terrain inhabited by Maikop communities, wheeled vehicles would have been of minor practical advantage in everyday life. The erection of hundreds or thousands of burial mounds, however, was a task where animal power was a major potential asset. If it was used, it at once shifted workloads from humans to animals and involved this new animal workforce and the human builders in an important ritual act. Putting up burial monuments was a focal activity in Maikop societies. Accordingly, incorporating animals into the labour side this process may have emphasised their symbolic importance and enhanced appropriation of this new aspect of animal exploitation. The abundance of mounds, most of which were probably built in competition, provides a reasonable scenario for a rapid spread of knowledge on how to train and involve draught animals in socially important work processes. Ploughing is another potential use for draught animals. At Maikop sites, classical domesticated species are present (Lebedeva 2011), but the intensity of agricultural activity is questionable (Knipper *et al.* forthcoming).

Yamnaya and Catacomb culture communities, on the other hand, lived a mobile life with their stocks. Life in the steppes called for more intensive interaction between humans and animals than the herding practices of settlers. As in Maikop communities, steppe dwellers erected burial mounds, so it seems likely that draught animals would have been pressed into service to transport the requisite material. However, the transportation aspect is likely to have been more important still in permanently mobile lifestyles. The mobility radii of those groups are still unclear, but even if the territorial scales of mobility were limited, this does not detract from the usefulness of transport vehicles. Domesticated plants were not part of the subsistence diet on the steppe, so ploughing can be excluded (Shishlina 2008).

At the community level, bioarchaeological investigations seriously challenge the hypothesis of large-scale mobility both in piedmont and in steppe environments. Given this background, intensification of contact through a greater volume of trade promoted by wheeled transport (Sherratt 1981) is rather unlikely. Other mechanisms for the transfer of knowledge need to be considered (Frachetti 2012). The small number of non-local individuals in the North Caucasian sample suggest an isolated exchange of individuals and the knowledge they brought with them.

It would indeed be tempting to see the male from Sharakhalsun 2/6, grave 18 not only as the driver of the oldest wooden vehicle dated so far but also as the trainer and master of the animals that once pulled his cart. Despite his frequent serious fractures, an anthropological examination cannot confirm or refute this hypothesis beyond all doubt. At the time, all individuals were very muscular due to constant heavy work and walking long distances.

Taken together, the findings do indeed advocate a very careful interpretation of the available data. Burials and the transportation involved need not necessarily imply mobile individuals or mobile communities. While the usefulness of wheeled transport for pastoral communities living in steppe environments is indisputable, this need not entail large-scale mobility or long-distance migrations. In fact, the bioarchaeological data currently available from the North Caucasus suggest that the opposite was the case. We should thus focus our attention on the symbolism of the objects related to early transport or animal labour. The appropriation of traction discussed here as a social act in two different symbolic traditions did not tremendously change the normal lifestyle of the communities involved. However, it does reveal a sharp difference between the communities that emphasised social difference and power relations in their societies, *e.g.* Maikop, and those that did not, *e.g.* Yamnaya. It is most probably the symbolic aspect of the activities for which draught animals were employed that prompt different representations of traction in the two symbolic systems.

During the early, innovative phase of appropriation, the link between humans, animals and their work largely seems to have been individually constructed. The heterogeneity of burials with animal offerings in the piedmont area and the heterogeneity in the deposition of vehicles in the steppe zone give us no clues to standardised ritual procedure.

This strongly suggests that the late 4th and the early 3rd millennia BC were indeed the appropriation period for animal labour in this area, where there were no standards regulating the proper use of the new techniques. Only in the following late 3rd and early 2nd millennia BC in the South Caucasus, Eastern Anatolia and Mesopotamia did wagon burials become formalised status markers for elite burials (Sagona 2013).

Acknowledgements

We should like to thank Svend Hansen for the opportunity to discuss the data from the recent bioarchaeological projects of the Eurasia Department. We are grateful to the excavation teams of Sharakhalsun and Mar'inskaya, the team of the Institute for Anthropology and the Institute for Organic Chemistry at Mainz University, the radiocarbon and isotope lab team at the Curt-Engelhorn-Centre for Archaeometry in Mannheim and the staff of the Anthropology group of the DAI for technical support, sample preparation and analysis. Financial support for the isotope analyses was granted by the Internal University Research Funding department at the University of Mainz; the anthropological study and the radiocarbon dating were financed by DAI Natural Scientific, the Eurasia Department and 'Nasledie'. We are also indebted to Jakov B. Berezin (Stavropol') and Tamerlan A. Gabuev (Moscow) for information on the unpublished excavations at Lysogorskay 6 in 2015. Likewise, we are grateful to Juri J. Piotrovsky (St Petersburg) for updates on his ongoing research on the Maikop burial.

Notes

1   Sabine Reinhold: Eurasia Department, German Archaeological Institute, Berlin; Julia Gresky: Department of Natural Sciences, German Archaeological Institute, Berlin; Natalia Berezina: Anuchin Research Institute and Museum of Anthropology, Lomonosov Moscow State University, Moscow; Anatoly R. Kantorovich: Department of Archaeology, Lomonosov Moscow State University, Moscow; Corina Knipper: Curt-Engelhorn-Centre for Archaeometry, Mannheim; Vladimir E. Maslov and Vladimira G. Petrenko: Institute of Archaeology, Russian Academy of Sciences, Moscow; Kurt W. Alt: Center for Natural and Cultural History of Teeth, Danube Private University, Krems-Stein; Andrey B. Belinsky: Ltd. 'Nasledie', 'Stavropol'.

2   Together with complexes from several other sites, the graves are the object of a project by the Eurasia Department and the Natural Science Department of the German Archaeological Institute, the local heritage organisation 'Nasledie' and the Curt-Engelhorn-Centre for Archaeometry in Mannheim.

References

AlQahtani, S. J., Hector, M. P. and Liversidge, H. M. (2010) Brief Communication, The London Atlas of Human Tooth

Development and Eruption. *American Journal of Physical Anthropology* 142, 481–490.

Ambrose, S. H. (1993) Isotopic Analysis of Paleodiets. Methodological and Interpretive Considerations. In M. K. Sandford (ed.), *Investigations of Ancient Human Tissue.* Pennsylvania, Gordon and Breach.

Antipina, E. E. and Lebedeva, E. Y. (2005) Opyt kompleksnych archeobiologicheskich issledovaniyazemledeliya i skotovodstva, modeli vzaimodeystviya. *Rossiyskaya Archeologiya* 4, 70–78.

Balabina, V. I. (2004) Glinyany'e modeli sanej kul'tu ry' kukuten'-tripol'e i tema puti. In V. V. Volkov (ed.), *Pamyatni ki archeologii i drevnego iskusstva evrazii.* Moscow, Institut Archeologii RAN.

Beaujard, P. (2011) Evolutions and Temporal Deliminations of Bronze Age World-Systems in Western Asia and the Mediterranean. In T. C. Wilkinson, S. Sherratt and J. Bennet (eds), *Interweaving Worlds, Systemic Interactions in Eurasia, 7th to 1st Millennia BC,* 7–26. Oxford, Oxbow Books.

Belinsky, A. B. and Kalmykov, A. A. (2004) Neue Wagenfunde aus Gräbern der Katakombengrabkultur im Steppengebiet des zentralen Vorkaukasus. In M. Fansa and S. Burmeister (eds), *Rad und Wagen – Der Ursprung einer Innovation. Wagen im Vorderen Orient und Europa.* Archäologische Mitteilungen aus Nordwestdeutschland Beiheft 40, 201–220. Mainz, Philipp von Zabern.

Bentley, R. A. (2006) Strontium Isotopes from the Earth to the Archaeological Skeleton: A Review. *Journal of Archaeological Method and Theory* 13/3, 135–187.

Bloch, M. (1967) Technical Change as a Problem of Collective Psychology. In M. Bloch (ed.), *Land and Work in Medieval Europe,* 124–135. Berkeley, CA, University of California Press.

Burmeister, S. (2000) Archaeology and Migration. Approaches to an Archaeological Proof of Migration. *Current Anthropology* 41/4, 539–67.

Burmeister, S. (2011) Innovationswege – Wege der Kommunikation, Erkenntnisprobleme am Beispiel des Wagens im 4. Jt. v. Chr. In S. Hansen, and J. Müller (eds), *Sozialarchäologische Perspektiven, gesellschaftlicher Wandel 5000–1500 v. Chr. zwischen Atlantik und Kaukasus,* 211–240. Mainz, Philipp von Zabern.

Cavalcanti, H. B. (2005) Human Agency in Mission Work, Missionary Styles and Their Political Consequences. *Sociology of Religion* 66/4, 381–398.

Childe, V. G. (1936) Man Makes Himself. London, Watts & Co.

Erdal, Y. S. and Erdal, O. D. (2012) Organized Violence in Anatolia. A Retrospective Research on the Injuries from the Neolithic to Early Bronze Age. *International Journal of Paleopathology* 2, 78–92.

Fansa, M. and Burmeister, S. (eds) (2004) *Rad und Wagen – Der Ursprung einer Innovation. Wagen im Vorderen Orient und Europa.* Archäologische Mitteilungen aus Nordwestdeutschland Beiheft 40. Mainz, Philipp von Zabern.

Frachetti, M. D. (2012) Multiregional Emergence of Mobile Pastoralism and Nonuniform Institutional Complexity Across Eurasia. *Current Anthropology* 53/1, 2–38.

Frangipane, M. (1997) A 4th-Millennium Temple/Palace Complex at Arslantepe-Malatya. North-South Relations and the Formation of Early State Societies in the Northern Regions of Greater Mesopotamia. *Paléorient* 23/1, 45–73.

Frangipane, M. (2012) 'Transitions' as an Archaeological Concept. Interpreting the Final Ubaid – Late Chalcolithic Transition in the Northern Periphery of Mesopotamia. In C. Marro (ed.), *After the Ubaid. Interpreting Changes from the Caucasus to Mesopotamia at the Dawn of Urban Civilization (4500–3500 BC).*

Furholt, M. and Müller, J. (2011) The Earliest Monuments in Europe – Architecture and Social Structures (5000–3000 cal BC). In M. Furholt, F. Lüth and J. Müller (eds), *Megaliths and Identities: Early Monuments and Neolithic Societies from the Atlantic to the Baltic,* 1–32. Bonn, Habelt.

Gei, A. N. (2000) *Novotitorovskaja kul'tura.* Moscow, Institut Archeologii RAN.

Govedarica, B. (2004) *Zepterträger – Herrscher der Steppen. Die frühen Ockergräber des älteren Äneolithikums im karpatenländischen Gebiet und im Steppenraum Südost- und Osteuropas.* Mainz, Philipp von Zabern.

Greene, K. (1999) V. Gordon Childe and the Vocabulary of Revolutionary Change. *Antiquity* 73, 97–109.

Greenfield, H. J. (2010) The Secondary Products Revolution: The Past, the Present and the Future. *World Archaeology* 42/1, 29–54.

Hägerstrand, T. (1967) *Innovation Diffusion as a Spatial Process.* Chicago, IL, University of Chicago Press.

Haggett, P. (1991) *Geographie: Eine moderne Synthese.* Stuttgart, Ulmer.

Halstead, P. and Isaakidou, V. (2011) Revolutionary Secondary Products: The Development and Significance of Milking, Animal-Traction and Wool-Gathering in Later Prehistoric Europe and the Near East. In T. C. Wilkinson, S. Sherratt and J. Bennet (eds), *Interweaving Worlds, Systemic Interactions in Eurasia, 7th to 1st Millennia BC,* 61–76. **Oxford**, Oxbow Books.

Hansen, S. (2011) Technische und soziale Innovationen in der zweiten Hälfte des 4. Jahrtausends v. Chr. In S. Hansen and J. Müller (eds), *Sozialarchäologische Perspektiven, gesellschaftlicher Wandel 5000–1500 v. Chr. zwischen Atlantik und Kaukasus,* 153–191. Mainz, Philipp von Zabern.

Hansen, S. (2014) The 4th Millennium: A Watershed in European Prehistory. In B. Horejs and M. Mehofer (eds), *Western Anatolia before Troy. Proto-Urbanisation in the 4th Millennium BC?* 243–259. Vienna, Austrian Academy of Science Press.

Häusler, A. (1982) Zur ältesten Geschichte von Rad und Wagen im nordpontischen Raum. *Ethnologisch-Archäologische Zeitschrift* 25, 629–682.

Hollund, H. I., Higham, T., Belinsky, A. and Korenevskij, S. (2010) Investigation of Palaeodiet in the North Caucasus (South Russia) Bronze Age Using Stable Isotope Analysis and AMS Dating of Human and Animal Bones. *Journal of Archaeological Science* 37, 2971–2983.

Ingold, T. (1994) From Trust to Domination: An Alternative History of Human Animal Relations. In A. Manning and J. A. Serpell (eds), *Animals and Human Society: Changing Perspectives,* 1–22. London, Routledge.

Kaiser, E. (2007) Wagenbestattungen des 3. vorchristlichen Jahrtausends in der osteuropäischen Steppe. In M. Blečić, M.Crešnar, B. Hänsel, A. Hellmuth, E. Kaiser and C. Metzner-Nebelsick (eds), *Scripta Praehistorica in honorem Biba Teržan,* 129–149. Ljubljana, Narodni Muzej Slovenije.

Kaiser, E. (2011) Egalitäre Hirtengesellschaft versus Nomadenkrieger? Rekonstruktion einer Sozialstruktur der

Jamnaja- und Katakombengrabkultur (3. Jt. v.Chr.). In S. Hansen and J. Müller (eds), *Sozialarchäologische Perspektiven, gesellschaftlicher Wandel 5000–1500 v. Chr. zwischen Atlantik und Kaukasus*, 193–210. Mainz, Philipp von Zabern.

Kantorovich, A. R., Maslov, V. E. and Petrenko, V. G. (2013) Pogrebenija majkopskoj kul'tury kurgana No. 1 mogil'nika Mar'inskaja 5. *Materialy po izucheniju istoriko-kul'turnogo nasledija Severnogo Kavkaza X*, 71–108. Moscow, Pamjatniki istoricheskoy mysli.

Knipper, C., Held, P., Fecher, M., Nicklisch, N., Meyer, C., Schreiber, H., Zich, B., Metzner-Nebelsick, C., Hubensack, V., Hansen, L., Nieveler, E. and Alt, K. W. (2015) Superior in Life – Superior in Death, Dietary Distinction of Central European Prehistoric and Medieval Elites. *Current Anthropology* 56/4, 579–589.

Knipper, C., Reinhold, S., Gresky, J., Belinskiy, A. and Alt, K. W. (forthcoming) Economic Strategies at Bronze Age and Early Iron Age Upland Sites in the North Caucasus: Archaeological and Stable Isotope Investigations. In A. R. Ventresca Miller and C. Makarewicz (eds), *Isotopic Investigations of Pastoralism in Prehistory*, 123–140. Abingdon, Routledge.

Kohl, P. and Trifonov, V. (2014) The Prehistory of the Caucasus: Internal Developments and External Interactions. In C. Renfrew and P. Bahn (eds), *The Cambridge World Prehistory* 3/7, 1571–1595. New York, Cambridge University Press.

Korenevsky , S. N. (2011) *Drevnejshij metall Predkavkaza*. Moscow, TAUS.

Korenevsky , S. N. (2012) *Rozhdenie kurgana*. Moscow, TAUS.

Lebedeva, E. Y. (2011) Pervye rezultaty arkheobotanicheskikh issledovanii na arkheologicheskikh pamyatnikakh Adygei. *Analiticheskie issledovaniya laboratorii estestvennohauchnykh metodov* 2, 244–257.

Limberis, N. J. and Marchenko, I. I. (2002) Ein Kurgan der Novotitarovskaja-Kultur bei Novovelicovskaja, Kuban-Gebiet, Nordwestkaukasien. *Eurasia Antiqua* 8, 1–37.

Lyonnet, B. (2007) La Culture de Maïkop, la Transcaucasie, l'Anatolie orientale et le Proche-Orient. Relations et chronologie. In B. Lyonnet (ed.), *Les Cultures du Caucase (VIe–IIIe millénaires avant notre ère). Leurs relations avec le Proche-Orient*, 133–161. Paris, CNRS.

Mansfeld, G. (2013) *Der Held auf dem Wagen. Archäologische Belege zur technischen Entwicklung des Wagens*. Ruhpolding, Rutzen.

Mariotti, V., Facchini, F. and Giovanna Belcastro, M. (2007) The Study of Entheses: Proposal of a Standardised Scoring Method for Twenty-Three Entheses of the Postcranial Skeleton. *Collegium Antropologicum* 31/1, 291–313.

Marro, C. (2007) Upper-Mesopotamia and Transcaucasia in the Late Chalcolithic Period (4000–3500 BC). In B. Lyonnet (ed.), *Les Cultures du Caucase (VIe–IIIe millénaires avant notre ère). Leurs relations avec le Proche-Orient*, 77–94. Paris, CNRS.

Merpert, N. J. (1974) *Drevneyshe skotovody Vol'zhsko-Ural'skogo mezhdurech'ya*. Moscow, Nauka.

Mimokhod, R. A. (2013) *Lolinskaja Kul' tura, severo-zapadnyj Prikaspij na rubeže srednego i pozdnego periodov bronžovogo veka*. Moscow, Institut Archeologii RAN.

Müller, J. (2013) Demographic Traces of Technological Innovation, Social Change and Mobility, from 1 to 8 million Europeans (6000–2000 BCE). In S. Kadrow and P. Włodarczak (eds),

Environment and Subsistence, Forty Years after Janusz Kruk's 'Settlement studies...' 493–506. Rzeszów and Bonn, Rzeszów University and Habelt.

Müller, J. (2015) Eight Million Neolithic Europeans: Social Demography and Social Archaeology on the Scope of Changes – From the Near East to Scandianvia. In K. Kristiansen, L. Šmejda and J. Turek (eds), *Paradigm Found*. Oxford, Oxbow Books.

Munchaev, R. M. (1975) *Kavkaz na zare bronzovogo veka*. Moscow, Nauka.

Munchaev, R. M. (1994) Maikopskaya kul'tura'. In K. C. Kušnarëva and V. I. Markovin (eds), *Archeologija 5. Ėpocha bronzy Kavkaza I Srednej Azii – Rannjaja is srednjaja bronza Kavakza*, 158–225. Moscow, Nauka.

Nagler, A. (1996) *Kurgane der Mozdok-Steppe in Nordkaukasien*. Espelkamp, Marie Leidorf.

Nebieridze, L. (2010) *The Tsopi Chalcolithic Culture*. Tbilisi, SABC.

Nechaev, A. A. (1992) Domaikopskaya kul'tura Severnogo Kavkaza. *Archeologicheskie Vesti* 1, 76–96.

Oates, J., McMahon, A., Karsgaard, P., Al Quntar, S. and Ur, J. (2007) Early Mesopotamian Urbanism: A New View from the North. *Antiquity* 81/313, 585–600.

Piggott, S. (1992) *Wagon, Chariot and Carriage: Symbol and Status in the History of Transport*. London, Thames & Hudson.

Piotrovsky, J. J. and Bochkaryov, V. S. (eds) (2013) *Bronzovyj vek. Evropa bes granic. Bronzezeit. Europa ohne Grenzen. 4.–1. Jahrtausend v. Chr*. St. Petersburg, Chisty list.

Reinhold, S. (2012) Zur Konstruktion von Identität in der Bronzezeit Kaukasiens. In B. Horejes and I. Heske (eds), *Bronzezeitliche Identitäten und Objekte,* 83–106. Bonn, Habelt.

Rice, M. (1998) *The Power of the Bull*. London, Routledge.

Rogers, E. M. (2003) *Diffusion of Innovations*. New York, Free Press.

Rosenstock, E. and Masson, A. (2011) Das Rind in Vorgeschichte und traditioneller Landwirtschaft: Archäologische und technologisch-ergologische Aspekte. *Mitteilungen der Berliner Gesellschaft für Anthropologie, Ethnologie und Urgeschichte* 32, 81–106.

Russell, N. (2012) *Social Zooarchaeology. Humans and Animals in Prehistory*. Cambridge, Cambridge University Press.

Sagona, A. (2013) Wagons and Carts of the Trans-Caucasus. In O. Tekin, M. H. Sayar and E. Konyar (eds), *Tarhan Armağani. M. Taner Tarhan'a Sunulan Makaleler. Essays in Honour of M. Taner Tarhan*, 277–297. Istanbul, Ege Yayınlar.

Schultz, M. (1982) Krankheit und Umwelt des vor- und frühgeschichtlichen Menschen. In: H. Wendt and N. Loacker (eds), *Kindlers Enzyklopädie. Der Mensch*, 259–312. Zürich, Kindler.

Siebenga, J., Segers, M. J. M., Elzinga, M. J., Bakker, F. C., Haarman, H. J. and Patka, P. (2006) Spine Fractures Caused by Horse Riding. *European Spine Journal* 15/4, 465–471.

Sherratt, A. G. (1981) Plough and Pastoralism. Aspects of the Secondary Products Revolution. In I. Hodder, G. Isaac and N. Hammond (eds), *Pattern of the Past. Studies in Honor of David Clarke*, 261–305. Cambridge, Cambridge University Press.

Sherratt, A. G. (1983) The Secondary Products Revolution of Animals in the Old World. *World Archaeology* 15, 90–104.

Sherratt, A. G. (2004) Wagen, Pflug, Rind. Ihre Ausbreitung und Nutzung – Probleme der Quelleninterpretation. In M. Fansa and S. Burmeister (eds), *Rad und Wagen – Der Ursprung einer Innovation. Wagen im Vorderen Orient und Europa.* Archäologische Mitteilungen aus Nordwestdeutschland Beiheft 40, 409–428. Mainz, Philipp von Zabern.

Shishlina, N. (2008) *Reconstruction of the Bronze Age of the Caspian Steppes: Life Styles and Life Ways of Pastoral Nomads.* Oxford, Archaeopress.

Shishlina, N., Kovalev, D. and Ibragimova, E. (2013) Wagen der Katakombengrabkultur aus den Steppen Eurasiens. In J. J. Piotrovsky and V. S. Bočkarev (eds), *Bronzovyj vek. Evropa bes granic. Bronzezeit. Europa ohne Grenzen. 4.–1. Jahrtausend v. Chr.*, 118–126. St Petersburg, Chisty list.

Shishlina, N. I., Orfinskaya, O. V. and Golikov, V. P. (2003) Bronze Age Textile from the North Caucasus: New Evidence of Fourth Millennium BC Fibers and Fabrics. *Oxford Journal of Archaeology* 22/4, 331–344.

Steadman, S. R. (2005) Reliquaries on the Landscape, Mounds as Matrices of Human Cognition. In S. Pollock and R. Bernbeck (eds), *Archaeologies of the Middle East: Critical Perspectives*, 286–307. Malden, MA, Blackwell.

Stein, G. J. (2012) The Development of Indigenous Social Complexity in Late Chalcolithic Upper Mesopotamia in the 5th–4th Millennia BC – An Initial Assessment. *Origini* 34, 125–151.

Temple, D. H. (2010) Patterns of Systemic Stress during the Agricultural Transition in Prehistoric Japan. *American Journal of Physical Anthropology* 142/1, 112–124.

Trifonov, V. (1987) Nekotorye voprosy peredneasiatskich svjazey Maikopskoy kul'tury. *Kratkie Soobsheniya Institut Archeologii* 192, 18–26.

Trifonov, V. (2004) Die Majkop-Kultur und die ersten Wagen in der südrussischen Steppe. In M. Fansa and S. Burmeister (eds), *Rad und Wagen – Der Ursprung einer Innovation. Wagen im Vorderen Orient und Europa.* Archäologische Mitteilungen aus Nordwestdeutschland Beiheft 40, 167–176. Mainz, Philipp von Zabern.

Tucker, K., Berezina, N., Reinhold, S., Kalmykov, A., Belinskiy, A. and Gresky, J. (2017) An Accident at Work? Traumatic Lesions in the Skeleton of a 4th millennium BCE "Wagon Driver" from Sharakhalsun, Russia. *Homo* 68/4. *DOI: 10.1016/j. jchb.2017.05.004.*

Uerpmann, M. and Uerpmann, H.-P. (2011) Zug und Lasttiere zwischen Majkop und Trialeti. In S. Hansen, A. Haupmann, I. Motzenbäcker and E. Pernicka (eds), *Von Majkop bis Trialeti: Gewinnung und Verbreitung von Metallen und Obsidian in Kaukasien im 4.–2. Jt. v. Chr.* Kolloquien zur Vor- und Frühgeschichte 13, 237–261. Bonn, Habelt.

Ur, J. A. (2010) Cycles of Civilization in Northern Mesopotamia, 4400–2000 BC. *Journal of Archaeological Research* 18, 387–431.

Ward, C. (2006) Boat-Building and Its Social Context in Early Egypt: Interpretations from the First Dynasty Boat-Grave Cemetery at Abydos. *Antiquity* 80, 118–129.

Woolley, L. (1934) *Ur Excavations 2. The Royal Cemetery: A Report on the Predynastic and Sargonid Graves Excavated between 1926 and 1931.* London, British Museum.

Yakovlev, A. V. and Samoylenko, V. G. (2003) Novye pogrebeniya s Maikopskoy keramikoy na severo-vostoke Stavropol'ya. *Kratkie Soobsheniya Institut Archeologii* 214, 74–83.

Chapter 9

Innovation, Interaction and Society in Europe in the 4th Millennium BCE: The 'Traction Complex' as Innovation and 'Technology Cluster'

Maleen Leppek

In his seminal work, Childe proclaimed (Childe 1951) that the advent of wheeled vehicles should be dated to the mid-4th millennium, an opinion still commonly voiced today. This view was later supported by Piggott's carefully elaborated studies on the topic (1983; 1992) and most prominently by Sherratt's Secondary Products Revolution (1981). Since then, it has become one of the most intensely discussed topics in research on European prehistory and interest still seems to be growing rather than declining. Accordingly, research on these early wheeled vehicles itself reflects the research ventures undertaken and the theoretical approaches employed over the last few decades. Notably the quest for the geographical origins of the cart has claimed the attention of researchers from the outset, and various attempts have been made to identify where carts were used for the first time and to reconstruct potential diffusion routes. While Childe and later Sherratt clearly favoured South Mesopotamia as the place of origin with subsequent diffusion of the traction complex to Europe and Eurasia, later scholars argued in favour of other possible regions (*e.g.* Häusler 1992; Maran 2001; 2004) that had not been taken into consideration before. Processual archaeology gained in impact, and its supporters rejected classical diffusion models, advancing rival theories of autochthonous emergence in two or more regions. These theories were supported by a significant number of scholars (*cf.* Woytowitsch 1978; Vosteen 1996; Fansa 2004).

This concern to identify the earliest examples of wheeled vehicles eventually culminated in a debate on the rival claims of monocentrism and polycentrism that has lost none of its significance, notably because it so clearly reflects underlying assumptions on the existence of an early horizon of material evidence. While supporters of monocentric origins insisted that innovations spread from one single region of origin to another by way of various diffusion mechanisms, the 'polycentrists' believed in autochthonous emergence in two or more regions more or less independently of external influences. Leaving aside pure coincidence, the aim of the polycentrists was to identify the conditions paving the way for the evolution of the cart in different regions. The main candidates were subsistence and environmental factors (Maran 2001, 737–738; Kohl 2007, 4–5; Burmeister 2011, 233–234). These assumptions imply that wheeled vehicles first appeared in contexts where the environmental circumstances had created a certain need for them or at least constituted favourable conditions facilitating their 'invention'. Supporters of monocentric emergence, by contrast, did all they could to locate the place of initial emergence and to trace diffusion paths, frequently acting on the implicit assumption that the technological superiority of an innovation would automatically cause it to spread and be adopted elsewhere. Common to both these approaches was the desire to track down a region where all the components characteristic of wheeled vehicles were detectable in the archaeological evidence. Accordingly, they were both overly concerned with questions of origin and emergence.

Interestingly, the question of *chronological* emergence has been far less controversial and the spate of recent research on the subject appears to have produced a broad consensus about the positioning of the earliest wheeled vehicles on the timeline. Most supporters both of monocentric and polycentric emergence have agreed on the mid- to late 4th millennium as the most probable time for the emergence of the traction complex because all the material evidence points in that direction. They differ however in the questions they ask. While supporters of the monocentric hypothesis have asked themselves how an innovation could spread so far in such a short time, polycentrists have speculated on what had led to the emergence of this innovation in different areas at roughly the same time (Burmeister 2004, 16–17; 2011, 427)

As Sherratt wisely contends in his study (and as has been pointed out by Sherratt 1981 and Greenfield 2010), it can justifiably be said that the mid-4th millennium is probably far more a time of 'breakthrough' than of initial emergence. It is therefore not surprising, that there has been a tendency to date the real emergence of wheeled vehicles somewhat earlier, *i.e.* at the beginning of the 4th or the end of the 5th millennium, slightly prior to the first horizon of material evidence. Their dissemination was probably accelerated by existing exchange networks that had been firmly established during these millennia (Maran 2001; 2004b; Burmeister 2012). It has frequently been said that one of the main drawbacks bedevilling archaeological endeavour is the inability to identify the initial stages of something termed an innovation, largely because we hardly ever come across prototypes or evidence of abortive 'trial runs'. If we rely on 'oldest' finds to identify regions of origin, every new 'oldest' find can potentially shift that region somewhere else. The problem this causes is that we can never be sure of having found the oldest specimen of something: finds will always reflect how far research has progressed at a given point in time (Johannsen 2006, 43–44; Burmeister 2011, 211; 2013, 51). Nevertheless, every archaeological object that has somehow found its way into the archaeological find horizon bears witness to certain circumstances like environmental conditions or the particular attitude to the object by the community. This also means that if something figures in the material evidence, then it was usually well-established in the relevant society. So find contexts tell us much more about the value and the importance attached to something by a community than about actual processes of emergence or invention – always assuming that we are able to read these contexts correctly and distinguish coincidence and favourable environmental conditions from intentional practice. Thus, preserved wheels and axles from lakeside settlements buried under trackways (familiar from the Circum-Alpine region[1]) are likely to tell us far more about their significance for the community than to disclose anything about their actual emergence. The same applies to the Baden-associated wagon models of the Carpathian Basin, which are known from burial contexts (Bondár 2012),[2] or wheels and axles that have survived for millennia because they were buried in bogs. The current find situation is heterogeneous, assembling very different finds like iconographic evidence, preserved wheels and axles, and cattle bones displaying pathologies that are possibly indicative of

draught cattle usage.[3] It therefore seems reasonable to make use of this potential and focus on those aspects that can be addressed with the help of the material evidence we have. In connection with wheeled vehicles, I shall attempt to indicate how and why our attention should home in on the divergence of the material horizon, which is very likely a reflection of local appropriation processes.

Wheeled vehicles and the Secondary Product Revolution: the 'traction complex'

To address those local appropriation processes, we need to expand our perspective to wheeled vehicles as innovations that do not unfold their potential independently of everything else but rather interact with existing local contexts. Probably the most influential fillip for research on wheeled vehicles in their role as innovations came from Andrew Sherratt and his highly influential work on the so-called Secondary Product Revolution, which he believed represented the introduction of an entire range of new subsistence-related technologies that had a profound impact on societies in the course of the 4th millennium. According to Sherratt, the changes involved were bound up with the use of the so-called 'secondary products' provided by living animals, as opposed to the sole use of primary products requiring their death. This generated innovations like the use of milk products, wool production, the domestication and integration of formerly wild animals like donkeys and horses into the domestic sphere, and the exploitation of draught animals to pull ploughs and carts. In the course of the 4th millennium, all these innovations significantly extended the role played by of animals in prehistoric societies. They triggered a multitude of technological, economic and societal changes leading to increased mobility, more intensive contact between groups, and changes in subsistence and settlement structure (Sherratt 1981; 83; Greenfield 2010, 29–30). Intensified interaction between animals, plants and humans ultimately resulted in changing world views encouraging the emergence of new social structures in the societies of 4th millennium Eurasia (Sherratt 1981, 261–262; Vosteen 1996, 11–12; Greenfield 2010, 29–30). Taking his lead from Childe's concept of a 'Neolithic Revolution' (Childe 1936), Sherratt termed the horizon of the first significant and archaeologically visible use of these secondary products a 'revolution' that he believed had spread from the Mesopotamian proto-urban centres to the European 'periphery' (Sherratt 1981; 1997).

Ever since, reasoned doubts have been expressed not only about initial emergence in the early civilizations of the Near East but also about the existence of such a clearly circumscribed horizon of new technologies spreading as a package from one core region to the peripheries. The justified doubts came from various scholars, who split the package into its components and managed to prove the existence of some secondary products during earlier periods of the Neolithic at various places all over Eurasia (*e.g.* Greenfield 2010). This new wave of research inspired by Sherratt's ground-breaking concept also began to unravel the so-called traction complex by separating wheeled vehicles from the plough and treating them as distinct innovations calling for individual investigation.

Although, as later research has shown, Sherratt's initial assumption of close relations between the emergence and spread of the SPR components was probably erroneous (*e.g.* Vosteen 1996; Greenfield 2010), his theory contains remarkable ideas that go far beyond the emergence of a clutch of important innovations. First, it seems entirely plausible to point to the 4th millennium as a period when specific interactions between some of these components laid the foundations for fundamental changes in prehistoric communities (*e.g.* new subsistence methods) all over Eurasia. Second, he also imposed an image of wheeled vehicles as part of a larger innovation that can legitimately be referred to as the traction complex, thus rightly emphasising the strong interconnection between these vehicles and the plough. But it is Sherratt's specific notion of the *'transformative potential'* of interacting innovations exerting a strong influence on existing societies that is perhaps his most valuable contribution.

One thing important to bear in mind in connection with innovations is the fact that contact with, and the potential availability of, knowledge about, something that is new to a community does not automatically bring about its adoption in that community. Acceptance and perception by the appropriating communities plus the nature of the local context may in fact be the crucial aspects obstructing or supporting the adoption of an innovation (Meir 1988; Rogers 2003; Burmeister 2013, 49; Schier 2013, 5). These aspects may have been quite different in prehistoric communities of the 4th and early 3rd millennia, leading to various outcomes that may also be reflected in the heterogeneous archaeological find horizon, which has been described as reflecting different centres where early use was made of wheeled vehicles (Vosteen 2002; Burmeister 2012, 83). In turn, the regionally divergent material horizon of early evidence for the existence of wheeled vehicles might then faithfully reflect different appropriation processes shaped by their interaction with local contexts. Accordingly, if we want to shed new light on the nature of these local appropriation processes, the 'adoption environment' concept proposed by Avinoam Meir (1988) seems the most promising one to work with. It suggests that the local context into which new elements are integrated has a substantial influence on how these are appropriated and therefore on how innovations develop. This would mean that new elements appropriated by a society may run into a whole range of different conditions ranging from highly favourable to 'full of pitfalls'. An unpropitious adoption environment may also result in the complete rejection of a new technology.

Conversely, a positive adoption environment is required for new technologies to be appropriated successfully. This means that new elements like technologies must be susceptible of integration into existing structures like local systems, beliefs, social structures, *etc.* A successful innovation process can only take place if the new aspect is somehow compatible with existing preconditions, including all the factors operative in the social, spiritual and economic spheres. It is therefore very likely that successful innovation is the result of a reciprocal appropriation process that affects both the appropriating context and the new element.

In the case of wheeled vehicles, the assumption that the appropriating contexts had a major influence themselves on how certain innovation processes developed would eventually shift the 'transformative potential' from new individual technologies to the interaction between pre-existing local contexts and these upcoming new technologies. As a consequence, innovations should be studied in micro-regional contexts in order to reveal their specific local interactions. Research should encompass catalytic factors, transmission mechanisms and specific local appropriations plus interaction with pre-existing technologies, subsistence aspects and the constitution of the society in question. All these factors can play a role in the specific appropriation of the new technology by local communities and should therefore be carefully examined if we wish to understand the nature of this 'transformative potential' arising from the interaction of 'innovations' and the specific environments adopting them. This could be helpful in achieving a better understanding of later developments rooted in those very specific interactions. It could also facilitate interpretation of the heterogeneous material horizon associated with early wheeled vehicles, because this heterogeneity is in all probability a reflection of local developments in different areas on the way from the mid-4th to the early 3rd millennium BCE.

Technologies as networks vs. entangled innovations

At this point it is worth recalling that this view of technological 'innovations' is based on the assumption that technologies are never isolated phenomena but are invariably embedded in an environment with which they form an interactive network involving both human and non-human agents. Small changes in single parts of such a network can potentially cause a cascade of effects with far-reaching impacts beyond the part of the network originally affected (in line with Latourian concepts of the network: Latour 1991). Given the important role of local contexts in the appropriation of innovations, these impacts can legitimately be regarded as reciprocal in nature. This implies that innovations have the potential to affect local communities and, vice versa, the constitution of local contexts will almost inevitably influence the appropriation of a new technology. The interaction between them would then shape the outcome of the innovation process and requires cautious, meticulous examination because reciprocal relationships tend to blur causes and effects. This in its turn strongly implies that innovations cannot usefully be studied in isolation but only in their interactions and interconnections with all pre-existing aspects of subsistence, environment, social structures, beliefs, transmission mechanisms and any other aspect that may have played a role in obstructing or supporting the integration of an innovation.

It is also important to note that 'innovations' in this case should not be confused with 'inventions'. These words are not exact synonyms (Burmeister 2013, 50; Burmeister and Müller-Scheeßel 2013, 1–2). Innovations and inventions differ in the sense that while inventions refer to something entirely new, the term innovation focuses more on the dynamic process of change affecting both the appropriating local contexts and the new element that is being appropriated. This makes it irrelevant whether something is completely new in the absolute sense of the term. What matters is whether something is *new to the appropriating society* (Rogers 2003, 26; Burmeister 2013, 49). This is of particular importance in prehistoric contexts because we can never be certain that we have in fact found the oldest representatives of something making a new appearance in the archaeological context (Johannsen 2006, 43–44; Burmeister 2013, 51). Thus, the pertinent distinction between innovation and invention gives us a further justification, not to say an outright need, to shift the focus of our enquiries from 'earliest emergence' to 'specific local appropriation'.

The traction complex as a technology cluster

In the present case, it is important to point out that vehicles (like many other innovations) are not single elements. They should be understood as 'technology clusters' (Rogers 2003, 14)[4] created by the successful combination of a whole range of individual technologies that are so inextricably interconnected that they are hardly perceived as distinct elements (Rogers 2003, 14; Burmeister 2004, 21; 2012, 85–86). As a result, it is easy to forget that there is no immanent necessity to combine them in this specific way. In fact, all these technological elements could be used independently of one another in entirely different combinations. Sherratt was very close to this idea when he spoke of wheeled vehicles as part of the traction complex, thus closely connecting them to the plough. Both are in fact dependent on the same technologies required to link them to the draught cattle pulling them. As we shall see, this perception of wheeled vehicles as technology clusters could have a substantial impact on the interpretation of the material evidence from the different contexts.

Researchers had in a way already countenanced this perception of wheeled vehicles as 'technology clusters' – but in the light of a quite different background. As part of the quest for either monocentric or polycentric origins, various attempts were made to trace from the archaeological evidence the underlying technological principles that in combination lead to wheeled vehicles and also constitute the traction complex. Theories emerged about how these could have been derived from other technologies and newly combined to construct innovative wheeled vehicles (Höneisen 1989, 7–10; Burmeister 2004, 21; 2011, 233). Although their purposes were different, this was equally essential for the champions of monocentric and polycentric emergence. From a monocentric perspective, it was important where these elements were first combined before the whole complex spread, while the polycentrists were more concerned about identifying contexts where all principles were present in the course of the 4th millennium and how these contexts had enabled wheeled vehicles to emerge without the assistance of 'classical' diffusion (Burmeister 2004, 16–17).

Stefan Burmeister has indicated several elements that in combination triggered the emergence of early wheeled vehicles but most probably existed prior to that juncture (Burmeister 2004, 21; 2012, 85–86). If we extend our purview to the traction complex, we find that they are connected to each other by additional technological elements as well. To understand their reciprocal impacts and their interconnections with local contexts, it may initially be helpful to disentangle them and regard them as distinct technological elements.

These factors are (1) the rotation principle in the form of the rotating wheel or rotating axle, (2) the device enabling transportation of heavier loads, and (3) the application of animal traction, which freed humans of the necessity to carry or pull heavy loads themselves. In terms of the adoption environment, further aspects for consideration would be the topographical environment, subsistence, social organisation, the structure of society, a certain willingness to adopt new technologies, concrete ideas about appropriate fields of application, and pre-existing technologies required for concrete realisation (*e.g.* elaborate metal tools appropriate for building highly sophisticated multipartite disc wheels: Childe 1951; Vosteen 1999; Burmeister 2004).

Looking first at animal traction, most prehistoric communities had had potential access to domesticated cattle since the beginning of the Neolithic, which meant access to an animal that could carry or pull heavier loads than humans could. We do not know when people first used cattle for this job but it is probable that it was quite early on in prehistory (Boroffka 2004, 467–468). The use of draught cattle to pull heavy loads without a transportation device or just with the help of simple padding like leather or fur could have occurred quite early in prehistory. It was considered for the building of the huge Neolithic longhouses known from *e.g.* LBK contexts that made it necessary to move and raise tree trunks of enormous size and weight (Bakker 2004, 283–284; Greenfield 2010, 38–3). Even today, the transport of heavy tree trunks from densely forested areas, especially in steep mountain ranges, is often done with the help of horses dragging the timber by their sheer muscular strength without the use of any transportation device. I have personally witnessed this in south-western Bulgaria.

This immediately leads us on to the next small but very important element of the 'traction complex' technology cluster, as *e.g.* horses could not be used for pulling heavy loads until a yoking device was invented that engineered the appropriate distribution of force to the horse's neck. The yoking techniques required for cattle are very different (Mainberger 2002, 85). Both ethnological and archaeological evidence indicates that the methods of yoking cattle may be quite divergent, ranging from the total absence of a yoke to the use of sophisticated and highly efficient techniques like double yokes harnessing two animals (Johannsen 2006; Rosenstock and Masson 2011). Material evidence pointing to draught cattle use in the 4th and even 5th millennium is relatively common and can be found all over Eurasia (*cf.* Boroffka 2004, 468; Matuschik 2002). It comprises iconographical evidence and preserved finds dating from the late 5th to the early 3rd millennium. Examples of such evidence are legion, it will suffice to refer to the famous copper models of cattle harnessed by a double yoke and a number of clay models showing one or two animals equipped with harnessing elements (Matuschik 2002; Bondár 2012). Likewise, finds of preserved yokes from wetland contexts such as the object from Arbon Bleiche 3, Switzerland (Leuzinger 2002, 107–108) or the roughly cut object from Chalain 1, France (Petréquin *et al.* 2002) tell us that certain prehistoric communities were definitely familiar with these technologies. Nevertheless, we should not forget that the use of harnessing techniques is not restricted to transportation. The efficient use of the ox-drawn plough also requires appropriate yoking techniques, as does the use of sledges, slides or any other instance of the use of animals for their traction power. This only goes to show the potential of yoking techniques as multi-purpose innovations and how certain techniques can be connected interdependently. Some researchers have therefore legitimately posed the question whether it was perhaps the advent of double yoking techniques rather than the wheel that accelerated the widespread occurrence of wheeled vehicles during the 4th millennium (Sherratt 1981;[5] Masson and Rosenstock 2011). On the other hand, the example of the yoke and animal power is an eloquent example of the ambiguity of material evidence when such questions are addressed with the help of archaeological methods only. The presence of yoking techniques is not a clinching argument for the use of wheeled vehicles by prehistoric communities as these techniques could equally well be indicative of ploughs or wheel-less transport.

The use of suitable yoking techniques like the harnessing of a team of cattle with a double yoke makes transporting heavy loads or ploughing more efficient. Conversely, the successful use of yokes and harnessing technologies needs draught animals to be trained at an appropriate age, which could turn into a time-consuming activity and require a lot of effort on the part of the people entrusted with the task. It is readily conceivable that individuals or whole prehistoric communities may have pursued different approaches and had varying degrees of knowledge about techniques for training and using cattle. This knowledge may involve such things as the ideal age for castration, animal handling, proper feeding, and the selection of particular individuals for breeding (Masson and Rosenstock 2011). It is not unlikely that these measures enhanced the economic value of efficiently trained draught animals as they were not as easy to replace as animals only kept for their meat. Besides this, such new practices may have served as a catalyst for changes in the relationship between humans and animals, possibly with an impact on the role and perception of cattle in spiritual matters as well (Bogucki 1993). In turn, specials skills and knowledge in connection with training and handling cattle is likely to have been a valuable species of knowledge that could be exchanged with other individuals or communities and may have played a role in interregional exchange and contact. It could also have provided certain individuals or groups within a community with a kind of special knowledge or ability that guaranteed them a certain status and attendant economic advantages.

Another aspect of the traction complex that needs to be addressed is the transportation device itself. As we have seen, not just carts or wagons can be used for transportation, even very simple devices like pieces of fur or leather can reduce friction adequately for the purpose. Wheeled vehicles of course are ideal in this respect. Accordingly, they are generally perceived as representing a much higher state of development than other means of transport like slides and sledges, which are often regarded as more primitive predecessors (Piggott 1983, 16; Mainberger 2002). Researchers often automatically assume that adding wheels to an animal powered transportation device was the last step in a series of gradual improvements from wheel-less devices to the fully developed cart. This implies a subconscious notion of wheeled vehicles as superior transportation devices that need to be adequately identified in the archaeological context, thus making wheels and axles diagnostic features of wagons clearly distinguishing them from wheel-less means of transport like sledges or slides (Burmeister 2004, 21).

As a result, special attention has been paid to the principle of wheel and axle. It is important to note that the earliest material evidence shows that there are two different ways of applying the rotation principle that have often been associated with either two- or four-wheeled vehicles. Wheels can be either connected to axles by rectangular axle holes, which means the whole wheel-and-axle construction would turn underneath the vehicle, or the wheels are connected to axles by round holes enabling them to turn around the axle. This is an argument frequently drawn upon by the supporters of polycentric emergence (*cf.* Woytowitsch 1978; Vosteen 1996; Fansa 2004). But recent finds of wheel models from Olzreuter Ried, Bad Schussenried discovered in the direct vicinity of real preserved wooden disc wheels indicate that the lines cannot always be drawn so clearly. The finds suggesting the presence of both principles in one context indicate the familiarity of the respective Neolithic community with the two principles (Schlichtherle 2010). This indicates that the preference for one of these options might be a carefully considered decision favouring one alternative over the other. Various theories have been advanced on how these wheel and axle technologies may have emerged. They range from pure coincidence and 'observation and translation' to active transfer from other technologies to the cart (Höneisen 1989, 8). Reconstructing the translation of such a principle from one technology to the other is a plausible approach, but in the case of the wheel it has not always been convincing. This, I believe, is particularly true of the translation of the rotation principle from technologies like the potter's wheel or spindle whorls to wheeled vehicles. Here I prefer to follow Piggott, who has rightly emphasised that these are not really aspects of the same technology, and I would further add that the rotation principle is not the only factor at stake here (Piggott 1983, 17; 1992, 14). This can best be illustrated by considering spindle whorls, which are in fact flyweights (*Schwungkörper*) influencing the coarseness and thickness of a thread, while the wheel, as we saw earlier, substitutes rolling resistance for friction resistance. Nevertheless, spindle whorls are a very important when we are considering the material evidence for the presence of wheeled vehicles in certain contexts as they play a dual role in research history. They are of course evidence for the rotation principle, but much more frequently they are considered potential wheel models for model wagons. Such wheel models definitely existed, as the wagon-model finds from the Carpathian Copper and Bronze Age (*cf.* Bondár 2012) indicate, but there is justified doubt about whether all the objects considered to be wheel models should actually be regarded as such. This is an especially important point when are looking at very old finds that may indicate an earlier horizon, examples being the discovery of wheel models in Gumelniţa contexts in Bulgaria and Romania (Dinu 1981; 143; Häusler 1992, 18; Vosteen 2002, 6).[6]

The image of these early wheeled vehicles proposed by Childe and Sherratt has changed little over the last decades, and reconstructions often refer to them as big, heavy, wooden and extremely cumbersome (Hayen 1990). Wheels, transportation and draught animals become an inseparable complex with wheels as diagnostic elements of crucial importance. This perception of wheels and axles as the basis for superior means of transport also restricts the role of sledges and slides to that of (primitive) predecessors of wagons, although the advantages of wheels are in fact restricted to quite specific environmental conditions and topographical givens (Mainberger 2002, 86–87). The advantage of wheeled vehicles – reduced friction means heavier loads can be transported with less effort – is null and void in sandy, swampy or hilly areas, where reduced friction actually turns into a disadvantage. It is hardly surprising, therefore, that there are a significant number of regions where up to the present-day people have used sledges rather than wheels. This is often the case when there are no paved roads and the ground is sandy, wet or soft and wheeled vehicles would just sink in and get stuck. The same applies to communities located in very steep, hilly areas (*e.g.* Alpine regions and the mountain ranges of Svanetia, Georgia, where sledges where used until very recently) (Mainberger 2002, 87–88). Here the reduced friction of wheels is a real disadvantage because vehicles move downwards much faster and more easily, a potential drawback that is in fact exacerbated by rotation (Rosenstock and Masson 2011). The list of unfavourable environments is long, densely forested areas would fall into the same category. Unless communities in such environments took intensive measures to create more 'wheel-friendly' conditions, wheels had no advantages over sledges or slides, which may actually have been a much better choice.

This may seem rather too positivistic *vis-à-vis* the material evidence, as it would definitely make sledge models (as known from Trypillia giant settlements in the north-west Pontic region, (*cf.* Gusev 1998) and related iconographic evidence not explicitly featuring wheels a little more ambiguous. However, it could also crucially widen our perspective on sledges (and slides). It would be equally conceivable to regard them as predecessors of actual vehicles, contemporaneous means of transport and the results of local appropriation processes. We should at least try to see them not only as simple precursors whose full potential was not realised until they were equipped with wheels but also as a genuine alternative with respect to local conditions. Although it is very likely that wheels are indeed crucially important in certain cases, this question can only be dealt with in specific contexts rather than making universal or generalised assumptions.

The same applies to the powerful combination of animals and wheeled vehicles that constitutes the traction complex as it is usually envisaged. Vehicle reconstructions correctly based on existing wheel finds are therefore often strongly suggestive and support the combination of the vehicles with draught animals and certain yoking technologies. They include a lot of guesses about other parts of carts that are seldom preserved and usually reconstructed on the basis of the size and weight of the preserved wheels, which are indeed often massive (*cf.* the well-known, very sophisticated specimen from Stare Gmajne Slovenia: Velušček *et al.* 2009). This notion is also deeply rooted in present-day ideas on what the first wheeled vehicles may have looked like and as such is highly suggestive. Indeed, the sheer size and weight of these reconstructions make it indeed hardly conceivable that they could have been used without the help of animal traction. In this context it is worth referring to a remarkable reconstruction from the Federsee Museum, Bad Buchau which suggests that mighty composite wheels usually serving as a basis for reconstructing huge vehicles could in fact belong to very different constructions (Fig. 9.1)[7] This small, humble and certainly much more manoeuvrable vehicle is based on the big wheels found in the Neolithic settlement of Seekirch 'Achwiesen' (Schlichtherle 1989) and gives the impression that they could easily have functioned without draught cattle. The same is shown by a similar reconstruction (Fig. 9.2). They shall serve here as a gentle reminder that big wheels do not necessarily have to be part of huge wagon constructions and could also be misleading when drawn upon as the only indication for the existence of the fully developed traction complex in certain contexts.

Wheeled vehicles as local technology clusters

This also gives rise to questions about the actual importance of the single elements that make up wheeled vehicles. There seems to be no general consensus on which of these aspects were critically important. Can the use of animal traction for both transport and ploughing be termed a breakthrough, as Sherratt (1981) proclaimed? Or was the wheel added to existing but simpler modes of transport as a technology that significantly reduced the traction power needed? Or did perhaps entirely different aspects like changes in local conditions, subsistence, specific knowledge and new ideas transmitted along communication channels, or other technological 'improvements' establish new fields of application? When we go through the material evidence with a fine-tooth comb to unearth all the elements we regard as necessary for the construction of wheeled vehicles, we might ask ourselves whether in fact we are not merely replacing old biases with new ones. Those elements may just not be there.

This very positivist view of the material could be rather insufficient as the finds of secondary validity may comprise hints for favourable preconditions as well as certain changes that could be connected to the emergence of wagons. From a

Figure 9.1: Reconstruction of a simple wheeled vehicle based on the finds from Seekirch 'Achwiesen' (courtesy Federseemuseum Bad Buchau).

Figure 9.2: Hypothetical reconstruction of a hand-drawn transportation device (courtesy Federseemuseum Bad Buchau).

present-day perspective, it may simply seem too self-evident that the full potential of certain innovations like wheeled vehicles as we imagine them today can only be realised by interaction with a whole range of other elements, these being traction animals, specific environmental conditions and a certain state of society enabling such realisations in the first place or translating them into certain kinds of appropriate application. The crucial point may be precisely this 'appropriate application', which in some cases may have involved compromises with local contexts, including the eventuality that aspects of the traction complex or early 'wheeled' vehicles may have occurred in quite a wide range of varieties – perhaps even without wheels. These forms of appropriate application would then be the result of processes including the removal of obstacles to the integration of wheeled vehicles into the local context and changing those vehicles to adapt them to local conditions and the needs of the community. On the other hand, does the notion of wheeled vehicles as technology clusters

definitely require realignments in the perception of them that will also affect their relations with other (pre-existing) technologies? We might therefore do well to expend more attention on investigating how the different components are reflected in the material evidence. Do they really point to a coherent innovation or are we merely dealing with specific local variants in the combination of technological elements?

In my opinion, a major obstacle militating against a clearer picture of the role and involvement of early wheeled vehicles in local contexts is that we do not have an adequately differentiated picture of what we are actually looking for. The traction complex may be much less standardised than is often assumed. As the discussion on rotating and fixed axles has shown, variations are possible and the absence of certain elements of the traction complex may not always reflect unfavourable preservation conditions or selective deposition practices. It may well be the case that parts of the complex were either consciously rejected because they did not fit in with local needs or that people were just never confronted with them. Either that, or the appropriation of certain features may have involved a huge effort with major repercussions on both the environment and the new element itself. Sherratt referred to the use of the plough as one of the major innovations of the Secondary Products Revolution and closely connected to carts because ploughing and carts are associated with the same technology. Here he was not just pointing out their close connection. To a certain degree, this connection also reflects the impossibility of clearly assigning yoke finds or certain pathologies on cattle bones to either wheeled vehicles, sledges, slides or even ploughs. It is hard to resist overstraining the potential of the archaeological evidence.

I still have no intentions of applying a pure positivistic approach because such a narrow view on the material would be rather insufficient as the finds of secondary validity may comprise valuable hints for favourable preconditions as well as certain changes that could well be connected to the emergence of wagons. Accordingly, I would opt for a more specific look at the archaeological evidence in micro-regional contexts that treats ambiguous evidence as ambiguous evidence. Otherwise, important information about the very local appropriation of innovations could get lost as a result of accidental generalisation. I would eschew assumptions that leave no room for a wider range of potential explanations for the way innovations were appropriated in local contexts. It seems fair to assume that the confrontation with something both entirely new and highly complex is not the same as being confronted with something constituting either a slight or even a vast improvement in something that is already familiar or clearly boasting the potential for combination with pre-existing technologies. This may well make us more receptive to the wide range of manifestations in which a particular innovation may have occurred and the way in which pre-existing aspects of society like subsistence strategies and established technologies or the social constitution and individual decisions may have influenced these appropriation processes, leaving their mark both on the innovation itself and on the local context.

Consequences of research on wheeled vehicles

It is clear that innovations may be much less pre-defined than is often assumed and may have occurred in a multitude of different guises. As our discussion has shown, we should perhaps also consider other models than the spread or emergence of a whole technology cluster that was inseparably connected. Instead we should try to shed more light on the notion of early wheeled vehicles as local manifestations of interconnected technologies that interacted strongly with local contexts. This also implies that we may need to rethink our position on material horizons, as there may have been no clearly defined horizon of earliest wheeled

vehicles and the material evidence we can lay our hands on is more likely to reflect different states of innovation processes in their local manifestations. Accordingly, even the different means of transport, like sledges and slides, and the two different concepts of wheeled vehicles, like four-wheeled and two-wheeled carts, could equally well be the results of local appropriation processes adopting different technologies on the basis of entirely different preconditions. They may therefore have emerged as a result of the combination of different technologies in accordance with local contexts.

This still leaves enough scope for the assumption that, due to the advent of changes in ideology, new technologies, new contacts, new groups or new individuals, the meaning of a certain technology or a complex like early wagons changed over time. All those factors could readily have changed the meaning of something that was apparently firmly established in the community, which in its turn implies that appropriation processes may never be fully complete because theoretically both technological improvements and new ideas/beliefs could change the perception of earlier innovations and bestow a new meaning on them. This makes it obvious that the question of polycentric or monocentric origins for innovations has to be regarded as much more complex than is often assumed, with multidirectional interactions, ideas, technologies *etc.* emerging more probably as a result of certain combinations than of the unique dissemination of an innovation regarded as universally beneficial in all appropriating contexts, regardless of preconditions like already existing technologies or subsistence strategies.

In favour of a micro-scale approach

I would therefore argue in favour of a micro-scale approach enabling us to trace the local concatenations of an innovation and thus to gain insights into concrete aspects of its local appropriation and the transformative potentials generated by the interaction of certain 'novelties' with aspects and elements of existing local contexts. Thus, new technologies can be seen as not just independently unfolding their potential in the societies adopting them but as embedded in local contexts and interacting with the technological, societal and environmental factors operative in those societies. It is this that enables them both to initiate a transformation of the existing contexts and to change character themselves. Accordingly, defining broader contexts like a chronologically narrow horizon of earliest wheeled vehicles can only be the end of research, not the beginning. This may be more a by-product of the examination of these micro-regional contexts and the transformations that have occurred via the interaction of new technologies and local contexts, but it could also help us gain a better understanding both of long-term developments and of sudden changes.

Notes

1   For more detailed information see Schlichterle 2004.
2   For example, the famous wagon-shaped cup from Budakalász stems from a cenotaph, so it is very likely that the model itself was buried and did not serve as a grave good (Maran 2004; Bondár and Raczky 2009).
3   For a detailed image of the current find situation and chronological relations, *cf.* Burmeister 2011.
4   'A technology cluster consists of one or more distinguishable elements of technology that are perceived as being closely interrelated.' (Rogers 2003, 14).
5   *e.g.* Sherratt had originally assumed that although cattle had been domesticated at least since the 6th millennium BCE, exploitation of animal traction had not started before the mid-4th millennium because of the absence of appropriate harnessing technologies. This prompted him to contemplate the eventuality that it was in fact team-harnessing techniques that accelerated the widespread dissemination of wheeled vehicles (Sherratt 1981, 163).
6   But I agree with Stefan Burmeister that for some of these objects the interpretation of them as wheel models is indeed convincing (*cf.* Vosteen 2002, 143; Burmeister 2011, 224–225).
7   I would like to thank the Federseemuseum Bad Buchau for providing the pictures.

References

Bakker, A. (2004) Die neolithischen Wagen im nördlichen Mitteleuropa. In M. Fansa and S. Burmeister (eds), *Rad und Wagen – Der Ursprung einer Innovation: Wagen im Vorderen Orient und Europa*. Archäologische Mitteilungen aus Nordwestdeutschland Beiheft 40, 283–294. Mainz, Philipp von Zabern.

Bogucki, P. (1993) Animal Traction and Household Economies in Neolithic Europe. *Antiquity* 67, 492–503.

Bondár, M. and Raczky, P. (ed.) (2009) *The Copper Age Cemetery of Budakalász*. Budapest, Pytheas.

Bondár, M. (2012) *Prehistoric Wagon Models in the Carpathian Basin (3500–1500 BC)*. Archaeolingua, Series Minor 32. Budapest, Archaeolinigua.

Boroffka, N. (2004) Nutzung der tierischen Kraft und Entwicklung der Anschirrung. In M. Fansa and S. Burmeister (eds), *Rad und Wagen – Der Ursprung einer Innovation: Wagen im Vorderen Orient und Europa*. Archäologische Mitteilungen aus Nordwestdeutschland Beiheft 40, 467–480. Mainz, Philipp von Zabern.

Burmeister, S. (2004) Der Wagen im Neolithikum und in der Bronzezeit: Erfindung, Ausbreitung und Funktion der ersten Fahrzeuge. In M. Fansa and S. Burmeister (eds), *Rad und Wagen – Der Ursprung einer Innovation: Wagen im Vorderen Orient und Europa*. Archäologische Mitteilungen aus Nordwestdeutschland Beiheft 40, 13–40. Mainz, Philipp von Zabern.

Burmeister, S. (2011) Innovationswege – Wege der Kommunikation. Erkenntnisprobleme am Beispiel des Wagens im 4. Jahrtausend v. Chr. In S. Hansen and J. Müller (eds), *Sozialarchäologische Perspektiven: Gesellschaftlicher Wandel 5000–1500 v. Chr.*

zwischen Atlantik und Kaukasus. Archäologie in Eurasien 24, 211–240. Mainz, Philipp von Zabern.

Burmeister, S. (2012) Der Mensch lernt fahren – Zur Frühgeschichte des Wagens. *Mitteilungen der Anthropologischen Gesellschaft in Wien* 142, 81–100.

Burmeister, S. (2013) Migration – Innovation – Kulturwandel. Aktuelle Problemfelder archäologischer Investigation. In E. Kaiser and W. Schier (eds), *Mobilität und Wissenstransfer in diachroner und interdisziplinärer Perspektive*, 35–58. Berlin and Boston, MA, de Gruyter

Burmeister, S. and Müller-Scheeßel, N. (2013) Innovation as a Multi-Faceted Social Process: An Outline. In S. Burmeister, S. Hansen, M. Kunst and N. Müller-Scheeßel (eds), *Metal Matters. Innovative Technologies and Social Change in Prehistory and Antiquity*. Menschen – Kulturen – Traditionen. Studien aus den Forschungsclustern des Deutschen Archäologischen Instituts 12, 1–11. Rahden/Westf., Marie Leidorf.

Childe, V. G. (1936) *Man Makes Himself*. London, Watts & Co.

Childe, V. G. (1951) The First Wagons and Carts – From the Tigris to the Severn. *Proceedings of the Prehistoric Society* 17, 177–194.

Dinu, M. (1981) Clay Models of Wheels Discovered in Copper Age Cultures of Old Europe Mid-Fifth Millennium B. C. *Journal of Indo-European Studies* 9, 1–14.

Fansa, M. (2004) Einleitung. In M. Fansa and S. Burmeister (eds), *Rad und Wagen – Der Ursprung einer Innovation: Wagen im Vorderen Orient und Europa*. Archäologische Mitteilungen aus Nordwestdeutschland Beiheft 40, 9–12. Mainz, Philipp von Zabern.

Greenfield, H. J. (2010) The Secondary Products Revolution. The Past, the Present and the Future *World Archaeology* 42/1, 29–54.

Gusev, S. A. (1998) K voprosu o transportnych sredstvach tripol'skoj kul'tury [On the Issue of the Tripolye Culture Transportation Means]. *Rossijskaja Archeologija* 1998, 15–28.

Hayen, H. (1990) *Ein Vierradwagen des dritten Jahrtausends v. Chr. Rekonstruktion und Nachbau*. Experimentelle Archäologie in Deutschland, Beiheft 4, 172–190. Oldenburg, Isensee.

Häusler, A. (1992) Der Ursprung des Wagens in der Diskussion der Gegenwart. *Archäologische Mitteilungen Nordwestdeutschland* 15, 179–190.

Höneisen, M. (1989) Die jungsteinzeitlichen Räder der Schweiz. Die ältesten Europas. In B. A. Schüle (ed.), *Das Rad in der Schweiz vom 3. Jt. v. Chr. bis um 1850*. Katalog Schweizerisches Landesmuseum, Zürich, 22 August–26 November 1989, 13–22. Zürich, Schweizerisches Landesmuseum.

Johannsen, N. N. (2006) Draught Cattle and the South Scandinavian Economies of the 4th Millennium BC. *Environmental Archaeology* 11, 35–48.

Johannsen, N. N. (2011) Past and Present Strategies for Draught Exploitation of Cattle. In U. Albarella and A. Trentacoste (eds), *Ethnozooarchaeology. The Present and Past of Human-Animal Relationships*, 13–19. Oxford, Oxbow Books.

Kohl, P. (2007) *The Making of Bronze Age Eurasia*. Cambridge, Cambridge University Press.

Latour, B. (1991) Technology is Society Made Durable. In J. Law (ed.), *A Sociology of Monsters? Essays on Power, Technology and Domination*, 103–131. London and New York, Routledge.

Leuzinger, U. (2002) Das vermutete Joch von Arbon–Bleiche 3, Schweiz. In J. Köninger, M. Mainberger, H. Schlichterle and M.

Vosteen (eds), *Schleife, Schlitten, Rad und Wagen. Zur Frage früher Transportmittel nördlich der Alpen.* Hemmenhofener Skripte 3, 107–108. Freiburg, Janus.

Mainberger, M. (2002) Sommerschlitten, Ackerrutschen, Pflugschleifen. Rezente radlose Transportfahrzeuge und die Schleife von Reute-Schorrenried. In J. Köninger, M. Mainberger, H. Schlichterle and M. Vosteen (eds), *Schleife, Schlitten, Rad und Wagen. Zur Frage früher Transportmittel nördlich der Alpen.* Hemmenhofener Skripte 3, 83–92. Freiburg, Janus.

Maran, J. (1998) Die Badener Kultur und der ägäisch-anatolische Bereich: Eine Neubewertung eines alten Forschungsproblems. *Germania* 76, 497–525

Maran, J. (2001) Zur Westausbreitung von Boleráz-Elementen in Mitteleuropa. In P. Roman (ed.), *Proceedings of the International Symposium 'Cernavoda III-Boleráz'. Ein vorgeschichtliches Phänomen zwischen Oberrhein und der Unteren Donau, Mangalia 1999,* 733–746. Bucharest, Vavila Edinf.

Maran, J. (2004) Kulturkontakte und Wege der Ausbreitung der Wagentechnologien im 4. Jahrtausend v. Chr. In M. Fansa and S. Burmeister (eds), *Rad und Wagen – der Ursprung einer Innovation. Wagen im Vorderen Orient und Europa.* Archäologische Mitteilungen aus Nordwestdeutschland Beiheft 40, 429–442. Mainz, Philipp von Zabern.

Masson, A. Rosenstock E. (2011) Das Rind in Vorgeschichte und traditioneller Landwirtschaft, *Mitteilungen der Berliner Gesellschaft für Anthropologie, Ethnologie und Urgeschichte* 32, 81–106.

Matuschik, I (2002) Kupferne Rindergespann-Darstellungen der mitteleuropäischen Kupferzeit. In J. Köninger, M. Mainberger, H. Schlichtherle and M. U. Vosteen (eds), *Schleife, Schlitten, Rad und Wagen. Zur Frage früher Transportmittel nördlich der Alpen.* Hemmenhofener Skripte 3, 111–122. Freiburg, Janus.

Matuschik, I. (2006) Invention et diffusion de la roue dans l'Ancien Monde: L'apport de l'iconographie. In P. Pétrequin, R.-M. Arbogast, A.-M. Pétrequin, S. van Willigen and M. Bailly (eds), *Premiers chariots, premiers araires. La diffusion de la traction animale en Europe pendant les IVe et IIIe millénaires avant notre ère.* Centre National de la Recherche Scientifique Monographies 29, 279–297. Paris, CNRS Éditions.

Meir, A. (1988) Adoption Environment and Environmental Diffusion Processes: Merging Positivistic and Humanistic Perspectives. In P. J. Hugill and D. B. Dickson (eds), *The Transformation and Transfer of Ideas and Material Culture,* 233–347. College Station, TX, Texas A&M University Press.

Pétrequin, P., Arbogast, R.-M., Viellet, A., Pétrequin, A.-M. and Maréchel, D. (2002) Eine neolithische Stangenschleife vom Ende des 31. Jhs. v. Chr. in Chalain (Fontenu, Jura, Frankreich). In J. Köninger, M. Mainberger, H. Schlichtherle and M. U. Vosteen (eds), *Schleife, Schlitten, Rad und Wagen. Zur Frage früher Transportmittel nördlich der Alpen.* Hemmenhofener Skripte 3, 55–65. Freiburg, Janus.

Piggot, S. (1983) *The Earliest Wheeled Transport. From the Atlantic Coast to the Caspian Sea.* London, Thames & Hudson.

Piggot, S. (1992) *Wagon, Chariot and Carriage. Symbol and Status in the History of Transport.* London, Thames & Hudson.

Rogers, E. M. (2003) *Diffusion of Innovations.* New York *et al.,* Free Press.

Schier, W. (2013) Mobilität und Wissenstransfer in prähistorischer und interdisziplinärer Perspektive. In E. Kaiser and W. Schier (eds), *Mobilität und Wissenstransfer in diachroner und interdisziplinärer Perspektive.* Topoi, Berlin Studies of the Ancient World 9, 1–10. Berlin and Boston, MA, de Gruyter.

Schlichtherle, H. (1989) Neue Fundstellen im Federseemoor bei Bad Buchau, Oggelshausen, Alleshausen und Seekirch, Kreis Biberach. *Archäologische Ausgrabungen Baden-Württemberg,* 57–62.

Schlichtherle, H. (2002) Die jungsteinzeitlichen Radfunde vom Federsee und ihre kulturgeschichtliche Bedeutung. In J. Köninger, M. Mainberger, H. Schlichtherle and M. U. Vosteen (eds), *Schleife, Schlitten, Rad und Wagen. Zur Frage früher Transportmittel nördlich der Alpen.* Hemmenhofener Skripte 3, 9–34. Freiburg, Janus.

Schlichtherle, H. (2004) Wagenfunde aus den Seeufersiedlungen im zirkumalpinen Raum. In M. Fansa and S. Burmeister (eds), *Rad und Wagen – Der Ursprung einer Innovation: Wagen im Vorderen Orient und Europa,* 295–314. Mainz, Philipp von Zabern.

Schlichtherle, H. (2010) Als die ersten Räder rollten… Räder der Jungsteinzeit aus dem Olzreuter Ried bei Bad Schussenried. *Denkmalpflege in Baden-Württemberg* 3, 140–144.

Sherratt, A. (1981) Plough and Pastoralism: Aspects of the Secondary Products Revolution. In I. Hodder, G. Isaac and N. Hammond (eds), *Pattern of the Past,* 261–306. Cambridge, Cambridge University Press.

Sherratt, A. (2004) Wagen, Pflug, Rind: Ihre Ausbreitung und Nutzung – Probleme der Quelleninterpretation. In M. Fansa and S. Burmeister (eds), *Rad und Wagen – Der Ursprung einer Innovation. Wagen im Vorderen Orient und Europa.* Archäologische Mitteilungen aus Nordwestdeutschland Beiheft 40, 409–428. Mainz, Philipp von Zabern.

Velušček, A., Čufar, K. and Zupančič, M. (2009) Prazgodovinsko leseno kolo z osjo s kolišča Stare gmajne na Ljubljanskem barju. Prehistoric Wooden Wheel with an Axle from the Pile-Dwelling Stare gmajne at the Ljubljansko barje. In A. Velušček (ed.), *Koliščarska naselbina Stare gmajne in njen čas. Ljubljansko barje v 2. polovici 4. tisočletja pr. Kr. Stare gmajne Pile-Dwelling Settlement and Its Era. The Ljubljansko barje in the 2nd Half of the 4th Millennium BC.* Opera Instituti Archaeologici Sloveniae 16, 197–222. Ljubljana, Inštitut za arheologijo ZRC SAZU.

Vosteen, M. (1996) *Unter die Räder gekommen. Untersuchungen zu Sherratts 'Secondary Products Revolution'.* Archäologische Berichte 7. Bonn, Holos.

Vosteen, M. U. (1999) *Urgeschichtliche Wagen in Mitteleuropa. Eine archäologische und religionswissenschaftliche Untersuchung neolithischer bis hallstattzeitlicher Befunde.* Freiburger Archäologische Studien 3. Rahden/Westf., Marie Leidorf.

Vosteen, M. U. (2002) Die fünffache Erfindung von Rad und Wagen. In J. Köninger, M. Mainberger, H. Schlichtherle and M. U. Vosteen (eds), *Schleife, Schlitten, Rad und Wagen. Zur Frage früher Transportmittel nördlich der Alpen.* Hemmenhofener Skripte 3, 143–148. Freiburg, Janus.

Woytowitsch, E. (1985) Die ersten Wagen der Schweiz: Die ältesten Europas. *Helvetia Archaeologica* 16, 2–45.

Chapter 10

Wheels of Change: The Polysemous Nature of Early Wheeled Vehicles in 3rd Millennium BCE Central and Northwest European Societies

Joseph Maran

For Igor Manzura in friendship on the occasion of his 60th birthday

Introduction

The present volume is based on the contention that if we want to shed new light on the various aspects of the Secondary Products Revolution and the introduction of bronze metallurgy in the Old World, then we will need a shift of perspective from the issues connected with origins and dispersal routes that have dominated research so far to the contextualization of technologies and their transformative effects on the lifeworlds of people. In this article, I shall be using early wheeled vehicles as an instructive example of the importance of such a shift in perspective. As is well known, starting at the middle of the 4th millennium BCE we have almost simultaneous evidence for wheeled vehicles in a vast zone ranging from northern Germany to southern Mesopotamia, while before that date evidence for acquaintance with this particular technology is lacking in almost all of the sub-regions of this geographical zone (Maran 2004a; 2004b; Schier 2015). This has provoked an intense debate about whether to interpret the apparently rapid adoption of wheeled vehicles as the result of a polycentric invention occurring independently in various areas or of a monocentric invention in which this technology arose in a specific region and then spread to other areas (Fansa and Burmeister 2004; Pétrequin 2006; Burmeister 2011; 2012). Based on our perception of wheeled vehicles as an important means of enhancing mobility and facilitating

tedious labour, it was taken for granted in these discussions that the usefulness of this innovation would be immediately obvious to people and therefore lead to a uniform appraisal of its practical advantages. But once we focus our attention on the significance of such vehicles in the various sub-regions of their appearance, we realise that the attempts to define potential areas of invention and mechanisms of diffusion have distracted us from one striking fact: starting with the earliest phases of their appearance, the perceptions and connotations associated with such vehicles seem to have differed markedly in the various zones in which they are encountered. This applies especially to aspects that were not primarily of an economic nature but were rather related to ideological constructs. In the following, I shall explore the polysemous character of wheeled vehicles in central and northwest Europe, which in cultural practice intimately and often intriguingly linked economic, religious, and social aspects.[1]

The nature of our sources for the earliest horizons of the use of wheeled vehicles between *c.* 3500 and 2500 BCE to the north of the Alps is extremely uneven. In the centuries between roughly 3500 and 3100 BCE, our knowledge of the existence of such vehicles derives almost exclusively from iconographic sources such as terracotta models or depictions on pottery and stone. However, when we move to the centuries between 3100 and 2500, iconographical representations become extremely rare in the zone north of the Alps. During these centuries our knowledge is largely based on numerous finds of wooden parts of actual vehicles (Hayen 1986; Burmeister 2004a; 2011). I have interpreted the relatively high number of iconographical attestations in the centuries from the middle of the 4th millennium BCE onwards as an indication that during this period such vehicles were still very rare and were accordingly regarded as particularly prestigious items (Maran 1998, 521; 2001, 741; 2004a, 277–278). Also, I have linked the ensuing decrease in the depictions of wheeled vehicles to the wider availability of this technology, which may have resulted in a decline of their symbolic significance (Maran 1998, 521). As this article shows, I now believe that the latter conclusion should be rejected and that the available evidence requires a different interpretation.

Both the deposition and discovery contexts and the construction of components of Neolithic wheeled vehicles display marked regional differences. In the northern regions of Europe, such objects stem from uninhabited parts of the wetlands and were discovered by chance during the extraction of peat. A number of such objects also came to light during excavations of bog trackways. The finds exclusively represent vehicles with single-piece disc wheels rotating around the axle. By contrast, in the Circum-Alpine zone such vehicle parts were encountered in excavations of wetland settlements, and they belong to vehicles with composite wheels and an axle rotating with the wheel under the cart or wagon.[2] Concerning the vehicle parts from Neolithic wetland settlements, there is clear research consensus that they should be interpreted in a profane way as parts either stored for future use or piled up somewhere as useless scrap wood (Vosteen 1999, 50–51). Characteristically, Helmut Schlichtherle (2002, 25; 2004, 297) has stated that such finds of wooden parts of vehicles in wetland settlements are 'not dissimilar to the rotting vehicle pool that we nowadays frequently encounter in the surroundings of agricultural estates' (my translation). Initial doubts about the validity of such a view arise when we ask ourselves (a) why such substantial vehicle parts should not have been used as firewood instead of being left to rot, and (b) why it is always specific parts of wheeled vehicles that are encountered in settlements, while other large agricultural tools, such as scratch ploughs (ards), sledges, pitchforks, shovels *etc.*, are underrepresented or do not figure at all in the archaeological record. As we will see, there are other reasons to challenge Schlichtherle's seemingly commonsensical interpretation.

In contrast to the profane interpretation proposed for the vehicle parts from Late and Final Neolithic wetland settlements, a ritual interpretation of such parts from northern central European and northwest European bogs is firmly established in research. In 1964, Johannes Diderik

Van der Waals was the first to outline certain patterns in the chronological distribution and deposition of wooden parts of wheeled vehicles in Dutch wetlands that to him clearly indicated that the finds were of a ritual nature (Van der Waals 1964, 47–50). He noted that there was a chronological cluster of objects dating from the final stages of the Neolithic, that only specific parts of the vehicles appear to have been selected for deposition, and that wheels were sometimes deposited in pairs although they did not necessarily belong to the same vehicle. In his study on the hoards of the Funnel Beaker and Single Grave Cultures Manfred Rech concurred with Van der Waals and proposed a cultic interpretation of the vehicle parts found in the bogs of northern Germany (Rech 1979). Markus Vosteen also argued in favor of a sacral character of some of the depositions of vehicle parts found in bogs, and he juxtaposed them with the profane nature of the occurrences of similar objects in wetland settlements (Vosteen 1999, 40–42, 159–160). But such views have not gone unchallenged, and the occurrences of wooden vehicle parts in the uninhabited wetlands of northern central and northwest Europe have also been explained in a profane way as damaged, discarded or lost objects. In 1972, Hajo Hayen refuted the possibility of a ritual interpretation for such vehicle parts (Hayen 1972). While he did not explicitly address Van der Waals' views, it is clear that Hayen rejected the latter's conclusions out of hand. He argued that archaeologists have to be very cautious before invoking cult as an explanation and that 'four wheels lying flat in the peat are not enough to justify a cultic interpretation' (Hayen 1972, 84: my translation). Hayen insists that to warrant such an interpretation, archaeological objects have to fulfill the following criteria: The objects' morphological traits and find associations have to be such that they can only be interpreted in a cultic way, the objects must be unsuited for practical purposes, or an analogous body of material must unequivocally point to a cultic interpretation. In Hayen's opinion, the vehicle parts found in bogs were either discarded because they were broken or, in the case of undamaged and even apparently unused objects, were parts that had been taken to the respective bog in the course of the production process for the wood to be 'watered' and then, for unknown reasons, left there (see also Rostholm 1977, 200).

In the hitherto most comprehensive assessment of the occurrences of vehicle parts in uninhabited wetlands, Stefan Burmeister (2004b, 334–335) acknowledged that certain indications seem to suggest a deposition of some of the finds in the course of religiously motivated practices. In most cases, however, he found a profane interpretation to be more likely. All in all, the ambivalent nature of many of the finds of wooden vehicle parts in uninhabited wetlands prompted him to adopt a cautious stance, concluding that the 'cultic character of such finds is not beyond all doubt' (Burmeister 2004b, 335).

Common deposition patterns of Neolithic vehicle parts in European wetlands

The controversy about the meaning to be attributed to the appearance of wooden parts of wheeled vehicles in central and northwest European wetlands bears a certain resemblance to the discussion of Bronze Age metal hoards in the same geographical zone, for which both sacral and profane interpretations have been advanced. The criteria defined by Hayen for refuting a cultic interpretation of vehicle parts found in uninhabited wetlands are based on the assumption that it should be possible to neatly assign objects either to a 'functional' or a 'cultic' context. In opposition to the approach chosen by previous research on depositions of bronze objects, David Fontijn (2001/2002, 20–22) has stated that we have to free ourselves from simplistic dichotomies and should rather look at each bronze hoard separately and investigate whether it displays possible deposition patterns that might enable us to infer ritual practices. Such a practice-oriented approach is indeed required, since in rituals various economic, political, social, and religious aspects are often so inextricably linked that clear distinction becomes impossible (Fontijn 2001/2002, 273–279; Wentink and Van Gijn 2008, 30). Van der Waals (1964, 44–46) was the first to stress the necessity of looking for common patterns underlying different deposition occurrences in an attempt to establish whether we are dealing with interrelated cases based on deliberate decisions to perform rituals or rather with cases of randomly discarded or lost objects. In using our archaeological sources to infer possible patterns of practice, what is of particular interest is the distinction between ritual practices and practices that follow from the routines of everyday life, such as cooking, tool-making, harvesting, hunting, the disposal of waste, or the storage of spare parts. Daily routines may also encompass ritualized elements, sometimes even of a religious character, but they differ from ritual practices in that the latter are marked off from daily life and are carried out at specific places or times. They channel emotion, are based on a cognitive script, and are meaningful to the participants, who see themselves as engaging in practices that are out of the ordinary (Tambiah 1985, 128; Snoek 2006, 11–13; Brosius *et al.* 2013, 13–15).

Returning to the interpretation by Van der Waals, it is remarkable that most of the patterns he describes also demonstrably apply to the vehicle parts from Neolithic wetland settlements that were still unknown when he published his ground-breaking contribution. The composite wheels with a rectangular axle hole that proved to be so typical of the Circum-Alpine zone were first recognized

in the 1970s on the basis of the finds from Zürich-Pressehaus and other sites along the shores of Lake Zurich (Ruoff 1978). This paved the way for the identification of wooden fragments of such wheels from earlier excavations (Woytowitsch 1985; Winiger 1987; Höneisen 1989a). The applicability of the patterns described by Van der Waals to the finds from wetland settlements suggests that at least some of the latter may also derive from out-of-the-ordinary practices. The fact that research has not yet considered the possibility of a ritual motivation for the deposition of wooden vehicle parts from settlements is likely to be rooted in two factors. I believe the first probable factor is the absence of find associations and/or find circumstances providing additional indications for an unequivocal ritual motivation of the deposited vehicle parts, which have usually been found embedded in an occupation layer without any associated objects. The second, and perhaps equally important, factor for ruling out ritual motives for the appearance of wooden vehicle parts in Late/Final Neolithic settlements is to my mind that in our scale of values, wood is a cheap raw material. Accordingly, we have automatically assumed that wooden components must have lost their importance as soon as they were detached from the vehicle (see also Burmeister 2003, 53).

Turning to those arguments suggesting a ritual deposition of wooden parts of wheeled vehicles, three patterns can be discerned that are common to many of those objects, irrespective of whether they were found in uninhabited wetlands or in wetland settlements. The first pattern is chronological distribution. Van der Waals himself recognized that most of the finds from the Netherlands date back to the 3rd millennium BCE. His attribution to the Single Grave Culture has been confirmed by additional finds (Drenth and Lanting 1997, 59; Fontijn 2011, 440 fig. 7) and probably also applies to most of the Neolithic wooden components of wheeled vehicles from bogs in northern Germany and southern Scandinavia (Burmeister 2004b, 328–330; Hesse 2011), though some of the finds are slightly older and belong to the late Funnel Beaker culture (Burmeister 2004b, 329; Johannsen and Laursen 2010, 16). If we now look at the wooden parts of wheeled vehicles found in wetland settlements in Slovenia, Switzerland, and southwestern Germany, the centuries between the end of the 4th and the middle of the 3rd millennium BCE again stand out as the period that all such finds date back to (Schlichtherle 2004; Burmeister 2011; Schier 2015, 108–111). This is particularly surprising when we recall that to my knowledge the numerous Swiss and south German Bronze Age wetland settlements from the centuries roughly between 1900 and 1500 BCE have not yielded a single example of a wheel or other parts of such vehicles (Schlichtherle 2004, 308). Since during these centuries wheeled vehicles must have been at least as common as 1000 years earlier, the question arises why the allegedly profane habit of storing or discarding waste wood in settlements was not also practiced in the Early and Middle Bronze Ages of central and northern Europe? Tellingly, such wooden vehicle parts only reappear frequently as of the Late and Final Bronze Ages (Hayen 1986, 114–123; Höneisen 1989b; Schlichtherle 2004, 308–312), during a period when practices associated with the ritual deposition of bronze objects both in settlements and outside them had reasserted themselves. Among these are the magnificent bronze spoked wheels of Hassloch and related wheels of the so-called Coulon group (Pare 1987, 49–55). Here no one has ever questioned the ritual nature of the deposition of parts of ceremonial wagons, simply because these wheels consist of metal and look much more impressive than the outwardly humble wooden wheels. Nevertheless, such bronze wheels may simply be a different manifestation of a practice that also encompassed wooden vehicle parts.

The second deposition pattern revolves around the choice of wooden components of such vehicles, since finds of wheels are clearly overrepresented in settlements and bogs. While among the Neolithic wheels from wetland settlements the composite type with rectangular axle hole predominates (Schlichtherle 2002; 2004; Schier 2015, 109–110), the disc wheels from the northern wetlands are all of the single-piece type with projecting nave and cylindrical axle hole. In the north, even rough-outs of such wheels dating to the Neolithic and Bronze Age are known, although they could never have been in use (Van der Waals 1964, 41, fig. 15, pl. I; Burmeister 2004b, 330, 334). Axle finds also come from settlements and from uninhabited wetlands (Burmeister 2004b; 2011; Schlichtherle 2004). By contrast, there are only a few examples of yokes (Winiger 1986, 108, fig. 20; Fansa and Schneider 1994, 32; Leuzinger 2002; Pétrequin *et al.* 2006, 374, fig. 13), and possible parts of the draught pole or the box of a cart or wagon have only rarely been discovered (Hayen 1990, 179–183; Fansa and Schneider 1994, 32; Schlichtherle 2015, 96, fig. 53.3–4). A predominance of wheels is to be expected among the wooden objects discovered by chance during peat-cutting because these are the parts of a vehicle most easily recognized by laypersons (Burmeister 2004b, 323). Strikingly, however, wheels and axles are also by far the most common vehicle components from excavated wetland settlements and bog trackways. For instance, along an approx. 400-metre segment of a Neolithic bog trackway dating between *c.* 2490 and 2160 BCE ('Pfahlweg XV [Le]') excavated by Hayen and Reinhard Schneider in the *Meerhusener Moor* in Lower Saxony, numerous fragments of axles and wheels as well as pieces of two draught poles and possibly of the box of a vehicle were found (Fig. 10.1) (see below; Fansa and Schneider 1994; Burmeister 2004b, 322–323; Both and Fansa 2011, 64–66). The predominance of specific parts of wheeled vehicles in wetland settlements and uninhabited wetlands contradicts the notion that the

objects were discarded as waste and points instead to selective deposition.

The third deposition pattern has to do with the combination of objects. Often pairs of wheels or even more than two wheels are encountered, either from the same or from different vehicles, sometimes combined with an axle (Burmeister 2004b; Schlichtherle 2004). This again points to a deliberate choice of objects for deposition and resembles the way in which bronze items of a certain shape and number were selected and combined in hoards of the central European Bronze Age.

Beside these three patterns common to finds from uninhabited wetlands and wetland settlements, regularities in the deposition sites can also be observed, although in this case it is necessary to distinguish between finds from uninhabited wetlands and from wetland settlements. The likelihood of the majority of wheel depositions in bogs resulting from specific ritual practices rather than mere scrap disposal is upheld by the fact that there appear not to be any similar concentrations of finds from other European wetlands, including those of the Circum-Alpine zone. In fact, such vehicle parts are restricted to uninhabited wetlands in a narrowly circumscribed and contiguous zone of Europe encompassing parts of the Netherlands, northern Germany and southern Scandinavia. Outside of this zone, I know of only two occurrences of wooden disc wheels, both found in bogs and both of unknown date, the first from the vicinity of Braniewo in north-eastern Poland (Schneider 1952, 14, fig. 2; Hayen 1986, 111 and n. 3),[3] the second from Aulendorf in Upper Swabia (Schneider 1952). To fully appreciate just how unusual this concentration of Neolithic vehicle parts in the aforementioned zone of northern Europe actually is, we need to recall that without the extraction of peat from wetlands the likelihood of discovering wooden vehicle parts is very low. But to my knowledge, the most important peat-producing regions in Europe – the vast wetlands of Belarus, Russia, the Baltic States and the Ukraine – do not appear to have yielded comparable find concentrations,[4] although the funerary record suggests a wide distribution of wheeled vehicles in eastern Europe at least from the early 3rd millennium BCE onwards (Belinskij and Kalmykov 2004; Gej 2004; Tureckij 2004; Kaiser 2007; 2010). To my knowledge, also no Neolithic vehicle parts have come to light in wetlands of the British Isles, in spite of their proximity to the northwestern and northern parts of continental Europe, where such objects frequently occur. In fact, the find-spots of vehicle parts from bogs of the Netherlands, northern Germany and Denmark are all situated within the distribution zone of the western Funnel Beaker and Single Grave cultures. All this clearly points to specific patterns of ritual deposition and militates against the 'commonsensical' interpretation of the finds as wooden parts discarded due to damage or accidentally left in the bog in the process of 'watering' wood. Interpretation of these

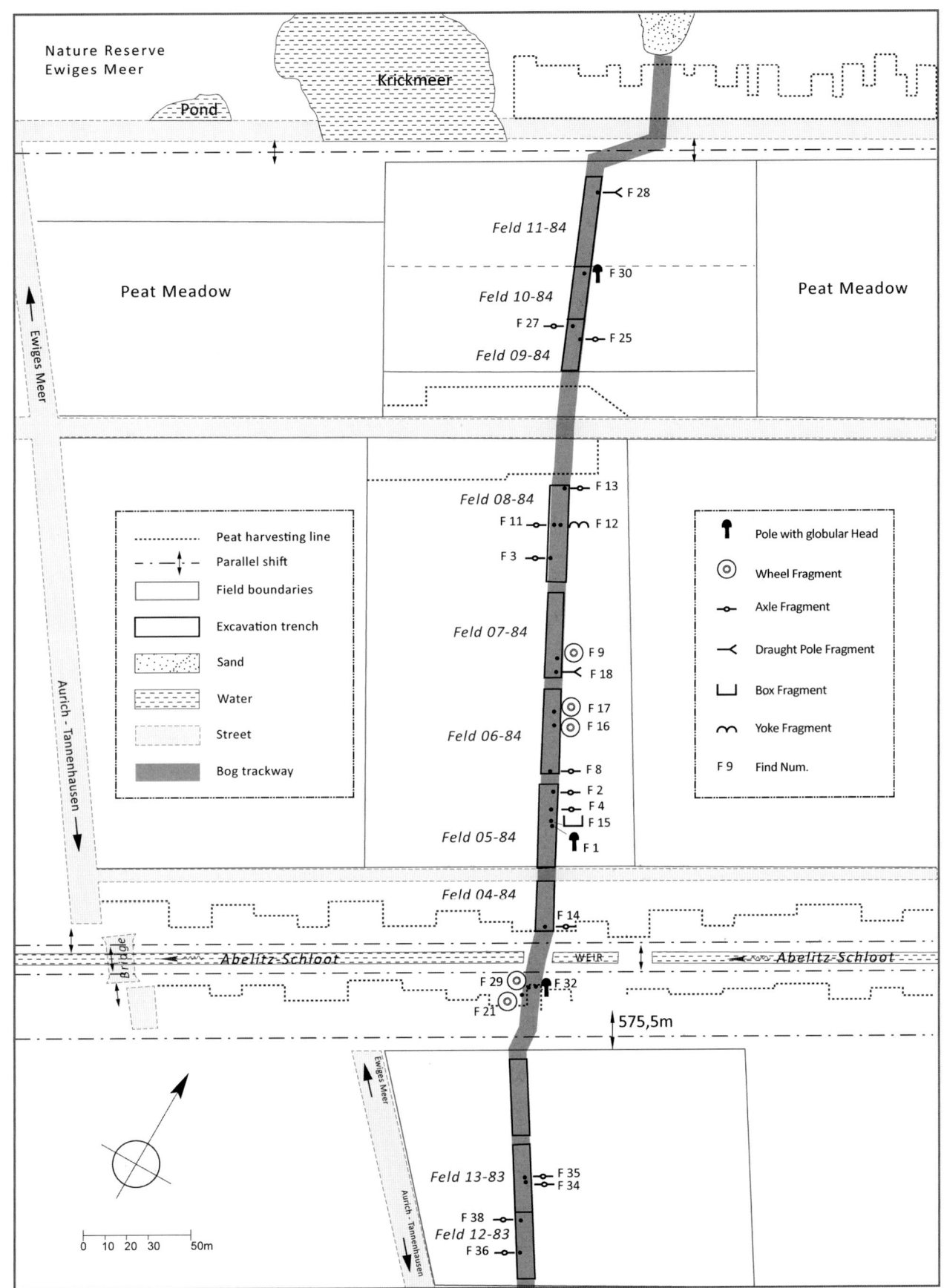

Figure 10.1: Meerhusener Moor, Lower Saxony. Plan of bog trackway XV (Le) with find-spots of parts of wheeled vehicles (redesigned after Fansa and Schneider 1994, fig. 11; graphics by M. Kostoula).

finds as a specific depositional practice is also suggested by Karsten Wentink's observation (2006, 105; fig. 11.2) that the find sites of wooden wheels in Dutch wetlands are often very close to places where stone axes of the Single Grave Culture have been discovered. Some of the wheels have been found deep inside bogs, a phenomenon interpreted some time ago by Rech (1979, 58) as indicating ritual deposition (but see Burmeister 2004b, 334 for a different interpretation).

Contextual evidence for the ritual significance of deposited Neolithic vehicle parts

Beside the general patterns we have just outlined, which suggest a ritual interpretation of at least many of the Late and Final Neolithic vehicle parts found in wetlands north of the Alps, there is additional contextual evidence corroborating such a conclusion.

A number of observations pertaining to the contexts of vehicle parts found along bog trackways contradict the standard interpretation as damaged components left in the bog after the repair of a vehicle. Although, ultimately, Burmeister (2003) also opted for the interpretation of vehicle parts from excavated bog trackways in Lower Saxony as discarded trash, he drew upon Hayen's detailed excavation documentation to distinguish distinct practices associated with the careful deposition of fragments of wheels and axles (Burmeister 2003, 52). His analysis shows that some pieces were evidently deposited beneath the trackway,[5] while others were rammed into the ground or even buried deep down. It was Van der Waals (1964, 47) who noted that broken vehicle parts found along trackways may also have been ritually deposited, just as damaged objects are a frequent constituent of bronze hoards. Indeed, the unusual patterns of practice described by Burmeister could very well point to deliberate ritual activity rather than the disposal of damaged objects.

Additional evidence contradicting the notion that practices of an exclusively profane nature were associated with Neolithic bog trackways is provided by wooden poles of a seemingly anthropomorphic shape found in 1984 at three different places associated with exactly the same segment of 'Pfahlweg XV (Le)' that has yielded so many fragments of axles and wheels (Fig. 10.1). The best-preserved pole (Fig. 10.2 [pole 1]), with an estimated length of roughly two metres and a width of about 11 cm, was found in excavation sector ('Feld') 05–84 (Figs 10.1, 10.5 [F1]). It had a round section and an offset globular head which, although it was discovered sticking head-down in the sediment, showed signs of heavy weathering (Fansa and Both 2011, 66, fig. 7). The special morphological features of the pole together with the traces of weathering on its upper part prompted Hayen to identify it as a 'cult figure' ('Kultfigur') or a 'cult pole' ('Kultpfahl') that had originally been standing upright beside the trackway (Hayen 1990, 183; Fansa and Both 2011, 66). According to Hayen, when the trackway ceased to be used, the pole was

removed from its original position and rammed head-down into the lane of the trackway for ritual concealment purposes (Fansa and Schneider 1994, 26–27, fig. 6; Fansa and Both 2011, 66).[6] The second pole (Fig. 10.3 [pole 2]), also with an offset globular head, has a preserved length of 36 cm and a width of 6.5 cm. It was found lying next to the trackway about 213 metres to the south of pole 1 in excavation sector 12–84 (Fig. 10.4 [F32]). The third pole (Fig. 10.6 [pole 3]) with a globular head was found about 126 metres to the north of pole 1 in excavation sector 10-84. It has a preserved length of 62 cm and a width of 7.5 cm.

Strangely, Hayen did not draw a connection between the 'cult poles' and the parts of vehicles found close by, probably because he had committed himself to an exclusively 'profane' interpretation of the latter. It is a fact, however,

Figure 10.2: Meerhusener Moor, Lower Saxony. Bog trackway XV (Le), excavation sector ('Feld') 05-84. Upper part of pole with globular head (F1) (courtesy Landesmuseum Natur und Mensch Oldenburg).

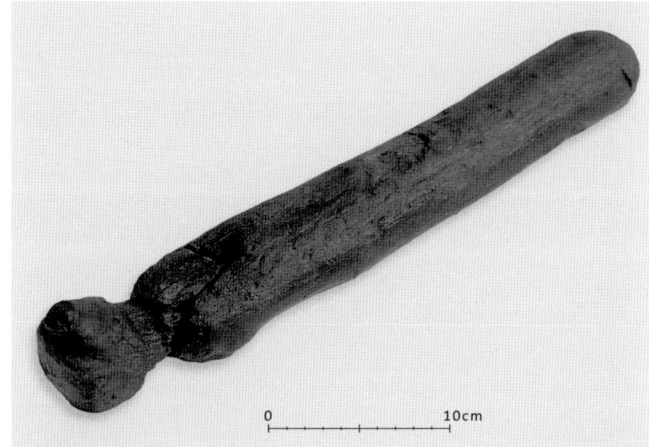

Figure 10.3: Meerhusener Moor, Lower Saxony. Bog trackway XV (Le), excavation sector ('Feld') 12-84. Pole with globular head (F32) (courtesy Landesmuseum Natur und Mensch Oldenburg).

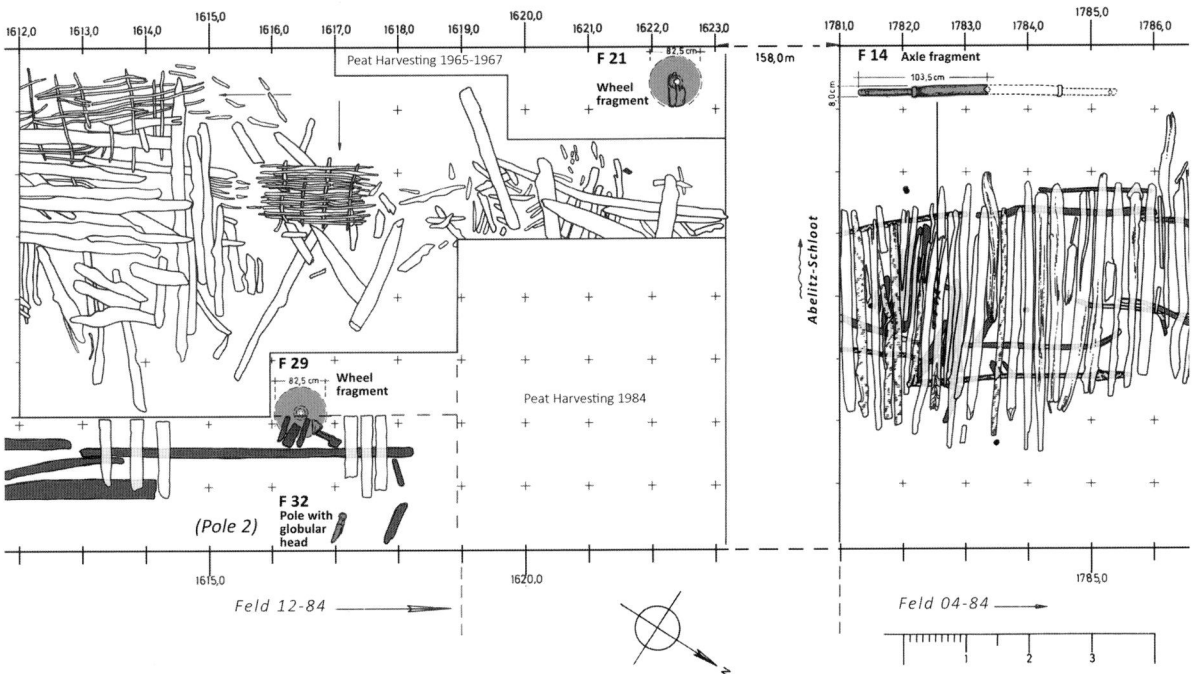

Figure 10.4: Meerhusener Moor, Lower Saxony, bog trackway XV (Le). Plan of excavation sector ('Feld') 12-84 and 04-84 with find-spots of parts of wheeled vehicles and fragment of pole with globular head. Scale approx. 1:130 (redesigned after Fansa and Schneider 1994, fig. 15; graphics by M. Kostoula).

that most of the vehicle parts found during the excavation of 'Pfahlweg XV (Le)' were discovered in the segments of the trackways situated between the three poles with globular heads, and some of the parts were encountered together with other objects in the immediate vicinity of such a pole (Figs 10.1, 10.4–10.7). At a distance of less than two metres to the north of the findspot of pole 1, a perforated plank from the bottom of the box of a cart or wagon was found (Fig. 10.5 [F15]; Hayen 1990, 183, fig. 9), and roughly 7 metres and 11 metres respectively to the north of it an axle preserved to about three-quarters of its length (F4) and a round stone with smooth surface (F6) (Fig. 10.5; Hayen 1990, 183). The missing piece (F2) of the axle F4 came to light at a distance of about 11.5 metres to the north-west of pole 1 (Fig. 10.5). At distances of *c.* 1.5 m and 8.5 m respectively to the north-west and west of the head of pole 2, two fragments of one and the same disc wheel were encountered (Fig. 10.4 [F29 and F21]), while 23 m and 29 m respectively to the south of pole 3 two axles preserved to about three-quarters of their length were found (Fig. 10.6 [F27 and F25]). In addition, a segment of a flint blade (F26) and a forked piece of worked wood (F33) came to light at distances of 19 m and 14 m to the south of pole 3 (Figs 10.6–10.7).

Even if we demur at Hayen's designation 'cult figures' as prematurely privileging a specific interpretation, the originally upright position of at least one of the wooden poles with a globular head, the allusion to an anthropomorphic body, and the concentrations of objects found in the vicinity make it likely

that the poles served to mark places of special significance that became focal points of ritual activity. One motive for depositing vehicle parts under, in, or along trackways may have been the desire to magically protect people traveling overland by cart or wagon against accidents in an environment that was regarded as particularly hazardous.[7]

Turning to the wooden parts of wheeled vehicles found in settlements, Markus Höneisen (1989a, 17–19) and Helmut Schlichtherle (2002, 25; 2004, 297) have observed that they were often found near landward rows of houses or close to fences or palisades. However, this find position does not necessarily favour an interpretation as a 'rotting vehicle pool'. These may have been areas that, as transitional zones from and to the outside world, were perceived as potentially threatened by outside forces and as liminal points were thus in need of a ritual protection. Future research on the positioning of such vehicle parts and possible associations with other objects may shed additional light on the interpretation of the individual case examples.

The ideological significance of wheeled vehicles in settlements of the first half of the 3rd millennium BCE is further corroborated by some recent finds. In 2009, a variety of wooden objects were encountered on a wooden floor in a Late Neolithic settlement dating to around 2900 BCE in the so-called 'Olzreuter Ried' in Upper Swabia. The finds comprise one entirely preserved wheel, fragments of three additional wheels, a fragment of an axle, and a miniature wooden wheel with a round axle hole (Schlichtherle 2010;

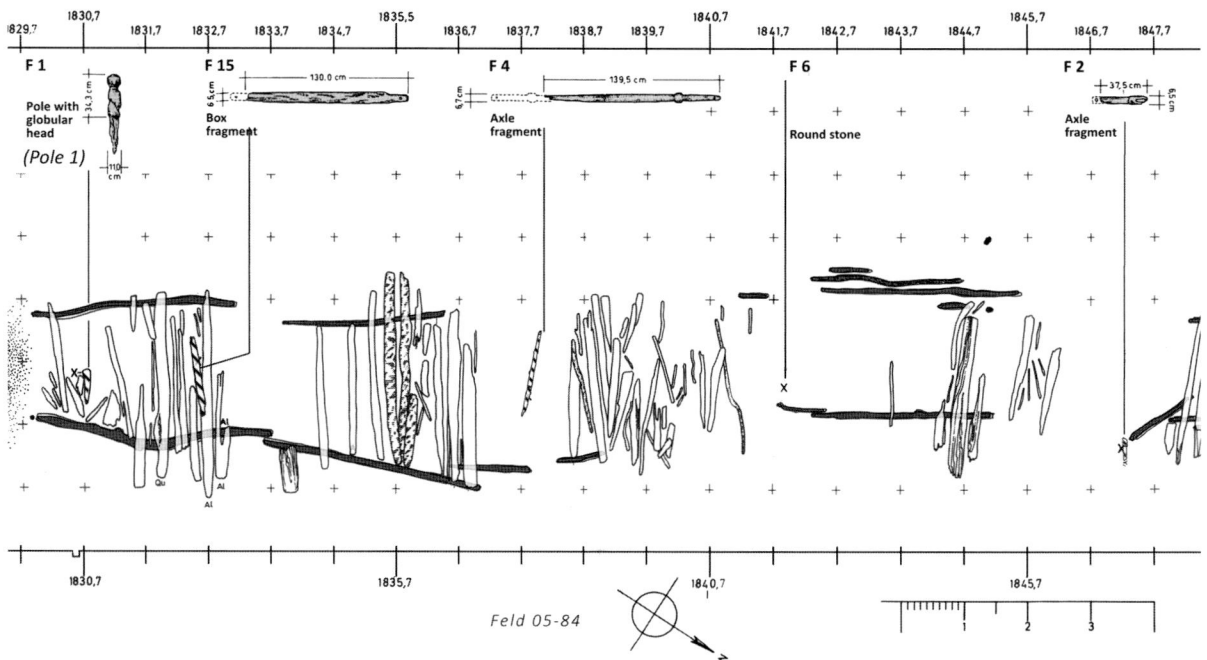

Figure 10.5: Meerhusener Moor, Lower Saxony, bog trackway XV (Le). Plan of excavation sector ('Feld') 05-84 with find-spots of parts of wheeled vehicles, fragment of pole with globular head and round stone. Scale approx. 1:130 (redesigned after Fansa and Schneider 1994, fig. 15; graphics by M. Kostoula).

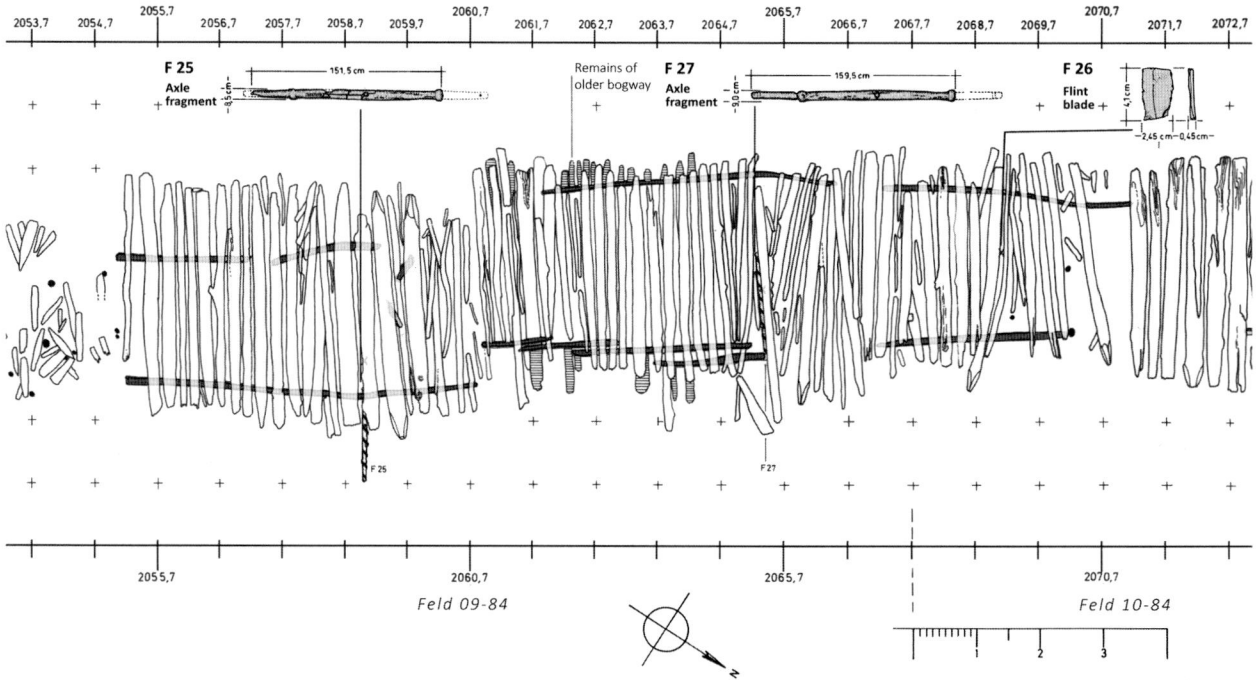

Figure 10.6: Meerhusener Moor, Lower Saxony, bog trackway XV (Le). Plan of excavation sector ('Feld') 09-84 and 10/84 with find spots of parts of wheeled vehicles and flint blade. Scale approx. 1:130 (redesigned after Fansa and Schneider 1994, fig. 18; graphics by M. Kostoula).

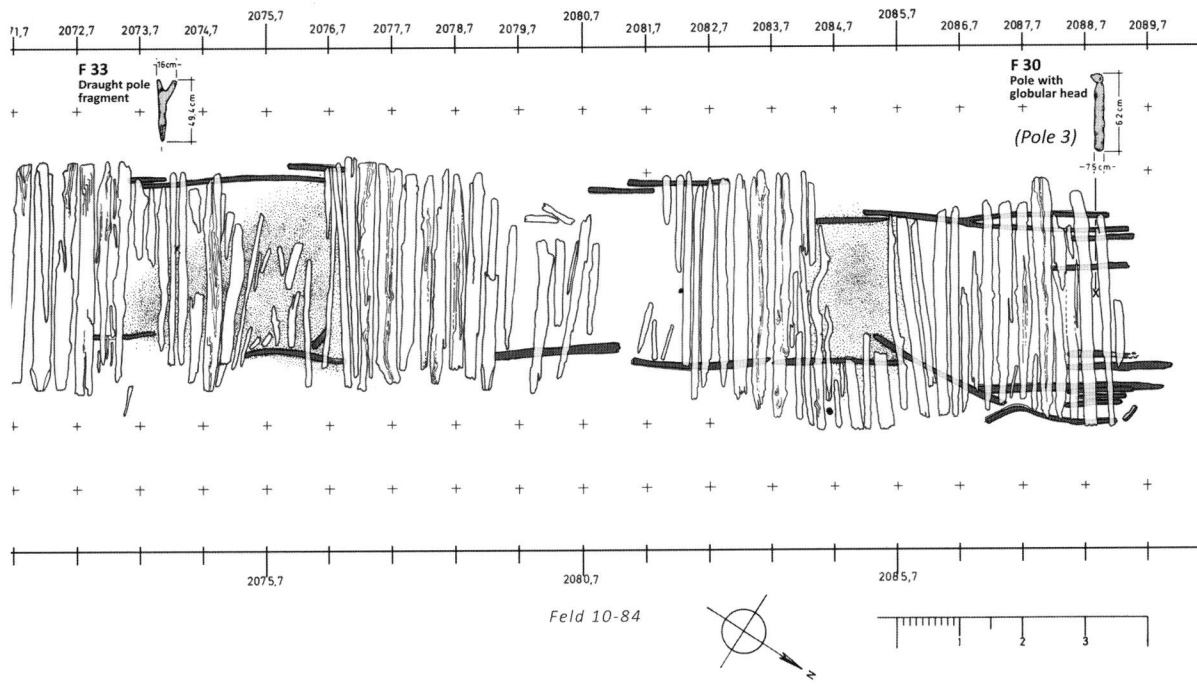

Figure 10.7: Meerhusener Moor, Lower Saxony, bog trackway XV (Le). Plan of excavation sector ('Feld') 10/84 with find spots of part of wheeled vehicle and fragment of pole with globular head. Scale approx. 1:130 (redesigned after Fansa and Schneider 1994, fig. 18; graphics by M. Kostoula).

2015; 2016). In 2015, three additional miniature wheels, two with a round, the other with a square axle hole came to light (Schlichtherle 2015; 2016). Based on traces of use-wear, Schlichtherle has convincingly attributed the miniature wheel found in 2009 to a cart model used for ceremonial purposes. Thus, irrespective of whether the deposition of the four regular-sized wheels derives from a practice of daily life or from a ritual, the find complex attests for the first time in regions north of the Alps the usage of models of wheeled vehicles with wooden wheels that up till now were known only from eastern central Europe (Schlichtherle 2010, 143–144; Bondár and Székely 2011; Bondár 2012, 23).

Excavations at the Final Neolithic settlement site of Wattendorf-Motzenstein in Upper Franconia have yielded a number of unusual clay objects, among them models of battle axes and pierced clay discs. Based on their size and shape, Timo Seregély (2004; 2008, 61–62) has convincingly proposed an interpretation as models of disc wheels rather than spindle whorls. The objects were discovered under find circumstances that suggest a ritual context, and the miniatures may have been intentionally broken as part of these rituals (Müller and Seregély 2008, 182). Evidently, in the Final Neolithic societies of southern Germany, battle-axes and vehicle wheels were the only objects of the lifeworld deemed important enough to be reproduced as clay miniatures.

Let us now turn to the funerary sector. In contrast to the situation in the Northern Pontic region with its cart burials

or the Jutland Peninsula with the stone heap graves of the late Funnel Beaker culture, there is no evidence that parts of wheeled vehicles or entire vehicles were deposited in Final Neolithic tombs of central and northwestern Europe. This is particularly noteworthy in the light of the hundreds of graves of the Single Grave and Corded Ware Cultures that are known in this part of Europe. If one further considers that copper battle-axes of the so-called Eschollbrücken type are never found in tombs but only as ritual depositions in wetlands (Maran 2008), the selective deposition of wooden wheels fits into an overall picture foregrounding the outstanding symbolic significance of certain items whose deposition was restricted to other ritual spheres than the funerary sector. Finally, the deposition of axes and wheeled vehicle parts during the Late and Final Neolithic displays great similarity with the choice of objects with economic and military functions described by Fontijn (2001/2002, 276) as typical for the Early Bronze Age.

Conclusions: The inextricable link between the sacred and the profane

To recapitulate: It seems that the decrease of iconographical attestations for wheeled vehicles in central Europe in the centuries after 3100 BCE does not imply that such vehicles had lost their symbolic significance, but rather that this significance was expressed in different ways than in the centuries before. The occurrences of wooden parts of wheeled vehicles from uninhabited wetlands in the centuries between *c.* 3100 and 2500 BCE are probably much more closely related to those from contemporary wetland settlements than was hitherto deemed possible. One reason why this linkage has not yet been recognized is the fact that the finds from the northern wetlands have rarely been discussed together with those from the Circum-Alpine zone. In addition, the perception of wood as an allegedly cheap raw material and a simplistic concept of neatly separable realms of the 'sacred' and the 'profane' prevented research from realizing that, in both cases, we are dealing with the result of out-of-the-ordinary practices. Although some of the occurrences of wooden vehicle parts may indeed derive from everyday practices, the aforementioned patterns interlinking the finds from uninhabited wetlands and wetland settlements point to the performance of special rituals in which specific parts of wheeled vehicles were dedicated as *pars pro toto* depositions both in settlements and outside them. On the current evidence, we can juxtapose two geographically distinct patterns of deposition. In the distribution area of the western Funnel Beaker and Single Grave Cultures, we find wooden vehicle parts deposited in uninhabited wetlands, where they either appear in association with trackways or as single finds without any associated building structures, but at least in some cases in the immediate vicinity of other dedicated objects. The pattern of depositing vehicle parts in Northern wetlands appears to survive in this zone after 2000 BCE and into the Nordic Bronze Age, although probably on a markedly reduced level (Burmeister 2004b, 330).[8] By contrast, in the Late and Final Neolithic of the northern and eastern Circum-Alpine zone we encounter depositions of wooden vehicle components in settlements only.[9] This deposition pattern seems to disappear at the end of the Corded Ware Culture. The considerable range of connotations that may have been ascribed to the vehicles and the differences in deposition sites suggest that the ritual practices were guided by a variety of ideas and intentions. The available evidence suggests that the magical protection of communities and people traveling overland against outside threats was one important motive for carrying out rituals in settlements and along trackways.

It is the aim of this article to show that the Late and Final Neolithic wooden wheels and other parts of wheeled vehicles found in uninhabited parts of northern Europe are closely related to those from Circum-Alpine wetland settlements. Both groups of finds are likely to provide evidence for the execution of ritual practices, in which specific parts of wheeled vehicles were dedicated as *pars pro toto* depositions both in settlements and outside them. While the deposition of such vehicle parts in wetland settlements of south-western Germany and Switzerland seems to be a Late and Final Neolithic phenomenon dating roughly to the centuries between 3100 and 2500 BCE, the practice of depositing such objects in bogs in the Netherlands,

northern Germany and Denmark is attested from the beginning of the 3rd millennium BCE and continued at least occasionally into the Bronze Age. During the Late and Final Neolithic, the practice of depositing wooden vehicle parts seems to have had a geographical focus in parts of central, north-western and northern Europe and was shared by communities that have been archaeologically assigned to a variety of different culture groups, such as Goldberg III and Horgen as well as the Funnel Beaker and Corded Ware/Single Grave Cultures. Accordingly, this practice does not seem to have been culture-specific but rather region-specific, and was ultimately based on ideological constructs that had emerged within societies living in these regions in the centuries after the first introduction of wheeled vehicles around 3500 BCE.

The example of early wheeled vehicles in central Europe emphasizes that a technology is always much more than that. The vehicles were polysemous in the sense that they were integrated into cultural practices uniting economic, ideological and social aspects, which gave them the potential to affect the ways in which societies imagined themselves. As a means for transport and communication they were important for the economy and especially for facilitating overland travel and time-consuming agricultural activities, such as harvesting and the procurement of building materials; as objects whose production and maintenance required specialist knowledge and considerable investment they were suited as markers of social distinction (Bogucki 1993; 1999, 227–230); and because of these properties they were readily associated with meanings bound up with fertility, affluence and movement that predestined them to be 'vehicles' of religious ideas. For these reasons, it is fair to expect that the rituals during which the vehicle parts were deposited were not only performed to communicate with supernatural powers, they were also guided by the intention of shaping and influencing the daily life, the economy, the social space and/or other aspects of the lifeworld of members of Neolithic communities that we would associate with the realm of the profane.

Notes

1 I would like to thank Maleen Leppek and Dr Stefan Burmeister for valuable information and helpful discussions. The director of the Landesmuseum Natur und Mensch Oldenburg, Dr Peter-Rene Becker, has kindly permitted me to use the photographs in Figures 10.2 and 10.3, for which I am most grateful. I am indebted to an anonymous reviewer of this article for important suggestions. Special thanks are due to Dipl.-Arch. Maria Kostoula for her expert preparation of the illustrations figuring in this article.

2 The wooden disc wheel from Aulendorf in Upper Swabia (Schneider 1952; Van der Waals 1964, 64, 75; Hayen 1996, 110–111) was destroyed during the Second World War and accordingly cannot be dated via radiocarbon analysis. If the

wheel originated from the Late or Final Neolithic, it would constitute the only evidence for the type of disc wheel rotating around the axle well known from the northern wetlands, but not otherwise attested in the Circum-Alpine zone, apart from the miniature wheels from the Olzreuter Ried (Schlichtherle 2015).

3   In the literature, the discovery site of this wheel is referred to with the pre-World War II German designation 'Schönsee, administrative district Braunsberg, East Prussia' (Gaerte 1929, 130, fig. 94; Schneider 1952; Burmeister 2004b, 323).

4   I would like to thank Maleen Leppek for information on wooden objects found in east European wetlands. For the distribution of global peat resources, see the website of the International Peat Society: http://www.peatsociety. org/peatlands-and-peat/global-peat-resources-country (last accessed 3 July 2017).

5   For similar depositions of objects under bog trackways, see Van der Waals 1964, 47–50.

6   I am grateful to Dr Stefan Burmeister for sending me an excerpt from the entry in Hajo Hayen's excavation diary describing the discovery of pole 1. Hayen refers to the extreme signs of weathering on the entire outer face of the globular head, which led him to conclude that the pole must have originally stood head-up on the trackway. In the same segment of the diary, Hayen also conjectures that the pole must have been ritually concealed ('rituelles Verbergen') by ramming it head-down into the middle of the trackway. In this way, the lower part of the pole was then also exposed to extreme weathering.

7   For medieval times the fears of supernatural beings threatening overland travelers are described in Dinzelbacher (1996) 13–14.

8   If Klavs Randsborg (1991; 2010) is right in interpreting the enigmatic Gallemose copper rods as parts of a yoke and a draught pole, we would even have evidence for the deposition of highly unusual metal vehicle parts in the north in the early 2nd millennium BCE.

9   It is unclear whether the disc wheel from Aulendorf, which may have dated to the Neolithic but was unfortunately destroyed during World War II, came from a wetland settlement or was found in an uninhabited bog.

References

Belinskij, A. B., and Kalmykov, A. A. (2004) Neue Wagenfunde aus Gräbern der Katakombengrab-Kultur im Steppengebiet des zentralen Vorkaukasus. In M. Fansa and S. Burmeister (eds), *Rad und Wagen – Der Ursprung einer Innovation: Wagen im Vorderen Orient und Europa*, 201–220. Mainz, Philipp von Zabern.

Bogucki, P. (1993) Animal Traction and Household Economies in Neolithic Europe. *Antiquity* 67, 492–503.

Bogucki, P. (1999) *The Origins of Human Society*. Oxford and Malden, MA, Blackwell.

Bondár, M. (2012) *Prehistoric Wagon Models in the Carpathian Basin (3500–1500 BC)*. Archaeolingua, Series Minor 32. Budapest, Archaeololingua.

Bondár, M., and Székely, G.V. (2011) A New Early Bronze Age Wagon Model from the Carpathian Basin. *World Archaeology* 43, 538–553.

Both, F., and Fansa, M. (2011) Die Moorwege im Weser-Ems-Gebiet. In M. Fansa and F. Both (eds), *'O, schaurig ist's, übers Moor zu gehen': 220 Jahre Moorarchäologie. Begleitschrift zur Ausstellung im Landesmuseum Natur und Mensch Oldenburg 2011*, 61–188. Mainz, Philipp von Zabern.

Brosius, C., Michaels, A., and Schrode, P. (2013) Ritualforschung heute – ein Überblick. In C. Brosius, A. Michaels, and P. Schrode (eds), *Ritual und Ritualdynamik*, 9–24. Göttingen, Vandenhoeck & Ruprecht.

Burmeister, S. (2003) 'Don't Litter' – Müll am steinzeitlichen Wegesrand. In M. Fansa and S. Wolfram (eds), *Begleitschrift zur Sonderausstellung 'Müll – Facetten von der Steinzeit bis zum Gelben Sack'*. Schriftenreihe des Landesmuseums für Natur und Mensch 27, 47–54. Mainz, Philipp von Zabern.

Burmeister, S. (2004a) Der Wagen im Neolithikum und in der Bronzezeit: Erfindung, Ausbreitung und Funktion der ersten Fahrzeuge. In M. Fansa and S. Burmeister (eds), *Rad und Wagen – Der Ursprung einer Innovation: Wagen im Vorderen Orient und Europa*, 13–40. Mainz, Philipp von Zabern.

Burmeister, S. (2004b) Neolithische und bronzezeitliche Moorfunde aus den Niederlanden, Nordwestdeutschland und Dänemark. In M. Fansa and S. Burmeister (eds), *Rad und Wagen – Der Ursprung einer Innovation: Wagen im Vorderen Orient und Europa*, 321–340. Mainz, Philipp von Zabern.

Burmeister, S. (2011) Innovationswege – Wege der Kommunikation. Erkenntnisprobleme am Beispiel des Wagens im 4. Jt. v. Chr. In S. Hansen and J. Müller (eds), *Sozialarchäologische Perspektiven: gesellschaftlicher Wandel 5000–1500 v. Chr. zwischen Atlantik und Kaukasus; Internationale Tagung 15.–18. Oktober 2007 in Kiel*. Archäologie in Eurasien 24, 211–240. Darmstadt, Philipp von Zabern.

Burmeister, S. (2012) Der Mensch lernt fahren – Zur Frühgeschichte des Wagens. *Mitteilungen der Anthropologischen Gesellschaft Wien* 142, 81–100.

Dinzelbacher, P. (1996) *Angst im Mittelalter: Teufels- Todes- und Gotteserfahrung, Mentalitätsgeschichte und Ikonographie*. Paderborn, Schöningh.

Drenth, E., and Lanting, A. E. (1997) On the Importance of the Ard and the Wheeled Vehicle for the Transition from the TRB West Group to the Single Grave Culture in the Netherlands. In P. Siemen (ed.), *Early Corded Ware Culture: The A-Horizon – Fiction Fact? International Symposium in Jutland, 2nd–7th May 1994*. Arkæologiske Rapporter Esbjerg Museum 2, 53–73. Esbjerg, Esbjerg Museum.

Fansa, M., and Burmeister, S. (eds) (2004) *Rad und Wagen – Der Ursprung einer Innovation: Wagen im Vorderen Orient und Europa*. Mainz, Philipp von Zabern.

Fansa, M., and Schneider, R. (1994) Steinzeitlicher Pfahlweg XV (Le) im Meerhusener Moor zwischen Aurich-Tannenhausen, Landkreis Aurich, im Südosten und dem Ewigen Meer, Landkreis Wittmund, im Nordwesten. *Archäologische Mitteilungen aus Nordwestdeutschland* 17, 15–37.

Fontijn, D. (2001–2002) Sacrificial Landscapes: Cultural Biographies of Persons, Objects and 'Natural' Places in the Bronze Age of the Southern Netherlands, *c.* 2300–600 BC. *Analecta Praehistorica Leidensia* 33/34, 1–392.

Fontijn, D. (2011) The 'Ritual' Fabric of Prehistoric Landscape: Funerary Places and Deposition Sites in the Low Countries, *c.* 5000–1500 cal BC. In S. Hansen and J. Müller (eds),

Sozialarchäologische Perspektiven: Gesellschaftlicher Wandel 5000–1500 v. Chr. zwischen Atlantik und Kaukasus. Internationale Tagung, 15.–18. Oktober 2007 in Kiel, 429–447. Archäologie in Eurasien 24. Mainz, Philipp von Zabern.

Gaerte, W. (1929) *Urgeschichte Ostpreußens*. Königsberg, Gräfe und Unzer.

Gej, A.N. (2004) Die Wagen der Novotitarovskaja-Kultur. In M. Fansa and S. Burmeister (eds), *Rad und Wagen – Der Ursprung einer Innovation: Wagen im Vorderen Orient und Europa*, 177–190. Mainz, Philipp von Zabern.

Hayen, H. (1972) Vier Scheibenräder aus dem Vehnemoor bei Glum (Gemeinde Wardenburg, Landkreis Oldenburg). *Die Kunde* 23, 62–86.

Hayen, H. (1986) Der Wagen in europäischer Frühzeit. In W. Treue (ed.), Achse, Rad und Wagen: Fünftausend Jahre Kultur- und Technikgeschichte, 109–138. Göttingen, Vandenhoeck & Ruprecht.

Hayen, H. (1990) Ein Vierradwagen des dritten Jahrtausends v. Chr. – Rekonstruktion und Nachbau. In M. Fansa (ed.), *Experimentelle Archäologie in Deutschland*. Archäologische Mitteilungen aus Nordwestdeutschland, Beiheft 4, 172–191.

Hesse, S. (2011) Ein neues Datum für ein altes Rad. Archäologische Funde von Rad- und Wagenteilen aus dem Teufelsmoor zwischen Gnarrenburg und Karlshöfen. *Rotenburger Schriften* 91, 235–244.

Höneisen, M. (1989a) Die jungsteinzeitlichen Räder der Schweiz: Die ältesten Europas. In B. A. Schüle, D. Studer and C. Oechslin (eds), *Das Rad in der Schweiz vom 3. Jt. v. Chr. bis um 1850*, 13–22. Zürich, Schweizerisches Landesmuseum.

Höneisen, M. (1989b) Die bronzezeitlichen Räder der Schweiz. In B. A. Schüle, D. Studer and C. Oechslin (eds), *Das Rad in der Schweiz vom 3. Jt. v. Chr. bis um 1850*, 23–30. Zürich, Schweizerisches Landesmuseum.

Johannsen, N., and Laursen, S. (2010) Routes and Wheeled Transport in Late 4th–Early 3rd Millennium Funerary Customs of the Jutland Peninsula: Regional Evidence and European Context. *Prähistorische Zeitschrift* 85, 15–58.

Kaiser, E. (2007) Wagenbestattungen des 3. vorchristlichen Jahrtausends in der osteuropäischen Steppe. In M. Blečić, M. Črešnar, B. Hänsel, A. Hellmuth, E. Kaiser and C. Metzner-Nebelsick (eds), *Scripta Praehistorica in honorem Biba Teržan*. Situla 44, 129–149. Ljubljana, Narodni muzej Slovenije.

Kaiser, E. (2010) Wurde das Rad zweimal erfunden? Zu den frühen Wagen in der eurasischen Steppe. *Prähistorische Zeitschrift* 85, 137–158.

Leuzinger, U. (2002) Das vermutete Joch von Arbon-Bleiche 3, Schweiz. In J. Köninger, M. Mainberger, H. Schlichtherle, and M. Vosteen (eds), *Schleife, Schlitten, Rad und Wagen: Zur Frage früher Transportmittel nördlich der Alpen*. Rundgespräch Hemmenhofen 10. Oktober 2001, 107–108. Freiburg, Janus-Verlag

Maran, J. (1998) Die Badener Kultur und der ägäisch-anatolische Bereich: Eine Neubewertung eines alten Forschungsproblems. *Germania* 76, 497–525

Maran, J. (2001) Zur Westausbreitung von Boleráz-Elementen in Mitteleuropa. In P. Roman (ed.), *Proceedings of the International Symposium 'Cernavoda III – Boleráz. Ein vorgeschichtliches Phänomen zwischen Oberrhein und der Unteren Donau', Mangalia 1999*, 733–746. Bucharest, Vavila Edinf.

Maran, J. (2004a) Die Badener Kultur und ihre Räderfahrzeuge. In M. Fansa and S. Burmeister (eds), *Rad und Wagen – Der Ursprung einer Innovation: Wagen im Vorderen Orient und Europa*, 265–282. Mainz, Philipp von Zabern.

Maran, J. (2004b) Kulturkontakte und Wege der Ausbreitung der Wagentechnologie im 4. Jahrtausend v. Chr. In M. Fansa and S. Burmeister (eds), *Rad und Wagen – Der Ursprung einer Innovation: Wagen im Vorderen Orient und Europa*, 429–442. Mainz, Philipp von Zabern.

Maran, J. (2008) Zur Zeitstellung und Deutung der Kupferäxte vom Typ Eschollbrücken. In F. Falkenstein, S. Schade-Lindig, and A. Zeeb-Lanz (eds), *Kumpf, Kalotte, Pfeilschaftglätter. Zwei Leben für die Archäologie. Gedenkschrift für Annemarie Häußer und Helmut Spatz*. Internationale Archäologie – Studia honoraria 27, 173–187. Rahden/Westf., Marie Leidorf.

Müller, J., and Seregély, T. (2008) Die schnurkeramische Siedlungsweise in Mitteleuropa. In J. Müller and T. Seregély (eds), *Endneolithische Siedlungsstrukturen in Oberfranken* II. *Wattendorf- Motzenstein: Eine schnurkeramische Siedlung auf der Nördlichen Frankenalb. Naturwissenschaftliche Ergebnisse und Rekonstruktion des schnurkeramischen Siedlungswesens in Mitteleuropa*. Universitätsforschungen zur prähistorischen Archäologie 155, 175–188. Bonn, Habelt.

Pare, C. F. E. (1987) Der Zeremonialwagen der Bronze- und Urnenfelderzeit: Seine Entstehung, Form und Verbreitung. In F. E. Barth (ed.), *Vierrädrige Wagen der Hallstattzeit: Untersuchungen zu Geschichte und Technik*. Römisch-Germanisches Zentralmuseum Mainz, Monographien 12, 25–67. Bonn, Habelt.

Pétrequin, P. (ed.) (2006) *Premiers chariots, premiers araires: La diffusion de la traction animale en Europe pendant les IVe et IIIe millénaires avant notre ère*. Monographie du Centre de Recherches Archéologiques 29. Paris, CNRS Éditions.

Pétrequin, P., Pétrequin, A.-M., and Bailly, M. (2006) Vues du Jura français: les premières tractions animals au Néolithique en Europe occidentale. In P. Pétrequin (ed.), *Premiers chariots, premiers araires: La diffusion de la traction animale en Europe pendant les IVe et IIIe millénaires avant notre ère*. Monographie du Centre de Recherches Archéologiques 29, 361–398. Paris, CNRS Éditions.

Randsborg, K. (1991) Gallemose: A Chariot from the Early Second Millennium BC in Denmark? *Acta Archaeologica* 62, 109–122.

Randsborg, K. (2010) Bronze Age Chariots: From Wheels and Yoke to Bridles, Goad and Double-Arm Knob. *Acta Archaeologica* 81, 251–269.

Rech, M. (1979) *Studien zu Depotfunden der Trichterbecher- und Einzelgrabkultur des Nordens*. Neumünster, Wachholtz.

Rostholm, A. H. (1977) Neolitiske skivehjul fra Kideris og Bjerregårde i Midtjylland. *Kuml*, 185–222

Ruoff, U. (1978) Die schnurkeramischen Räder von Zürich-Pressehaus. *Archäologisches Korrespondenzblatt* 8, 175–183.

Schier, W. (2015) Central and Eastern Europe. In C. Fowler, J. Harding, and D. Hofmann (eds), *The Oxford Handbook of Neolithic Europe*, 99–120. Oxford, Oxford University Press.

Schlichtherle, H. (2002) Die jungsteinzeitlichen Radfunde vom Federsee und ihre kulturgeschichtliche Bedeutung. In J. Köninger, M. Mainberger, H. Schlichtherle, and M. Vosteen (eds), *Schleife, Schlitten, Rad und Wagen: Zur Frage früher Transportmittel nördlich der Alpen*.

Rundgespräch Hemmenhofen 10. Oktober 2001, 9–34. Freiburg, Janus-Verlag.

Schlichtherle, H. (2004) Wagenfunde aus den Seeufersiedlungen im zirkumalpinen Raum. In M. Fansa and S. Burmeister (eds), *Rad und Wagen – Der Ursprung einer Innovation: Wagen im Vorderen Orient und Europa*, 295–314. Mainz, Philipp von Zabern.

Schlichtherle, H. (2010) Als die ersten Räder rollten … Räder der Jungsteinzeit aus dem Olzreuter Ried bei Bad Schussenried. *Denkmalpflege in Baden-Württemberg* 3, 140–144.

Schlichtherle, H. (2015) Zwei endneolithische Straßendörfer im Olzreuter Ried: Georadar, Pegelmessung, neue Funde zu Rad und Wagen. *Archäologische Ausgrabungen in Baden-Württemberg*, 94–97.

Schlichtherle, H. (2016) Im Olzreuter Ried – Räder, Räder, nochmal Räder. In H. Schlichtherle, M. Heumüller, F. Haack, and B. Theune-Großkopf (eds), *4.000 Jahre Pfahlbauten*, 411–413. Ostfildern, Thorbecke.

Schneider, G. (1952) Das vorgeschichtliche Wagenrad von Aulendorf. *Vorzeit am Bodensee*, 13–17.

Seregély, T. (2008) *Endneolithische Siedlungsstrukturen in Oberfranken I. Wattendorf-Motzenstein: Eine schnurkeramische Siedlung auf der Nördlichen Frankenalb. Studien zum dritten vorchristlichen Jahrtausend in Nordostbayern.* Universitätsforschungen zur prähistorischen Archäologie 154. Bonn, Habelt.

Seregély, T. (2004) Radmodell und Votivaxt: Außergewöhnliche Funde der Kultur mit Schnurkeramik von der Nördlichen Frankenalb. In M. Fansa and S. Burmeister (eds), *Rad und Wagen – Der Ursprung einer Innovation: Wagen im Vorderen Orient und Europa*, 315–320. Mainz, Philipp von Zabern.

Snoek, J. A. M. (2006) Defining 'Rituals'. In J. Kreinath, J. Snoek, and M. Stausberg (eds), *Theorizing Rituals: Issues, Topics, Approaches, Concepts*, 3–14. Leiden *et al.*, Brill.

Tureckij, M. A. (2004) Wagengräber der grubengrabzeitlichen Kulturen im Steppengebiet Osteuropas. In M. Fansa and S. Burmeister (eds), *Rad und Wagen – Der Ursprung einer Innovation: Wagen im Vorderen Orient und Europa*, 191–200. Mainz, Philipp von Zabern.

Tambiah, S. J. (1985) A Performative Approach to Ritual. In S. J. Tambiah (ed.), *Culture, Thought, and Social Action: An Anthropological Perspective*, 123–166. Cambridge, MA *et al.*, Harvard University Press.

Van der Waals, J. D. (1964) *Prehistoric Disc Wheels in the Netherlands*. Groningen, J.B. Wolters.

Vosteen, M. (1999) *Urgeschichtliche Wagen in Mitteleuropa: Eine archäologische und religionswissenschaftliche Untersuchung neolithischer bis hallstattzeitlicher Befunde*. Freiburger Archäologische Studien 3. Rahden/Westf., Marie Leidorf.

Wentink, K. (2006) *Ceci n'est pas une hache: Neolithic Depositions in the Northern Netherlands*. Leiden, Sidestone Press.

Wentink, K. and Van Gijn, A. (2008) Neolithic Depositions in the Northern Netherlands. In C. Hamon and B. Quilliec (eds), *Hoards from the Neolithic to the Metal Ages: Technical and Codified Practices. Session of the XIth Annual Meeting of the European Association of Archaeologists.* British Archaeological Reports, International Series 1758, 22–43. Oxford, Archaeopress.

Winiger, J. (1987) Das Spätneolithikum der Westschweiz auf Rädern. *Helvetia Archaeologica* 18, 78–109.

Woytowitsch, E. (1985) Die ersten Wagen der Schweiz: Die ältesten Europas. *Helvetia Archaeologica* 16, 2–45.

Chapter 11

Appropriating Draught Cattle Technology in Southern Scandinavia: Roles, Context and Consequences

Niels N. Johannsen

This paper discusses the introduction, roles and consequences of draught cattle technology in Neolithic communities of southern Scandinavia during the 4th and early 3rd millennia BCE. I start with an overall discussion of the significance of draught cattle technologies and then review the main lines of empirical evidence. I then discuss what this evidence tells us about the use of particular draught cattle technologies, including ard/scratch plough agriculture, wheeled transport and non-vehicular transport. The discussion identifies patterns of variation in both time and space, including general chronological developments and what seem to be more specific differences between various parts of the region. It is notable that even within a region of this relatively limited size, there appears to be some variation in the trajectories taken by the use of draught cattle technologies in different areas. This apparent variation may in part be related to differences between the landscapes inhabited by the regional groups in question; however, it may equally, at least to some extent, be the product of taphonomic and cultural factors affecting the composition and representativeness of the empirical record. Finally, I discuss how the evidence on draught cattle technology from southern Scandinavia ties in with a broader understanding of cultural developments in this region during the TRB period, 4000–2700 BCE, and what this tells us the about the link between technological developments and broader socioeconomic and ideological change in this case.

The appropriation of draught cattle technologies in Neolithic communities of southern Scandinavia during the 4th and early 3rd millennia BCE was embedded in, and contributed to, a series of significant transformations of prehistoric societies in this region. This paper gives an overview of the various lines of evidence providing information on this overall process during the *Trichterbecher* (TRB) period, 4000–2700 BCE, and on the use of particular draught cattle technologies (such as ard/scratch ploughing and wheeled transport). I then go on to discuss both general trends and differences in the way draught cattle technologies became significant in different parts of the region. This involves juxtaposing the evidence for draught cattle technologies with a wide range of other evidence. It also necessitates discussing the methodological challenges associated with comparing overall trends across different periods and different areas with uneven distributions of various data types – challenges that are potentially significant even when working on a region of relatively limited size that has been the site of relatively intense archaeological activity since the later 19th century. The paper concludes by considering the observations on technological change and its place in broader socioeconomic and ideological developments that we might feel justified in making on the basis of the southern Scandinavian material discussed. Before turning to the evidence from Neolithic southern Scandinavia, however, it will be useful to ask *why* draught cattle technology is worthy of consideration within a broader, contextually sensitive exploration of the appropriation and transformative potential of new technologies in past societies.

The significance of draught cattle technology

The draught utilisation of cattle is arguably the form of animal power utilisation (draught and carrying) that notwithstanding the success of horseback riding has had the biggest socioeconomic impact on world history – and it certainly ranks among the most significant of all secondary animal products. In many regions of Eurasia and in northern

Africa, draught cattle exploitation has been practised for millennia, whereas in other parts of the world it was introduced by European colonialism in the past few centuries. As recently as the early 1990s, estimates indicated that more than 200 million cattle were being utilised for traction on a world scale (Starkey 1991). Like most other major genres of technology, the introduction of cattle traction in a given cultural context not only moved the boundaries of human activity in relation to clearly delimited, practical tasks and purposes but also contributed much more broadly to shaping human socioeconomic strategies and lifeways. But while the historical success of draught cattle technology on a global scale is unquestionable, it seems important to outline the more concrete differences made by this technological genre in general and its more specific instantiations in particular. Before proceeding to do so, it may be useful to make the conceptual starting point of the discussion clear by providing a brief, general definition of the term 'technology' as I understand it and shall be using it here.

A technology may be defined as a specific, learned form of material activity or practice that most often involves extrasomatic artefacts, structures or media. Technologies are in general *purposive, i.e.* have an inbuilt directedness towards *something* (a more or less flexible but finite spectrum of objectives), and they always depend on a particular set of means necessary for their execution, including knowledge, corporeal skill and, usually, one or several external tools (Johannsen *et al.* 2014, 336). In the case of draught cattle technologies, the prerequisite set of corporeal skills and (in its basic sense) knowledge are of course distributed across human and bovine agents. As discussed further below, technologies involving cattle traction – like all other technologies, past or present – were and are situated in a general cultural system or matrix of activities and purposes, and in general they have to possess at least some degree of compatibility with a range of other, neighbouring elements of that whole system. The fact that virtually all technologies have more or less significant unintended consequences (to which we shall also return) does not detract from these observations concerning purposive fitness and systemic compatibility.

Several influential figures focusing on technology in 20th century archaeology and anthropology devoted special attention to draught cattle technology, including Gordon Childe, Leslie White and Andrew Sherratt. Childe's seminal 1936 volume *Man makes himself* was an account of Eurasian prehistory in which technological innovation in general played a central role. Among the most prominent technologies in Childe's synthesis were those based on animal traction, notably the employment of wheeled transport and (scratch) plough agriculture, which he perceived as important prerequisites for the accumulation of capital associated with the 'Urban Revolution' (Childe 1936, 137–142) – the emergence of the first Mesopotamian civilisations. But for Childe, these innovations were not just interesting as isolated

phenomena contributing to the change of specific prehistoric societies. In his eyes, they represented the crossing of the threshold to the long process of mechanisation that is still characteristic of human society today:

> Harnessing the strength of oxen or asses and the forces of the wind was man's first effective essay in making natural forces work for him. When he had succeeded, he found himself for the first time controlling and even directing continuous forces not supplied by his own muscles. He was on the right road to releasing his body from the more brutal forms of physical labour – the road that leads to the internal combustion engine and the electric motor, the steam hammer and the mechanical navvy. And at the same time he was learning new principles in mechanics and physics (Childe 1936, 137–138).

Childe thus had a distinct sense of the long-term causal unfolding and the repercussions of major technological changes, but even though his perspective transcended any specific historical setting, he also acknowledged the decisive role of social context in the implementation of new technology and focused to a high degree on the overarching links between technological change and broader cultural change (*e.g.* Childe 1936, 9–16).

Leslie White's intellectual framework was very different from Childe's but he still shared his basic assessment of the significance issue:

> A tremendous advantage of domestication over hunting lies in the continuous use of animals in the living form instead of the consumption of dead ones. Milk, eggs, and wool can be obtained again and again from animals without killing them. At certain levels of cultural development domestic animals may be used as forms of mechanical power, to pull sledges or travois, to carry burdens including human beings, to draw plows and carts [...] All the advantages that we have cited for domestication as compared with hunting can, however, be reduced to a single and simple statement: it is a means of producing more human need-serving goods and services per unit of human labor, and hence, per capita. Culture has advanced as a consequence of increase in the amount of energy harnessed per capita (White 1959, 46–47).

While White's framework was boldly deterministic, with a clear calorific focus, and though his general outlook was geared to universal cultural evolution, his sensitivity to the significance of what subsequently became known as 'secondary' animal products pointed in the direction of historical processes of a much more specific and culturally complicated nature. Outlining and understanding these historically specific processes was the task that Andrew Sherratt later set himself. Much like Childe, Sherratt suggested that:

> The plough increased production and made economic the cultivation of a range of poorer-quality soils; it thus resulted

in the colonisation of a wider area than had been possible under previous systems of cultivation. Both the ox-cart and the horse, as well as the pack-donkey, opened up new possibilities for bulk transport and reduced the friction of distance. They made economic a range of locations and settlement types, including systems with cities, which would otherwise have involved huge amounts of effort (Sherratt 1981, 262).

For Sherratt, such secondary animal exploitation explained not so much the similarities as the differences between the long-term cultural trajectories seen in different parts of the world.

> The contrast between the development of agriculture in the Old and New Worlds is an instructive one. In both, the domestication of a cereal crop allowed a massive increase in population, first in village communities and later in towns and cities. The major difference lay in the role of animal domesticates. [But] it is not without significance that the next threshold, that of industrialisation, was attained only in the Old World; for the employment of animal-power as the first stage in the successive harnessing of increasingly powerful sources of energy beyond that of human muscle was only possible where these animals were domestic, not wild. The critical difference between the utilisation of animals in the Old and New World lay less in their uses as a source of meat than in their emergent properties when other applications were explored – their 'secondary products' (Sherratt 1981, 261).

In the decades after Sherratt proposed his 'secondary products revolution', new chronological data started to chip away at the historical package he had originally interpreted as revolutionary. This led Sherratt to focus increasingly on the impact of draught cattle technology, specifically in the later prehistory of western Eurasia (Sherratt 2003; 2004). To some extent, Sherratt thus seemed to be moving in the direction suggested by Bogucki (1999, 228) of modifying his original long-term revolution to an 'animal traction revolution'.

These brief excerpts from three influential bodies of work display a striking similarity in the way animal traction technology is emphasised as a major innovation in human prehistory – despite the fact that the three perspectives indisputably evolved out of very different intellectual climates. If this congruence enhances plausibility, what conclusions can we draw from a current perspective? First of all, paired cattle traction probably *was* the first technology that provided Neolithic farmers with an external force combining the following two crucial attributes: 1) it far exceeded the capacity of the human body and 2) it could be controlled and directed in a different way from other external sources of energy (*e.g.* wind, fire or water). In this particular sense, as proposed by Childe and Sherratt, we may consider draught cattle technology to constitute the first reasonably broad and yet significantly potent mechanisation of the

practical realm in many prehistoric societies. As a relatively versatile force, draught cattle could be utilised in a range of economic activities, such as the tilling of agricultural soil, wheeled transportation of goods and people, threshing of crops using sledges and, not least, a wide spectrum of activities involving the movement of heavy objects across land, *e.g.* forest clearance and construction work (for dwellings, monuments *etc.*). We shall look more closely at evidence for these activities in the southern Scandinavian case below.

A second significant thing to note is that Childe, Sherratt and White all sensed the close connection between technological change and the broader societal context. Even if White in particular favoured a specific causal order (technology determines social organisation), all three attempted to grasp the relational position of technological change within sets of cultural prerequisites and consequences. Through the social roles and the status of the agents practising them, the economic and social domains they function within, and the two sets of material products that they respectively depend upon and produce, virtually all technologies are connected to large parts of the wider cultural matrix they are practised within (strictly speaking, of course, to all of it). This fact has engendered the proposition that technologies may usefully be conceptualised as networks, as indeed the editors of the present volume themselves suggest (Maran and Stockhammer in this volume). Even if the position of a given technology in the wider matrix is conceptualised in synchronous terms, it is no simple thing to grasp. But the factor that really makes understanding roles and consequences exceedingly complicated is that the interconnected causal context into which a given technology is introduced always changes dynamically over time (albeit at a varying pace), as does (potentially) the specific technology itself, since adjustments in both the wider context and the specific technology frequently occur as a result of their causal interplay. The full spectrum or extent of the relevant causal links is never transparent to the human agents actually contributing to this process. Accordingly, although technologies are deliberate human creations designed to serve particular needs, their transformative effects usually go far beyond those foreseen or intended. Many episodes of extensive socioeconomic reconfiguration are, essentially, triggered by a specific, apparently delimited instance of technological change that sets off minor or major 'cascades' of unintended change and deliberate adjustment throughout the wider cultural network or system (*cf.* Schiffer 2005). Despite the benefit of hindsight, the analytical complexity of dealing with such processes as archaeologists is proportionate to the strategic challenge faced by the prehistoric agents who simultaneously created and navigated them. Understanding the different ways in which these processes may unfold thus remains an important challenge to archaeology in particular and the social and human sciences in general, and addressing it requires the juxtaposition of case studies that expose (some of) the factors and the dynamics involved in concrete cultural contexts. Both in terms of its global historical significance and the variability of its roles and consequences, draught cattle technology constitutes one potentially informative technological genre on which to focus such work.

Evidence for draught cattle technology in Neolithic southern Scandinavia

Four main lines of more or less direct evidence tell us about the use of draught cattle technology in southern Scandinavia during the 4th and early 3rd millennia BCE:

1. ard marks (scratch plough agriculture)
2. wooden disc wheels and preserved tracks (wheeled transport)
3. funerary structures and their alignments (wheeled transport)
4. osteomorphological data (unspecified draught exploitation)

The first three of these refer to two specific genres of draught cattle technology, while the fourth type of data provides direct but non-specific evidence for draught exploitation of cattle in that it says nothing about the concrete activities in which these animals were employed. This last category is important because, as discussed below, significant areas of specific draught application may well be undiscernible in the archaeological record but nonetheless reflected in this non-specific line of evidence.

Ard marks

Ard marks dating back to the TRB period, 4000–2700 BCE, have been found in more than 50 instances distributed across practically all of southern Scandinavia (Fig. 11.1). Almost all of these marks have been found under barrows from the Middle TRB, *c.* 3500–3100 BCE and have been dated by reliable *terminus ante quem* relationships to the funerary structures covering them (Thrane 1991; Beck 2013; Mischka 2014). However, recent results have indicated that the earliest uses of the ard or scratch plough probably predated this period somewhat. Ard marks under the megalithic long barrow LA4 at Flintbek in Schleswig-Holstein appear to have been made before 3560 BCE (calibrated ^{14}C spectrum, 1 sigma: 3636–3565 BCE, Mischka 2014, 129), and an even earlier application is indicated by marks preserved under a non-megalithic long barrow at Højensvej 7 on the island of Funen (Fig. 11.2), which appear to have been made before 3640 BC (calibrated ^{14}C spectrum, 1 sigma: 3705–3645 BCE, Beck 2013, 102). There has been some debate as to the significance of these marks, some scholars suggesting that they do not reflect agriculture as such but ritual preparation of the ground on which barrows were to be constructed (*e.g.* Rowley-Conwy

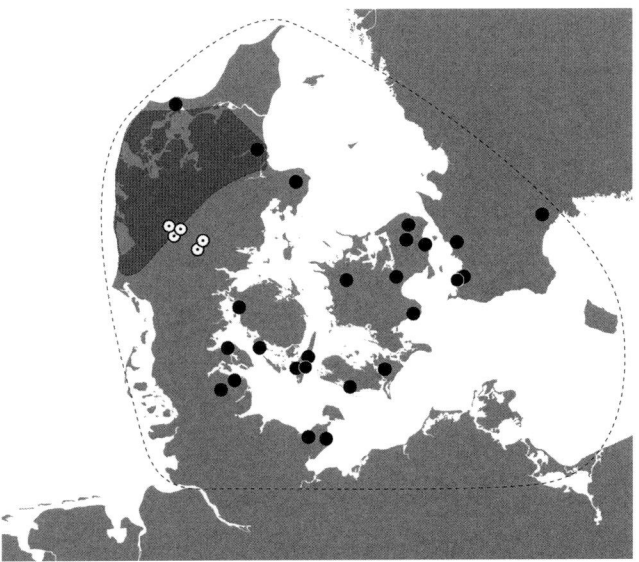

Figure 11.1: Map showing the distributions of different types of evidence for draught cattle utilization in southern Scandinavia during the TRB period, 4000–2700 BCE. The area in focus, referred to here as southern Scandinavia, is delimited approximately by the punctuated line. This line simultaneously indicates the distribution of ard marks preserved under barrows in the region (after Beck 2013; Mischka 2014; Thrane 1991 and the Danish Sites and Monuments Registry), since these finds are distributed across the whole region. The black dots indicate the distribution of sites yielding the bone assemblages on which the osteomorphological analyses of domestic cattle bones discussed in the text were based (after Johannsen 2006). The white and black icons clustering in central Jutland mark the sites where wooden disc wheels dating to the early 3rd millennium have been found (after Rostholm 1978 and the Danish Sites and Monuments Registry). The shaded area indicates the distribution area of the late TRB stone heap graves (after Johannsen et al. 2016).

1987). But while in these societies there were probably ritual aspects to ploughing (and most other activities), the evidence in favour of seeing the ard marks as reflections of agricultural tilling that was (also) economically motivated seems overwhelming. First of all, experimental ploughing with an animal-drawn ard has produced identical marks in the subsoil (Hansen 1969), and given that the patterns of furrows observed a) reflect an even application of substantial power for extended periods of time and b) are found on all soil types, including heavier soils (*cf.* Thrane 1991, 115), the instruments used in these Neolithic contexts were in all likelihood drawn by cattle. Furthermore, the criss-cross patterns in which ard marks are mostly found correspond very accurately to ploughing strategies commonly observed ethnographically in societies using ards/scratch ploughs today (*e.g.* Fig. 11.3), and similar strategies are described in some detail by several Roman authors, including Columella, Pliny, Varro and Virgil (Nielsen 1993, 115–118).

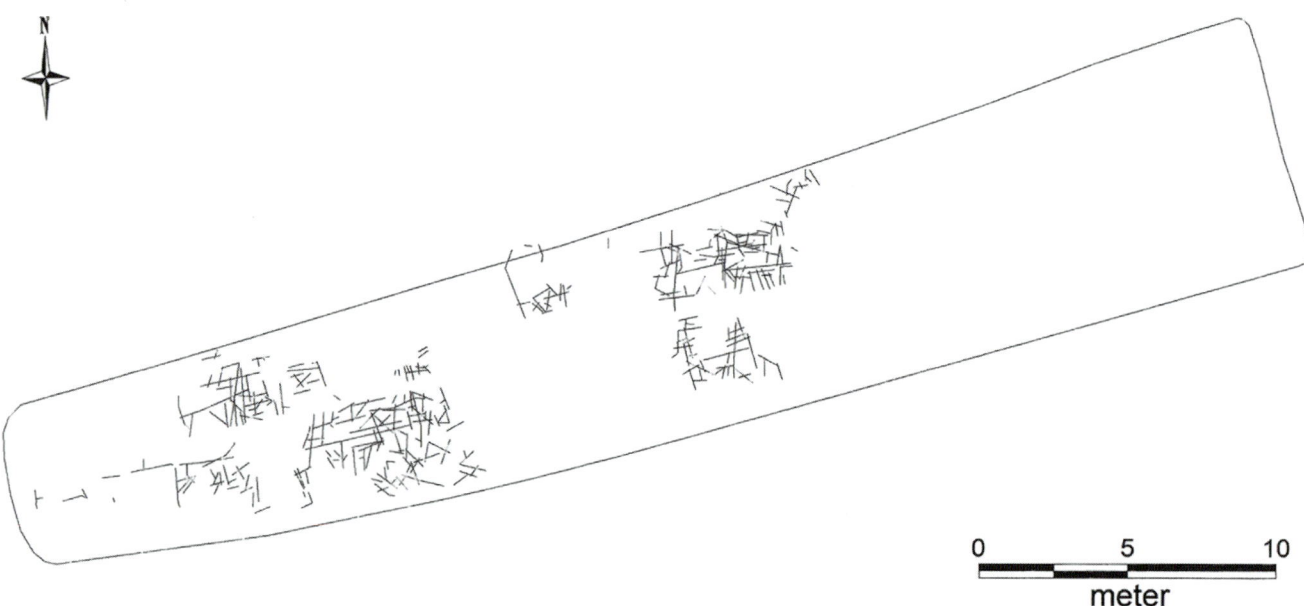

N

0   5   10
meter

Figure 11.2: Ard marks preserved under the long barrow Højensvej 7 on the island of Funen, Denmark (after Beck 2013, 70).

Figure 11.3: Peruvian farmer ploughing with a scratch plough drawn by a team of oxen near the shores of Lake Titicaca, 2002 (photo by the author).

Disc wheels and preserved tracks

The presumed cart or wagon tracks preserved under a long barrow at Flintbek at the base of the Jutland Peninsula constitute the earliest piece of evidence for wheeled transport in southern Scandinavia, indicating that such transport was already practised in the region around 3400 BCE (Bayesian modelling of 20 AMS results has produced a calibrated ^{14}C spectrum of 3420–3385 BCE, Mischka 2011, 750–755). However, as this data point is the only one predating 3100 BCE in southern Scandinavia, it remains difficult at present to evaluate the prevalence or significance of this technology in the region during the second half of the 4th millennium BCE. After 3100 BCE, the evidence for wheeled transport in this region becomes much more substantial. In several instances, tracks similar to those at Flintbek have been found in close proximity to

stone heap graves from the period 3100–2750, and though these tracks lack a correspondingly reliable *terminus* date, they are likely to reflect traffic contemporary with these graves (Johannsen and Egeberg forthcoming; Johannsen and Laursen 2010, 40–41). More unambiguous are the finds of wheeled vehicle parts, in all cases disc wheels, which appear during this period. Five wooden, single-piece disc wheels, all found in the central part of the Jutland Peninsula (Fig. 11.1; 11.4) have been dated to the early 3rd millennium, but due to the well-known plateau in the radiocarbon calibration curve for this period, it has long been difficult to narrow down the chronology of these finds more closely than 2950–2600 BCE (Rostholm 1978; 2007). More recently, however, two of these finds, both fragmented wheels from the Pilkmose site, have been dated dendrochronologically to 2935 BCE (wheel 1) and 2775 BCE (wheel 2), respectively (Christensen 2007, 231). Like ^{14}C-dates on wheels and axles found in the neighbouring Weser-Ems region (Burmeister 2004, 329), the more precise date of the oldest wheel from Pilkmose in particular provides clear evidence for the use of wheeled vehicles in this region during the Late TRB, 3100–2700 BCE. While it is possible that the deposition of these wheels in the bogs that preserved them was ritually motivated (*cf.* Maran this volume), extensive wear and, in one case, even repair of these wheels seem to indicate that they had (also) been part of vehicles that were put to extensive practical use.

Funerary structures and their alignments

One particular type of funerary structure provides indirect, but nonetheless significant, information on wheeled transport in Late TRB southern Scandinavia. Around 250 so-called stone heap graves have been found across 50 different sites located exclusively in the north-western part of the Jutland Peninsula (Fig. 11.1). These graves were constructed during the period 3100–2750 and appear to have been vehicle graves in which single individuals were interred either in a fully functional or more symbolic vehicle placed in a more or less rectangular structure and equipped with a full draught team of two oxen. The animals were placed with their bodies in the two associated oblong pits with their heads resting high on the adjacent surface, facing away from the quadrangular

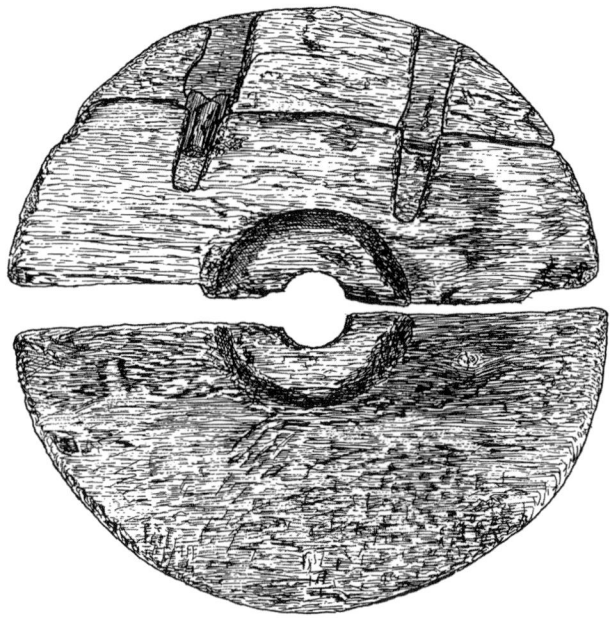

Figure 11.4: Drawing of one of the two disc wheels found at Kideris in 1940 (wheel 1). The wheel was made of oak and today measures 73.5 cm in diameter (minor post-recovery shrinkage has probably occurred). While the wheel was broken in two through the nave when found, an earlier breakage – visible in the upper part of the drawing – has been repaired by its users through the addition of two wooden wedges (after Rostholm 1978, 188).

structure, as if ready to pull the vehicle behind them. All of these features were subsequently covered by a large heap of stones (Fig. 11.5; Johannsen and Laursen 2010; Johannsen and Kieldsen 2014; Johannsen *et al.* 2016). The clear thematic focus of these graves is stressed by their alignment in small linear groups that in some places accumulated into large, linear cemeteries (Jørgensen 1977; Fabricius 1996) following the corridors of (vehicular) transport used in the area in question (Johannsen and Laursen 2010). In conjunction with the direct evidence provided by finds of vehicle parts and tracks, this more indirect line of evidence suggests that the use of wheeled transport was sufficiently extensive in these communities to be reflected in their way of thinking about a person's ultimate transition – or journey – from the state of being 'alive' to that of being 'dead'.

Osteomorphological data

To supplement the lines of evidence discussed above, which concern two specific genres of draught cattle technology, an effort has been made to develop osteological methods for identifying draught cattle utilisation independently of such specific applications and thus potentially facilitating a more precise understanding of the role and impact of draught cattle technology in general. Ethnographically and historically, draught cattle are known also to have been important in several other activities than soil processing and wheeled transport, but the only element involved in all applications is, of course, the draught animals themselves. In some cases morphological changes in the skeletons provide indications of work utilisation notably extensions of the articular surfaces found in the proximal (1st) phalanx and in the metapodia of the animals. The conclusion that these morphological changes probably reflect draught utilisation rests on the presence of such changes in the distal limb bones of modern draught cattle with known life/work histories and on their absence in the same bones of wild cattle (aurochs) and smaller samples of non-draught domestic cattle, plus more general knowledge about the response of mammalian

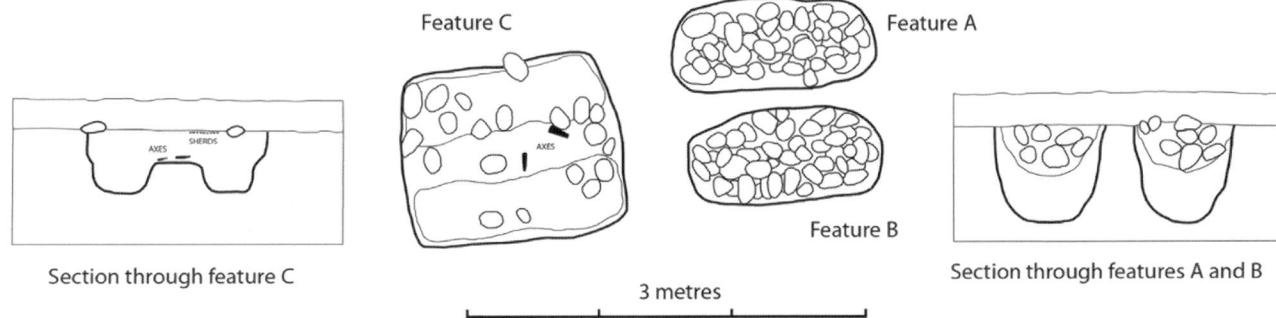

Figure 11.5: Stone heap grave at Vroue Hede II in plan and section (structure IX). The finds have been plotted on to the section through the quadrangular feature, C (after Jørgensen 1993, 113).

skeletal tissue to very specific or biologically abnormal activity patterns (see Bartosiewicz *et al.* 1997; Johannsen 2005; 2006 for extensive methodological discussion).

1033 distal limb bones of domestic cattle, reflecting a minimum of 122 individual cattle, have been studied from the TRB period in southern Scandinavia. As shown in Fig. 11.1, almost all this material was recovered in the eastern/south-eastern parts of the region where soil conditions are for the most part more favourable for the preservation of organic material than further west. The oldest morphological evidence for draught cattle utilisation in this sample was found in the metapodials of an individual ox skeleton dated to the mid-4th millennium BCE (calibrated ^{14}C spectrum, 2 sigma: 3650–3360 BCE, Johannsen 2006, 43). However, it is not until after 3300 BCE that such evidence becomes prevalent (Johannsen 2006, 40–43). Looking at the five largest assemblages from the later centuries of the TRB, a relatively uniform pattern emerges (Tab. 11.1). The combined distributions of pronounced and extreme extensions or 'lipping' (stages 3 and 4) of the proximal articular surface of the proximal/1st phalanx, which are likely to reflect draught exploitation, are very similar across these five assemblages. Despite the fact that one site, Hindby Mosse (5.8%), diverges somewhat from the values found in the other four assemblages (8.0–10.9%), the overall similarity in the occurrence of such changes in the five assemblages is statistically significant (Chi (p) = 0.626), and this points to a certain regularity across different sites in the way that cattle were managed and used.

When interpreting these figures, several methodological limitations must be taken into account. Only 50 bones (reflecting a minimum of nine individual cattle) out of the total sample of 1033 bones predate 3300 BCE, which means that the small sample size for the period predating 3300 BCE may in part explain why symptoms of draught

Table 11.1: Incidence of proximal lipping of the proximal/1st phalanx of domestic cattle in southern Scandinavian TRB assemblages where more than 30 specimens could be examined for this type of change. In the top row, site names are followed by approximate date of the assemblage and sample size. The bottom row shows the total percentage of proximal phalanges with pronounced or extreme lipping indicative of draught exploitation (stages 3 and 4; after Johannsen 2006, 40)

Proximal lipping of the proximal phalanx	Troldebjerg 3000–3200 BCE (N = 174)	Bundsø c. 3000 BCE (N = 55)	Hindby mosse c. 3000 BCE (N = 171)	Lindø c. 3000 BCE (N = 50)	Spodsbjerg 2900–2800 BCE (N = 37)
Stage 1	61.5%	60.0%	72.5%	80.0%	67.6%
Stage 2	28.7%	29.1%	21.6%	12.0%	21.6%
Stage 3	9.2%	10.9%	3.5%	8.0%	10.8%
Stage 4	0.6%	–	2.3%	–	–
Stages 3 & 4	*9.8%*	*10.9%*	*5.8%*	*8.0%*	*10.8%*

utilisation have only been found in one case between 4000 and 3300 BCE (Kærup). With regard to the prevalence of such symptoms in the five largest assemblages, all postdating 3300 BCE, it is important to note two things: 1) only regular draught exploitation for longer periods of time can provoke these relatively subtle skeletal adaptations, and this implies that animals used for a relatively short period and perhaps slaughtered young, are unlikely to be identified as draught animals; and 2) older animals, such as draught animals utilised for several years (and animals kept for other secondary products, or for reproduction), are in general underrepresented in archaeological assemblages because the turnover of animals killed when younger is faster (Johannsen 2006, 41, fig. 4). Accordingly, the figures in Table 11.1 are indications of the probable *minimum* ratio of draught animals in the cattle herds of these Neolithic communities, and actual ratios in the live herds are likely to have been somewhat higher. What the figures demonstrate with some reliability is that no later than 3300 BCE animals used extensively for traction seem to have been a standard component of most cattle herds in this region, at least in its eastern parts, from which almost all this osteological evidence is derived.

Summary and discussion

Four lines of evidence for the use of draught cattle technology during the TRB period (4000–2700 BCE) in southern Scandinavia have been reviewed above. Osteomorphological evidence indicates that draught utilisation of cattle was common by 3300 at the latest but says nothing about the specific applications of this tractive force. Strictly speaking, however, this evidence tells us almost exclusively about the situation in the eastern/south-eastern parts of the region, and it is uncertain whether projecting this conclusion onto the western/north-western part of the region is legitimate or not, given that the landscape differs to some extent (see below). In contrast, the evidence for ard/scratch plough agriculture is distributed across the whole region, and the earliest ard marks suggest that, at least locally, the presence of draught cattle technology in this region slightly predates the mid-4th millennium BCE. But while the two earliest data points lie between 3700 and 3500 BCE, the remaining body of this substantial line of evidence comes from the centuries after 3500 BCE. This indicates that ard agriculture was a significant part of the socioeconomic strategies pursued across this region from the mid-4th millennium BCE onwards, and it also suggests that the very limited osteological evidence for draught utilisation of cattle predating 3300 BCE is indeed the result of the small sample size for the period 4000–3300 BCE.

All the earliest evidence for wheeled transport has been found on the Jutland Peninsula. The Flintbek tracks, which reflect an exceptionally fortuitous situation of preservation, provide the earliest data point around 3400 BCE – but all other evidence for this genre of draught cattle technology in the TRB period is derived from its late phase 3100–2700. And strikingly, all of the latter comes from a limited area covering a north-western and central part of the peninsula. If this geographical concentration is taken to reflect that wheeled transport was particularly important in north-western Jutland (by comparison to other parts of southern Scandinavia) there is a potential circularity in the claim, particularly as the funerary evidence, which is quantitatively quite substantial, clearly reflects a deliberate symbolic focus on this technology – a focus that may have been absent in other parts of the region for many different reasons. On the other hand, while it is hard to draw negative conclusions regarding those parts of the region where these lines of evidence are absent, drawing positive conclusions for the area in which they have been found still seems justified. In particular, it is difficult to imagine that the stone heap grave custom would have come into being in a cultural context where wheeled transport was insignificant or culturally marginal. And given that at the onset of the Neolithic already, this was, by intraregional comparison, a very flat, less densely forested area (see discussion below), the natural preconditions in this area would have been more favourable to wheeled transport than in many other areas. Overall, the evidence permits us to conclude that wheeled transport appears to have been present in southern Scandinavia from around 3400 BCE, at least sporadically, but also that it is not until the period starting around 3100 that we have evidence indicating that wheeled transport was socioeconomically important in the westernmost part of the region, potentially also further east.

Roles, context and consequences of draught cattle technology

As already suggested in the first part of this paper, attaining a reasonably useful understanding of the roles and consequences of a particular technology or technological genre in relation to locally varying conditions and circumstances is no simple matter. Below I shall attempt to provide the contours of such an understanding concerning the appropriation of draught cattle technology during the Neolithic in southern Scandinavia, though given the constraints of space this account will necessarily be superficial.

The evidence available at present indicates that while at least sporadic applications of draught cattle technology predate the middle of the 4th millennium BCE, its importance clearly accelerated during the centuries following 3500 BCE. A number of changes in the wider archaeological and palaeoenvironmental record are of interest here. First of all, the appearance of widespread evidence for ard agriculture after 3500 BCE broadly coincides with the classical 'landnam' sequence documented in the pollen record of central and eastern/south-eastern parts of southern Scandinavia (Aaby 1986; 1988; Andersen 1991; Berglund

et al. 1991; Berglund and Kolstrup 1991; Gaillard and Göransson 1991; Regnéll 1991; Dörfler 2001; Kalies and Meurers-Balke 2001; Rasmussen 2005; Feeser *et al.* 2012). From the onset of the Neolithic in this region, land-use strategies seem to have involved elements of slash-and-burn agriculture in a pattern where small groups, presumably individual households, relocated relatively frequently (Madsen and Jensen 1982; Larsson 1985; Andersen 1992a; Kalies and Meurers-Balke 2001). However, around 3500 BCE the level and character of human impact on the landscape changed significantly. High percentages of birch pollen are succeeded by a prolonged peak in hazel combined with an increased presence of grasses and herbs, notably ribwort (*Plantago lanceolata*). This stage appears to represent a general intensification of human land-use and the regional establishment of a new type of economy, which combined intensive agricultural exploitation in delimited, open areas with extensive open forest management that included grazing and widespread coppicing (Christensen 1997; Kalies and Meurers-Balke 2001; Westphal 2009), while the use of fire still seems to have played a role in land management (Andersen 1992b). During the Late TRB phase, 3100–2700 BCE, the extent of the hazel-dominated cultural landscape was reduced, and in some areas the mixed high-canopy forest started to regenerate away from the increasingly intensively exploited parts of the landscape. Parallel to these developments in land-use, a general development towards larger settlements with intensive exploitation of the local area took place. This applies not least to the Late TRB, after 3100 BCE, during which thick layers caused by erosion and deposition of cultural material became a characteristic feature of many sites covering several hectares (Skaarup 1985; Madsen 1990; 1998, Nielsen 1999; Artursson *et al.* 2003). Furthermore, the apparent transition in land use occurring during the mid-4th millennium BCE coincides with a notable intensification of ritual activity from around or slightly before 3500 BCE, which includes the construction of thousands of megalithic tombs (Ebbesen 2011), the construction of a significant number of very large causewayed enclosures of clearly supra-local significance (Andersen 1997; Klassen 2014) and a pronounced peak in sacrificial wetland depositions, including polished stone axes, ceramic vessels, domestic animals and, occasionally, humans (Ebbesen 1993; Koch 1998).

It has previously been suggested that a change in the agricultural regime of Neolithic communities in southern Scandinavia, as reflected not least by the evidence for ard agriculture, played a major causal role in the developments just described (Johannsen 2006). This interpretation of the socioeconomic sequence was inspired in no small part by the earlier observation that the intensification of ritual activity taking place from around 3500 BCE might well reflect a cultural need for relatively strict regulation of socioeconomic relationships at several different scales in societies going through an extended phase of socioeconomic growth, which possibly included considerable demographic expansion (Madsen 1990). In comparison with hoe agriculture, ard agriculture often does not increase the crop yield per area unit (in some cases it even decreases) but it significantly increases the area that can be cultivated by a household and hence the potential total yield in most environments – presuming that the requisite amount of suitable land is available (Clark and Haswell 1970; Goody 1976; Webster and Wilson 1980; McCann 1984; Halstead 1995; Savadogo *et al.* 1998; Singh 1988). In other words, ard agriculture may have contributed to an economic and demographic feedback dynamic that simultaneously increased the demand for good arable land *and* made an expansion in overall productivity possible if this demand was met. This dynamic is in turn likely to have contributed to new notions of territoriality, land ownership or use-rights, inheritance and marital systems (*cf.* Goody 1976; Kjærby 1983; Sosovele 1994; Farnham 1997). Recently, the idea of a causally significant change in the agricultural regime has been substantiated further by the archaeobotanical identification of a marked shift in the crop spectrum in this region: a drastic reduction in the prevalence of naked wheat/bread wheat (tetraploid free-threshing wheat) from around 3500 BCE (Kirleis and Fischer 2014). This change is likely to be directly related to the rising significance of ard cultivation, in the sense that these two developments were part of the same reconfiguration of the agricultural system. In many ways, the scenario outlined above appears as a classic case of the 'cascading processes' mentioned earlier, in which an instance of technological change leads to a series of intended or unintended changes and corresponding readjustments through much of the socioeconomic system (*cf.* Schiffer 2005).

But draught cattle technology was probably not only involved in creating a need for ritual regulation of socioeconomic relations – it probably also contributed to making the necessary investment of resources in this domain possible. Estimates have indicated that perhaps as many as 25,000 megalithic tombs were constructed between 3500 and 3100 BCE in present-day Denmark alone (Ebbesen 1985, 37–40). In addition to some 100 to 1000 cubic metres of earth, the builders of such a monument had to move the large stones mostly weighing 4–8 tons, but in some cases reaching 20 tons (Ebbesen 2011, 279). As discussed extensively by Sjögren (2003, 237–251), the use of draught cattle would have reduced the human labour requirement significantly in such a demanding project, and detailed analyses of monumental construction work in a different cultural context, the well-documented Late Bronze Age of the Argos Plain in present-day Greece, have provided similar indications of the substantial quantitative difference that the availability of draught cattle is likely to have made in this type of activity (Brysbaert 2013). Turning back to southern Scandinavia, the period when megalithic monuments were built also displays another spectacular,

large-scale investment of labour in the construction of the causewayed enclosures mentioned earlier. An estimated 3350 tons of earth and some 3700 tree trunks weighing at least 600 tons, were used/moved in the construction of just one of these enclosures, Sarup 1 (Andersen 1988; 1997). As with the megalithic tombs, draught cattle would have been highly valuable for these construction projects, not least for hauling the large number of tree trunks used in the massive palisades of these enclosures from the place where they were felled to the site of the enclosure.

The discussion of monumental construction projects undertaken during the second half of the 4th millennium BCE in southern Scandinavia illustrates an important point: draught cattle probably played a major role in a broad range of additional activities other than agricultural tilling and wheeled transport. For many of these activities it is likely that the only artefactual equipment involved was the yoke and the ropes attached to whatever it was that was being transported, sometimes probably supplemented with wooden rollers, sliders, sledges or travois. Given that the circumstances would have to be truly exceptional for such artefacts to be preserved *and* recognised archaeologically (for an example, see Pétrequin *et al.* 2002), such activities may not have left any artefactual signature that we are likely to identify. This point in its turn underlines the importance of considering all available lines of evidence on draught cattle technology, including such data as osteomorphological changes and the information they may supply on the general significance of such technology. Also, it stresses that the impact of draught cattle technology in these societies has to be evaluated in relation to its overall effect in and on the socioeconomic system.

As I have emphasised, with the exception of ard marks, which are found throughout the region, the brief account of socioeconomic developments through the TRB period in southern Scandinavia provided above is very much based on evidence from the central and eastern parts of the region, and it thus applies primarily to these parts. As we have seen, however, the evidence for wheeled transport during (the late part of) this period comes specifically from those north-western and central parts of the Jutland peninsula that are *not* included in that purview (Fig. 11.1), and the evidence for this specific genre of draught cattle technology is thus clustered in a part of the region where the natural landscape and vegetation differed somewhat from the rest of the region. In large parts of north-western and central Jutland, the landscape is exceedingly flat, and it is furthermore dominated by very light, sandy soils on which forest cover was naturally less dense than on the heavier, clayey deposits of the Weichselian glaciation that cover most of the region further to the east. Although the pollen record of this western area shows significant variation between sequences for different localities, the dominant picture is a trend towards an increasing opening of the landscape with large areas of pasture and, in some places, *Calluna* heath that can be been to start no later than the Late TRB period, 3100–2700 BCE, and then accelerate in some localities with the advent of the first single Grave or Corded Ware groups around 2800 BCE (*cf.* Odgaard 1988; 1991; 1994; Odgaard and Rostholm 1988; Andersen 1993; 1998; Hübner 2005, 702–707). This development is accompanied by scattered, small-scale evidence for domestic activity and settlement that both in terms of character and specific location in the landscape (Rostholm 1977; 1982; 1986; Davidsen 1978; Hansen 1986; Simonsen 1986; Hübner 2005; Siemen 2008; Johannsen *et al.* 2016) exhibits a remarkable degree of continuity throughout the transition between the Late TRB and early Single Grave/Corded Ware period.

It is likely that the pollen and settlement data together reflect the predominance of a particular socioeconomic strategy in north-western and central Jutland during the Late TRB and the following period. The relatively open landscape in this area would have been highly suitable for herding domestic ruminants (cattle, sheep and goats), and the grazing/browsing of these animals on the sandy soils would then promote further opening of the landscape. Also, the situation of the late TRB sites offered good access to different types of landscape. When plotted on a soil classification map, sites with stone heap graves can be shown to be located predominantly in areas with different types of sandy soil, combined with relative vicinity to the coast/fjords and/or major river valleys (Johannsen *et al.* 2016, fig. 9). With good access to shore meadows, river valleys and drier parts of the inland, a pattern of settlement and mobility centred around these sites would provide suitable foraging environments for the animals at different times of the year (Johannsen *et al.* 2016). The correct term for at least a part of this predominantly pastoral and potentially seasonally mobile economy may well be 'transhumance', but no doubt animal husbandry was supplemented by other wild and domestic resources. Incidentally, rather similar conclusions have been reached concerning the socioeconomic strategies of so-called Globular Amphora groups inhabiting parts of the continental lowlands to the south-east of southern Scandinavia (Szmyt 2002; Czebreszuk and Szmyt 2011; Woidich 2014), and there were clearly connections between these groups and the late TRB groups of north-western Jutland, as reflected not least in the custom of cattle burial and the exchange of stone battle axes (Johannsen and Laursen 2010, 25–28).

If this brief sketch of socioeconomic strategies in north-western and central Jutland has any truth to it, it may – despite the biases discussed earlier – neither be coincidental nor entirely misleading for us to find the bulk of the evidence for wheeled transport in this part of southern Scandinavia. Not only did the landscape here offer comparatively favourable natural conditions for using wheeled vehicles, it also invited socioeconomic strategies in which the ability to move tools, foodstuffs, sheltering equipment and perhaps passengers such as young children or the elderly without having to follow the main water courses, would have been highly valued –

strategies that in turn promoted open land, thus making such vehicles even more useful (Johannsen *et al.* 2016). What we can say with great certainty is that this dynamic process between mobility and animal husbandry started during the Neolithic and founded long-term trajectories in land-use and landscape impact the subsequent stages of which we can identify in later periods of prehistory and in historical times (Holst and Rasmussen 2013).

Concluding remarks

In broad terms, the introduction of draught cattle technology in southern Scandinavia may well share an overall chronology, but this technological genre – or its more specific applications – seem to have been appropriated rather differently in different parts of this region. We must acknowledge that, to a certain degree, some of the apparent variation between the empirical records of the eastern and western parts of the region may be the product of taphonomically and culturally induced biases, but I have argued above that such potential distortion is unlikely to provide an adequate explanation of the fundamental differences we see. In general, draught cattle technology seems to have been every bit as impactful in the societies in question as envisioned by Childe, White and Sherratt, all of whom, to one extent or another, considered this technological genre sufficiently fundamental for socioeconomic activities to constitute a cultural 'game changer'. But again, the brief analysis presented above only supports a qualified version of this view. True, draught cattle technology seems to have contributed to shaping cultural developments across southern Scandinavia, but the profile of this significance varies considerably between the eastern/central parts and the western part of the region. In the former area, the specific employment of draught cattle in ard agriculture seems to have played a central role in a major, potentially turbulent restructuring of the socioeconomic system, which was apparently negotiated by the communities involved through various forms of extensive ritual regulation – which most likely also involved draught cattle technology – and which was followed by a gradual, localised intensification of the system established as a result of this process. In the latter area, where base conditions were less favourable to arable agriculture and more favourable to (mobile) animal husbandry, such a development towards large settlements with long duration and very high local economic intensity is not visible in the palaeoenvironmental record or the settlement pattern. On the contrary, in this area developments in the Late TRB seem to promote increasing mobility for small social units about the size of families/households, at least for some parts of the year. And here we also find substantial indications in the archaeological record that wheeled transport was important to people living there, both in practical and in ideological terms. The stone heap grave custom in particular reflects this cultural integration, providing what seems to be a good illustration of the way in which technological choices can bring (unforeseen) ideological consequences with them (*cf.* Johannsen 2010).

The initial appropriations of draught cattle technology in southern Scandinavia played out differently and thus introduced several different long-term trajectories or path dependencies in the socioeconomic systems of this region. None of the individual agents or groups involved could have grasped these nascent trajectories fully, but innumerable agents inheriting the outcomes of their choices had to navigate and appropriate these conditions themselves, according to gradually changing circumstances. In this sense, the appropriation of draught cattle technology in southern Scandinavia was a process that only really ended when this technological genre was finally abandoned in the region during the 20th century AD. But during the decades and centuries directly after the TRB period already, the difference between developments in areas with the apparently different technological trajectories described above is remarkable. In western and central Jutland, the so-called Single Grave Culture emerged as a distinct, regional manifestation of the Corded Ware phenomenon around 2850 BCE, overlapping with the latest TRB groups, and successfully spread across much of the peninsula within decades. Further to the east in southern Scandinavia, the impact of the Corded Ware current was both delayed and, more importantly, culturally much less substantial (Hübner 2005; Iversen 2014; Johannsen *et al.* 2016). If the data discussed above have been interpreted correctly, then this difference was not only caused by the same variation in natural landscape preconditions across the region that TRB communities had faced, but also by the ways that these communities had already worked and lived within these different preconditions, shaping landscapes, technologies and socioeconomic modes as well as ideals of how to live in different directions. And in this process, draught cattle technology and cattle husbandry in different constellations probably contributed significantly, and quite literally, to clearing the ground for regionally distinct economic and cultural histories in southern Scandinavia.

References

Aaby, B. (1986) Trees as Anthropogenic Indicators in Regional Pollen Diagrams from Eastern Denmark. In K. E. Behre (ed.), *Anthropogenic Indicators in Pollen Diagrams*, 73–93. Rotterdam, Balkema.

Aaby, B. (1988) The Cultural Landscape as Reflected in Percentage and Influx Pollen Diagrams from Two Danish Ombrotrophic Mires. In H. H. Birks, H. J. B. Birks, P. E. Kaland and D. Moe (eds), *The Cultural Landscape – Past, Present and Future*, 209–228. Cambridge, Cambridge University Press.

Andersen, N. H. (1988) *Sarup: Befæstede kultpladser fra bondestenalderen*. Højbjerg, Jutland Archaeological Society.

Andersen, N. H. (1997) *The Sarup Enclosures*. Sarup 1. Højbjerg, Jutland Archaeological Society.

Andersen, S. T. (1991) Natural and Cultural Landscapes Since the Ice Age Shown by Pollen Analyses from Small Hollows in a Forested Area in Denmark. *Journal of Danish Archaeology* 8, 188–199.

Andersen, S. T. (1992a) Pollen Spectra from Two Early Neolithic Lugged Jars in the Long Barrow at Bjørnsholm, Denmark. *Journal of Danish Archaeology* 9, 59–63.

Andersen, S. T. (1992b) Early- and Middle-Neolithic Agriculture in Denmark: Pollen Spectra from Soils in Burial Mounds of the Funnel Beaker Culture. *Journal of European Archaeology* 1, 153–180.

Andersen, S. T. (1993) Early Agriculture. In S. Hvass and B. Storgaard (eds), *Digging Into the Past: 25 Years of Archaeology in Denmark*, 88–91. Copenhagen and Aarhus, The Royal Society of Northern Antiquaries and Jutland Archaeological Society.

Andersen, S. T. (1998) Pollen Analytical Investigations of Barrows from the Funnel Beaker and Single Grave Cultures in the Vroue Area, West Jutland, Denmark. *Journal of Danish Archaeology* 12, 107–132.

Artursson, M., Linderoth, T., Nilsson, M.-L. and Svensson, M. (2003) Byggnadskultur i södra & mellersta Skandinavien. In M. Svensson (ed.), *I det neolitiska rummet*, 40–171. Lund, Riksantikvarieämbetet UV Syd.

Berglund, B. E., Hjelmroos, M. and Kolstrup, E. (1991) The Köpinge Area: Vegetation and Landscape through Time. In B. E. Berglund (ed.), *The Cultural Landscape during 6000 Years in Southern Sweden – The Ystad Project*. Ecological Bulletins 41, 109–112. Copenhagen, Munksgaard.

Berglund, B. E. and Kolstrup, E. (1991) The Romele Area: Vegetation and Landscape through Time. In B. E. Berglund (ed.), *The Cultural Landscape during 6000 Years in Southern Sweden – The Ystad Project*. Ecological Bulletins 41, 247–249. Copenhagen, Munksgaard.

Bartosiewicz, L., Van Neer, W. and Lentacker, A. (1997) *Draught Cattle: Their Osteological Identification and History*. Annales Sciences Zoologiques 281. Tervuren, Musée Royal de l'Afrique Centrale.

Beck, M. R. (2013) Højensvej Høj 7 – en tidligneolitisk langhøj med flere faser. *Aarbøger for Nordisk Oldkyndighed og Historie* 2011–12, 33–117.

Bogucki, P. (1999) *The Origins of Human Society*. Oxford, Blackwell.

Brysbaert, A. (2013) Set in Stone? Socio-Economic Reflections on Human and Animal Resources in Monumental Architecture of Late Bronze Age Tiryns in the Argos Plain, Greece. *Arctos* 47, 49–96.

Burmeister, S. (2004) Neolithische und bronzezeitliche Moorfunde aus den Niederlanden, Nordwestdeutschland und Dänemark. In M. Fansa and S. Burmeister (eds), *Rad und Wagen: Der Ursprung einer Innovation. Wagen in Vorderen Orient und Europa*. Archäologische Mitteilungen aus Nordwestdeutschland 40, 321–340. Mainz, Philipp von Zabern.

Childe, V. G. (1936) *Man Makes Himself*. London, Watts & Co.

Christensen, K. (1997) Wood from Fish Weirs – Forestry in the Stone Age. In L. Pedersen, A. Fischer, and B. Aaby (eds), *The Danish Storebælt Since the Ice Age – Man, Sea and Forest*, 147–156. Copenhagen, A/S Storebælt Fixed Link.

Christensen, K. (2007) Forhistorisk dendrokronologi i Danmark. *Kuml*, 217–236.

Clark, C. and Haswell, M. (1970) *The Economics of Subsistence Agriculture*. 4th ed. London, Macmillan.

Czebreszuk, J. and Szmyt, M. (2011) Identities, Differentiation and Interactions on the Central European Plain in the 3rd millennium BC. In S. Hansen and J. Müller (eds), *Sozialarchäologische Perspektiven: Gesellschaftlicher Wandel 5000–1500 v. Chr. zwischen Atlantik und Kaukasus*. Archäologie in Eurasien 24, 269–291. Darmstadt, Philipp von Zabern.

Davidsen, K. (1978). *The Final TRB Culture in Denmark: A Settlement Study*. Arkæologiske Studier 5. Copenhagen, The Royal Society of Northern Antiquaries.

Dörfler, W. (2001) Von der Parklandschaft zum Landschaftspark: Rekonstruktion der neolithischen Landschaft anhand von Pollenanalysen aus Schleswig-Holstein. In R. Kelm (ed.), *Zurück zur Steinzeitlandschaft*. Albersdorfer Forschungen zur Archäologie und Umweltgeschichte 2, 39–55. Heide, Boyens & Co.

Ebbesen, K. (1985) *Fortidsminderegistrering i Danmark*. Copenhagen, The National Agency for the Protection of Nature, Monuments and Sites.

Ebbesen, K. (1993) Sacrifices to the Powers of Nature. In S. Hvass and B. Storgaard (eds), *Digging Into the Past: 25 Years of Archaeology in Denmark*, 122–125. Copenhagen and Aarhus, The Royal Society of Northern Antiquaries and Jutland Archaeological Society.

Ebbesen, K. (2011) *Danmarks Megalitgrave 1*, 1. Copenhagen, Attika.

Fabricius, K. (1996) Tragtbægerkulturens mellemneolitiske stendyngegrave. In K. Fabricius and C. J. Becker (eds), *Stendyngegrave og Kulthuse: Studier over Tragtbægerkulturen i Nord- og Vestjylland*. Arkæologiske Studier 11, 13–276. Copenhagen, Akademisk Forlag.

Farnham, D. A. (1997) *Plows, Prosperity, and Cooperation at Agbassa*. Selinsgrove, PA, Susquehanna University Press.

Feeser, I., Dörfler, W., Averdieck, F-R. and Wiethold, J. (2012) New Insight into Regional and Local Land-Use and Vegetation Patterns in Eastern Schleswig-Holstein during the Neolithic. In M. Hinz and J. Müller (eds), *Siedlung, Grabenwerk, Grosssteingrab: Studien zu Gesellschaft, Wirtschaft und Umwelt der Trichterbechergruppen im nördlichen Mitteleuropa*, 159–190. Bonn, Habelt.

Gaillard, M.-J. and Göransson, H. (1991) The Bjäresjö Area: Vegetation and Landscape through Time. In B. E. Berglund (ed.), *The Cultural Landscape during 6000 Years in Southern Sweden – The Ystad Project*. Ecological Bulletins 41, 167–174. Copenhagen, Munksgaard.

Goody, J. (1976) *Production and Reproduction: A Comparative Study of the Domestic Domain*. Cambridge Studies in Social Anthropology 17. Cambridge, Cambridge University Press.

Halstead, P. (1995) Plough and Power: The Economic and Social Significance of Cultivation with the Ox-Drawn Ard in the Mediterranean. *Bulletin on Sumerian Agriculture* 8, 11–22.

Hansen, M. (1986) Enkeltgravskulturens bopladsfund fra Vesthimmerland og Ribe-området. In C. Adamsen and K. Ebbesen (eds), *Stridsøksetid i Sydskandinavien*. Arkæologiske Skrifter 1, 286–291. Copenhagen, Forhistorisk Arkæologisk Institut.

Hansen, H.-O. (1969) Experimental Ploughing with a Døstrup Ard Replica. *Tools and Tillage* 1/2, 67–92.

Holst, M. K. and Rasmussen, M. (2013) Herder Communities: Longhouses, Cattle and Landscape Organization in the Nordic Early and Middle Bronze Age. In S. Bergerbrant and S. Sabatini (eds), *Counterpoint: Essays in Archaeology and Heritage Studies in Honour of Professor Kristian Kristiansen*. British Archaeological Reports, International Series 2508, 99–110. Oxford, Archaeopress.

Hübner, E. (2005) *Jungneolithische Gräber auf der Jütischen Halbinsel: Typologische und chronologische Studien zur Einzelgrabkultur* 1. Nordiske Fortidsminder, Series B 24/1. Copenhagen, The Royal Society of Northern Antiquaries.

Iversen, R. (2014) Transformation of Neolithic Societies: An East Danish perspective on the 3rd millennium BC. Unpublished thesis. University of Copenhagen, Copenhagen.

Johannsen, N. N. (2005) Palaeopathology and Neolithic Cattle Traction: Methodological Issues and Archaeological Perspectives. In J. Davies, M. Fabiš, I. Mainland, M. Richards, and R. Thomas (eds), *Health and Diet in Past Animal Populations*, 39–51. Oxford, Oxbow Books.

Johannsen, N. N. (2006) Draught Cattle and the South Scandinavian Economies of the 4th Millennium BC. *Environmental Archaeology* 11, 33–46.

Johannsen, N. N. (2010) Technological Conceptualization: Cognition on the Shoulders of History. In L. Malafouris and C. Renfrew (eds), *The Cognitive Life of Things: Recasting the Boundaries of the Mind*, 59–69. Cambridge, McDonald Institute for Archaeological Research.

Johannsen, N. and Egeberg, T. (forthcoming) Orientations and Linearity of Neolithic Burial Monuments on the Jutland Peninsula: Causes and Consequences of Monumental Alignments.

Johannsen, N. N. and Kieldsen, M. (2014) En stendyngegrav ved Kvorning: fund, kontekst og betydning. *Kuml*, 9–28.

Johannsen, N. N. and Laursen, S. (2010) Routes and Wheeled Transport in Late 4th – Early 3rd Millennium Funerary Customs of the Jutland Peninsula: Regional Evidence and European Context. *Praehistorische Zeitschrift* 85, 15–58.

Johannsen, N. N., McGraw, J. and Roepstorff, A. (2014) Introduction: Technologies of the Mind. *Journal of Cognition and Culture* 14/5, 335–343.

Johannsen, N. N., Nielsen, S. K. and Jensen, S. T. (2016) North-Western Jutland at the Dawn of the 3rd Millennium: Navigating Life and Death in a New Socioeconomic Landscape? In M. Furholt, R. Grossmann and M. Szmyt (eds), *Transitional Landscapes? The 3rd Millennium BC in Europe*, 35–51. Bonn, Habelt.

Jørgensen, E. (1977) *Hagebrogård – Vroue – Koldkur. Neolitische Gräberfelder aus Nordwest-Jütland*. Arkæologiske Studier 4. Copenhagen, Akademisk Forlag.

Kalies, A. J. and Meurers-Balke, J. (2001) Zur Landnutzung der Trichterbecherkultur in der norddeutschen Jungmoränenlandschaft. In R. Kelm (ed.), *Zurück zur Steinzeitlandschaft*. Albersdorfer Forschungen zur Archäologie und Umweltgeschichte 2, 56–69. Heide, Boyens & Co.

Kirleis, W. and Fischer, E. (2014) Neolithic Cultivation of Tetraploid Free Threshing Wheat in Denmark and Northern Germany: Implications for Crop Diversity and Societal Dynamics of the Funnel Beaker Culture. *Vegetation History and Archaeobotany* 23, Supplement 1, 81–96.

Kjærby, F. (1983) *Problems and Contradictions in the Development of Ox-Cultivation in Tanzania.* SIAS Research Reports 66. Uppsala, Scandinavian Institute of African Studies.

Klassen, L. (2014) *Along the Road: Aspects of Causewayed Enclosures in South Scandinavia and Beyond.* Aarhus, Aarhus University Press.

Koch, E. (1998) *Neolithic Bog Pots from Zealand, Møn, Lolland and Falster.* Nordiske Fortidsminder, Series B, 16. Copenhagen, The Royal Society of Northern Antiquaries.

Larsson, M. (1985) *The Early Neolithic Funnel-Beaker Culture in South-West Scania, Sweden.* British Archaeological Reports, International Series 264. Oxford, BAR.

Madsen, T. and Jensen, H. J. (1982) Settlement and Land Use in Early Neolithic Denmark. *Analecta Praehistorica Leidensia* 15, 63–86.

Madsen, T. (1990) Changing Patterns of Land Use in the TRB Culture of South Scandinavia. In D. Jankowska (ed.), *Die Trichterbecherkultur. Neue Forschungen und Hypothesen* 1, 27–41. Poznań, Instytut Prahistorii Uniwersytetu im. Adama Mickiewicza w Poznaniu.

Madsen, T. (1998) Die Jungsteinzeit in Südskandinavien. In J. Preuss (ed.), *Das Neolithikum in Mitteleuropa 1/2: Teil B: Übersichten zum Stand und zu Problemen der archäologischen Forschung*, 423–450. Weissbach, Beier & Beran.

McCann, J. (1984) *Plows, Oxen, and Household Managers: A Reconsideration of the Land Paradigm and the Production Equation in Northeast Ethiopia.* African Studies Center Working Papers 95. Boston, MA, Boston University.

Mischka, D. (2011) The Neolithic Burial Sequence at Flintbek LA 3, North Germany, and its Cart Tracks: A Precise Chronology. *Antiquity* 85, 742–758.

Mischka, D. (2014) The Significance of Plough Marks for the Economic and Social Change and the Rise of the First Monumental Burial Architecture in Early Neolithic Northern Central Europe. In B. Gaydarska and B. S. Paulsson (eds), *Neolithic and Copper Age Monuments.* British Archaeological Reports, International Series 2625, 125–137. Oxford, Archaeopress.

Nielsen, P. O. (1999) Limensgård and Grødbygård: Settlements with House Remains from the Early, Middle and Late Neolithic on Bornholm. In C. Fabech and J. Ringtved (eds), *Settlement and Landscape*, 149–165. Aarhus, Jutland Archaeological Society.

Nielsen, V. (1993) *Jernalderens Pløjning: Store Vildmose.* Hjørring, Vendsyssel Historiske Museum.

Odgaard, B. V. (1988) Heathland History in Western Jutland, Denmark. In H. H. Birks, H. J. B. Birks, P. E. Kaland and D. Moe (eds), *The Cultural Landscape – Past, Present and Future*, 311–319. Cambridge, Cambridge University Press.

Odgaard, B. V. (1991) Cultural Landscape Development through 5500 Years at Lake Skånsø, Northwestern Jutland as Reflected in a Regional Pollen Diagram. *Journal of Danish Archaeology* 8, 200–210.

Odgaard, B. V. (1994) *The Holocene Vegetation History of Northern West Jutland, Denmark.* Opera botanica 123. Copenhagen, Council for Nordic Publications in Botany.

Odgaard, B. V. and Rostholm, H. (1988) A Single Grave Barrow at Harreskov, Jutland. Excavation and Pollen Analysis of a Fossil Soil. *Journal of Danish Archaeology* 6, 87–100.

Pétrequin, P., Arbogast, R.-M., Viellet, A., Pétrequin, A.-M. and Maréchal, D. (2002) Eine neolithische Stangenschleife vom Ende des 31. Jhs. v. Chr. in Chalain (Fontenu, Jura, Frankreich). In J. Köninger, M. Mainberger, H. Schlichtherle and M. Vosteen (eds), *Schleife, Schlitten, Rad und Wagen.* Hemmenhofener Skripte 3, 55–65. Gaienhofen-Hemmenhofen, Landesdenkmalamt Baden-Württemberg.

Rasmussen, P. (2005) Mid- to Late-Holocene Land-Use Change and Lake Development at Dallund Sø, Denmark: Vegetation and Land-Use History Inferred from Pollen Data. *The Holocene* 15/8, 1116–1129.

Regnéll, J. (1991) The Krageholm Area: Vegetation and Landscape through Time. In B. E. Berglund (ed.), *The Cultural Landscape during 6000 Years in Southern Sweden – The Ystad Project.* Ecological Bulletins 41, 221–224. Copenhagen, Munksgaard.

Rostholm, H. (1977) Nye fund fra yngre stenalder fra Skarrild Overby og Lille Hamborg. *Hardsyssels Årbog*, 91–112.

Rostholm, H. (1978) Neolitiske skivehjul fra Kideris og Bjerregårde i Midtjylland. *Kuml* 1977, 185–222.

Rostholm, H. (1982) A Grave Complex of the Early Single Grave Culture at Skarrild Overby, Central Jutland. *Journal of Danish Archaeology* 1, 35–38.

Rostholm, H. (1986) Lustrup og andre bopladsfund fra Herning-egnen. In C. Adamsen and K. Ebbesen (eds), *Stridsøksetid i Sydskandinavien.* Arkæologiske Skrifter 1, 301–317. Copenhagen, Forhistorisk Arkæologisk Institut.

Rostholm, H. (2007) Stenalderhjul til behandling. *Midtjyske fortællinger*, 63–70.

Rowley-Conwy, P. (1987) The Interpretation of Ard Marks. *Antiquity* 61, 263–266.

Savadogo, K., Reardon, T. and Pietola, K. (1998) Adoption of Improved Land Use Technologies to Increase Food Security in Burkina Faso: Relating Animal Traction, Productivity, and Non-Farm Income. *Agricultural Systems* 58, 441–464.

Schiffer, M. B. (2005) The Devil is in the Details: The Cascade Model of Invention Processes. *American Antiquity* 70/3, 485–502.

Sherratt, A. G. (1981) Plough and Pastoralism: Aspects of the Secondary Products Revolution. In I. Hodder, G. Isaac and N. Hammond (eds), *Pattern of the Past. Studies in Honour of David Clarke*, 261–305. Cambridge, Cambridge University Press.

Sherratt, A. G. (2003) The Baden (Pécel) Culture and Anatolia: Perspectives on a Cultural Transformation. In E. Jerem and P. Raczky (eds), *Morgenrot der Kulturen: Frühe Etappen der Menschheitsgeschichte in Mittel- und Südosteuropa*, 415–429. Budapest, Stiftung Archaeolingua.

Sherratt, A. G. (2004) Wagen, Pflug, Rind: Ihre Ausbreitung und Nutzung – Probleme der Quelleninterpretation. In M. Fansa and S. Burmeister (eds), *Rad und Wagen: Der Ursprung einer Innovation. Wagen in Vorderen Orient und Europa.* Archäologische Mitteilungen aus Nordwestdeutschland Beiheft 40, 409–428. Mainz, Philipp von Zabern.

Siemen, P. (2008) Settlements from the 3rd Millennium BC in Southwest Jutland. In W. Dörfler and J. Müller (eds), *Umwelt – Wirtschaft – Siedlungen im dritten vorchristlichen Jahrtausend Mitteleuropas und Südskandinaviens.* Offa-Bücher 84. 67–82. Neumünster, Karl Wachholtz.

Simonsen, J. (1986) Nogle nordvestjyske bopladsfund fra Enkeltgravskulturen og deres topografi. In C. Adamsen and K. Ebbesen (eds), *Stridsøksetid i Sydskandinavien.* Arkæologiske

Skrifter 1, 292–300. Copenhagen, Forhistorisk Arkæologisk Institut.

Singh, R. D. (1988) *Economics of the Family and Farming Systems in Sub-Saharan Africa: Development Perspectives*. Boulder, CO, Westview Press.

Sjögren, K.-G. (2003) *'Mångfalldige uhrminnes grafvar...' Megalitgravar och samhälle i Västsverige*. GOTARC Series B No. 27. Gothenburg, University of Gothenburg, Institute for Archaeology.

Skaarup, J. (1985) *Yngre stenalder på øerne syd for Fyn*. Rudkøbing, Langelands Museum.

Sosovele, H. (1994) Transfer of Animal Traction Technology: Cultural and Social Issues in Tarime District, Tanzania. In P. Starkey, E. Mwenya and J. Stares (eds), *Improving Animal Traction Technology*, 318–320. Wageningen, Technical Centre for Agricultural and Rural Cooperation.

Starkey, P. (1991) Draught Cattle World Resources, Systems of Utilization and Potential for Improvement. In C. G. Hickman (ed.), *Cattle Genetic Resources*. World Animal Science B 7, 153–200. Amsterdam, Elsevier.

Szmyt, M. (2002) Kugelamphoren-Gemeinschaften in Mittel- und Osteuropa: Siedlungsstrukturen und soziale Fragen. In J. Müller (ed.), *Vom Endneolithikum zur Frühbronzezeit: Muster sozialen Wandels?* Universitätsforschungen zur prähistorischen Archäologie 90, 195–233. Bonn, Habelt.

Thrane, H. (1991) Danish Plough-Marks from the Neolithic and Bronze Age. *Journal of Danish Archaeology* 8, 111–125.

Webster, C. C. and Wilson, P. N. (1980) *Agriculture in the Tropics*. 2nd ed. London, Longman.

White, L. A. (1959) *The Evolution of Culture: The Development of Civilization to the Fall of Rome*. New York, McGraw-Hill.

Westphal, J. (2009) Vidnesbyrd om organiseringen af tragtbægerkulturens landskab i pollenprøver fra megalitgrave. In A. Schülke (ed.), *Plads og rum i tragtbægerkulturen*. Nordiske Fortidsminder, Series C, 6, 105–138. Copenhagen, The Royal Society of Northern Antiquaries.

Woidich, M. (2014) *Die westliche Kugelamphorenkultur. Untersuchungen zu ihrer raumzeitlichen Differenzierung, kulturellen und anthropologischen Identität*. Topoi, Berlin Studies of the Ancient World 24. Berlin, De Gruyter.

Chapter 12

Key Techniques in the Production of Metals in the 6th and 5th Millennia BCE: Prerequisites, Preconditions and Consequences

Svend Hansen

Introduction

Human society could not have existed without techniques. The production of tools has become an essential defining criterion for humankind, in contradistinction to the animal kingdom. The history of humankind is unimaginable without hunting techniques, animal husbandry or the achievements of the Neolithic age. Such techniques give a structure to social processes and relationships. The ancient Greek term *technē* referred to art or artisanship based on experience and the kind of knowledge that can be acquired by learning. The modern term 'technology', by contrast, refers mostly to engineering or large-scale infrastructures.

How techniques come into being and the social repercussions they involve was a topic in archaeology very early on. Gordon Childe called the complex ensemble of innovations including domestication of animals, pottery-making, house-building and polished stone tools the 'Neolithic Revolution' (Childe 1936; Çilingiroğlu 2005). Further, he regarded the sailing boat, the wagon, the plough and metalworking as prerequisites for the 'Urban Revolution' (Childe 1950; see Sherratt 1997; Hansen 2011 on these innovations). A closer study of these fundamental innovations, however, was long hindered by the absence of a reliable dating method. It took the establishment of ^{14}C chronology to provide a precise mapping of the development and dissemination of these technical accomplishments. Likewise, increasing interest has been aroused by the interrelationship between technical innovations and social conditions. Were new technologies the precondition for a growth in production involving surplus manufacture and the concentration of economic power only in a few hands? Or did innovations flourish under the favourable conditions of peace and liberty?

Innovations rest upon inventions, new methods and technical solutions. They are effective as such when they are integrated into production and when the products ultimately are brought onto the market. This process is also referred to as 'diffusion' (Müller-Prothmann and Dörr 2009, 7). Yet, in the archaeological context a more pragmatic use of the term 'innovations' is advisable.[1] Innovations are acknowledged when they appear in archaeological contexts, *i.e.* integrated into social practices like artistic depictions or funeral rites. But this of course tells us nothing about the time of invention or the process of diffusion. This article endeavours to cast light on the early stages of two important innovations by looking at two selected instances: alloying of copper and casting in the lost-wax form.

The beginnings of metallurgy

To understand innovations, it is important to have a description of the specific time at which they materialised and the state of knowledge representing the prerequisites for the development of new technologies. As early as the 18th century, artisanal techniques were regarded as precursors of scientific knowledge. Most notably, the invention of metallurgy and the workmanship it involved was recognised as the prime cause for initial artistic endeavour, craftsmanship and science (Plessing 1787, 182). Metals were identified as the driving force behind practical inventions (Orell 1786, 497).

The fact that modern industry would not have emerged without the invention of metallurgy is evident. Accordingly, the origin of metallurgy is one of the most important research sectors in Eurasian archaeology. Objects made of copper and bronze are among the most comprehensively published groups of finds in archaeology. Aside from typological categorisation, the technological development of metallurgy has become an increasingly central concern in interdisciplinary research. The significance of metallurgy for social organisation is a matter of controversy. By contrast, another no less important aspect, the 'magical' dimensions of metallurgy as a process of transformation, has scarcely been touched upon, although this aspect was a source of considerable fascination for early metallurgists (Eliade 1980).

In comparison with the Americas, Africa and Australia, one specific model stands out: the Eurasian model of Neolithisation and its rapid dynamism, extending to civilisations with writing. In Eurasia, the relationship between technical innovations and the processes generating social hierarchies was particularly close, leading to the first cities and the emergence of the state some 250 generations after Neolithisation. One element in these historical processes was the development of metallurgy. Gordon Childe even considered the metal industry to be an essential prerequisite for the formation of the first city states (Childe 1950). While the state did not necessarily arise directly from metallurgy, the Mesopotamian city states would surely not have emerged in their particular form without metalworking.

In the late 1960s, Colin Renfrew adopted an influential anti-diffusionist viewpoint (Renfrew 1969). He dated the cemetery at Varna, Bulgaria, to the time around 4000 BCE, drawing upon the ^{14}C datings available at the time to contend that metallurgy appeared on the Balkan Peninsula at an earlier date than in the Near East. This prompted him to argue in favour of autonomous development in metalworking. Metallurgy was not invented in economic centres but on the periphery and evidently had little effect on social development. Nor did it have anything to do with the emergence of the state. This autochthonous concept of culture received enthusiastic support, but it was unable to provide a conceptual explanation for the emergence of metallurgy. The early death of I. Ivanov, the excavator of the Varna cemetery, meant that there is no concluding publication on the excavations and this has placed restrictions on the socio-historical evaluation of the site. Accordingly, the study of Varna seems doomed to remain investigative and assertive rather than analytic and well-substantiated.

Unfortunately, there is a lamentable literary *topos* in a number of recent publications suggesting a 'still existing predominance of diffusionist models and other Childean theories of elite dominance, core-periphery dynamics and specialized craftspeople' (Thornton and Roberts 2009, 181; *cf.* also Chapman *et al.* 2007; Higham *et al.* 2007, 651). Tobias Kienlin attempts to 'challenge evolutionist assumptions in our notions of technological "progress",' because 'sharply defined technological stages tend to become blurred by new discoveries' (Kienlin 2014, 447). All in all, the 'evolutionist grand narratives linking perceived technological progress to the emergence of hierarchical society' have come under fire (Kienlin 2014, 466). After all, Renfrew's autonomy model has been taken into consideration once more (Radivojević and Rehren 2015).

Metals of power/the power of metals

It has become evident today through new ^{14}C-datings, that the first exploitation of ores by mining and the initial metal casting began in the early 5th millennium BCE on the Balkan Peninsula. No doubt, this is only an intermediate result, which can, however, eventually be expected to compare with the improved state of research in eastern Anatolia, the Caucasus and Iran. Evidence of the earliest metals, mostly in the form of simple flat axes, is of course not very abundant. Only shortly before and around the middle of the 5th millennium BCE did copper objects and a small number of gold items start appearing in settlements in the west Pontic area. The cemetery I graves in Varna on the Bulgarian Black Sea coast are spectacular, because it was here that individuals were buried for the first time with an opulent supply of metal goods, indicating that they belonged to persons from the highest echelons in the hierarchical social structure (*cf.* also Chapman *et al.* 2007; Higham *et al.* 2007, 651).

The recent research dating the rich graves in Varna I to the 45th and 44th centuries BCE and the excavations in Pietrele on the lower Danube have occasioned an entirely new conception of the Copper Age in southeast Europe. The differences in wealth found at these two sites were the result of a development that took place during the first half of the 5th millennium BCE, a 'Late Neolithic' process that long received scarcely any attention.

It is a well-known fact that next to the famous cemetery I in Varna there is an older cemetery – Varna II – that has not been completely excavated yet. The burials in cemetery II were lavishly furnished with bead ornaments and weapons (Fig. 12.1) (Ivanov 1978). Hence, a tradition already existed at that time in Varna for marking differences in wealth and status. In this regard, the first jadeite axes recently identified in graves 1 and 3 in the older cemetery are of great significance (Pétrequin *et al.* 2012). Only a few years ago, P. Petréquin succeeded in identifying the sources of jadeite in the French Alps, going on from there to catalogue jadeite finds in Europe and so to reconstruct a principally west European network. In the course of this research, it was possible to identify several jadeite axes found in present-day Bulgaria. According to the results, long-distance connections for the exchange of exotic objects already existed in the time of Karanovo V (that is, before 4600 BCE). These trade relations extended as far as the

Figure 12.1: Varna II. Richly furnished grave (photo by Museum Varna).

Alps.[2] The jadeite axes in two of the graves in Varna II are socially highly significant. Especially spectacular is the coeval 28-piece hoard found in Svoboda near Plovdiv with its different, supraregionally distributed jadeite axe types.[3] We also encounter the placement of jadeite axes in graves 4 and 43 of in Varna I, which have the richest depositions of gold and grave goods in the entire cemetery (Fig. 12.2).

Numerous studies have been dedicated to the European dimensions of the Varna phenomenon (Petréquin *et al.* 2002; 2005), as clearly emphasised by the jadeite axes overlooked hitherto.[4] The deposition of a jadeite axe in grave 43, lying between the legs of the deceased, is an even more graphic demonstration of the high symbolic status of the axe in Varna. Decisive for the 'biography' of the site is the fact that the Alpine axe had been refashioned to produce a typical KGK-VI axe, a modification known from a number of settlement mounds of the Gumelniţa culture and designed to integrate the axe into the local system of signs. These goods from foreign lands did not represent any real value for their owners, who, in such societies, were distinguished rather by their progeny. Nevertheless, the possession of such axes probably enhanced their position and reputation within their own society and thus helped them to major economic success (Helms 1992, 157–159).

The importance of metal becomes apparent in both graves (Fol and Lichardus 1988).[5] Some 1.5 kilograms of gold were deposited in each burial. The value represented by the gold is of course difficult to estimate. One sensible approach to figuring this out, however, is to look at the working time needed to procure the metal. Using mining of gold deposits as a gauge, we can estimate a measurement, five grams of gold per ton of rock.[6] To mine and smelt the gold found in both graves, 300 tons of rock would have had to be removed, an estimate that makes it safe to presume that the deceased did not do this work themselves.[7]

Among the grave goods (Fig. 12.2) there are several weapons made of copper: shafthole axes, adzes, a spearhead and a pick axe. This weaponry clearly made a greater effect than corresponding objects made of stone or deer antlers (Hansen 2014b, 246 fig. 3). Both of the deceased received a stone axe, whose wooden shaft was sheathed in sheet gold. As a whole, these shafted axes can be understood as a symbol of power (like a sceptre). Some of the flint blades are extremely long, one of them measuring 40 cm. Bracelets were made of *Spondylus*, which came from the Mediterranean. The golden bracelets on both wrists and the luxurious necklaces of gold and carnelian beads are also symbols of power. Various gold discs, some quite large, were sewn onto the clothing of the deceased, so that the corpses of both the interred were literally 'covered' with gold. The metal grave goods were used to express social differentiation and meant to be understood as a representation of power.

Innovations in metallurgy

Recent research has been able to prove that copper smelting and casting began on the Balkan Peninsula as early as the beginning of the 5th millennium BCE.[8] Yet, there will surely be no great desire to answer the question as to where then metallurgy developed, in the Orient or on the Balkan Peninsula with positivism basing on the now known oldest cast copper objects.[9] The answer to this question calls for a source-critical understanding of evidential archaeological finds. Quantitatively, the Balkan Peninsula seems to have quite an abundance of finds, especially compared to Anatolia and the Near East. However, the fact is that most of these massive metal objects stem either from graves or hoards, *i.e.* they represent intentional depositions. They were thus removed from the 'normal' circulation of metal objects, a process that usually ended with their being melted down to make new objects. As early as the 5th millennium BCE, the deposition landscapes crystallised, in which metals were used not only in graves, but above all for exchange with imagined powers – spirits and divinities (Hansen 2009). Therefore, one may not consider the absence of numerous metal objects in other regions, for example in Anatolia, as proof that the use of metals was unknown there or only took place on a lower level.

Cast metal was the result of a process of transformation, through which – simply said – stone was changed into metal. Humankind had already gathered experience with the conversion of matter for a longer time. Mircea Eliade writes: 'The alchemist like the smith and before him the potter is a master of the fire. By using fire he caused the transformation of matter from one condition into another' (Eliade 1980, 83).

The oldest preserved example of the transformation of matter is the Palaeolithic clay figurine found in Dolni Věstonice, Moravia. Long before Neolithisation the potentialities of fire had been confronted again and again. As early as the 7th millennium BCE, the first pottery kilns

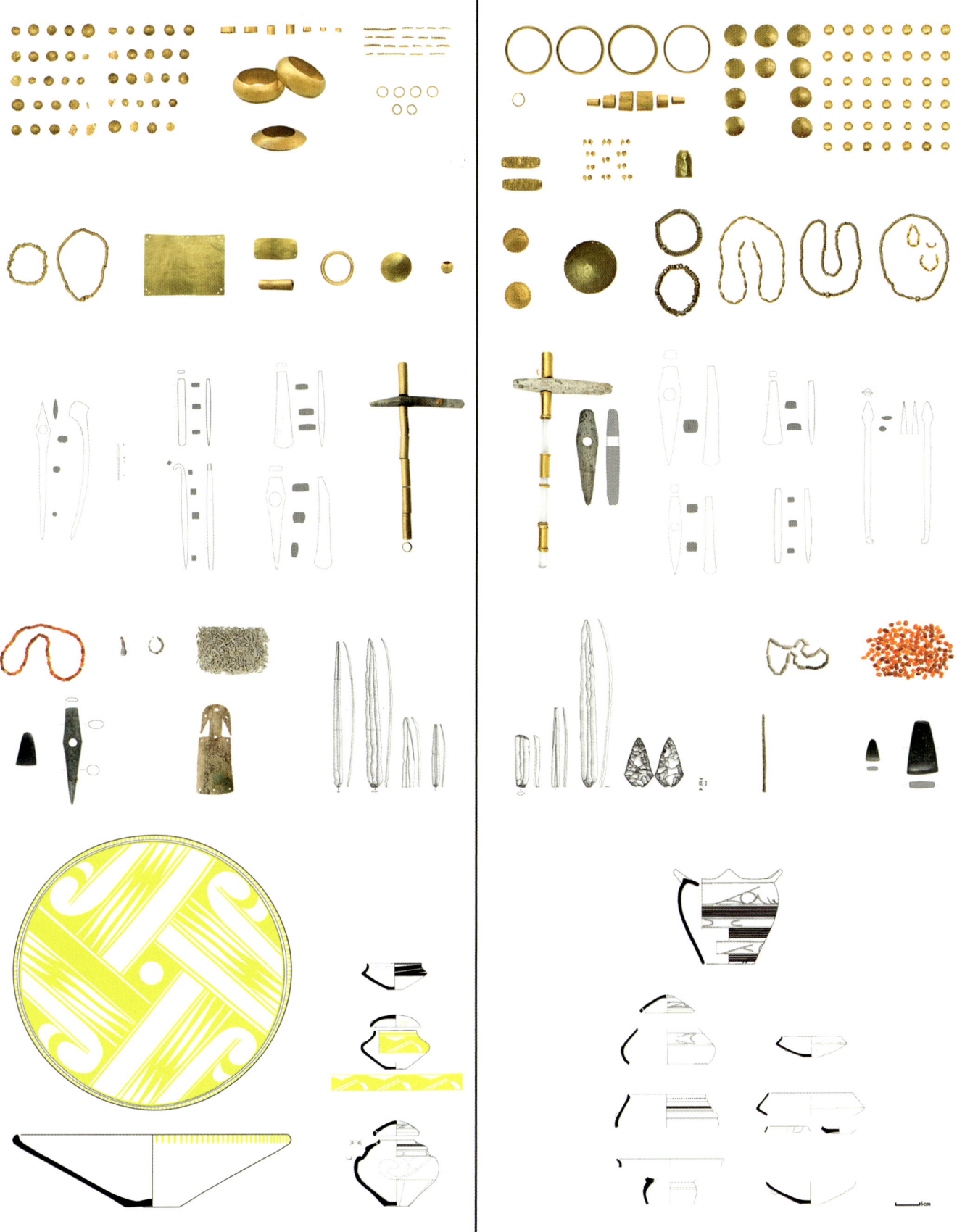

Figure 12.2: Overview of grave goods from graves 4 and 43 (after Fol and Lichardus 1988).

in West Asia were put to use, in order to gain better control over firing pottery. This resulted in the ability to achieve higher and more specific temperatures.

Many important innovations came about on the long path from the beginnings of mining and smelting the copper ore to industrial mining and the large steelworks of the modern age. But one of the most significant innovations occurred as far back as the later 5th and the 4th millennium BCE: the decisive step from using soft copper to hardened bronze by alloying copper with arsenic. Ultimately, we cannot explain how the idea dawned of adding another metal to copper. It is generally assumed that pieces of naturally occurring copper and arsenic were first fired and melted, producing alloyed copper (Pernicka *et al.* 2002, 124–125; Roberts *et al.* 2009, 1015; Weeks 2012, 30).

However, there are good arguments suggesting that other ingredients were intentionally added to copper which were meant to change the properties of the metal material (Lechtman 1996, 509). An intentional preference for polymetallic ores has also been proposed (Di Nocera 2010, 271).

The oldest alloyed product known so far was recently identified by Verena Leusch and Ernst Pernicka a disc-ring pendant from grave 271 in Varna I, consisting of 50% gold, 14% silver and 36% copper (Fig. 12.3) (Leusch *et al.* 2014, 175 fig. 11a). Leusch and Pernicka offer a plausible explanation for the enhanced silver content in gold beads from grave 43 in Varna as resulting from the natural occurrence of large amounts of silver in gold (Leusch *et al.* 2014, 177). The authors presume that experimentation with

different metals may have been behind the manipulation of colour and the change in properties.

Nonetheless, the question still remains unanswered as to whether this achievement came by accident or was the result of a specific experiment. The mental step of mixing different sorts of metal with one another is quite comprehensible in view of the Neolithic state of ceramic know-how. Thus, one can presume that the metallurgists adopted the method of tempering clay for producing pottery as a model.[10] For instance, the statuette from Dolni Věstonice was already made of tempered clay. Indeed, the addition of various organic or mineral agents can render a successful firing of pottery, hindering cracks and fissures during firing or drying. The kind of temper, however, changed the properties of the particular vessel: the weight of the vessel, the porosity of the walls, or the colour of the clay could be manipulated. Further, the production of appropriate crucibles as well as operable clay forms was essential for successful bronze casting.[11]

So, alloying copper was ultimately the adaption of a concept in pottery-making for use in metal production. The addition of another element can change the properties of copper. By adding arsenic, copper gains a silvery colour, whereas the addition of tin to copper lends a golden hue to the object.[12] By means of the corresponding alloy the otherwise soft copper gains hardness, while brittleness and elasticity can be altered. The flow of the molten metal is greatly improved, because the additional elements act as antioxidants that reduce the formation of bubbles in the metal, and in that way aid in producing a homogenous, solid object. Alloying copper was the groundbreaking technical achievement in metalworking, from the production of 'prestigious objects' to making practical tools and weapons.

Hence, metalworking did not start from a fictive zero point; instead this craft took recourse to older funds of knowledge and existing techniques and craftsmanship. Early metallurgists closely noted the chemical reactions and physical properties of the finished products, and early metallurgy undoubtedly lay in the hands of specialists and experts (Tadmor *et al.* 1995, 98). The rapid development of the associated techniques is otherwise hard to explain. Craft specialisation in the 5th millennium BCE is not only to be found in the sphere of metallurgy. In the Copper Age settlement of Pietrele on the lower Danube, we have evidence of households specialising in various occupations such as textile production and fishing. In one case, the tradition of hunting and fishing could be traced back six house-generations, or 250 years (Benecke *et al.* 2013). Newly specialised groups formed in the population. Specialisation as farmers and stock-raisers had been around for a long time. Now they were joined by potters, producers of long flint blades, brewers and metal casters (Chapman *et al.* 2007, 160).

Figure 12.3: Varna Grave 271. Pendant (photo by B. Armbruster/V. Leusch/Vl. Slavchev).

Nahal Mishmar

In 1961, a large number of metal alloys were found in Nahal Mishmar, located west of the Dead Sea (Bar-Adon 1980; Goren 2008; Ussishkin 2014). A total of 429 objects including 416 metal items wrapped in a reed mat had been concealed in a cave that could only be entered by a rope ladder. The majority of the objects (251) were simple, round metal maceheads. The second largest group was made up of 92 'standards' (Fig. 12.4–12.6). There were also singletons such as 'crowns', a sceptre with two ibex figures and two metal vessels. The simple flat axes in the metal assemblage should also be mentioned. In addition, there were rhinoceros horns and a container made of a hollowed-out elephant tusk. The find from Nahal Mishmar was assessed anew and dated by Florian Klimscha, also more recently by Isaac Gilead and Milena Gošić, to the last quarter of the 5th millennium BCE (Klimscha 2013, 37; Gilead and Gošić 2014).

In a series of analyses of 28 objects, Miriam Tadmor's team were able to distinguish three kinds of metal: objects made of pure copper, copper objects with high arsenic and antimony content and objects with high nickel and arsenic content. Only a small number of pure copper objects were identified, including simple tools like axes and hammers. A few objects contain high amounts of nickel. But by far the most numerous objects in Nahal Mishmar are antimony-arsenic bronzes with between 1% and 25% antimony and 0.4% to 15% arsenic (Tadmor *et al.* 1995, 129–131 tab. 2; *cf.* Shalev and Northover 1993, 43 tabs 1–2).

To judge by production residues, the simple tools of pure copper were made in settlements like Abu Matar or Shiqmim, whereas there is no indication of the production sites for casting the highly-alloyed copper objects, so their origin must be regarded as unknown (Kerner 2010, 190). Nevertheless, detailed examination of one mace-head found in Shiqmim shows that it was alloyed and cast in a complicated manner, in this case the lost-wax form with a stone core (Shalev *et al.* 1992, 66 fig. 2). Petrographic examination of the remains of the clay lost-wax forms for 70 metal objects in Nahal Mishmar led to the conclusion that the material for the casting forms probably came from the none too distant environs of the cave (Goren 2008, 391).

Therefore, it is quite probable that the metal, but not the finished products, originated from different mining districts. Lead isotope analyses of the copper objects indicate that the copper came from Fenan (Tadmor *et al.* 1995, 137 fig. 32). The search for ore deposits containing large amounts of arsenic and antimony pointed to eastern Anatolia and the Caucasus, although corresponding ore deposits have not actually been located (Hauptmann 2000, 186–187; for ore deposits in the Caucasus in general, *cf.* Twaltschrelidze 2001). Therefore, it is questionable whether or not the theory that the objects found in Nahal Mishmar represent natural alloys is correct (Hauptmann 2000, 186). Namely, it cannot be excluded that antimony and arsenic were used as alloys for copper. It is an assumption that early metallurgists could precisely distinguish the various metals just by their colour. An 'unintentional' or uncontrolled alloying of copper, as found in such a quantitative measure as in Nahal Mishmar, is hardly thinkable (Shugar and Gohm 2011, 133, on the simple

Figure 12.4: Nahal Mishmar (after Bar Adon 1980).

Figure 12.5: Nahal Mishmar (after Bar Adon 1980).

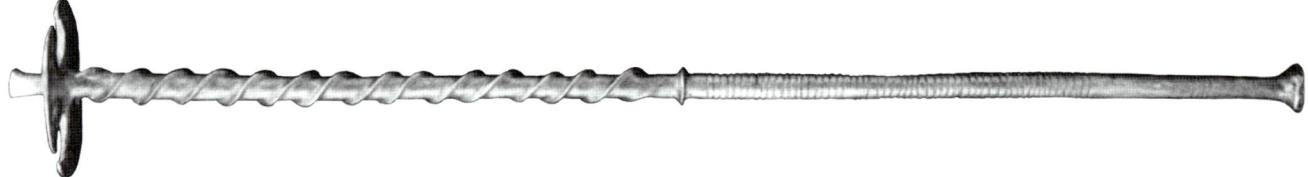

Figure 12.6: Nahal Mishmar (after Bar Adon 1980).

Figure 12.7: Aruchlo (Republic of Georgia). Copper bead (photo by Daniel Steininger).

recognition of arsenic). The basic problems are associated with the 'prestigious objects' in Nahal Mishmar is (1) that metal finds of the 5th millennium BCE in the Levant and Caucasus are extremely rare and (2) that objects made of copper alloyed with large amounts of arsenic and antimony are later in date (Tadmor *et al.* 1995, 140–141; Hauptmann 2000, 186–187). At present, therefore, the objects found in Nahal Mishmar cannot be assigned to a broad spectrum of experiments with different metals in the 5th millennium BCE. It remains singular.

Recent examinations of slag from Arisman, Iran, have substantiated the production of 'arsenspeiss' in the early 3rd millennium BCE. This at least proves the technical capability of adding a specific amount of arsenic as an alloy in the Early Bronze Age (Rehren *et al.* 2012). Whether or not this capability had existed earlier we cannot say. The one known antimony mine exploited in the Bronze Age is located in the Caucasus, more specifically in Racha, Republic of Georgia. Here copper and arsenic deposits occur next to one another. The oldest known traces of ore exploitation, however, date to the 2nd millennium BCE at the earliest (Maisuradse and Gobedschischwili 2001; new research in this sector region is a desideratum). This matches opportunely copper-antimony alloys as well as beads made of pure antimony that are known to the Late Bronze Age (Gambaschidze *et al.* 2001, 276–278 nos 87–92, 94–102; Hauptmann and Gambaschidze

2001, 150; Meliksetian *et al.* 2011, 205). A gold mine located in the Mašavera valley southwest of Tbilisi (Republic of Georgia) and dated to the 4th millennium BCE has been investigated in the past few years (Stöllner *et al.* 2010; Stöllner, 2014) and we can expect confirmation of much more comprehensive mining activities in the Caucasus as early as the Early Bronze Age and possibly the Late Copper Age. Probable native copper was already exploited and worked on in the 6th millennium BCE in Neolithic settlements like Aruchlo (Republic of Georgia) (Fig. 12.7). In Armenian Aratashen, 57 beads of arsenical copper were discovered dating to the Neolithic (Badalyan *et al.* 2007; Courcier 2014, 591, fig. 22.5). These find contexts are older than, or coeval with, recently discussed finds from Serbia (*cf.* Radivojević and Rehren 2015). Evidence in northern Mesopotamia points to copper metalworking practised within a clearly more complex system of craft specialisation, economic differentiation, administrative techniques and possibly public buildings between 5300 and 4500 BCE (Stein 2012, 146; for political organisation, *cf.* Özbal 2010, 43; Tell Brak: Oates *et al.* 2007). A recently published awl find in a grave from the late 6th millennium BCE in Tel Tsaf (Israel) can be added here to the early evidence of metallurgy in the large area between the Levant and the Caucasus (Garfinkel *et al.* 2014). According to ^{14}C dates, the well-known cast axe from Mersin-Yumuktepe can be dated to around 5000

BCE (Yalçin 2000, 22; Gülçür 2012, 218; Hansen 2013, 139 fig. 1).

In light of revised radiocarbon datings, the earliest copper casting in the geographic area between Iran and the Balkans can be dated to the beginning of the 5th millennium BCE. The metal objects in Nahal Mishmar do not mark the advent of metallurgy. Their diversity and technical complexity are the results of longer previous developments in metalworking.

Experiments and standard alloys

To a certain extent, experimentation in mixing metals is a distinctive feature of the second half of the 5th millennium BCE. The small wheels found in Mehrgarh, Pakistan (late 5th millennium BCE) can be added to the finds from Nahal Mishmar and Varna described above. They were made of a mixture of copper and lead, the lead content being 30–40% (Fig. 12.8; Roux *et al.* 2013, 65–68). Experimentation continued in the 4th millennium BCE. The royal grave at Arslantepe near Malatya (dated around 3000 BCE) yielded up to 28 objects for various uses made of copper alloyed with a large amount of silver (Hauptmann *et al.* 2002). Comparable alloys were established in a series of objects from Mesopotamia, the Caucasus, western Turkey, Bosnia and Romanian Moldova, without, however, being able to determine the reason for the addition of silver (up to 50%) (Hansen and Helwing 2016). A pin with a rolled head, part of the hoard found at Tülintepe (*c.* 3000 BCE), may represent one of the first tin bronzes. The copper contained 5.27% tin (Yalçın and Yalçın 2008, 114–117). The sword and the spearheads found in Tülintepe were also coated with tin. In the quest for the best alloy, copper alloyed with arsenic prevails in the 4th millennium BCE. Evgenij Chernych views the copper-arsenic-nickel combination as particularly characteristic of the Maikop culture (Черных 1966, 98–101 tabs 1–2; Chernykh 1992, 74, 145). He presumes the source of this metal to be Anatolia or Iran, whereas the source of copper with low nickel content is probably the southern Caucasus. The amount of arsenic in metal objects can vary considerably between 0.3% and 13%, as in finds from Novosvobodnaya-Klady (northern Caucasus) (Резепкин 2012, 64–66 tabs 1–3). Controlled alloying with arsenic thus signifies not only an extension of the aesthetic repertoire in metal craftsmanship via manipulation of colour from reddish copper to silvery arsenical bronze, it also brought about an improvement in material properties. Arsenical bronze is harder and has greater elasticity than pure copper (Lechtman 1996, 494). Finally, the addition of arsenic to the molten metal reduced the formation of bubbles in the casting process. The prevention of cavities in the cast objects was an advance in quality improvement. Cavities in cast massive axes are ignorable, but by contrast represent a great problem in thin

Figure 12.8: Mehrgarh (Pakistan). Copper and lead wheel (photo by Benoit Mille).

dagger blades. For when a blade is resharpened, the cavities cause pits in the surface. Alloying copper with arsenic was therefore a technical innovation of great consequence to the production of dagger- and sword-blades. The formation of cavities was less a problem for massive axes and flat axes, for which reason the simple axes in Nahal Mishmar like many Early Bronze Age shafthole axes of the later 4th and early 3rd millennium BCE continued to be produced out of pure copper (Szeverényi 2013).

The increase in the length of blades was considerable. Daggers with a length of 34.7 cm were already familiar in the first half of the 4th millennium BCE (Piotrovskij 2013, 312 fig. 20.30). In the last third of the 4th millennium BCE, swords more than 60 cm in length were produced both in

Arslantepe in eastern Anatolia and in Novosvobodnaya in the northern Caucasus. At the turn of the 4th to the 3rd millennium BCE, long functional blades made of arsenical bronze were widespread throughout parts of eastern and southern Europe. These new weapons facilitated the infliction of violence on others and thus represent a decisive innovation for further developments in history. The palace-like building upon the Arslantepe in which such swords were found, or the royal graves at Maikop, Novosvobodnaya and Arslantepe, which held an arsenal of weapons, show that the powerholders of the time were aware of the opportunities afforded by such technical developments. Thus it seems quite plausible that, as heads of early states, they exercised control over metal production.

Casting in lost-wax form

Developments in alloying stood in direct association with techniques in casting metal, as adding alloys could change conditions in casting. This applied to the lowered melting point of the metal mixture, the formation of bubbles and flow of molten metal, among other things. Therefore, it is certainly no coincidence that evidence of casting in the lost-wax form has been found at three sites where alloying is attested. This process represents a complex sequence in production, as opposed to casting over an open hearth. It enabled almost any imaginable form to be made (on this process, *cf.* Born 2001; Levy 2007; Goren 2008; Davey 2009). For this, a wax model was made of the object to be cast later. The wax form was covered with clay and then fired, so that the clay would bake and meanwhile the wax would melt. The clay form thus gained was then filled with molten metal poured through a funnel. It is the only process, by means of which the production of complex forms like the crown with bird protomes or the standard with ibex protomes in Nahal Mishmar was possible (Bar-Adon 1980, 28 no. 8, 42 no. 17).

The casters in Nahal Mishmar were indeed masters of their craft: they succeeded in casting the 58.5-cm long sceptre with torsion decoration (Bar-Adon 1980, 92 no. 127). Most of the other objects in the hoard were produced in the lost-wax method as well (Tadmor *et al.* 1995, 101–103; 124–126). Hence, Nahal Mishmar provides us with the most comprehensive testimony for this casting method in the 5th millennium BCE. The only hitherto single evidence for the use of this process in the Balkans is a gold bead found in Varna grave 43 (Pernicka 2013, 72; Leusch *et al.* 2014, 175 fig. 10b). This casting technique had already been postulated in earlier research, but until now it could not be verified (Schubert and Schubert 1999, 671). A further example of the lost-wax casting method process is represented by the spoked metal wheels from Mehrgarh in Baluchistan/Pakistan (Fig. 12.8). They are only 2 cm in diameter, each wheel with six spokes, and date back to the late 5th millennium BCE (Roux *et al.* 2013, 65–68).

This means that the lost-wax casting method spread from Pakistan and the Levant to the west Black Sea coast as early as the 5th millennium BCE. In the 4th millennium BCE this technology then found its way to central Europe (Mille *et al.* 2004, 264–270; Hansen 2014a, 405 fig. 25).

Conclusion

According to the present state of research, metal casting commenced in the early 5th millennium BCE, perhaps on the Balkan Peninsula, though possibly also in adjacent areas to the east extending as far as Persia. From more or less the same zone we have the oldest available evidence for lost-wax casting, its most impressive manifestation being the hoard in Nahal Mishmar. The multi-faceted and technical complexity of the objects found there far exceed anything that has come down to us from the Balkan Copper Age metallurgy of the 5th and 4th millennia BCE. Developments in this period are characterised rather by stagnation at the level of the finds from Varna.

The development of lost-wax casting was linked with an improvement in pouring capacity, for which experiments were made with a variety of alloy materials. Both innovations – alloying copper and lost-wax casting – were probably closely contingent upon one another. At least casting in lost-wax form was based upon Neolithic techniques in pottery production. Perhaps adding alloys to copper was suggested conceptually from adding temper to clay.

Experiments in alloying were nonetheless constrained by a variety of conditions. The procurement of arsenic ore (or speiss), antimony (ore) or lead involved travelling long distances. Successful reproduction producing the same result was by no means guaranteed. All these factors played a role over a time span of several hundred years, a time in which the knowledge that had been gained could be forfeited locally at any time due to events like sudden disasters, epidemics, warfare, etc. The broad, 'synchronous' dissemination of technical methods is indicative of the transfer of knowledge by various and different 'cultures' and political systems, a process that is not to be taken for granted. In whatever ways this transfer may have been effected in individual cases, the result was that it helped to preserve the knowledge that had been gained through experimentation.

The few metal finds of the 5th millennium BCE indicate the creativity of early metallurgists, but also the fallacies they operated under, the blind alleys they explored and the failures they suffered before their success with arsenical bronzes. The metallurgical innovations of the 5th millennium BCE did not materialise out of nowhere, they built upon a Neolithic body of knowledge. They were not isolated inventions but were related through the way they took shape, and in their turn they formed the preconditions for the exploration of completely new avenues in metal use in the 4th millennium BCE.

Acknowledgements

I would like to thank the organisers of the conference, Joseph Maran and Philipp W. Stockhammer, for their kind invitation to participate and also for their consummate patience with regard to my schedule problems.

Another vote of thanks goes to Daniel Steiniger and Petar Zidarov for their suggestions and discussions on this topic. Barbara Armbruster, Verena Leusch, Benoit Mille, Ernst Pernicka and Vladimir Slavchev provided me with photographs. Mehmet Karaucak prepared Figure 12.3. Anke Reuter was responsible for the graphics. Emily Schalk translated my German text. My thanks to all of them.

Notes

1   This is the basis of a research project aiming at a 'digital atlas of innovations', a project conducted since 2013 in the framework of the excellence cluster 'Topoi' of the Eurasia Department (DAI) and the Max Planck Institute for the History of Science in Berlin.
2   In terms of absolute chronology, I refer here to the ^{14}C sequence for Pietrele. Petréquin *et al.* 2012 assume somewhat earlier datings.
3   (Types Bégude, Chelles, Durrington) Petréquin *et al.* 2012, 1263.
4   Kienlin 2014, 453, differs on this point, seeing the 'systems of elite exchange supposedly stretching as far as from Brittany to Varna' as a 'considerable extrapolation from the archaeological data'.
5   Fol and Lichardus 1988, 189–203. The individual buried in grave 43 was a 40–50-year-old male, whereas grave 4 did not contain any skeletal remains and is thus considered a symbolic grave.
6   For the calculation of the time involved in the exploitation of gold, see Stöllner 2014, 93–95. On alluvial gold, *cf.* Albiez 1950, who views 20 grams of gold per year for a gold panner as realistic.
7   Bulgaria has an abundance of gold, and in the 5th millennium BCE fair-sized gold nuggets were surely common.
8   In Belovode, Serbia: Radivojević *et al.* 2010; in Tepe Yhaya, Iran: Thornton 2010, 33; the dating to around 4300 BCE is not adequately substantiated. The find could be older.
9   'Belovode predates anything comparable in the Near East' (Kienlin 2014, 450) is a misinterpretation.
10 Hansen Streily 2000, 17–25.
11 For a theoretical model *cf.* the 'community of practice' concept (Wenger 1998).
12 The significance of colour aesthetics in the spread of tin bronze is explained by Ernst Pernicka 1998.

References

Albiez, G. (1950) Rheingold. *Badische Heimat* 30, 125–129.

Badalyan, R., Lombard, P., Avetisjan, P., Chataigner, C., Chabot, J., Vila, E., Hovsepjan, R., Wilcox, G. and Pessin, H. (2007) New Data on the Late Prehistory on the Southern Caucasus. The Excavations at Aratashen (Armenia): Preliminary Report. In B. Lyonnet (ed.), *Les cultures du Caucase (VIe–IIIe millénaires avant notre ère). Leurs relations avec le Proche-Orient,* 37–62. Paris, CNRS.

Bar-Adon, P. (1980) *The Cave of the Treasure. The Finds from the Caves in Nahal Mishmar.* Jerusalem, Israel Exploration Society.

Benecke, N., Hansen S., Nowacki, D., Reingruber, A., Ritchie, K. and Wunderlich, J. (2013) Pietrele in the Lower Danube Region: Integrating Archaeological, Faunal and Environmental Investigations. *Documenta Praehistorica* 40, 175–193.

Born, H. (2001) Die Herstellungstechnik der Helme und Waffen. In H. Born and S. Hansen (eds), *Helme und Waffen Alteuropas.* Sammlung Axel Guttmann 9, 167–268. Mainz, Philipp von Zabern.

Childe, V. G. (1936) *Man Makes Himself.* London, Watts & Co.

Childe, V. G. (1950) The Urban Revolution. *The Town Planning Review* 21, 3–17.

Chapman, J., Higham, T., Slavchev, V., Gaydarska, B. and Honch, N. (2007) Social Context of the Emergence, Development and Abandonment of the Varna Cemetery, Bulgaria. *European Journal of Archaeology* 9, 157–181.

Chernykh, E. N. (1992) *Ancient Metallurgy in the USSR. The Early Metal Age.* Cambridge, Cambridge University Press.

Courcier, A. (2014) Ancient Metallurgy in the Caucasus from the Sixth to the Third Millennium BCEE. In B. W. Roberts and C. P. Thornton (eds), *Archaeometallurgy in Global Perspective. Methods and Syntheses.* New York *et al.*, Springer.

Davey, C. J. (2009) The Early History of Lost-Wax Casting. In J. Mei and T. Rehren (eds), *Metallurgy and Civilisation: Eurasia and Beyond,* 147–154. London, Archetype.

Eliade, M. (1980) *Schmiede und Alchemisten.* 2nd ed. Stuttgart, Klett-Cotta.

Fol, A. and Lichardus, J. (eds) (1988) *Macht, Herrschaft und Gold. Das Gräberfeld von Varna (Bulgarien) und die Anfänge einer neuen europäischen Zivilisation.* Saarbrücken, Moderne Galerie des Saarland-Museums.

Gambaschidze, I., Hauptmann, A., Slotta, R. and Yalcin, Ü. (eds) (2001), *Georgien. Schätze aus dem Land des goldenen Vlies. Katalog der Ausstellung des Deutschen Bergbau-Museums Bochum in Verbindung mit dem Zentrum für Archäologische Forschungen der Georgischen Akademie der Wissenschaften Tbilissi vom 28. Oktober 2001 bis 19. Mai 2002.* Bochum, Deutsches Bergbau-Museum.

Garfinkel, Y., Klimscha, F., Shalev, S. and Rosenberg, D. (2014) The Beginning of Metallurgy in the Southern Levant: A Late 6th Millennium CalBC Copper Awl from Tel Tsaf, Israel. *PLOS ONE.* DOI: 10.1371/journal.pone.0092591.

Gilead, I. and Gošić, M. (2014) Fifty Years Later: A Critical Review of the Stratigraphy, Chronology and Context of the Nahal Mishmar Hoard Mitekufat Haeven. *Journal of the Israel Prehistoric Society* 44, 226–239.

Goren, Y. (2008) The Location of Specialized Copper Production by the Lost Wax Technique in the Chalcolithic Southern Levant. *Geoarchaeology: An International Journal* 23, 374–397.

Gülçür, S. (2012) The Chalcolithic period in Central Anatolia Aksaray-Niğde region. *Origini* 24, 10–20.

Hansen, S. (2009) Kupferzeitliche Äxte zwischen dem 5. und 3. Jahrtausend in Südosteuropa. In L. Dietrich, O. Dietrich, B. Heeb and A. Szentmiklosi (eds), *Aes aeterna. Festschrift für Tudor Soroceanu zum 65. Geburtstag.* Analele Banatului, S. N. Arheologie – Istorie 17, 141–160. Timişoara, Mirton.

Hansen, S. (2011) Technische und soziale Innovationen in der zweiten Hälfte des 4. Jahrtausends v. Chr. In S. Hansen and J. Müller (eds), *Sozialarchäologische Perspektiven: Gesellschaftlicher Wandel 5000–1500 v. Chr. zwischen Atlantik und Kaukasus*. Archäologie in Eurasien 24, 153–191. Mainz, Philipp von Zabern.

Hansen, S. (2013) Innovative Metals: Copper, Gold and Silver in the Black Sea Region and the Carpathian Basin During the 5th and 4th Millennia BCE. In S. Burmeister, S. Hansen, M. Kunst and N. Müller-Scheessel (eds), *Metal Matters: Innovative Technologies and Social Change in Prehistory and Antiquity*, 137–170. Rahden/Westf., Marie Leidorf.

Hansen, S. (2014a) Gold and Silver in the Maikop Culture. In Meller, H., Risch, R. and Pernicka, E. (eds), *Metalle der Macht – Frühes Gold und Silber. Metals of Power – Early Gold and Silver. 6. Mitteldeutscher Archäologentag vom 17. bis 19. Oktober 2013 in Halle (Saale)*. Tagungen des Landesmuseums für Vorgeschichte Halle 11/1, 389–410. Halle (Saale), Landesamt für Denkmalpflege und Archäologie Sachsen-Anhalt and Landesmuseum für Vorgeschichte.

Hansen, S. (2014b) The 4th Millennium: A Watershed in European Prehistory. In B. Horejs and M. Mehofer (eds), *Western Anatolia Before Troy. Proto Urbanisation in the 4th Millennium BCE? Proceedings of the International Symposium Held at the Kunsthistorisches Museum Wien. Vienna, Austria, 21–24 November 2012*, 243–260. Vienna, Austrian Ac. of Sciences Press.

Hansen, S. and Helwing, B. (2016) Die Anfänge der Silbermetallurgie in Eurasien. In M. Bartelheim, B. Horejs and R. Krauß (eds), *Von Baden bis Troia; Ressourcennutzung, Metallurgie und Wissenstransfer. Eine Jubiläumsschrift für Ernst Pernicka*, 41–58. Rahden/Westf., Marie Leidorf.

Hansen Streily, A. (2000) Bronzezeitliche Töpferwerkstätten in der Ägäis und Westanatolien. Unpublished thesis, University of Mannheim.

Hauptmann, A. (2000) *Zur frühen Metallurgie des Kupfers in Fenan/Jordanien*. Bochum, Deutsches Bergbau-Museum.

Hauptmann, A. and Gambaschidze, I. (2001) Antimon – Eine metallurgische Besonderheit aus dem Kaukasus. In I. Gambaschidze, A. Hauptmann, R. Slotta and Ü. Yalcin (eds), *Georgien. Schätze aus dem Land des goldenen Vlies. Katalog der Ausstellung des Deutschen Bergbau-Museums Bochum in Verbindung mit dem Zentrum für Archäologische Forschungen der Georgischen Akademie der Wissenschaften Tbilissi vom 28. Oktober 2001 bis 19. Mai 2002*, 150–154. Bochum, Deutsches Bergbau-Museum.

Hauptmann, A., Schmitt-Strecker, S., Begemann, F. and Palmieri, A. (2002) Chemical Composition and Lead Isotopy of Metal Objects from the 'Royal' Tomb and Other Related Finds at Arslantepe, Eastern Anatolia. *Paléorient* 28/2, 43–70.

Helms, M. (1992) Long Distance Contacts, Elite Aspirations, and the Age of Discovery. In E. M. Schortman and P. A. Urban (eds), *Resources, Power, and Interregional Interaction*, 157–174. New York, Plenum Press.

Higham, T., Chapman, J., Slavchev, V., Gaydarska, B., Honch, N., Yordanov, Y. and Dimitrova, B. (2007) New Perspectives on the Varna Cemetery (Bulgaria) – AMS Dates and Social Implications. *Antiquity* 81, 640–654.

Ivanov, I. (1978) Раннохалколитни гробоведо град Варна. *Известия на народния музей Варна* 15, 81–93.

Kerner, S. (2010) Craft Specialisation and Its Relation with Social Organisation in the Late 6th to Early 4th Millennium BCE of the Southern Levant. *Paléorient* 36, 179–198.

Kienlin, T. (2014) Aspects of Metalworking and Society from the Black Sea to the Baltic Sea from the Fifth to the Second Millennium BCE. In B. W. Roberts and C. P. Thornten (eds), *Archaeometallurgy in Global Perspective. Methods and Syntheses*, 447–472. New York, Springer.

Klimscha, F. (2013) Another Great Transformation: Technical and Economic Change from the Chalcolithic to the Early Bronze Age in the Southern Levant. *Zeitschrift für Orient Archäologie* 6, 82–112.

Lechtman, H. (1996) Arsenic Bronze: Dirty Copper or Chosen Alloy? A View from the Americas. *Journal of Field Archaeology* 23, 477–514.

Leusch, V., Pernicka, E. and Armbruster, B. (2014) Chalcolithic Gold from Varna – Provenance, Circulation, Processing, and Function. In. H. Meller, R. Risch and E. Pernicka (eds), *Metalle der Macht – Frühes Gold und Silber. Metals of Power – Early Gold and Silver. 6. Mitteldeutscher Archäologentag vom 17. bis 19. Oktober 2013 in Halle (Saale)*. Tagungen des Landesmuseums für Vorgeschichte Halle 11/1, 165–182. Halle (Saale), Landesamt für Denkmalpflege und Archäologie Sachsen-Anhalt, Landesmuseum für Vorgeschichte.

Levy, T. E. (2007) *Journey to the Copper Age. Archaeology in the Holy Land*. San Diego, CA, San Diego Museum of Man.

Maisuradse, B. and Gobedschischwili, G. (2001) Alter Bergbau in Ratscha. In I. Gambaschidze, A. Hauptmann, R. Slotta and Ü. Yalcin (eds), *Georgien. Schätze aus dem Land des goldenen Vlies. Katalog der Ausstellung des Deutschen Bergbau-Museums Bochum in Verbindung mit dem Zentrum für Archäologische Forschungen der Georgischen Akademie der Wissenschaften Tbilissi vom 28. Oktober 2001 bis 19. Mai 2002*, 130–135. Bochum, Deutsches Bergbau-Museum.

Meliksetian, K., Kraus, S., Pernicka, E., Avetissian, P., Devejian, S. and Petrosyan, L. (2011) Metallurgy of Prehistoric Armenia. In Ü. Yalcin (ed.), *Anatolian Metal V*. Der Anschnitt, Beiheft 24, 115–118. Bochum, Deutsches Bergbau-Museum.

Mille, B., Besenval, R. and Bougarit, D. (2004) Early 'Lost-Wax-Casting' in Baluchistan (Pakistan): The 'Leopards-Weight' from Shahi-Tump. In T. Stöllner, R. Slotta and A. Vatandoust (eds), *Persiens Antike Pracht. Bergbau, Handwerk, Archäologie. Katalog der Ausstellung des Deutschen Bergbau-Museums Bochum vom 28. November 2004 bis 29. Mai 2005*, 264–271. Bochum, Deutsches Bergbau-Museum.

Müller-Prothmann, T. and Dörr, N. (2009) *Innovationsmanagement. Strategien, Methoden und Werkzeuge für systemische Innovationsprozesse*. Munich, Carl Hanser Verlag GmbH & Co. KG.

Di Nocera, G. M. (2010) Metals and Metallurgy. Their Place in the Arslantepe Society between the End of the 4th and Beginning of the 3rd Millennium BCE. In M. Frangipane (ed.), *Economic Centralisation in Formative States. The Archaeological Reconstruction of the Economic System in 4th Millennium Arslantepe*. Studi di Preistoria Orientale 3, 255–274. Rome, Sapienza Univ. di Roma, Dipart. di Scienze Storiche Archeologiche e Antropologiche dell'Antichità.

Oates, J., McMahon, A., Karsgaard, P., Al Quntar, S. and Ur, J. (2007) Early Mesopotamian Urbanism: A New View from the North. *Antiquity* 81, 585–600.

Özbal, R. (2010) A Comparative Look at Halaf and Ubaid Period. Social Complexity and the Tell Kurdu Case. *TÜBA-AR* 13, 39–59.

Orell, J. H. (1786) *Vollständige theoretische und praktische Geschichte der Erfindungen. Oder Gedanken über die Gegenstände aller drey Naturreiche, die im menschlichen Leben teils zur Beschäftigung des Körpers, teils auch der Seele beygetragen haben.* Zürich, Füeßly.

Pernicka, E. (1998) Die Ausbreitung der Zinnbronze im 3. Jahrtausend. In B. Hänsel (ed.), *Mensch und Umwelt in der Bronzezeit Europas. Man and Environment in European Bronze Age. Die Bronzezeit: Das erste goldene Zeitalter Europas,* 135–147. Kiel, Oetker-Voges Verlag.

Pernicka, E. (2013) Die Ausbreitung der Metallurgie in der Alten Welt. In J. J. Piotrovskij (ed.), *Бронзовый Век Европа без Границ. Четвертое - первое тысячелетия до н. э. Bronzezeit. Europa ohne Grenzen. 4.–1. Jahrtausend v. Chr.,* 66–78. St Petersburg, Tabula Rasa.

Pernicka, E., Schmidt, K. and Schmitt-Strecker, S. (2002) Zum Metallhandwerk. In K. Schmidt (ed.), *Norşuntepe. Kleinfunde 2, Artefakte aus Felsgestein, Knochen und Geweih, Ton, Metall und Glas.* Archaeologica Euphratica 2, 115–145. Mainz, Philipp von Zabern.

Pétrequin, P., Cassen, S., Croutsch, C. and Herrera, M. (2002) La valorisation sociale des longues haches dans l'Europe néolithique. In J. Guilaine (ed.), *Matériaux, productions, circulations du néolithique à l'âge du Bronze,* 67–98. Paris, Éditions Errance.

Pétrequin, P., Pétrequin, A.-M., Errera, M., Cassen, S., Croutsch, C., Klassen, L., Rossy, M., Garibaldi, P., Isetti, E., Rossi, G. and Delcaro, D. (2005) *Beigua, Monviso e Valai. All'origine delle grandi asce levigate di origine alpina in Eurpa occidentale durante il V millennio.* Rivista di Scienze Preistoriche 55, 265–322.

Pétrequin, P., Cassen, S., Errera, M., Tsonev, T., Dimitrov, K., Klassen, L. and Mitkova, R. (2012) Les haches en 'jades alpins' en Bulgarie. In P. Pétrequin, S. Cassen, M. Errera, L. Klassen, A. Sheridan and A. M. Pétrequin (eds), *Jade. Grandes haches alpines du Néolithique européen Ve et Ive millénaires av. J.-C* 2, 1231–1279. Besançon, Presses Univ. de Franche-Comté *et al.*

Piotrovskij, J. J. (2013) Maikop-Kultur. In J. J. Piotrovskij (ed.), *Бронзовый Век Европа без Границ. Четвертое - первое тысячелетия до н. э. Bronzezeit. Europa ohne Grenzen. 4.–1. Jahrtausend v. Chr.,* 308–340. St Petersburg, Tabula Rasa.

Plessing, F. V. L. (1787) *Memnonium oder Versuche zur Enthüllung der Geheimnisse des Altertums* 1. Leipzig, Weygand.

Radivojević, M., Rehren, T., Pernicka, E., Šljivar, D., Brauns, M. and Borić, D. (2010) On the Origins of Extractive Metallurgy: New Evidence from Europe. *Journal of Archaeological Science* 37, 2775–2787.

Radivojević, M. and Rehren, T. (2015) Paint It Black: The Rise of Metallurgy in the Balkans. *Journal of Archaeological Method and Theory* 22, 1–38.

Rehren, T., Boscher, L. and Pernicka, E. (2012) Large Scale Melting of Speiss and Arsenical Copper at Early Bronze Age Arisman, Iran. *Journal of Archaeological Science* 39/6, 1717–1727.

Renfrew, C. (1969) The Autonomy of the South-East European Copper Age. *Proceedings of the Prehistoric Society* 35, 12–47.

Roberts, B., Thornton, C. P. and Pigott, V. C. (2009) Development of Metallurgy in Eurasia. *Antiquity* 83, 1012–1022.

Roux, V., Mille, B. and Pelegrin, J. (2013) Innovations céramiques, métallurgiques et lithiques au Chalcolithique: Mutations sociales, mutations techniques. In J. Jaubert, N. Fourment and P. Depaepe (eds), *Transitions, ruptures et continuité en Préhistoire 1, Évolution des techniques – Comportements funéraires – Néolithique ancient. Actes du XXVIIe Congrès préhistorique de France. Bordeaux – Les Eyzies, 31 mai–5 juin 2010,* 61–73. Paris, Société Préhistorique Française.

Schoop. U.-D. (1995) *Die Geburt des Hephaistos. Technologie und Kulturgeschichte neolithischer Metallverwendung im vorderen Orient.* Espelkamp, Marie Leidorf.

Schubert, E. and Schubert, F. (1999) Die Hammeräxte vom Typ Handlová. In F. R. Herrmann (ed.), *Festschrift für Günter Smolla* 2. Materialien zur Vor- und Frühgeschichte von Hessen 8, 657–671. Wiesbaden, Selbstverlag des Landesamtes für Denkmalpflege Hessen.

Shalev S., Goren, Y., Levy, T. E. and Northover, J. P. (1992) A Chalcolithic Mace Head from the Negev, Israel: Technological Aspects and Cultural Implications. *Archaeometry* 34, 63–71.

Shalev, S. and Northover, J. P. (1993) The Metallurgy of the Nahal Mishmar Hoard Reconsidered. *Archaeometry* 35, 35–47.

Sherratt, A. (1997) Sherratt, Economy and Society in Prehistoric Europe. Edinburgh. Edinburgh University Press.

Shugar, A. N. and Gohm, C. J. (2011) Developmental Trends in Chalcolithic Copper Metallurgy: A Radiometric Perspective. In J. L. Lovell and Y. M. Rowan (eds), *Culture, Chronology and the Chalcolithic: Theory and Transition.* Levant Supplementary Series 9, 133–148. Oxford, Oxbow Books.

Stein, G. (2012) The Development of Indigenous Social Complexity in Late Chalcolithic Upper Mesopotamia in the 5th–4th Millennia BCE – An Initial Assessment. *Origini* 34, 125–151.

Stöllner, T., Gambaschidze, I., Hauptmann, A., Mindiašvili, G. I., Gogočuri, G. and Steffens, G. (2010) Goldbergbau in Südostgeorgien. Neue Forschungen zum frühbronzezeitlichen Bergbau in Georgien. In S. Hansen, A. Hauptmann, I. Motzenbäcker and E. Pernicka (eds), *Von Majkop bis Trialeti. Gewinnung und Verbreitung von Metallen und Obsidian in Kaukasien im 4.–2. Jt. v.Chr. Beiträge des Internationalen Symposiums in Berlin vom 1.–3. Juni 2006.* Kolloquien zur Vor- und Frühgeschichte 13, 1–36. Bonn, Habelt.

Stöllner, T. (2014) Gold in the Caucasus: New Research on Gold Extraction in the Kura-Araxes Culture of the 4th Millennium and the Early 3rd Millennium. In. H. Meller, R. Risch and E. Pernicka (eds), *Metalle der Macht – Frühes Gold und Silber. Metals of Power – Early Gold and Silver. 6. Mitteldeutscher Archäologentag vom 17. bis 19. Oktober 2013 in Halle (Saale).* Tagungen des Landesmuseums für Vorgeschichte Halle 11/1, 71–110. Halle (Saale), Landesamt für Denkmalpflege und Archäologie Sachsen-Anhalt, Landesmuseum für Vorgeschichte.

Szeverényi, V. (2013) The Earliest Copper Shaft-Hole Axes in the Carpathian Basin: Interaction, Chronology and Transformations of Meaning. In A. Anders and G. Kulcsár (eds), *Moments in Time. Papers Presented to Pál Raczky on His 60th Birthday.* Ősrégészeti tanulmányok 1, 661–669. Budapest, L'Harmattan.

Tadmor, M., Kedem, D., Begemann, F., Hauptmann, A., Pernicka, E. and Schmitt-Strecker, S. (1995) The Nahal Mishmar Hoard from the Judean Desert: Technology, Composition, and Provenance. *Atiqot* 27, 95–148.

Thornton, C. (2010) The Rise of Arsenical Copper in Southeastern Iran. *Iranica Antiqua* 45, 31–50.

Thornton C. and Roberts, B. W. (2009) Introduction: The Beginnings of Metallurgy in Global Perspective. *Journal of World Prehistory* 22, 181–184.

Twaltschrelidze, A. G. (2001) Erzlagerstätten in Georgien. In I. Gambaschidze, A. Hauptmann, R. Slotta and Ü. Yalcin (eds), *Georgien. Schätze aus dem Land des goldenen Vlies. Katalog der Ausstellung des Deutschen Bergbau-Museums Bochum in Verbindung mit dem Zentrum für Archäologische Forschungen der Georgischen Akademie der Wissenschaften Tbilissi vom 28. Oktober 2001 bis 19. Mai 2002*, 78–89. Bochum, Deutsches Bergbau-Museum.

Ussishkin, D. (2014) The Chalcolithic Temple in Ein Gedi: Fifty Years after Its Discovery. *Near Eastern Archaeology* 77, 15–26.

Weeks, L. (2012) Metallurgy. In D. T. Potts (ed.), *A Companion to the Archaeology of the Ancient Near East* 1, 295–316. Chichester *et al.*, Wiley-Blackwell.

Wenger, E. (1998) *Communities of Practice: Learning, Meaning and Identity*. Cambridge, Cambridge University Press.

Yalçin, Ü (2000) Anfänge der Metallverwendung in Anatolien. In Ü. Yalçin (ed.), *Anatolian Metal I*, 17–30. Bochum, Deutsches Bergbau-Museum.

Yalçın, Ü. and Yalçın, H. G. (2008) Der Hortfund von Tülintepe, Ostanatolien. In Ü. Yalçın (ed.), *Anatolian Metal IV. Der Anschnitt*, Beiheft 21, 101–123. Bochum, Deutsches Bergbau-Museum.

Черных, Е. Н. (1966) *История Древнейшей Металлургии восточной Европы*. Moscow, Наука.

Резепкин, А. Д. (2012) *Новосвободненская культура (на основе материалам могильника Клады)*. Труды иимк ран 37. St Petersburg, Нестор-История.

Chapter 13

The Diffusion of Know-How within Spheres of Interaction: Modelling Prehistoric Innovation Processes between South-West Asia and Central Europe in the 5th and 4th Millennia BC

Florian Klimscha

Introduction

This paper deals with the problems archaeologists face when analysing innovation processes. For a long time, traditional archaeological modelling was necessarily an explanation *ex oriente*. With the advent of the radiocarbon method, this paradigm changed and has since been substituted by a multitude of local narratives.

This paper demonstrates the difficulties inherent in such approaches by highlighting the biased nature of the archaeological record and the resulting selectivity in defining key technologies. All too often the search for what is 'oldest' prompts prehistorians to disregard the intentionality of many archaeological sources or the necessity for a comparable state of research. The narratives thus created are based on a positivistic viewpoint, which is often not fully reflected.

My contention is that for the modelling of prehistoric innovations it is necessary to appreciate locally available know-how and its integration into production processes just as fully as the technical prerequisites and their social contexts. I shall argue that techniques in prehistory can often better be grasped as transformations of a collection of technical components. Only if we take these factors into account will we have any real chance to identify the spatial and chronological dimensions of prehistoric innovation processes.

Innovation and technique in prehistory

Studying the diffusion of innovations in prehistory will ultimately result in re-assessing the ground-breaking work of V. Gordon Childe. While modern studies seem to have left very little of Childe's original ideas standing, many parts of his underlying modelling are still useful, and his work remains essential to the understanding of the long-term relation between technological stages and social developments.

Childe's model was based on the invention of key technologies in an assumed Near Eastern core and their successive diffusion to the periphery. The combination of key techniques enabled societies to become more complex and finally achieve state level. In *Social Evolution*, his most important study dealing with technical innovations, Childe nevertheless points out in no uncertain terms that technology did not force society to change. Yet, he was still able to demonstrate the correlation between certain key techniques and the rise of complex societies in the Near East (Childe 1948; 1951).

During the radiocarbon revolution, these paradigms were confronted by many challenges as the scientific dating of many finds contradicted the age attributed to them in typological sequences. Colin Renfrew realised that this was an explosive issue for prehistory and used it to tackle the Childean models (Renfrew 1969; 1973). Because in some cases the periphery now offered older evidence than the presumed core, Renfrew argued that it may have been internal technical evolutions independent of Mesopotamia and Egypt that were responsible for the uptake of key innovations. The technical development of pottery kilns and the production of graphite pottery in south-eastern Europe, for instance, were thought to have been inspiring pyrotechnical experiments, and this, in turn, resulted in the smelting of copper ores. Thus, the connection between society and technology was loosened and human actors were given a much more prominent role in the scholarly discourse.

Diffusionist models have also been queried from other angles. The ethnographic data on which the (neo-) evolutionary stages were based have been thoroughly reviewed (Wolf 1982; Jung 2011) as has the concept of social complexity (Yoffee 2005; Pauketat 2007). Modern chronologies conflict with several of Childe's typological relations and sometimes even revert them. Does this mean that evolutionist and diffusionist models are so flawed that we should disregard them completely?

I would not say so. Instead, I shall argue here that, for all their shortcomings and modifications, they deserve to be defended against several key criticisms that have been levelled at them. The chronological discrepancies that loosened the possibility of diffusionist connections are partly due to the uneven state of research and the limited use of scientific dating methods outside central Europe.

Furthermore, while recent or sub-recent societies may have been influenced by Western colonial powers, this is not relevant for the archaeological issue at hand. The much broader timeframe that archaeology operates within still allows for answers to the question of how societies were able to create pristine, complex systems of relationships. Techniques had to be invented and developed to a stage where the innovation process could start to embed them in the socio-economic system. In the long run, a society with knowledge of the wheel or metallurgy or writing will develop differently from one without this knowledge. While freeing archaeological discourse from ethnographic data will definitely make the resulting models vaguer, it will also enable us to sidestep many of the criticisms mentioned above.

Long-term perspectives on the diffusion of techniques and the resulting social consequences are indeed necessary to fully understand the complexity of social change. Banning a diffusionist perspective will only help to transform local archaeological records into local narratives. The diffusion of techniques is much more than the passive ingestion of foreign know-how. It includes its adoption, recombination, transformation and ultimately further innovation-processes. This enables us to overcome the anachronistic division of prehistory into important centres and secondary or tertiary peripheral zones with a view to highlighting transformations within a complex decentralised network of information transfer.

Rethinking early wagons beyond diffusion

The introduction of early wheeled vehicles into western Eurasia is a good example of the problems inherent both in traditional diffusionist approaches and the anti-diffusionist stance. The evidence for early wagons is best summed up by Stefan Burmeister (2004a). There is no visible chronological gradient between wagons in the Near East and the North Sea coast. In fact, there is even evidence that the central European finds may be slightly older than those from Mesopotamia

(Bernbeck 2004, n. 1; Mischka 2011). Yet Burmeister clearly states that he favours neither the diffusion of the complete technical bundle nor of several independent inventions (*cf.* Vosteen 1999 for such a viewpoint) but instead argues for the diffusion of a mental template ('stimulus diffusion' *sensu* Kroeber 1940). A very quick diffusion of wagon technology has also been proposed by Jan Alber Bakker *et al.* (1999), but the dilemma with the archaeological record is that it is very difficult indeed to substantiate either of these positions (*cf.* also Maran 2004b).

Mapping all known finds from the 4th and early 3rd millennia improves our prospects of understanding of this phenomenon and opens up the possibility of modelling the adoption process (Fig. 13.1). Taking all its deficiencies into account, the archaeological record can still be understood as partly representative of the social appraisal accorded to wheeled vehicles. Despite the blurring effect of cultural and natural filters, I have argued that the simultaneous appearance of evidence for wagons in several different region-specific contexts can be understood as reflecting the social impact of the 'take-off' (Klimscha 2017).

To understand this point, it is helpful to briefly recall the general theory of the innovation process. Everett Rogers described it as the successive adoption of a given technique by a constantly rising number of users over time (Rogers 2003). Several stages can be identified within this diffusion. The most decisive one is the 'take-off. It is during this phase, which comes shortly after the initial introduction, that a critical number of opinion leaders adopt an innovation. This helps it succeed, even though it is still imperfect and may be at odds with traditions or rival technologies. If Rogers' model of the diffusion process is taken at face value, the take-off of wheeled vehicles has to be imagined as appearing in the third quarter of the 4th millennium (Fig. 13.2). Since all the components required for constructing wagons had been available since the Early Neolithic and the 'earliest' evidence of wheeled wagons in Europe displays not only a surprising degree of technical elaboration but also a range of completely different designs, it seems plausible to either argue for multiple inventions, as Markus Vosteen did (1999), or to interpret the evidence as proof of advanced stages of a technical evolution in which the technology had already been adapted to local social and environmental necessities (Klimscha 2017). From an innovation-theoretical viewpoint, I would argue that the different designs of wagons in the second half of the 4th millennium indicate a longer tradition of wagon usage. It is within the early 4th millennium that the archaeological data point to a new quality of connectivity between societies in the north European Plain, the Alps and the Black Sea, and this could be the social background for the quick diffusion of social appreciation for wagon technology (Klimscha 2017).

After *c.* 3500/3400 BC the amount of evidence quickly grows: wagons are known to have been used in the Wartberg

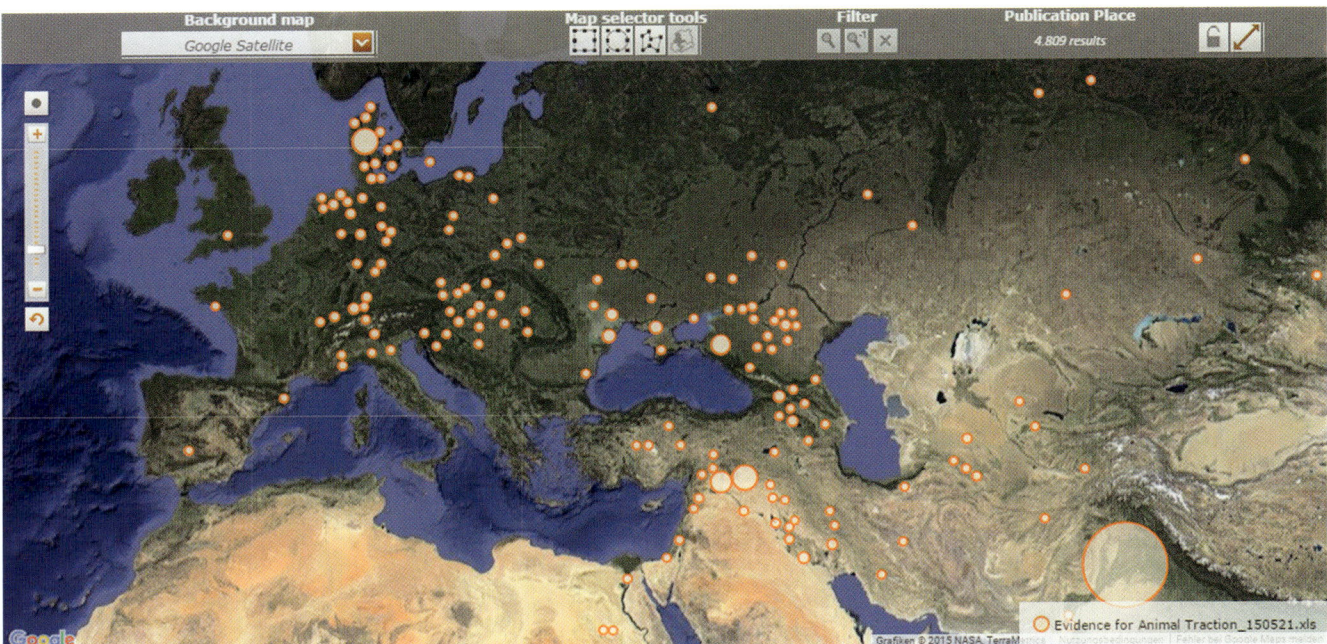

Figure 13.1: *Repartition of evidence for the use of cattle-traction* c. *5000–2800 BC (© S. Hansen, F. Klimscha, J. Renn, Digital Atlas of Innovations).*

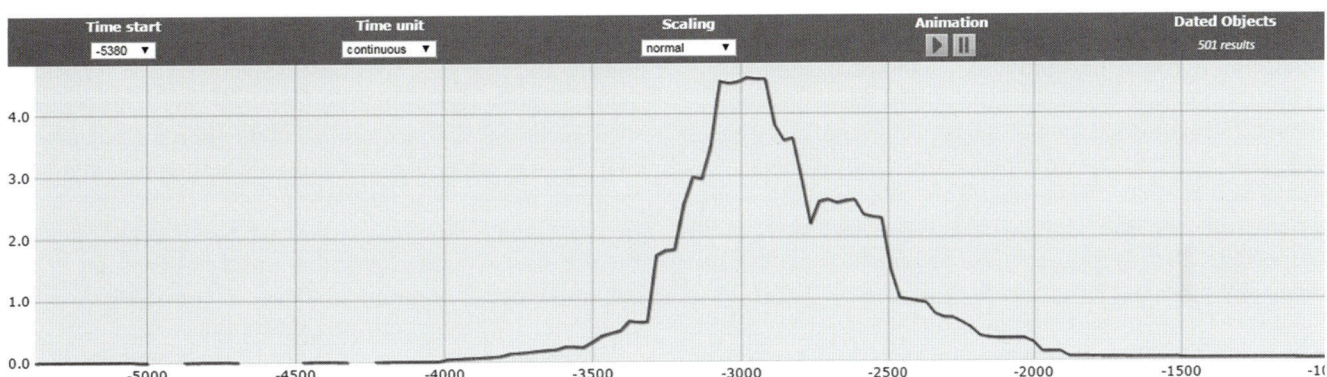

Figure 13.2: *Histogram of evidence for the use of cattle-traction* c. *5000–2800 BC (© S. Hansen, F. Klimscha, J. Renn, Digital Atlas of Innovations).*

Culture of Western and Southwestern Germany (Günther 1990), they figure as bog-finds from northern Germany and the Low Countries (Burmeister 2004b), as assumed cattle burials from Jutland (Johannsen and Laursen 2010), as depictions on pottery in the North European Plain including Poland (Bakker 2004), and copper figurines in eastern and central Europe (Bakker 2004), as wheeled drinking cups in the Carpathian Basin (Maran 2004a), as rock art and lake finds in the Circum-Alpine region (Schlichtherle 2004), in graves in the North Pontic steppes and the Caucasus (Gej 2004; Trifonov 2004; Turetskij 2004) and finally as pictograms and miniatures in Anatolia and Syria-Mesopotamia (Bernbeck 2004; Crouwel 2004).

The innovation process can only be understood from a long-term perspective taking into account the fact that the wagon was not an isolated technology but a combination of previously known techniques and modes of organisation. While most papers dealing with this problem stress the brevity of the diffusion, this is only true from a very abstract point of view. In fact, the 'horizon' of the earliest wagons extends across at least 500 years.

Besides the difficulty of dating them, one peculiarity of the early evidence for wheels and wagons is that it features regionally specific deposition patterns enabling us to distinguish various find groups (see above). What has not been sufficiently borne in mind is the connection of most find groups with funerary contexts. The wagon, or parts of it, may be placed in the grave itself as part of the inventory, which is characteristic of the northern Black Sea region (Gej 2004; Trifonov 2004; Turetskij 2004). Closely connected to this is a find group in which not the wagon itself but its means of traction, *i.e.* a pair of cattle, are deposited. This is known both from within the Baden culture and from central Germany (Behrens 1963; Stahlhofen and Kurzhals 1983, 157–160; Friederich and Hoffmann 2013) and, if one agrees with Johannsen's interpretation of the stone-heap graves, in Jutland as well (Johannsen and Laursen 2010). However, the depictions within the Wartberg culture are also only known from graves (Günther 1990; 1997; *cf.* also: Schunke 2013) and the same is true of the famous wheeled drinking vessels from the Baden culture (Maran 2004a) as well as miniature wheels from the Corded Ware culture (Seregély 2004). Wagon tracks (and plough marks) are only preserved when some structure is built over them soon after their materialisation (Zich 1992/1993; Mischka 2011; Mischka 2013). In most cases this is a barrow, so their preservation is also the result of a cultural bias favouring barrow constructions for graves.

There is some evidence independent of graves, but much of it is equally biased, like the evidence from hoards (Bakker 2004), or very difficult to date, like the rock art in the western Alps (de Lumley 2003), or only assumed to include wheels (Гусев 1998). But the Circum-Alpine lake dwellings and the bog discoveries in northern Germany and the Low Countries (Burmeister 2004b; Schlichtherle 2004) do feature some finds that are currently interpreted as non-intentional. While the northern find province might be thought to be peripheral to most Neolithic innovations, the alpine region should give us a good idea of the appearance of the wagon in Europe (even though, of course, this again is influenced by the non-continual settlement of the lake borders). Thus, the vast amount of evidence for early wheeled vehicles is connected with the social actions of prehistoric communities, especially the way they buried their dead.

In other words, the distribution map of early wagons in Europe reflects the role of wagon technology in regionally and culturally specific funeral rites but not the presence and absence of wagons *tout court*. Once the bias inherent in the map is understood, it becomes obvious what can actually be said about the diffusion of wagons.

When cattle burials are seen as *pars pro toto* for a wagon (in the case of the stone-heap graves it could even be argued that they are a *synecdoche* for a wagon), the same is true of the so-called wagon graves in the Pontic region. In the former the draught animals were buried, in the latter it was the chassis and/or wheels. Within the Wartberg culture, the depiction of wagons is also shortened and, apart from cattle drawing a two-wheeled wagon, cattle connected by a horizontal line or just two cattle next to each other are also known (Günther 1990; Bakker 2004, 285 fig. 4). The copper figurines that are usually also interpreted as wagons follow the same formula and just depict two cattle under a yoke (Bakker 2004, 284 fig. 2), and the same is true of the sculptured depiction on a handle from Krężnica Jara in Poland (Vosteen 1999, pl. 107, 62) or the vessels from Radošina (Maran 2004a, 271 fig. 2)

and Boglárlelle (Ecsedy 1982). In fact, the elliptical image can be found in an even wider area and enables us to posit a connection between the Black Sea/central Europe with regions as distant as the southern Levant, where a similar and contemporaneous sculpted depiction can be found on a bowl from Tell el-Farah (Dayagi-Mendels and Rozenberg 2010, 39 fig. 4). Given the absence of wagon graves, it is interpreted within a different narrative and seen as a *pars pro toto* depiction of a plough. Nonetheless, this is questionable because we also have depictions of sleds in Western Asia from the second half of the 4th millennium (Littauer and Crouwel 1979 fig. 2; Littauer and Crouwel 1990 fig. 1; Frangipane 1996 fig. 76a; Burmeister 2004b, 22 fig. 5 upper part), which triggers the assumption/proposition that the yoked pair of cattle in the bowl could also be a *pars pro toto* depiction of a sled pulled by cattle.

But how is it possible to distinguish cattle pairs used to pull vehicles from those used to pull ploughs? And is there any way of distinguishing wheeled from unwheeled vehicles in such contexts? Were the cattle pairs known in Europe unable to pull sleds (or a travois, which would also be a possibility; *cf.*: Pétrequin *et al.* 2002; Schlichtherle 2004, 302–303)?

I believe that such distinctions are impossible. Instead, we need to realise that it is the same technological component that is used for pulling ploughs, travois and wagons in Europe and ploughs, sleds (and wagons?) in the Near East.

The significance of the cattle pair for the functioning of these technical systems resulted in the shorthand depictions of just two cattle under a yoke. Whether this sign was invented independently or also diffused is difficult to say. A major point, however, is that we have strong evidence for the use of animal traction and possibly also of the wagon, which is often ignored due to its integration into a completely different scholarly discourse.

Discussion about the diffusion of the wagon must take into account the fact that the wagon itself is not replicated according to a standard template construct or exchanged as a finished artefact all over western Eurasia. What we see in the archaeological record is the local re-interpretation of several technical components according to social, economic and environmental requirements. This is why we have two-wheeled and four-wheeled wagons, travois, sleds, ploughs, different wheel-types as well as rotating axles and fixed axles within the same chronological horizon. All these are variations of the combination of three technical components: traction (including the yoke), a rotating axle (including two wheels) and a wooden chassis. The wagon uses all three components, the sled and the travois omit the rotary motion, and the plough just uses the traction required to pull a 'machine'. At a different level, individual components of these basic combinations may then be further differentiated, and wagons can have two or four wheels *etc*.

The most important part of all three technical systems is the traction by a team of two cattle. I would therefore like to suggest that to fully understand the innovation the wagon represents, it is necessary to refocus our perspective and see the wagon as the result of appropriating and reinterpreting animal traction.

It is impossible to determine whether there is an internal chronology hidden within the 500-year time span. However, we can assume that the very early stages of the innovation process are not visible in the archaeological record and that the invention of the wagon and all the technologies connected with it must have taken place earlier. Where exactly this happened is currently difficult to decide. While early evidence for the use of traction can be found in western Europe and the Levant, the earliest evidence for the wheel that we have at present does indeed come from the Northern Pontic zone (Klimscha 2017).

Innovating pyrotechnology: A new perspective on early metallurgy

Along with wheeled vehicles, the use of metals has been one of the most frequently cited examples in arguing for or against diffusionist models in prehistory. While in traditional models metals were first used in the Near East and then successively transmitted to the culturally 'less advanced' regions of Europe (*e.g.* Müller 1905), Renfrew reversed the picture (Renfrew 1969) when he stressed that the Copper Age cemetery of Varna on the Bulgarian Black Sea coast (Fol and Lichardus 1988) was older than anything comparable in the Near East, thus proving that the use of metallurgy was not invented in the Orient, nor did it lead, as Childe had argued (see above), to urbanism. Recent studies have made Varna even older (Higham *et al.* 2007) and new evidence from Vinča-Belo Vrda has made it possible to date the beginning of extractive metallurgy to around 5000 BC (Šljivar *et al.* 2006; Borić 2009; Radivojević *et al.* 2010). Varna is no longer the starting point for extractive metallurgy in the Balkan regions but instead the result of at least 500 years of continuous local experimentation with cast metals (Hansen 2013).

This, however, revived Renfrew's (1969) old narrative, and while one scholarly position still clings to models of monocentric diffusion (Roberts *et al.* 2009), the chronological primacy of smelting in the Western Balkans has recently reared its head once again (Radivojević and Rehren 2015).

To fully understand this process, it is not sufficient to just compare the 'oldest' dates of a single innovation in itself but to recognise smelting as the local reinterpretation of pyrotechnical know-how heavily filtered by taphonomic processes and social rules.

The importance of Varna has been discussed for a long time. Some colleagues have argued for one extreme in which influences can be seen reaching as far as Brittany (Cassen 1991; Cassen *et al.* 2011), others have favoured a model in which Varna is more or less unique and neighbouring societies stay egalitarian (Kienlin 2008). But similarly furnished and equally rich tombs are also known from

Hungary and Slovakia (Šiška 1964; Vizdal 1977; Zalai-Gaal *et al.* 2011), which suggests that similar hierarchies to those bodied forth in the Varna cemetery were indeed known across a much larger region.

However, these are just small spots on a big map. Other cemeteries display completely different grave-goods; here even the richest graves merely boast heavily reduced stone axes and nothing else (Angelova 1986). Is the sparsity of copper objects in cemeteries other than Varna (or to be more general: in some areas of the Balkans) indicative of less intensive copper use? The archaeological record is full of smallish copper tools for a variety of tasks, and the large number of small copper tools within the settlement layers of tell sites clearly contradicts such a notion (Hansen 2015, 279 fig. 17). But support also comes from closer analysis of the larger copper artefacts. The distribution of flat axes, adze axes and hammer axes made from copper is not arbitrary. Once the mapping is sorted by source type, regional preferences become apparent. Copper is only known from graves in the Western Pontic area and along the river Tisza, while in hoards it is found in parts of Transylvania, the Iron Gates, Oltenia and Moldova (Fig. 13.3). Depositions of copper in graves or hoards are mutually exclusive. Copper finds from the same context can be limited to specific regions. Identifying the preservation of larger copper artefacts as the result of region-specific deposition patterns is a very important argument for interpretation.

If distribution reflects *intentional* deposition patterns, then it cannot be connected with economic cycles. The absence of copper-rich graves does not at all signify a

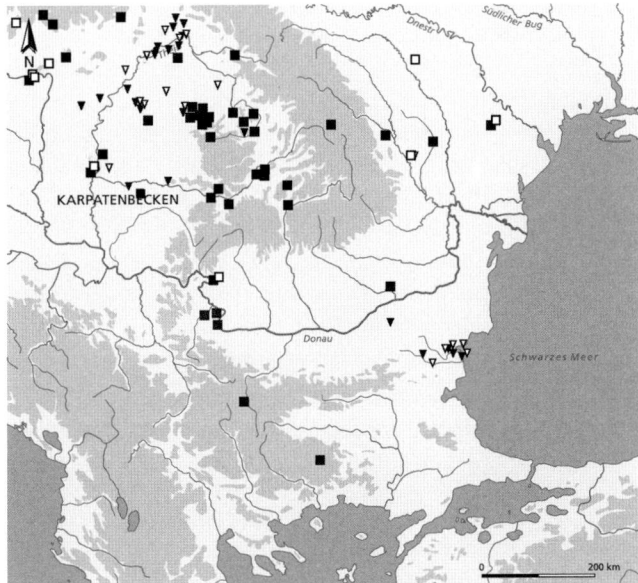

Figure 13.3: Repartition of copper axes in the 5th millennium in the Eastern Balkan region (from Todorova 1981; Vulpe 1975; Patay 1984; Dergačev 2002). Legend: black-white = flat axe; black = axe-adze or shafthole axe; triangle = grave-find; square = hoardfind.

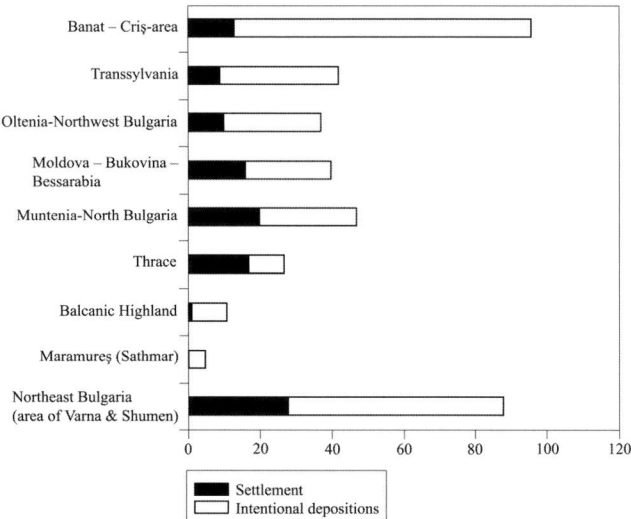

Figure 13.4: Comparison of the number of settlement-finds and intentionally deposited finds in graves and hoards during the 5th millennium in the Eastern Balkan region (from Todorova 1981; Vulpe 1975; Patay 1984; Dergačev 2002).

lack of copper (or an absence of social elites), it simply highlights different social constellations in which metals have prestige value (*cf.* Vandkilde 1996 for a similar perspective). Discrepancies in the distribution cannot be simply interpreted as reflecting the reality of the prehistoric availability. Regions like Thrace or Muntenia were certainly *not* devoid of copper, we know this for a fact. A comparison of the amount of copper from intentional deposition sources (graves and hoards) with copper from settlements changes the picture drastically (Fig. 13.4).

If distribution maps of south-eastern Europe cannot simply be read as showing hotspots of copper use and their periphery, what is the consequence for the 'bigger picture'?

I intend to argue that the minimalistic picture drawn by some scholars is inaccurate and that, a much wider availability of the know-how in the Balkans and the Carpathian region has to be assumed.

If the distribution maps of early copper items reflect intentional deposition patterns, then the visibility of copper items is strongly connected to their integration into the ideology of prehistoric societies and their use in rituals (which prescribed their deposition inside graves and hoards). If we are discussing the chronological primacy of smelting, a large number of copper artefacts is of limited use as a criterion, and this again leads to an interesting question. How valid is an argument for south-east European primacy in extractive metallurgy based only on a relatively small number of older radiocarbon dates? Or, to put it more provocatively, could it not be possible that cultural filters and the scarcity of graves in the Halaf and Ubaid periods in western Asia simply fool us?

One might argue against this line of inquiry by pointing out that especially the very early evidence preceding Varna

and the Karanovo VI-Gumelniţa cultures derives not from graves but from settlements. But the finds are not very numerous, and their importance was only realised after the intensive use of modern methods and large series of ^{14}C-datings (*cf.* Borić 2009; Hansen 2011). Moreover, in many parts of western Asia, the time span in question (the late 6th and early 5th millennia) has only been poorly researched so far or remains 'hidden' under, or possibly even 'destroyed' by, several metres of later cultural layers. The general archaeological record, however, is heavily influenced by the personal agendas of researchers, by chance, to a small degree by an interested public and above all by urban development combined with public funding. Neither in terms of intensity nor tradition is the research situation in Anatolia, Syria, Mesopotamia, Palestine, Iran or the Caucasus comparable to that of the Balkan peninsula. At the moment, any conclusions on chronological primacy are based on a heavily distorted research and publication situation.

Indeed, the published data suggests that we ought to be careful and not rule out the Near East completely. It is here that we find very long traditions of cold metalworking in existence since the 10th millennium (Yalçin 2000) alongside a long tradition of high-temperature kilns for pottery since the Late Neolithic (Hansen Streily 2000). New research, for instance at Tell Brak, also suggests that we must expect division of labour, craft specialisation and the beginnings of urban settlements at least from the late 5th millennium onwards (Oates *et al.* 2007). There is also an impressive metalworking tradition in Israel and northern Jordan dating back to before 4600 BC (Klimscha 2013; Garfinkel *et al.* 2014). The arsenic-copper objects in the hoard from the Cave of the Treasure in the Nahal Mishmar, which were cast using the lost-wax technique and can be dated to 4300–4100 BC, display a level of technical sophistication that far surpasses the known finds from the Balkans (Bar-Adon 1980; for the dating *cf.* Aardsma 2001; Klimscha 2014).

The extraordinary archaeological visibility of the Balkan copper industry is thus based on the fact that we see a larger *ritual network* in the archaeological record. In this network, copper is removed from the economic circuits to underline the status of elite groups. In contrast to the Balkans, Western Asia has a completely different state of modern urbanisation and therefore a smaller number of excavations (sometimes even the lack of institutionalised archaeological field work), which distorts the picture.

Spheres of interaction and diffusion processes

The discrepancies in the archaeological data presented in this paper make simple diffusion models difficult to uphold. Partly, this can be explained by research biases that give certain regions preferred by archaeologists or endowed with a higher budget on excavations, have a richer archaeological record, both quantitatively and qualitatively.

Nevertheless, we also face difficulties in dating stages of the innovation process precisely, and we must face up to the fact that the future prospects of resolving this problem at a super-regional level are remote (due to the soil-specific preservation of datable material and the uneven course of the radiocarbon curve).

In recent decades, the narrative of autochthonous evolutions was deduced several times out of the haziness of the archaeological sources. I would argue that in the case of technical innovations such a perspective can easily lead to flawed understandings of prehistory because it interprets heavily biased data as objective depictions of economic relations. There are no continuous sequences of metal objects or wagons as we have – or think we have – for pottery or flint tools. And our chances of ever being able to reconstruct them are very limited because such technologies are best discerned when they have either already passed the take-off stage or are have become part of the ideology of prehistoric communities and therefore been selected for preservation.

Accordingly, it is necessary to improve our understanding of the technical innovation process by comparing not the artefacts themselves but sequences from different *chaînes opératoires*. In this way, it is possible to identify the requisite knowledge and organisation behind every working step and hence also understand the connections between reinterpretations of the same technical component.

In a society, technical systems cannot evolve or be implemented without the know-how required for the production chains. Innovation (and the destruction of traditions) is an ongoing process of serial change (Eco 1989; 1990). Small changes are thus introduced into an otherwise unchanged technical system. In the long run, however, the successive implementation of these changes and their possible recombination(s) may lead to technologies that are perceived as completely new by users and consequently may trigger an innovation process in Rogers' sense of the term that leads to significant social changes.

The chronological solution and the porous record of archaeological sources, however, hides exactly this. Accordingly, Neolithic and Chalcolithic innovation processes will probably never be reconstructed in every single detail. With a fine disregard for the source-critical and theoretical limitations of archaeological modelling, we often take the archaeological record at face value. The disparate tradition suggests large distances between our various typological stages, which in turn seems to favour models thinking in terms of huge steps, revolutions and big changes.

Evolutionary relations can of course be modelled between the different technical systems encountered in the archaeological record. But when only single artefacts are analysed (like for instance in: Basalla 1989), these are significantly blurred by the filters I have described here. For a truer picture of diffusion and technical evolution, we need to understand the complete technical system. The more complex

such a system gets, the less plausible is the availability of the requisite know-how and the creation of the necessary social sub-structures merely by internal changes within a society. The combination of the know-how manifested in the stages of different *chaînes opératoires* with traditional archaeological data and new scientific methods may enable us to identify the way in which technical components are transferred, appropriated and transformed in the long term. Accordingly, close analysis of prehistoric technical systems will, in the long run, make for negotiation between simplifying narratives of cultural autochthony and hyper-diffusionism.

The *chaîne opératoire* of copper metallurgy is extremely complex. Not only do several of the working stages require highly specialised knowledge, even minor errors within the process will ruin the finished product. This is why most researchers favour one single invention (*e.g.* Hauptmann *et al.* 1993; Pernicka 1990; Yalçin 2000; Craddock 2001). The time period in question is very long and our data very sparse, especially around 5000 BC, which is currently assumed to represent the beginning of smelting. Nevertheless, we see evidence for smelting in the Balkans (Hansen 2013) and Iran (Helwing 2013) and possibly Anatolia (Hansen 2011), the Levant (Klimscha 2013; Garfinkel *et al.* 2014) and the Northern and Eastern Pontic zone (*cf.* the typological analogies brought forward by: Boroffka 2009) in the early 5th millennium.

Is it possible that all these areas invented smelting independently of each other? Or do we need to look to other crafts where the requisite knowledge was available? Barbara Helwing recently argued that new firing kilns in Iran made the wider availability of smelted copper items possible (Helwing 2013), while back in the 1960s Renfrew saw the high-temperature firing of pottery as one inspiration for Balkan metallurgy (Renfrew 1969). From this perspective, extractive metallurgy could also be understood as a combination of mining, pyrotechnical know-how and copper-working.

Significantly more metal items have been found in the Balkans and the Carpathian Basin than in any other region. But I would still contend that the technology used to produce these items was relatively simple (Boroffka 2009; *cf.* also the article by Svend Hansen in this volume); finds from other regions are much more sophisticated (*e.g.* Bar-Adon 1980).

In the Balkans, copper axes can be identified as status symbols or prestige items. Their exchange was essential for social reproduction (Klimscha 2009; 2012) and whoever controlled their production controlled this kind of communication as well. Heavy metal ruled the Copper Age and was therefore given a prominent place in burial rites and depositions. The distortion of the record thus seems to be mainly due to the fact that metal had a distinguishing effect in the prehistoric communities of south-eastern Europe (*e.g.* via the mechanisms described by Veblen 1899). And this, again, motivated the economic 'destruction' of huge amounts of wealth.

By contrast, the innovation process is much shorter for wheeled vehicles. The *chaîne opératoire* is quite simple, so wagons could be manufactured and used in wider parts of Eurasia and diverse social systems ranging from small hamlets in the Baltic to early states in Mesopotamia. But the wagon itself can be identified as a transformation of a more general technique also seen in action in parts of north Africa and western Asia. In these areas, we find sleds, ploughs, wagons and travois making use of animal traction and the yoke, but the wheel only appears in the context of pottery manufacture (potter's wheel) and the way goods and rooms were sealed (with the cylinder seal). A reservoir of technical elements comprising cattle traction, the yoke and discs rotating around an axis was transformed into a variety of technical systems and thereafter successively updated to such a degree of sophistication that the earliest wagons we can identify in the archaeological record at the moment are so well made and designed that it seems impossible that they should actually stand at the beginning of the innovation process (one such instance is the wheel with axle from the Stare Gmajne in Ljubljana; Schlichtherle 2004).

Conclusion

Prehistoric innovation processes can rarely be pinpointed to one specific place. They take place within spheres of interaction. These are characterised by the perpetual exchange of goods, people, gifts and know-how over time. Inside these spheres, knowledge and bodily techniques circulate, possibly in the form of gift giving, marital relations, apprenticeships, wandering craftsmen, violence cultures *etc.*

Most of the time, we get the best view of these new techniques when they have proved to be socially valuable (*cf.* Taylor 1999). This is the reason for the huge amount of metal in the Carpathian Basin and the Balkans in the 5th millennium and the wagon graves around the Black Sea in the 4th and 3rd millennia. In these time spans we can neither assume market-oriented selection mechanisms nor laboratory experiments, but it is dangerously close to circular logic to deduce from the data that it was only elite competition that triggered and promoted technical change because elite competition is the reason for the visibility of technical change.

I would propose two dimensions for modelling prehistoric innovation. First, local competition often triggers the appropriation, modification and ultimate recombination of techniques, leading to secondary or tertiary innovation processes. This, however, does *not* automatically result in autochthonous technical evolutions. An innovation needs to be socially embedded. Accordingly, conflict may ensue when it destroys traditions and thus perhaps undermines traditional foundations of social power.

For the 5th and 4th millennia, high mobility of stylistic knowledge can be demonstrated with regard to pottery, lithic technology or raw material distributions (Манолакакис

2002; Reingruber 2010; Klimscha 2012). The mobility of materials, styles, things and persons (see above, *cf.* also Wrobel Nørgaard 2014) brings people into contact and thus fuels the transmission of (technical) know-how (Frieman 2012). Last but not least, techniques from other crafts can also trigger change when they are integrated into a *chaîne opératoire*, and daily experimentation with a technique can also bring about small-scale changes. This know-how, in turn, is integrated into local technical traditions, thus making it possible to appropriate innovations, recombine available know-how and start up new innovations.

These factors certainly do not effect change continuously to the same degree. Rather, we might conceive of junctures where competition and experimentation intensify, during military or political conflict, for instance, and thus perhaps cause innovations to take off (Vandkilde 2007 suggests a similar model of 'hot times' (with cultural change) and 'cold times' (with intentional stagnation) referring to Lévi-Straus' model of 'hot' and 'cold' cultures). Also conceivable are periods in which skills are lost; ultimately, this might also cause the collapse of social systems. How technical systems survive such collapses is still anything but clear, but the mobility of people and know-how and the decentralised storage of that know-how in prehistoric communities might be fairly good explanations.

Therefore it is most striking that we can see long-term continuity in the way technical innovation processes of metallurgy and animal traction were appropriated. In the case of metal, the area north of the Alps and the Carpathians never fully adopted metallurgy before the Bronze Age – though there are shortish periods with a higher quantity of finds (Klassen 2000; 2004; Govedarica 2010). Wagon technology on the other hand seems to have been restricted to eastern and central Europe for more than a millennium after the innovation first took off.

This means that there must have been factors enabling connectivity and mobility within these zones in order to overcome conflicts, population shifts, diseases, drift and political collapse. These factors made for a constant flow of information and kept the innovation process going.

In the cases discussed in this paper, neither the surroundings nor the social system can be seen as responsible for that on their own. Both are extremely variable within the interaction spheres in which wagons and metallurgy were innovated, ranging from small hamlets in the North European Plain to large agglomerations of differentiated houses in the Near East.

Accordingly, there should be much more of a focus on the ability to host the technical substructures for innovations (*sensu* Rammert 2007) than is currently the case in archaeological discourse. A *chaîne opératoire* needs social institutions, environmental specifications and technical prerequisites if it is to work. Terrain and environment are not the only factors that can shape communication (in the sense of cultural landscapes/seascapes; *cf.* for instance: Frieman 2008). Equally important are available technical know-how and a society's ability to implement change in itself.

While all the prerequisites for the wagon were already in place in the Early Neolithic, this substructure enabled many societies to adopt it quickly, whereas the high specialisation and division of labour necessary for metallurgy limited it for a long time to those areas in which long-term experience with high-temperature firing was available and specialised production was valued and practised (Frieman 2012).

However, the decentralised storage of know-how within stylistic, technical, social and military interaction spheres is poorly understood, and further data on correlations between innovation processes and such chronospatial networks are very much a desideratum. Until we have freed our discourse from the social filters of the archaeological record, that discourse will be limited to the diffusion of ideology, not technology.

Acknowledgements

I should like to thank the organisers of the conference for their assistance, especially their help in actually getting me to the conference and thus enabling me to speak there. Catherine Frieman (Canberra) read this paper and made critical comments, for which I am very grateful to her.

References

Aardsma, G. E. (2001) New Radiocarbon Dates for the Reed Mat from the Cave of the Treasure, Israel. *Radiocarbon* 43/3, 1247–1254.

Angelova, I. (1986) Prehistoric Necropolis Near the Town of Targovište. Arheoloijski Institut i Muzeij na Ban. *Interdiscijplinarni Izsledvanija* 14 A, 49–66.

Bar-Adon, O. (1980) *The Cave of the Treasure. The Finds from the Caves in Naḥal Mishmar.* Jerusalem, Israel Exploration Society.

Bakker, J. A., Kruk, J., Lanting, A. E. and Milisauskas, S. (1999) The Earliest Evidence of Wheeled Vehicles in Europe and the Near East. *Antiquity* 73, 778–790.

Bakker, J. A. (2004) Die neolithischen Wagen im nördlichen Mitteleuropa. In M. Fansa and S. Burmeister (eds), *Rad und Wagen – Der Ursprung einer Innovation. Wagen im Vorderen Orient und Europa.* Archäologische Mitteilungen aus Nordwestdeutschland Beiheft 40, 283–294. Mainz, Philipp von Zabern.

Basalla, G. (1989) *The Evolution of Technology.* Cambridge, Cambridge University Press.

Behrens, H. (1963) Die Rindskelettfunde der Péceler Kultur und ihre Bedeutung für die Erkenntnis historischer Zusammenhänge. *Acta Archaeologica Academiae Scientiarum Hungaricae* 15, 33–36.

Bernbeck, R. (2004) Gesellschaft und Technologie im frühgeschichtlichen Mesopotamien. In M. Fansa and S. Burmeister (eds), *Rad und Wagen – Der Ursprung einer Innovation. Wagen im Vorderen Orient und Europa.* Archäologische Mitteilungen aus Nordwestdeutschland Beiheft 40, 59–68. Mainz, Philipp von Zabern.

Borić, D. (2009) Absolute Dating of Metallurgical Innovations in the Vinča Culture of the Balkans. In T. L. Kienlin and B. W. Roberts (eds), *Metals and Societies. Studies in Honour of Barbara S. Ottaway.* Universitätsforschungen zur Prähistorischen Archäologie 169, 191–245. Bonn, Habelt.

Boroffka, N. (2009) Simple Technology: Casting Moulds for Axe-Adzes. In T. L. Kienlin and B. W. Roberts (eds), *Metals and Societies. Studies in Honour of Barbara S. Ottaway.* Universitätsforschungen zur Prähistorischen Archäologie 169, 246–257. Bonn, Habelt.

Burmeister, S. (2004a) Der Wagen im Neolithikum und in der Bronzezeit: Erfindung, Ausbreitung und Funktion der ersten Fahrzeuge. In M. Fansa and S. Burmeister (eds), *Rad und Wagen – Der Ursprung einer Innovation. Wagen im Vorderen Orient und Europa.* Archäologische Mitteilungen aus Nordwestdeutschland Beiheft 40, 13–40. Mainz, Philipp von Zabern.

Burmeister, S. (2004b) Neolithische und Bronzezeitliche Moorfunde aus den Niederlanden, Nordwestdeutschland und Dänemark. In M. Fansa and S. Burmeister (eds), *Rad und Wagen – Der Ursprung einer Innovation. Wagen im Vorderen Orient und Europa.* Archäologische Mitteilungen aus Nordwestdeutschland Beiheft 40, 321–340. Mainz, Philipp von Zabern.

Cassen, S. (1991) Les débuts du Ive millénaire en Centre-Ouest: l'hypothèse du Matignons ancien. In A. Beeching, D. Binder, J.-C. Blanchet, C. Constantin, J. Dubouloz, R. Martinez, D. Mordant, J.-P. Thevenot and J. Vaquer (eds), *Identité du Chasséen. Actes du Colloque International de Nemours 17–19 mai 1989.* Mémoirs du Musée de Préhistoire d'Ile de France 4, 111–120. Nemours, APRAIF.

Cassen, S. S., Pétrequin, P., Boujot, C., Domínguez-Bella, S., Guiavarc'h, M. and Querré, G. (2011) Measuring Distinction in the Megalithic Architecture of the Carnac Region: From Sign to Material. In M. Furholt (ed.), *Megaliths and Identities. Early Monuments and Neolithic Societies from Atlantic to the Baltic. 3rd European Megalithic Studies Group Meeting. 13–15 May 2010 Kiel University*, 225–248. Bonn, Habelt.

Childe, V. G. (1948) *Man Makes Himself.* The Thinker's Library 87. London, Watts & Co.

Childe, V. G. (1951) *Social Evolution.* London, Watts & Co.

Craddock, P. T. (2001) From Hearth to Furnace. Evidence of the Earliest Metal Smelting Technologies in the Eastern Mediterranean. *Paléorient* 26/2, 151–165.

Crouwel, J. (2004) Der Alte Orient und seine Rolle in dr Entwicklng von Fahrzeugen. In M. Fansa and S. Burmeister (eds), *Rad und Wagen – Der Ursprung einer Innovation. Wagen im Vorderen Orient und Europa.* Archäologische Mitteilungen aus Nordwestdeutschland Beiheft 40, 69–86. Mainz, Philipp von Zabern.

Dayagi-Mendels, M. and Rozenberg, S. (2010) *Chronicles of the Land. Archaeology in the Israel Museum Jerusalem.* Jerusalem, Israel Museum.

Dergačev, V. (2002) *Die äneolithischen und bronzezeitlichen Metallfunde aus Moldavien.* Prähistorische Bronzefunde 20/9. Stuttgart, Steiner.

Eco, U. (1989) Serialität im Universum der Kunst und der Massenmedien. In U. Eco (ed.), *Im Labyrinth der Vernunft. Texte über Kunst und Zeichen*, 301–324. Leipzig, Reclam.

Eco, U. (1990) Die Innovation im Seriellen. In U. Eco (ed.), *Über Spiegel und andere Phänomene*, 155–180. Munich, dtv.

Ecsedy, I. (1982) Későrézkori leletek Boglárlelléről. Late Copper Age Finds from Boglárlelle. *Communicationes Archaeologicae Hungariae* 26, 15–27.

Fol, A. and Lichardus, J. (eds) (1988) *Macht, Herrschaft und Gold. Das Gräberfeld von Varna (Bulgarien) und die Anfänge einer neuen europäischen Zivilisation.* Saarbrücken, Moderne Galerie des Saarland-Museums.

Frangipane, M. (1996) *La Nascita dello Stato nel Vicino Oriente. Dai lignaggi alla burocrazia nella Grande Mesopotamia.* Quadrante 85. Rome, Laterza.

Friederich, S. and Hoffmann, V. (2013) Die Rinderbestattungen von Profen – mit Rad und Wagen. In H. Meller (ed.), *3300 BC. Mysteriöse Steinzeittote und ihre Welt. Sonderausstellung vom 14. November 2013 bis 18. Mai 2014 im Landesmuseum für Vorgeschichte Halle,* 83–87. Mainz, Nünnerich-Asmus.

Frieman, C. (2008), Islandscapes and Islandness: The Prehistoric Isle of Man in the Irish Sea-Scape. *Oxford Journal of Archaeology* 27/2, 135–151.

Frieman, C. (2012) Flint Daggers, Copper Daggers, and Technological Innovation in Late Neolithic Scandinavia. *European Journal of Archaeology* 15/3, 2012, 440–464,

Garfinkel, Y., Klimscha, F., Rosenberg, D. and Shalev, S. (2014) The Beginning of Metallurgy in the Southern Levant: A Late 6th Millennium CalBC Copper Awl from Tel Tsaf, Israel. *PLOS ONE.* DOI: 10.1371/journal.pone.0092591.

Gej, A. (2004) Der Wagen in der Novotitarovskaja-Kultur. In M. Fansa and S. Burmeister (eds), *Rad und Wagen – Der Ursprung einer Innovation. Wagen im Vorderen Orient und Europa.* Archäologische Mitteilungen aus Nordwestdeutschland Beiheft 40, 177–190. Mainz, Philipp von Zabern.

Govedarica, B. (2010) Spuren von Fernbeziehungen in Norddeutschland während des 5. Jahrtausends. *Das Altertum* 56, 1–12.

Гусев, С. А. (1998) К Вопросу о Транспортных Средствах Трипольской. *Россиыская Аркхеология,* 15–28.

Günther, K. (1990) Neolithische Bildzeichen an einem ehemaligen Megalithgrab bei Warburg, Kreis Höxter (Westfalen). *Germania* 68, 39–65.

Günther, K. (1997) *Die Kollektivgräber-Nekropole Warburg I-V.* Bodenaltertümer Westfalens 34. Mainz, Philipp von Zabern.

Hansen, S. (2011) Innovation Metall. Kupfer, Gold und Silber in Südosteuropa während des fünften und vierten Jahrtausends v. Chr. *Das Altertum* 56, 275–314.

Hansen, S. (2013) Innovative Metals: Copper, Gold and Silver in the Black Sea Region and the Carpathian Basin during the 5th and 4th Millennium BC. In S. Burmeister, S. Hansen, M. Kunst and N. Müller-Scheeßel (eds), *Metal Matters. Innovative Technologies and Social Change in Prehistory and Antiquity.* Menschen – Kulturen – Traditionen. Studien aus den Forschungsclustern des Deutschen Archäologischen Instituts 12, 137–167. Rahden/Westf., Marie Leidorf.

Hansen, S. (2015) Pietrele. A Lakeside Settlement, 5200–5250 BC. In: S. Hansen, P. Raczky, A. Anders and A. Reingruber (eds), *Neolithic and Copper Age between the Carpathians and the Aegean Sea. Chronologies and Technologies from the 6th to the 4th Millennium BCE. International Workshop Budapest 2012.* Archäologie in Eurasien 31, 273–294. Mainz, Habelt.

Hansen Streily, A. (2000) Early Pottery Kilns in the Middle East. *Paléorient* 26/2, 69–81.

Hauptmann, A., Lutz, J., Pernicka, E. and Yalcin, Ü. (1993) Zur Technologie der frühesten Kupferverhüttung im östlichen Mittelmeerraum. In M. Frangipane, H. Hauptmann, M. Liverani, P. Matthiae and M. Mellink (eds), *Between the Rivers and Over the Mountains: Archaeologica Anatolica et Mesopotamica Alba Palmieri Dedicate,* 541–572. Rome, Dipartimento di Scienze dell'Antichità, Università di Roma La Sapienza.

Helwing, B. (2013) Early Metallurgy in Iran – An Innovative Region as Seen from the Inside. In S. Burmeister, S. Hansen, M. Kunst and N. Müller-Scheeßel (eds), *Metal Matters. Innovative Technologies and Social Change in Prehistory and Antiquity.* Menschen – Kulturen – Traditionen. Studien aus den Forschungsclustern des Deutschen Archäologischen Instituts 12, 105–135. Rahden/Westf., Marie Leidorf.

Higham, T., Chapman, J., Slavchev, V., Gaydarska, B., Houch, N., Yordanov, Y. and Dimitrora, B. (2007) New Perspectives on the Varna Cemetery (Bulgaria). AMS Dates and Social Implications. *Antiquity* 81, 640–654.

Johannsen, N. and Laursen, S. (2010) Routes and Wheeled Transport in Late 4th–Early 3rd Millennium Funerary Customs of the Jutland Peninsula: Regional Evidence and European Context. *Prähistorische Zeitschrift* 85, 15–58.

Jung, M. (2011) Der 'Big Man'. Die Verselbständigung eines theoretischen Konstrukts und ihre Adaption in der Archäologie. *Das Altertum* 56, 187–240.

Kienlin, T. (2008) Tradition and Innovation in Copper Age Metallurgy. Results of a Metallographic Examination of Flat Axes from Eastern Central Europe and the Carpathian Basin. *Proceedings of the Prehistoric Society* 74, 79–107.

Klassen, L. (2000) *Frühes Kupfer im Norden. Chronologie, Herkunft und Bedeutung der Kupferfunde der Nordgruppe der Trichterbecherkultur.* Højbjerg and Aarhus, Aarhus University Press.

Klassen, L. (2004) *Jade und Kupfer. Untersuchungen zum Neolithisierungsprozess im westlichen Ostseeraum unter besonderer Berücksichtigung der Kulturentwicklung Europas 5500–3500 BC.* Jutland Archaeological Society Publications 47. Højbjerg and Aarhus, Aarhus University Press.

Klimscha, F. (2009) Studien zu den Steinernen Beilen und Äxten der Kupferzeit des Ostbalkanraumes (5. und 4. Jahrtausend). Unpublished thesis, Freie Universität Berlin.

Klimscha, F. (2012) 'Des goûts et des couleurs on ne discute pas'. Distinction sociale et échange des idées dans l'âge de cuivre en Europe de Sud-est. Datation, répartition et valeur sociale des haches en silex de la culture Gumelniţa. In P. Pétrequin, S. Cassen, M. Errera, L. Klassen, A. Sheridan and A.-M. Pétrequin (eds), *Jade. Grandes haches alpines du Néolithique européen. Ve et IVe millénaires av. J.-C* 2. Collection Les cahiers de la MSHE Ledoux 17. Série Dynamiques territoriales 6, 1208–1229. Besançon, Presses Université de Franche-Comté.

Klimscha, F. (2013) Another Great Transformation. Technical and Economical Change from the Chalcolithic to the Early Bronze Age in the Southern Levant. *Zeitschrift für Orient Archäologie* 6, 82–112.

Klimscha, F. (2014) Innovations in Chalcolithic Metallurgy in the Southern Levant during the 5th and 4th Millennium BC. Copper-Production at Tall Hujayrat al-Ghuzlan and Tall al-Magass, Aqaba Area, Jordan. In S. Burmeister, S. Hansen,

M. Kunst and N. Müller-Scheeßel (eds), *Metal Matters. Innovative Technologies and Social Change in Prehistory and Antiquity*. Menschen – Kulturen – Traditionen. Studien aus den Forschungsclustern des Deutschen Archäologischen Instituts 12, 31–64. Rahden/Westf., Marie Leidorf.

Klimscha, F. (2017) Transforming Technical Know-how in Time and Space. Using the Digital Atlas of Innovations to Understand the Innovation Process of Animal Traction and the Wheel. *eTOPOI. Journal for Ancient Studies* 6, 16–63. DOI: 10.17169/FUDOCS_document_000000026267.

Kroeber, A. L. (1940) Stimulus Diffusion. *American Anthropologist* 42/1, 1–20.

Littauer, M. A. and Crouwel, J. (1979) Wheeled Vehicles and Ridden Animals in the Ancient Near East. Leiden, Brill.

Littauer, M. A. and Crouwel, J. (1990) Ceremonial Threshing in the Ancient Near East. *Iraq* 52, 15–19.

Lumley, H. de (2003) *Région du mont Bego. Gravures protohistoriques et historiques*. Tende, Alpes-Maritimes 14. Secteur des Merveilles. Zone du Grand Capelet. Zone XII. Groupes I á VI. Aix-en-Provence, Edisud.

Манолакакис, Л. (2002) Фунĸзияата на големите пластини от Варненсĸия неĸропол. *Археология* (София) 43/3, 5–17.

Maran, J. (2004a) Die Badener Kultur und ihre Räderfahrzeuge. In M. Fansa and S. Burmeister (eds), *Rad und Wagen – Der Ursprung einer Innovation. Wagen im Vorderen Orient und Europa*. Archäologische Mitteilungen aus Nordwestdeutschland Beiheft 40, 265–282. Mainz, Philipp von Zabern.

Maran, J. (2004b) Kulturkontakte und Wege der Ausbreitung der Wagentechnologie im 4. Jahrtausend. In M. Fansa and S. Burmeister (eds), *Rad und Wagen – Der Ursprung einer Innovation. Wagen im Vorderen Orient und Europa*. Archäologische Mitteilungen aus Nordwestdeutschland Beiheft 40, 429–442. Mainz, Philipp von Zabern.

Mischka, D. (2011) The Neolithic Burial Sequence at Flintbek LA 3, North Germany, and Its Cart Tracks: A Precise Chronology. *Antiquity* 85, 742–758.

Mischka, D. (2013) Die sozioökonomische Bedeutung von Pflugspuren im Frühneolithikum des nördlichen Mitteleuropas. In I. Heske, H.-J. Nüsse and J. Schneeweiß (eds), *Landschaft, Besiedlung und Siedlung. Archäologische Studien im nordeuropäischen Kontext. Festschrift Karl-Heinz Willroth zu seinem 65. Geburtstag*. Göttinger Schriften zur Vor- und Frühgeschichte 33, 295–306. Neumünster, Wachholtz.

Müller, S. (1905) *Urgeschichte Europas. Grundzüge einer prähistorischen Archäologie*. Strasbourg, Trübner.

Oates, J., McMahon, A., Karsgaard, P., Al Quntar, S. and Ur, J. (2007) Early Mesopotamia Urbanism: A New View from the North. *Antiquity* 81, 585–600.

Patay, P. (1984) *Kupferzeitliche Meißel, Beile und Äxte in Ungarn*. Prähistorische Bronzefunde 9/15. Munich, Beck.

Pauketat, T. R. (2007) *Chiefdoms and other Archaeological Delusions*. Lanham, MD, AltaMira Press.

Pernicka, E. (1990) Gewinnung und Verbreitung der Metalle in prähistorischer Zeit. *Jahrbuch des Römisch-Germanischen Zentralmuseums* 31/1, 21–129.

Pétrequin, P., Arbogast, R.-M., Viellet, A., Pétrequin, A.-M. and Maréchal, D. (2002) Eine neolithische Stangenschleife vom Ende des 31. Jhs. v. Chr. In Chalain (Fontenu, Jura, Frankreich). In J. Köninger, M. Mainberger, H. Schlichtherle and M. U.

Vosteen (eds), *Schleife, Schlitten, Rad und Wagen. Zur Frage früher Transportmittel nördlich der Alpen*. Hemmenhofener Skripte 3, 55–65. Gaienhofen-Hemmenhofen, Janus.

Radivojević, M. and Rehren, T. (2015) Paint It Black: The Rise of Metallurgy in the Balkans. *Journal of Archaeological Method and Theory*. DOI: 10.1007/s101816-014-9238-3.

Radivojević, M., Rehren, T., Pernicka, E., Šljivar, D., Brauns, M. and Borić, D. (2010) On the Origins of Extractive Metallurgy. New Evidence from Europe. *Journal of Archaeological Science* 37/11, 2775–2787.

Rammert, W. (2007) *Technik – Handeln – Wissen. Zu einer pragmatistischen Technik- und Sozialtheorie*. Wiesbaden, VS Verlag für Sozialwissenschaften.

Reingruber, A. (2010) Keramische Hausinventare des 5. Jahrtausends v. Chr. aus Pietrele, Rumänien. In P. Kalábková, B. Kovár, P. Pavúk and J. Šuteková (eds), *PANTA RHEI. Studies in Chronology and Cultural Development of South-Eastern and Central Europe in Earlier Prehistory Presented to Juraj Pavúk on the Occasion of his 75. Birthday*. Studia Archaeologica et Mediaevalia 11,131–142. Bratislava, Comenius University.

Renfrew, C. (1969) The Autonomy of the South-East European Copper Age. *Proceedings of the Prehistoric Society* 35, 12–47.

Renfrew, C. (1973) *Before Civilization. The Radiocarbon Revolution and Prehistoric Europe*. London, Cape.

Roberts, B. W., Thornton, C. P. and Pigott, V. C. (2009) Development of Metallurgy in Eurasia. *Antiquity* 83, 1012–1022.

Rogers, E. (2003) *Diffusion of Innovations*. 5th ed. New York, Free Press.

Schlichtherle, H. (2004) Wagenfunde aus den Seeufersiedlungen im zirkumalpinen Raum. In M. Fansa and S. Burmeister (eds), *Rad und Wagen – Der Ursprung einer Innovation. Wagen im Vorderen Orient und Europa*. Archäologische Mitteilungen aus Nordwestdeutschland Beiheft 40, 295–314. Mainz, Philipp von Zabern.

Schunke, T. (2013) Klasdy – Göhlitzsch. Vom Kaukasus nach Mitteldeutschland oder umgekehrt? In H. Meller (ed.), *3300 BC. Mysteriöse Steinzeittote und ihre Welt. Sonderausstellung vom 14. November 2013 bis 18. Mai 2014 im Landesmuseum für Vorgeschichte Halle*, 151–155. Mainz, Nünnerich-Asmus.

Seregély, T. (2004) Radmodell und Votivaxt – außergewöhnliche Funde der Kultur mit Schnurkeramik von der Nördlichen Frankenalb. In M. Fansa and S. Burmeister (eds), *Rad und Wagen – Der Ursprung einer Innovation. Wagen im Vorderen Orient und Europa*. Archäologische Mitteilungen aus Nordwestdeutschland Beiheft 40, 315–320. Mainz, Philipp von Zabern.

Šiška, S. (1964) Phrebisko tiszapolgárskej kultúry v Tibave, *Slovenská Archeológia* 12/2, 293–356.

Šlijvar, D., Kuzmanović-Cvetković, J. and Jacanović, D. (2006) Belovode-Pločnik. New Contributions Regarding the Copper Metallurgy in the Vinča Culture. In N. Tasić and C. Grozdanov (eds), *Homage to Milutin Garašanin*, 254–266. Belgrade, Serbian Academy of Sciences and Arts.

Stahlhofen, H. and Kurzhals, A. (1983) Neolithische Rinderbestattungen bei Derenburg, Kr. Wernigerode. *Ausgrabungen und Funde* 28, 157–160.

Taylor, T. (1999) Evaluing Metal. Theorizing the Eneolithic 'Hiatus'. In R. Young, A. M. Pollard, P. Budd and R. A. Ixer (eds), *Metals in Antiquity*. British Archaeological Reports, International Series 792, 22–32. Oxford, Archaeopress.

Todorova, H. (1981) *Die kupferzeitlichen Äxte und Beile in Bulgarien.* Prähistorische Bronzefunde 9/14. Munich, Beck.

Trifonov, V. (2004) Die Majkop-Kultur und die ersten Wagen in der südrussischen Steppe. In M. Fansa and S. Burmeister (eds), *Rad und Wagen – Der Ursprung einer Innovation. Wagen im Vorderen Orient und Europa.* Archäologische Mitteilungen aus Nordwestdeutschland Beiheft 40, 167–178. Mainz, Philipp von Zabern.

Turetskij, M. (2004) Wagengräber der grubengrabzeitlichen Kulturen im Steppengebiet Osteuropas. In M. Fansa and S. Burmeister (eds), *Rad und Wagen – Der Ursprung einer Innovation. Wagen im Vorderen Orient und Europa.* Archäologische Mitteilungen aus Nordwestdeutschland Beiheft 40, 190–201. Mainz, Philipp von Zabern.

Vandkilde, H. (1996) *From Stone to Bronze. The Metalwork of the Late Neolithic and Earliest Bronze Age in Denmark.* Aarhus, Aarhus University Press.

Vandkilde, H. (2007) *Culture and Change in Central European Prehistory. 6th to 1st Millenium BC.* Aarhus, University Press.

Veblen, T. (1899) *The Theory of the Leisure Class: An Economic Study of Institutions.* New York, Macmillan.

Vizdal, J. (1977) *Tiszapolgárske pohrebisko vo vel'kých raškovciach.* Košice, Východoslovenské Vydavatel'stvo.

Vosteen, M. (1999) *Urgeschichtliche Wagen in Mitteleuropa. Eine archäologische und religionswissenschaftliche Untersuchung neolithischer bis hallstattzeitlicher Befunde.* Freiburger Archäologische Studien 3. Rahden/Westf., Marie Leidorf.

Vulpe, A. (1975) *Die Äxte und Beile in Rumänien* 2. Prähistorische Bronzefunde 9/5. München, Beck.

Wolf, E. (1982) *Europe and the People without History.* Berkeley, CA, University of California Press.

Wrobel Nørgaard, H. (2014) Are Valued Craftsmen as Important as Prestige Goods? Ideas about Itinerant Craftsmanship in the Nordic Bronze Age. In S. Reiter, H. Wrobel Nørgaard, Z. Kölcze and C. Rassmann (eds), *Rooted in Movement: Aspects of Mobility in Bronze Age Europe.* Jutland Archaeological Society Publications 83, 37–52. Aarhus, Jutland Archaeological Society.

Yalçin, Ü. (2000) Anfänge der Metallverwendung in Anatolien. In Ü. Yalçin (ed.), *Anatolian Metal I.* Veröffentlichungen aus dem Deutschen Bergbau-Museum Bochum 92. Der Anschnitt 13, 17–30. Bochum, Deutsches Bergbau-Museum.

Yoffee, N. (2005) *Myths of the Archaic State. Evolution of the Earliest Cities, States and Civilization.* Cambridge, Cambridge University Press.

Zalai-Gaál, I., Gál, E., Köhler K. and Osztás, A. (2011) Das Steingerätedepot aus dem Häuptlingsgrab 3060 der Lengyel-Kultur von Alsónyék, Südtransdanubien. In H.-J. Beier, R. Einicke and E. Biermann (eds), *Varia Neolithica 7. Dechsel, Axt, Beil & Co. Werkzeug, Waffe, Kultgegenstand? Aktuelles aus der Neolithforschung. Beiträge der Tagung der Arbeitsgemeinschaft Werkzeuge und Waffen im Archäologischen Zentrum Hitzacker 2010 und Aktuelles.* Beiträge zur Ur- und Frühgeschichte Mitteleuropas 63, 85–103. Langenweißbach, Beier & Beran.

Zich, B. (1992–1993) Die Ausgrabung chronisch gefährdeter Hügelgräber der Stein- und Bronzezeit in Flintbek, Kreis Rendsburg-Eckernförde. Ein Vorbericht. *Offa* 49–50, 15–31.

Chapter 14

A Comparative View on Metallurgical Innovations in South-Western Asia: What Came First?

Barbara Helwing

Leaving aside anecdotal occurrences in Neolithic contexts, the advent of metal as of the 5th millennium BCE is generally considered a highly important turning point in history. The development of a new and complex technology made it possible to turn stones into a malleable, shimmering material. Archaeologists have assigned high explanatory value to the very occurrence of the material, and the archaeological terminology of the Three Age system since the days of Christian Jurgensen Thomsen has corroborated this assumption further by calling two of its major periods after metals (Thomsen 1837; Hansen 2001). Today, the rise of metallurgy is still regarded by most archaeologists as crucial in the formation of early complex societies (Renfrew 1978).

This assumption, now widely accepted, had already been canonised by Gordon Childe (Childe 1948; Childe 1952). For him, metallurgy was one of the key technologies behind the subsequent urban revolution, and it is from these hubs of condensed urban life that itinerant smiths will subsequently have set out on their travels, carrying with them a fund of knowledge that was later to spread throughout the Old World. This verdict has since been reiterated many times and has taken root in archaeo-metallurgical research (Muhly 1988, 16) and beyond, as is obvious from the frequent references to it in this volume (*cf.* the contributions by L. Rahmstorf and S. Hansen). But as so correctly pointed out recently by Christopher Thornton and Benjamin Roberts (Thornton and Roberts 2014), Childe had formulated his theory practically without any backing from scientific data and at a time when no absolute datings were yet possible. Small wonder, then, that with the availability of more precise dating methods and analytical tools, and especially since the radiocarbon revolution, the internal consistency of the model has been challenged. This is best exemplified by the debate

on the autonomy of the European Copper Age sparked off by Colin Renfrew (Renfrew 1969). Thanks to the gradual addition of new find-sites and to chronological rectification, these controversies are now closer to settlement (Pernicka 1990; Radivojević *et al.* 2010).

But beyond such data-bound argumentation, metallurgy has today lost its status as the key innovation leading to early complexity through technological superiority. Metallurgy forms part of a package of innovations spreading over wide areas at unprecedented speed in the 5th and 4th millennia BCE (Burmeister 2011; Hansen 2011; Kienlin 2014). Also, early metallurgy appears as a small-scale and prestige-related activity and not as production aiming at a high output and hence favouring the 'industrial' production of more efficient tools and weapons with all the competitive advantages this would have implied (Radivojevic *et al.* 2013). In fact, quite the opposite is the case. Metallurgy seems to have spread within particular contexts of social demand and must therefore be considered a socially embedded innovation rather than simply a new technology. Early instances of the phenomenon as of the mid-5th millennium BCE have come in for a great deal of attention recently. Accordingly, we now know of an early horizon of selective metal use and accumulation depending on depositional customs in the different regions and manifesting itself either in spectacular hoards – best known from Nahal Mishmar, but also from south-eastern Europe (Hansen 2009; Klimscha 2013) – or in grave contents, such as the Varna cemetery (Fol and Lichardus 1988) or the Susa I graveyard (de Morgan and de Morgan 1912; Canal 1978; Hole 1990). A high variability of metal technology is discernible between different centres of production, and this has inspired models of local development during these early periods in what

Aslıhan Yener calls a 'balkanized' mode of production (Yener 2000, 20). Along similar lines, Tobias Kienlin has recently argued with regard to the European examples that metallurgical production should be conceptualised as a self-organising activity, a specialisation growing from a set of craft skills already existing in other fields and developed locally within the communities, rather than being dependent on, or controlled by, central locations (Kienlin 2014, 367).

The transmission of knowledge and skills is a complex issue. There are various kinds of large-scale knowledge diffusion that differ according to the nature of the contacts involved, the technology transmitted and travel and communication resources (Hyman and Renn 2012). Depending on the level of complexity involved, specific learning pathways are required, ranging from simple observation to lengthy periods of apprenticeship. A technology as complex as metal processing has prompted many authors to insist on the necessity of properly learning to 'work' metal in the course of a period of apprenticeship (Thornton 2009; Ottaway 2015). Michael Schiffer has variously elaborated on multi-phase models of transmission that proceed from the transfer of information proper to phases of experimentation and redesign among the new users before these users become skilled enough to replicate the object in question (Schiffer 2004).

By contrast, the approach proposed by Kienlin (Kienlin 2014) and the idea of cross-craft interaction advocated by Ann Brysbaert (Brysbaert 2011) emphasise the fact that any new technology, including metallurgy, is usually rooted in earlier domestic technologies. In line with these models, the spread of this innovation should not be thought of in terms of frameworks of full technology transfer requiring long-term apprenticeship and the systematic transmission of particular skills. They foreground the concept that a new idea could be picked up by different communities within a broad geographical exchange network if these communities themselves had a firmly ensconced tradition of metalworking and smelting themselves and could thus re-interpret or 'transpose' the original idea. The outcome of these processes differs depending on the background onto which they were grafted. I have chosen to call this process 'engrafted innovation' to underline the fact that it is superimposed on a pre-existing technology.

The purpose of this paper is to take a closer look at the centuries that follow upon this early horizon of metal use, a period subsequent to a threshold of rapid changes culminating in Mesopotamia in the formation of the first state-like systems in the second half of the 4th millennium (Nissen 1988). Starting from Kienlin's model of precedent self-organising specialisation, I argue that after this threshold of early state formation, technological development continued to unfold on parallel tracks and led to processes of *engrafted innovation* in communities at long distances from each other. This was only possible within the framework of the early states ensuring the stability and long-distance connectivity that was a necessary precondition for the rapid spread of technical innovation. I thus maintain

that the upsurge in metallurgical production in the later 4th millennium BCE is not a pre-condition for the formation of early states but rather a direct result of this process.

To pursue this enquiry, I shall be contrasting two innovations identifiable in the archaeological record of the late 4th/early 3rd millennium BCE. The first is the introduction of wind-powered smelting furnaces referred to by Paul Craddock as the first pyrotechnical structures built specifically for metallurgy – an innovation that significantly boosted copper output (Craddock 1995). Such furnaces occur at various points in south-western Asia at about the same time in the late 4th millennium BCE. The second instance is the appearance of metal weapons in graves as of the beginning of the 3rd millennium BCE. This case serves as a test of whether a more efficient practical technology – the availability of smelted copper to produce important artefacts – may indeed have been a competitive advantage.

Background

Around 3500 BCE, fundamental changes in social organisation took place in some areas of the ancient world. Populations in south-western Asia had begun to coalesce into urban agglomerations beginning in the late 5th/early 4th millennium BCE (Nissen 1988; Wilkinson *et al.* 2014). These centres were subsequently the focus of social reorganisation: political power was personified by the figure of a ruler at the summit of a state organisation (Selz 1998) and central institutions were established that exercised control over goods and manpower (Pollock 1999). This development is certainly linked to the necessity of organising hygiene in large urban settlements and settling tension and violence among its populations (McMahon and Stone 2012), but this process is not relevant to our topic. Once established, we find a rapid spread of state organisation and central institutions across south-western Asia around 3350 BCE (most recently, Schrakamp 2013). It is this new form of organisation that also forms the background for the reorganisation of craftsmanship. Textual sources on copper production remain restricted to the Mesopotamian lowlands and are highly eclectic in nature, but the archaic texts from Uruk and Ur attest to the central administration of copper and silver production (Reiter 1997, 150 and n. 4; 155 on copper; 83–84 and n. 34 on silver; Englund 1998, 98 and n. 217). Mesopotamia was, however, devoid of primary sources for the production of metal. Craftsmen reworked and refined copper and silver there, but the regions where people experimented with raw materials were the highland regions where the natural sources of ore are located.

Nevertheless, once the early states were established, new technologies started to be introduced in distant areas as well. In the case of metal technology, the almost contemporary introduction of wind-powered furnaces and large-scale smelting operations is observable in various regions a long way away from each other. These innovative practices favoured large-scale production of copper and copper alloys and heightened output. Although not primary metal producers themselves, the early Mesopotamian states provided a social platform for contact and exchange that enabled the knowledge pertaining to these innovations to 'travel', for example information about wind-powered furnace technology. An overview of the relevant furnace-using regions was presented by Craddock some time ago (Craddock 2000), but since then, important new evidence has come to light.

Copper-smelting furnaces in highland Iran

The Iranian Plateau with its rich deposits of ore has long been regarded as one of the heartlands of metallurgy, to use a phrase coined by V. Pigott (Pigott 1999). Throughout the 4th millennium BCE, it was the site of sophisticated and

ingenious experimentation in connection with the extraction of copper and silver (Helwing 2013a; Thornton 2014). On the western Central Plateau a *coiné* of copper production materialised; representative sites included Ghabrestan, Ma'mourin, Sialk and Arisman, where craftsmen used crucibles for the smelting of copper in a sort of 'cottage industry'. The heavy tools produced there seem to be very much in line with finds from the Chalcolithic Susa I graveyard (Helwing 2011). Regions further east, around Yahya (Thornton *et al.* 2002), or north, like Tappe Hesar (Thornton 2013), developed their own regional traditions.

In the so-called proto-Elamite period beginning after 3350 BCE (Helwing 2013b; Petrie 2013), the mode and scale of production changed significantly and wind-powered furnaces were introduced, as the evidence from Arisman confirms (Fig. 14.1). Copper smelting took place outside the settlement

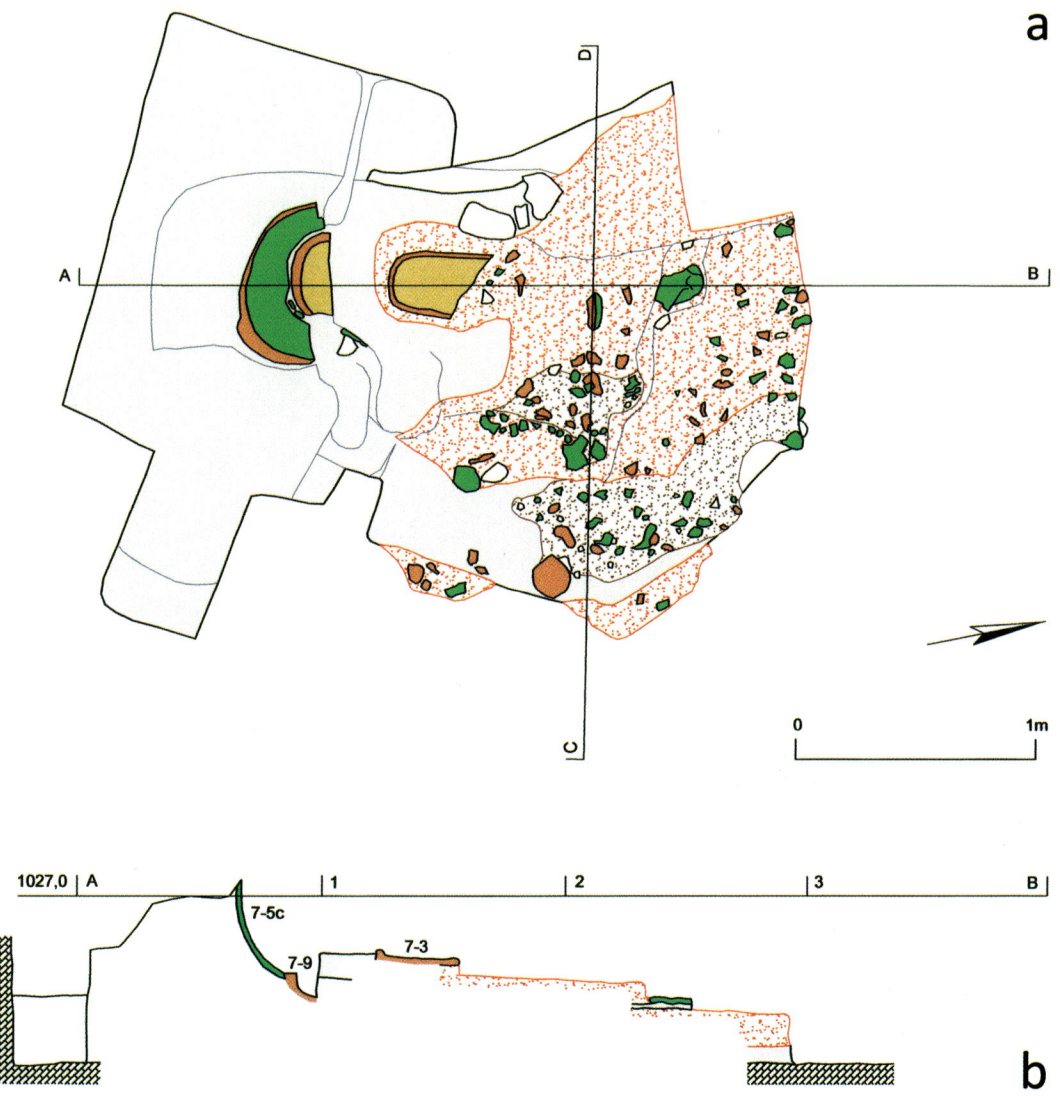

*Figure 14.1: Arisman, western Central Iran. Wind-powered furnace excavated in area A, proto-Elamite period (*c. 3000 BCE) (© DAI Eurasia Department, Arisman Excavation Archive).*

itself, while refining and casting was done in workshops set up in ruined houses in the former settlement area.

The smelting furnaces from Arisman (Fig. 14.1) are made entirely of mudbricks and clay. They take the form of a small mudbrick platform with an integrated depression in which the smelted material assembled. The upper part of the furnace chamber was lined with clay plaster, as was probably the front, but this had to be broken to get at the smelted material. No *tuyères* were found, so a wind-draft system is presumed to have been in place, possibly on a seasonal basis, as the winds descending from the Karkas Mountains are seasonal as well.

With the introduction of the furnaces, production sky-rocketed, leaving a total of 180 tons of slag. Metal artefacts have been rarely found on the site, either in graves from the late proto-Elamite period or in scrap-metal hoards, but the morphology of axe moulds used in Arisman matches the shape of axes found in contexts of the Jamdat Nasr to ED I period in northern Mesopotamia (Helwing 2011) and the trace-element composition of these Mesopotamian objects does not rule out the eventuality of their having been Iranian in origin. What we see here is the emergence of a new technology – the updraft furnace – and the stepping-up of output bound up with it taking place within the broader context of large-scale interaction with North Mesopotamian consumers.

Another production centre on the Iranian Plateau is Tappe Hesar on the northern fringe of the Central Desert. Crucible smelting of arsenical copper is known to have taken place here as well, and as in Arisman, a significant shift in smelting technology took place around 3000 in Hesar II. Although no direct evidence of furnace usage has survived, the production attested to in both domestic and industrial contexts indicates the existence of sophisticated smelting processes including the use of fluxes (Thornton 2014). With regard to artefacts, Hesar has yielded a wealth of material from tombs, but they date mainly from the Early Bronze Age and not before.

In the Iranian highlands and beyond, there were other sites involved in metal processing, for example Tappe Yahya, Shahr-e Sukhte (Hauptmann *et al.* 2003; Artioli *et al.* 2005) or Shahdad (Hakemi 1992). Further east in the Makran region, sophisticated smelting technology is known to have existed even earlier, *cf.* the famous leopard's weight from Shahi Tump (Mille *et al.* 2005). Unfortunately, research coverage is still extremely patchy and nowhere have genuine workshop areas been documented in such a way as to make a comparison between the sites a meaningful undertaking.

Copper-smelting furnaces in the southern Levant

A second crossroads of early metallurgy is the southern Levant around the Wadi Arabah, a region intensively investigated and long considered prototypic for the development of early smelting technologies worldwide, as pointed out by Thornton and Roberts (Thornton 2009). After a period of

open-cast copper mining dating back to the Chalcolithic period, large-scale copper processing began there shortly after 3500 BCE (EBA I in local terminology) and became a major factor from EB II onwards (Hauptmann 2007). South Levantine production stands out for its sophisticated spatial organisation, with specialised settlements such as Khirbet Hamra Ifdan and decentralised processing. From EBA I onwards, underground mining is attested to in Wadi Faynan, with ores smelted in wind-powered furnaces close to the mines, probably on a seasonal basis. Experimental reconstruction of such furnaces has confirmed that they were fully functional due to the currents forming inside the vessel (Fig. 14.2). The furnaces of Wadi Faynan were set on top of a rocky crest and occur in two major types. Some consist of little chambers of stone slates with the front open on the wind side. Others were made of clay, an option apparently depending on the availability of stone slates. These slates betray the frequent re-use and re-application of the furnace lining. Upright fingers of clay served to keep the charge in place without blocking the wind draft.

Copper-smelting furnaces in Egypt

In Egypt, the use of copper is known from the settlements of the north Egyptian Maadi-Buto sites but seems to have got off to a rather slow start in the 4th millennium BCE (Hartung 2013), possibly due to the lack of genuine raw material sources in the region. For that reason, northern Egypt seems to have relied largely on materials imported from the southern Levant since the beginning of metal usage (Killick 2014a, 510), and copper consumption was at first rather sporadic. Only at the end of the millennium did copper implements begin to replace flint tools. In the more southerly Naqada region, the copper supply was filtered off even further by the northern Maadi-Buto sites.

As of the early 3rd millennium BCE (Craddock 2000), evidence of early copper exploitation comes from the Eastern Desert in the Wadi Dara and from Wadi Umm Balad. Most likely, the furnaces used were greatly influenced by south Levantine copper production. They are similar in

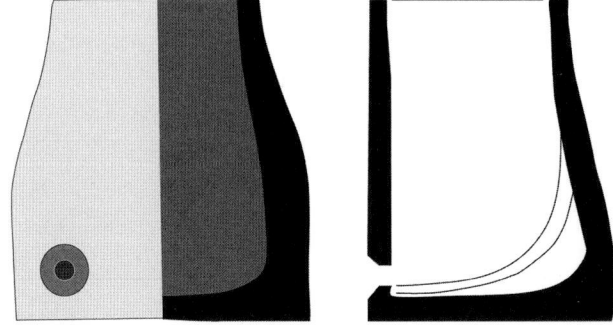

Figure 14.2: Reconstruction of a wind-powered furnace based on findings from Wadi Faynan (after Koelschbach 1999, fig. 7.9).

construction to the Wadi Faynan furnaces and consist of small chambers made from stone slabs, the interior of which will have been lined with clay.

Copper smelting in eastern and central Anatolia

A third source region for early metallurgy is the highlands of Anatolia and the Taurus Mountains, where – as in Iran – rich metal minerals had been exploited since prehistory. Evidence of smelting practice stems from the 4th millennium BCE in the form of moulds and crucible fragments. Among the smelting installations there is evidence of pit furnaces (Pernicka *et al.* 2002, 115–116; Schoop 2011). Indirect evidence of smelting also exists, for example the increase in the percentage of iron in the copper since the 4th millennium BCE (Killick 2014b, 33). As in Iran, the various centres of metalworking do not appear to be standardised. Each of them follows its own strategies of exploitation and processing, a pattern referred to by Yener as a 'balkanized' model (Yener 2000).

Copper smelting in western Anatolia and the Aegean

In the Aegean world, metal smelting is known to have existed from the 4th millennium BCE onwards, and a considerable corpus of metal artefacts has been documented and analysed (Pernicka *et al.* 2003). We have no evidence of true furnace construction there pre-dating 3000 BCE. The main indications of copper smelting are slags and crucible fragments dating back to the Late Chalcolithic, for example at Bakla Tepe (Şahoğlu and Tuncel 2014, 71), at Liman Tepe VII and at other sites like Yiali on Nisyros (Kouka 2014, 57). The recent excavations at Çukuriçi in Western Anatolia indicate that horseshoe-shaped fireplaces were used for smelting, but they did not avail themselves of the wind-powered furnace technique (Mehofer 2014). On Crete and the Cycladic islands, smelting sites are recorded from the 3rd millennium BCE onwards at sites like Chrysokamino (Betancourt and Armpis 2006), Kephala/Petras (Catapotis *et al.* 2011) and various sites on the island of Seriphos (Philaniotou *et al.* 2011). In Chrysokamino, we know of the use of a pierced ceramic container that held ore and charcoal and functioned like an up-draft chimney, but this postdates the western Asian prototypes by several centuries. The Seriphos smelting installations are pits carved out of natural rock, and the location close to the crest of natural mounds suggests that their builders were aware of the principles behind wind-powered furnaces.

Copper smelting on the Arabian Peninsula

Following a re-orientation of the Mesopotamian supply system to maritime routes via the Persian Gulf (Weeks 2003, 15–16; Helwing 2006), the exploitation of the productive copper sources on the Arabian Peninsula only began towards the end of the 4th or in the early 3rd millennium BCE. During the first half of the 3rd millennium BCE, furnaces for copper smelting seem to have existed at Maysar I in Oman, among other sites. These have been reconstructed as round containers, but due to the preservation situation no actual furnace could be documented in the field (Hauptmann 1985). Despite growing proof of the importance of the south-eastern Arabian Peninsula as a copper-supplying region as of the 3rd millennium BCE, we have no substantial evidence of the earlier existence of copper-smelting furnaces in that area (Weeks 2012, 302).

Evidence of copper smelting: a comparative discussion

The various cases we have just looked at (Fig. 14.3) fall within a narrow time range between the second half of the 4th and the beginning of the 3rd millennium BCE, the period of early state formation in many regions of south-western Asia and their subsequent collapse. The early states had introduced systems for the centralised control of goods and labour and ensured a period of relative stability and security. This is the background against which reliable overland contact and trade networks were established enabling the swift and accurate transmission of technical innovations such as the wind-powered furnaces known to have existed in places as far away as Iran and Jordan. How does the evidence briefly surveyed here fit into the established theoretical models on technology transfer outlined above, models proposing either apprenticeship or the transfer of ideas into a pre-existing corpus of knowledge?

In direct comparison, the various wind-powered furnaces we have been looking at display a remarkable variety of local/regional technological traditions. The examples from Arisman (Fig. 14.1) and Faynan (Fig. 14.2), the two best documented ones, display important differences in construction: both types are constructed largely from clay and consist of a chamber with a closed outer lining. The Arisman furnace was set into a platform in which a depression served to accommodate the molten metal. No *tuyères* were observed. The Faynan examples are located on the crest of a mountain summit and have a grill-like opening in the lower part to enable wind to enter and fan the fire. Both furnaces rely on the same principle of chamber construction, but the constructions themselves differ. The Egyptian examples are very close to the Faynan type. The Cycladic examples adhere to the same principle of wind power but were built considerably later than the Faynan examples.

I interpret this pattern as an indication that, although both furnace-construction traditions documented in Jordan and Iran adhered to the same principle of wind power, the individual form chosen differed as each community taking up the principle of furnace smelting already possessed the knowledge required for smelting copper and other metals.

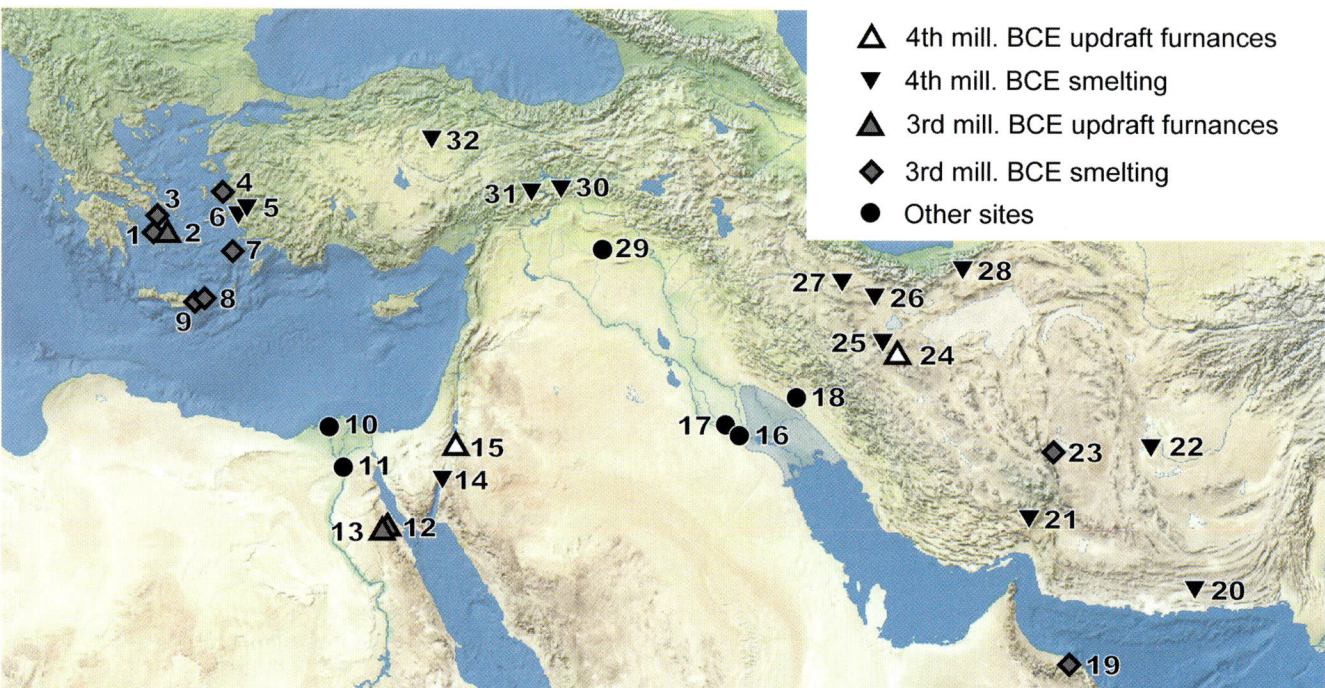

Figure 14.3: Map of south-western Asia, indicating major sites with evidence for smelting in the 4th and 3rd millennium BCE. (1) Kephala Seriphos; (2) Avissalos; (3) Kephala; (4) Liman Tepe; (5) Çukurici; (6) Bakla Tepe; (7) Giali; (8) Petras; (9) Chrysokamino; (10) Buto; (11) Maadi; (12) Wadi Dara; (13) Wadi Um Balad; (14) Hujayrat al-Ghuzlan; (15) Faynan; (16) Ur; (17) Uruk; (18) Susa; (19) Maysar; (20) Shahi Tump; (21) Yahya; (22) Shahr-e Sukhte; (23) Shahdad; (24) Arisman; (25) Sialk; (26) Ma'morin; (27) Qabrestan; (28) Hesar; (29) Brak; (30) Norşun; (31) Arslantepe; (32) Çamlıbel Tarlası.

The furnace technology was hence introduced into a fully functioning craft system. Each individual community differed in furnace construction, which may be a hint that although the principal idea may indeed have 'travelled', realisation in the form of actual built structures followed individual conceptual pathways. This pattern suggests an adaptive model in which the regional variability of copper technology can be explained through its continuous development from local roots plus the integration of new ideas into that pre-existing knowledge. In the two examples just discussed, the local tradition was advanced enough to integrate new knowledge via a process of 'engrafted innovation'.

In this case, the almost coeval occurrence of furnace technologies in regions remote from one another, such as highland Iran and the Wadi Faynan, took place without there being any proof of the existence of agents facilitating the adoption of the new technology. In fact, the design of the furnaces is different in the two regions. There are other regions no further away, such as Anatolia, where although sophisticated smelting had taken place in settlements there since the 4th millennium BCE at the latest, furnaces still seem not to have been adopted. This pattern in itself may thus indicate a special situation where metal-working communities came into contact with a new idea – wind-powered furnaces – but without any direct participation via observation or learning. The basic principles were recognised and transmitted vaguely, by hear-say or some other means of information

transfer, and subsequently were translated and integrated into local technology. Such a transfer would not have worked if the communities concerned had not already been in command of general metal-working skills enabling them to adopt the novel concept of the wind-powered furnace into their toolkit. But the circumstances favouring this theoretical transfer stem from the establishment of a large and stable interaction sphere among the early states of south-west Asia, which provided the necessary security for long-distance traffic and the untrammelled circulation of ideas.

Copper consumption in the late 4th/early 3rd millennium BCE

A different pattern is operative in the adoption of new modes of burial deposition occurring at the turn of the 4th to the 3rd millennium BCE, although the record is admittedly rather biased as we lack funerary evidence for most of the older sites. Individual burials appear on the scene from the beginning of the 3rd millennium BCE onwards, either with simple or with rich burial endowments. One outstanding group is that of graves with weapons – axes, daggers and swords – evidenced from the 30th century BCE onwards. These 'warrior burials' in south-western Asia have been discussed in detail by various authors (Philip 1995; Rehm 2003; Helwing 2012; see also Frangipane in this volume), so I do not intend to go into greater detail here. But it is important to note that a new custom – equipping individual burials with weaponry – occurred in various settings over most of south-western Asia at more or less the same time (Fig. 14.4). Evidently, this pattern is representative of another type of rapid and widespread innovation.

In the case of the 'warrior burials', the visible proof of metal used for weaponry and elaborate equipment postdates the introduction proper of a refined smelting technology that made increased copper output possible in the first place. Instead of the material availability of metal, what is important is the necessity for certain members of a given community to defend themselves and their community in inter-communal conflicts. Here we see conformity with a socially constructed innovation that rapidly spread between the Caucasus and Mesopotamia. The innovation hence reflects the changed life circumstances at the heart of the diffusion of this new custom. Chronologically, the appearance of warrior burials coincides with a point in time when the archaic states were on the verge of transformation or demise.

The spread of warrior burials thus seems to reflect a different type of innovative diffusion directly coinciding with a period marked by the demise of central authority. Accordingly, I contend that the dynamics behind the diffusion of this innovation are not linked to technology but to life circumstances and are hence ultimately rooted in social life as it was at the time. It is not the easy availability of metal that leads to its utilisation for the production of weaponry but the disappearance of central state institutions which, although the authority they imposed on the people may have been irksome, had at least been able for some time to guarantee a degree of stability and security for individual communities.

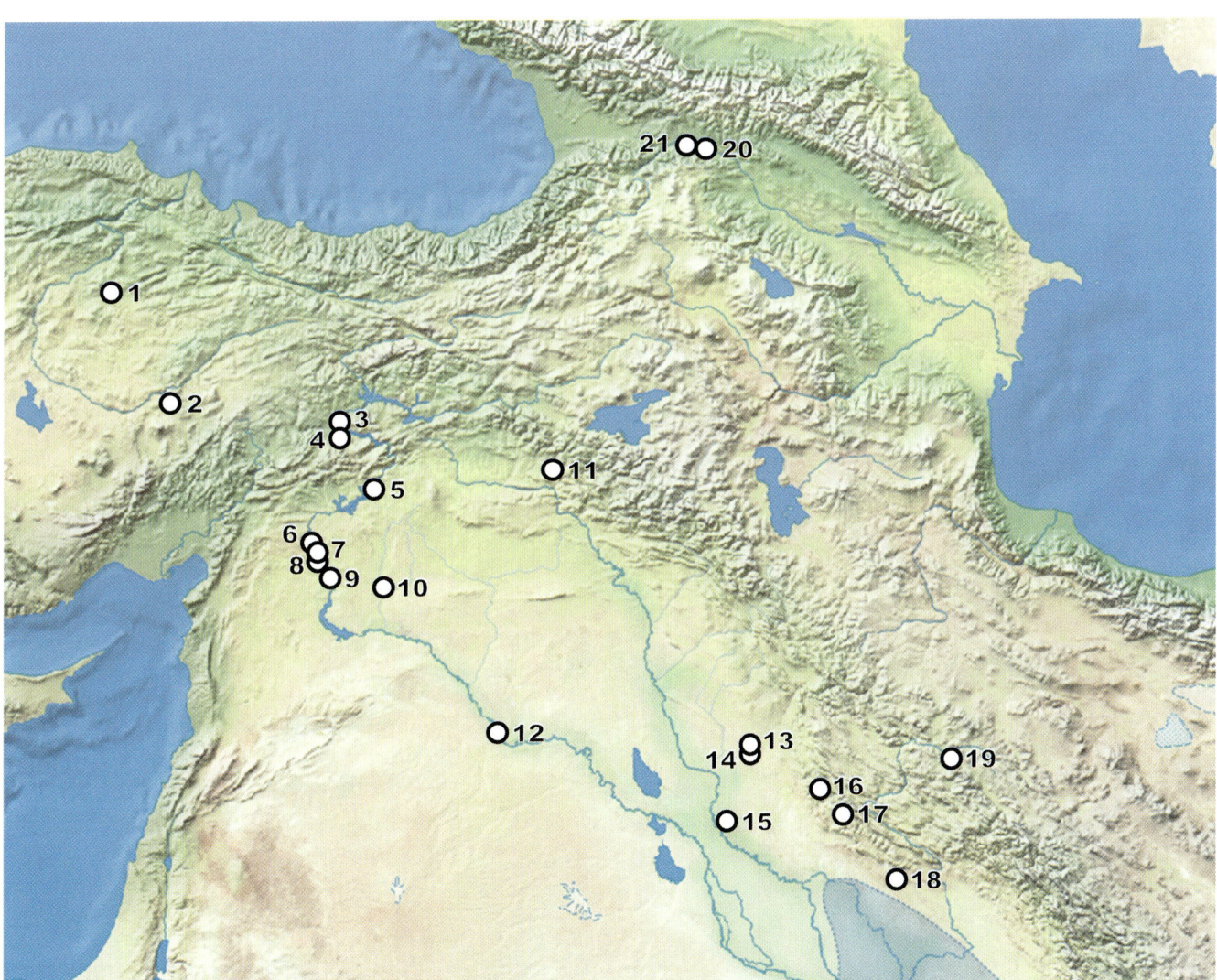

Figure 14.4: Map of south-western Asia indicating distribution of stone cist burials, some with weaponry. (1) Alacahöyük; (2) Kültepe; (3) Suyatağı; (4) Arslantepe; (5) Hassek; (6) Hacınebi; (7) Karkemish; (8) Jerablus Tahtani; (9) Qara Quzak; (10) Hammam et-Turkman; (11) Başur Höyük; (12) Mari; (13) Kheit Qasim; (14) Ahmat al-Hattu; (15) Khafaji; (16) Kalleh Nisar; (17) Gululal-i Galbi; (18) Aliabad; (19) Tappe Giyan; (20) Gudabertka; (21) Kvatskhela.

Conclusion

In Greater Mesopotamia, the spread of two innovations linked to metal, the wind-powered furnace technology of the late 4th millennium BCE and the metal weaponry occurring since the beginning of the 3rd millennium BCE, follow different patterns. In the furnace case, the integration of a new technological principle into pre-existing technologies represents an example of engrafted innovation adopted by communities with a pre-existing command of basic smelting technology. This happened at a moment in history when the emergence of centrally controlled institutions in the early states of the mid-4th millennium BCE paved the way for a re-organisation of metal production on a large scale. Furthermore, the (relatively) more stable relations between the early states favoured an extension and consolidation of long-distance trade and with it the rapid diffusion of technical innovations. The output from this production, which can now rightly be called an industry, was first absorbed by the central institutions, where it served to adorn buildings and to produce objects of material or symbolic value. While these certainly enhanced the symbolic representation of the state institutions, their emergence is not bound up with the availability of metal. In fact, the opposite is the case.

Subsequently, the adoption of new types of weapon and the new custom of having them figure in burial inventories spread via direct confrontation between neighbouring communities. It represents a species of arms race taking place some time after 3000 BCE, at a moment when the first states were about to collapse. Metal had long since been available. It only became a decisive commodity when the state as a social stability factor ceased to exist.

Both cases of innovation are hence embedded in specific social situations and can only be understood in terms of those contexts. This observation confirms the initial observation that we must ultimately regard innovations as social rather than technological phenomena.

References

Artioli, D., Giardino, C., Guida, G., Lazzari, A. and Vidale, M. (2005) The Exploitation and Transformation of Copper Ores at Shahr-i Sokhta (Iran): Preliminary Results. In U. Franke-Vogt and H.-J. Weisshaar (eds), *South Asian Archaeology 2003. Proceedings of the Seventeenth International Conference of the European Association of South Asian Archaeologists. 7–11 July 2003, Bonn.* Forschungen zur Archäologie Außereuropäischer Kulturen 1, 179–184. Aachen, Linden Soft.

Betancourt, P. P. and Armpis, E. (eds) (2006) *The Chrysokamino Metallurgy Workshop and Its Territory.* Hesperia, Supplement 36. Athens and Oxford, American School of Classical Studies at Athens and Oxbow Books.

Brysbaert, A. (2011) Introduction: Tracing Social Networks through Studying Technologies. In A. Brysbaert (ed.), *Tracing Prehistoric Social Networks through Technology. A Diachronic Persepctive on the Aegean.* Routledge Studies in Archaeology 3, 1–11. New York and London, Routledge.

Burmeister, S. (2011) Innovationswege – Wege der Innovation. Erkenntnisprobleme am Beispiel des Wagens im 4. Jt. v. Chr. In S. Hansen and J. Müller (eds), *Sozialarchäologische Perspektiven: Gesellschaftlicher Wandel 5000–1500 v. Chr. zwischen Atlantik und Kaukasus: Internationale Tagung, 15.–18. Oktober 2007 in Kiel.* Archäologie in Eurasien 24, 211–240. Mainz, Philipp von Zabern.

Canal, D. (1978) La haute terrasse de l'acropole de Suse. *Paléorient* 4, 169–176.

Catapotis, M., Bassiakos, Y. and Papadatos, Y. (2011) Reconstructing Early Cretan Metallurgy: Analytical Evidence from Kephala Petras, Siteia. In P. Betancourt and S. Ferrence (eds), *Metallurgy: Understanding How, Learning Why. Studies in Honor of James D. Muhly.* Prehistory Monographs 29, 69–78. Philadelphia, PA, INSTAP.

Childe, V. G. (1948) *Man Makes Himself.* The Thinker's Library 87. London, Watts.

Childe, V. G. (1952) *Stufen der Kultur.* Stuttgart, Kohlhammer.

Craddock, P. T. (1995) *Early Metal Mining and Production.* Edinburgh, Edinburgh University Press.

Craddock, P. T. (2000) From Hearth to Furnace: Evidences for the Earliest Metal Smelting Technologies in the Eastern Mediterranean. *Paléorient* 26/2, 151–165.

Englund, R. K. (1998) Texts from the Late Uruk Period. In J. Bauer, R. K. Englund and M. Krebernik (eds), *Annäherungen 1: Mesopotamien. Späturuk-Zeit und Frühdynastische Zeit.* Orbis Biblicus et Orientalis 160, 15–233. Freiburg and Göttingen, Universitätsverlag and Vandenhoek & Ruprecht.

Fol, A. and Lichardus, J. (1988) *Macht, Herrschaft und Gold. Das Gräberfeld von Varna und die Anfänge einer neuen europäischen Zivilisation.* Saarbrücken, Saarland-Museum.

Hakemi, A. (1992) The Copper Smelting Furnace of the Bronze Age at Shahdad. In J.-F. Jarrige (ed.), *South Asian Archaeology 1989. Papers from the Tenth International Conference of South Asian Archaeologists in Western Europe, Musée National des Arts Asiatiques – Guimet, Paris, France, 3–7 July 1989.* Monographs in World Archaeology 14, South Asian Archaeology 10, 119–132. Madison, WI, Prehistory Press.

Hansen, S. (2001) Von den Anfängen der prähistorischen Archäologie: Christian Jürgensen Thomsen und das Dreiperiodensystem. *Prähistorische Zeitschrift* 76, 10–23.

Hansen, S. (2009) Kupfer, Gold und Silber im Schwarzmeerraum während des 5. und 4. Jahrtausends v. Chr. In J. Apakidze, B. Govedarica and B. Hänsel (eds), *Der Schwarzmeerraum vom Äneolithikum bis in die Früheisenzeit (5000–500 v. Chr.): Kommunikationsebenen zwischen Kaukasus und Karpaten: Internationale Fachtagung von Humboldtianern für Humboldtianer im Humboldt-Kolleg in Tiflis, Georgien, 17.–20. Mai 2007.* Prähistorische Archäologie in Südosteuropa 25, 11–50. Rahden/Westf., Marie Leidorf.

Hansen, S. (2011) Innovation Metall. Kupfer, Gold und Silber in Südosteuropa während des fünften und vierten Jahrtausends v. Chr. *Das Altertum* 56, 275–314.

Hartung, U. (2013) Raw Material Supply and Social Development in Egypt in the 4th Millennium BC. In S. Burmeister, S. Hansen, M. Kunst and N. Müller-Scheessel (eds), *Metal Matters. Innovative Technologies and Social Change in Prehistory and Antiquity.* Menschen – Kulturen – Traditionen

12. ForschungsCluster 2, 13–30. Rahden/Westf., Marie Leidorf.

Hauptmann, A. (1985) *5000 Jahre Kupfer in Oman 1: Die Entwicklung der Kupfermetallurgie vom 3. Jahrtausend bis zur Neuzeit.* Bochum, Deutsches Bergbau-Museum.

Hauptmann, A. (2007) *The Archaeometallurgy of Copper. Evidence from Faynan, Jordan.* Berlin *et. al.*, Springer Publications.

Hauptmann, A., Rehren, T. and Schmitt-Strecker, S. (2003) Early Bronze Age Copper Metallurgy at Shahr-i Sokhta (Iran), Reconsidered. In T. Stöllner, G. Körlin, G. Steffens and J. Cierny (eds), *Man and Mining – Mensch und Bergbau. Studies in Honour of Gerd Weisgerber on Occasion of his 65th Birthday.* Der Anschnitt, Beiheft 16, 197–213. Bochum, Deutsches Bergbau-Museum.

Helwing, B. (2006) The Rise and Fall of Bronze Age Centers Around the Central Iranian Desert – A Comparison of Tappe Hesar II and Arisman. *Archäologische Mitteilungen aus Iran und Turan* 38, 35–48.

Helwing, B. (2011) Conclusions: The Arisman Copper Production in a Wider Context. In A. Vatandoust, H. Parzinger and B. Helwing (eds), *Early Mining and Metallurgy on the Western Central Iranian Plateau. Report on the First Five Years of Research of the Joint Iranian–German Research Project.* Archäologie in Iran und Turan 9, 523–531. Mainz, Philipp von Zabern.

Helwing, B. (2012) An Age of Heroes? Some Thoughts on Early Bronze Age Funerary Customs in Northern Mesopotamia. In H. Niehr, P. Pfälzner, E. Pernicka and A. Wissing (eds), *(Re-)constructing Funerary Rituals in the Ancient Near East. Proceedings of the First International Symposium of the Tübingen Post-Graduate School 'Symbols of the Dead' in May 2009.* 1st ed. Qaṭna Studien Supplementa 1, 47–58. Wiesbaden, Harrassowitz.

Helwing, B. (2013a) Early Metallurgy in Iran – An Innovative Region as Seen from the Inside. In S. Burmeister, S. Hansen, M. Kunst and N. Müller-Scheessel (eds), *Metal Matters. Innovative Technologies and Social Change in Prehistory and Antiquity.* Menschen – Kulturen – Traditionen 12. ForschungsCluster 2, 105–135. Rahden/Westfalen, Marie Leidorf.

Helwing, B. (2013b) Some Thoughts on the Mode of Culture Change in the Fourth-Millennium BC Iranian Highlands. In C. A. Petrie (ed.), *Ancient Iran and Its Neighbours. Local Developments and Long-Range Interactions in the Fourth Millennium BC.* Institute of Persian Studies Archaeological Monographs Series 3, 93–105. Oxford and Oakville, PA, Oxbow Books.

Hole, F. (1990) Cemetery or Mass Grave: Reflections on Susa I. In F. Vallat (ed.), *Contribution à l'histoire de l'Iran. Mélanges offerts à Jean Perrot,* 1–14. Paris, Editions Recherche sur les Civilisations.

Hyman, M. D. and Renn, J. (2012) Chapter 3. Survey: From Technology Transfer to the Origins of Science. In J. Renn (ed.), The Globalization of Knowledge in History. *Max Planck Research Library for the History and Development of Knowledge, Studies* 1, 75–104.

Kienlin, T. L. (2014) Aspects of Metalworking and Society from the Black Sea to the Baltic Sea from the Fifth to the Second Millennium BC. In B. W. Roberts and C. P. Thornton (eds), *Archaeometallurgy in Global Perspective,* 447–472. New York, Springer.

Killick, D. (2014a) Cairo to Cape: The Spread of Metallurgy through Eastern and Southern Africa. In B. W. Roberts and C. P. Thornton (eds), *Archaeometallurgy in Global Perspective,* 507–528. New York, Springer.

Killick, D. (2014b) From Ores to Metal. In B. W. Roberts and C. P. Thornton (eds), *Archaeometallurgy in Global Perspective,* 11–46. New York, Springer.

Klimscha, F. (2013) Innovations in Chalcolithic Metallurgy in the Southern Levant during the 5th and 4th Millennium BC. Copper Production at Tall Hujayrat al-Ghuzlan and Tall al-Magass, Aqaba Area, Jordan. In S. Burmeister, S. Hansen, M. Kunst and N. Müller-Scheessel (eds), *Metal Matters. Innovative Technologies and Social Change in Prehistory and Antiquity.* Menschen – Kulturen – Traditionen 12. ForschungsCluster 2, 31–63. Rahden/Westf., Marie Leidorf.

Kouka, O. (2014) Past Stories – Modern Narratives: Cultural Dialogues between East Aegean Islands and the West Anatolian Mainland in the 4th Millennium BC. In B. Horejs and M. Mehofer (eds), *Proceedings of the Symposion Western Anatolia before Troy : Proto-Urbanisation in the 4th Millennium BC, Kunsthistorisches Museum Wien, 22–24 November 2012.* Oriental and European Archaeology 1, 43–64. Vienna, Verlag der Österreichischen Akademie der Wissenschaften.

McMahon, A. and Stone, A. (2012) The Edge of the City: Urban Growth and Burial Space in 4th Millennium BC Mesopotamia. *Origini* 34, 59–77.

Mehofer, M. (2014) Metallurgy during the Chalcolithic and the Beginning of the Early Bronze Age in Western Anatolia. In B. Horejs and M. Mehofer (eds), *Proceedings of the Symposion Western Anatolia before Troy: Proto-Urbanisation in the 4th Millennium BC, Kunsthistorisches Museum Wien, 22–24 November 2012.* Oriental and European Archaeology 1, 463–490. Vienna, Verlag der Österreichischen Akademie der Wissenschaften.

Mille, B., Bourgarit, D. and Besenval, R. (2005) Metallurgical Study of the 'Leopard's Weight' from Shahi-Tump (Pakistan). In C. Jarrige and V. Lefèvre (eds), *South Asian Archaeology 2001. Proceedings of the Sixteenth International Conference of the European Association of South Sian Archaeologists, Held in Collège de France, Paris, 2–6 July 2001,* 237–244. Paris, Éditions Recherche sur les Civilisations.

Morgan, J. de (1912) Observations sur les couches profondes de l'acropole de Suse. In E. Pottier, J. de Morgan and R. de Mecquenem (eds), *Mémoires de la délégation en Perse 13.* Recherches archéolgiques 5ème série, 6–26. Paris, Éditions Ernest Leroux.

Muhly, J. D. (1988) The Beginnings of Metallurgy in the Old World. In R. Maddin (ed.), *The Beginning of the Use of Metals and Alloys,* 2–20. Cambridge, MA, MIT Press.

Nissen, H. J. (1988) *The Early History of the Ancient Near East 9000–2000 B.C.* Chicago, IL and London, The University of Chicago Press.

Ottaway, B. (2015) Experiments in Archaeometallurgy. In A. Hauptmann and D. Modarressi-Tehrani (eds), *Archaeometallurgy in Europe III.* Der Anschnitt, Beiheft 26, 337–346. Bochum, Deutsches Bergbau-Museum.

Pernicka, E. (1990) Gewinnung und Verbreitung der Metalle in prähistorischer Zeit. *Jahrbuch des Römisch-Germanischen Zentralmuseums* 37, 21–129.

Pernicka, E., Eibner, C., Öztunalı, O. and Wagner, G. A. (2003) Early Bronze Age Metallurgy in the North-East Aegean.

In G. A. Wagner, E. Pernicka and H.-P. Uerpmann (eds), *Troia and the Troad: Scientific Approaches.* Natural Science in Archaeology, 143–172. Berlin and Heidelberg, Springer.

Pernicka, E., Schmidt, K. and Schmitt-Strecker, S. (2002) Anhang I. Zum Metallhandwerk. In K. Schmidt (ed.), *Norşuntepe. Kleinfunde II, Artefakte aus Felsgestein, Knochen und Geweih, Ton, Metall und Glas.* Archaeologica Euphratica 2, 115–137. Mainz, Philipp von Zabern.

Petrie, C. A. (2013) Ancient Iran and Its Neighbours: Emerging Paradigms and Future Directions. In C. A. Petrie (ed.), *Ancient Iran and Its Neighbours. Local Developments and Long-Range Interactions in the Fourth Millennium BC.* Institute of Persian Studies Archaeological Monographs Series 3, 385–411. Oxford and Oakville, PA, Oxbow Books.

Philaniotou, O., Bassiakos, Y. and Georgakopoulou, M. (2011) Early Bronze Age Copper Smelting on Seriphos (Cyclades, Greece). In P. Betancourt and S. Ferrence (eds), *Metallurgy: Understanding How, Learning Why: Studies in Honor of James D. Muhly.* Prehistory Monographs 29, 157–164. Philadelphia, PA, INSTAP.

Philip, G. (1995) Warrior Burials in the Ancient Near-Eastern Bronze Age: The Evidence from Mesopotamia, Western Iran and Syria-Palestine. In S. Campbell and A. Green (eds), *The Archaeology of Death in the Ancient Near East,* 140–154. Oxford, Oxbow Books.

Pigott, V. C. (1999) A Heartland of Metallurgy. Neolithic/Chalcolithic Metallurgical Origins on the Iranian Plateau. In A. Hauptmann, E. Pernicka, T. Rehren and Ü. Yalçın (eds), *The Beginnings of Metallurgy. Proceedings of the International Conference 'The Beginnings of Metallurgy', Bochum 1995.* Der Anschnitt, Beiheft 9, 107–120. Bochum, Deutsches Bergbau-Museum.

Pollock, S. (1999) *Ancient Mesopotamia, The Eden That Never Was.* Cambridge, Cambridge University Press.

Radivojević, M., Rehren, T., Kuzmanović-Cvetković, J., Jovanović, M. and Northover, J. P. (2013) Tainted Ores and the Rise of Tin Bronzes in Eurasia, *c.* 6500 Years Ago. *Antiquity* 87, 1030–1045.

Radivojević, M., Rehren, T., Pernicka, E., Šljivar, D., Brauns, M. and Borić, D. (2010) On the Origins of Extractive Metallurgy: New Evidence from Europe. *Journal of Archaeological Science* 37/11, 2775–2787.

Rehm, E. (2003) *Waffengräber im Alten Orient. Zum Problem der Wertung von Waffen in Gräbern des 3. und frühen 2.Jahrtausends v. Chr. in Mesopotamien und Syrien.* Oxford, Archaeopress.

Reiter, K. (1997) *Die Metalle im Alten Orient unter besonderer Berücksichtigung altbabylonischer Quellen.* Alter Orient und Altes Testament 249. Münster, Ugarit-Verlag.

Renfrew, C. (1969) The Autonomy of the South East European Copper Age. *Proceedings of the Prehistoric Society* 35, 12–47.

Renfrew, C. (1978) Varna and the Social Context of Early Metallurgy. *Antiquity* 52, 199–203.

Şahoğlu, V. and Tuncel R. (2014) New Insights into the Late Chalcolithic of Coastal Western Anatolia: A View from Bakla Tepe, Izmir. In B. Horejs and M. Mehofer (eds), *Proceedings of the Symposion Western Anatolia before Troy : Proto-Urbanisation in the 4th Millennium BC, Kunsthistorisches Museum Wien, 22–24 November 2012.* Oriental and European Archaeology 1, 65–82. Vienna, Verlag der Österreichischen Akademie der Wissenschaften.

Schiffer, M. B. (2004) Studying Technological Change: A Behavioral Perspective. *World Archaeology* 36/4, 579–585.

Schoop, U.-D. (2011) Çamlıbel Tarlası, ein metallverarbeitender Fundplatz des vierten Jahrtausends v. Chr. im nördlichen Zentralanatolien. In Ünsal Yalçın (ed.), *Anatolian Metal V.* Der Anschnitt, Beiheft 24, 53–68. Bochum, Deutsches Bergbau-Museum.

Schrakamp, I. (2013) Die 'Sumerische Tempelstadt' heute. Die sozioökonomische Rolle eines Tempels in frühdynastischer Zeit. In K. Kaniuth, A. Löhnert, J. L. Miller, A. Otto, M. Roaf and W. Sallaberger (eds), *Tempel im Alten Orient: 7. Internationales Colloquium der Deutschen Orient-Gesellschaft, 11.–13. Oktober 2009, München.* Colloquien der Deutschen Orient-Gesellschaft 7, 445–465. Wiesbaden, Harrassowitz.

Selz, G. J. (1998) Über mesopotamische Herrschaftskonzepte. Zu den Ursprüngen mesopotamischer Herrscherideologie im 3. Jahrtausend. In M. Dietrich/O. Loretz (eds), *Dubsar anta-men: Studien zur Altorientalistik* Festschrift für Willem H. Ph. Römer zur Vollendung seines 70. Lebensjahres 283–344. Münster: Ugarit-Verlag.

Thomsen, C. J. (1837) *Leitfaden zur nordischen Alterthumskunde.* Translated from: Ledetraad til nordisk Oldkyndighed, Kopenhagen 1936. Gesellschaft für Nordische Alterthumskunde (ed.). Copenhagen, Im Secretariat der Gesellschaft.

Thornton, C. P. (2009) The Emergence of Complex Metallurgy on the Iranian Plateau: Escaping the Levantine Paradigm. *Journal of World Prehistory* 22/3, 301–327.

Thornton, C. P. (2013) The Bronze Age in Northeastern Iran. In D. T. Potts (ed.), *The Oxford Handbook of Ancient Iran,* 181–204. Oxford, Oxford University Press.

Thornton, C. P. (2014) The Emergence of Complex Metallurgy on the Iranian Plateau. In B. W. Roberts and C. P. Thornton (eds), *Archaeometallurgy in Global Perspective,* 665–696. New York, Springer.

Thornton, C. P. and Roberts, B. W. (2014) Introduction. In B. W. Roberts and C. P. Thornton (eds), *Archaeometallurgy in Global Perspective,* 1–10. New York, Springer.

Thornton, C. P., Lamberg-Karlovsky, C. C., Liezers, M. and Young, S. M. M. (2002) On Pins and Needles: Tracing the Evolution of Copper-Base Alloying at Tepe Yahya, Iran, via ICP-MS Analysis of Common-Place Items. *Journal of Archaeological Science* 29/12, 1451–1460.

Weeks, L. (2003) *Early Metallurgy of the Persian Gulf. Technology, Trade, and the Bronze Age World.* American School of Prehistoric Research 2. Boston, MA, Brill.

Weeks, L. (2012) Metallurgy. In D. T. Potts (ed.), *A Companion to the Archaeology of the Ancient Near East* 1, 295–316. Chichester *et. al.,* Wiley-Blackwell.

Wilkinson, T. J., Philip, G., Bradbury, J., Dunford, R., Donoghue, D., Galiatsos, N., Lawrence, D., Ricci, A. and Smith, S. L. (2014) Contextualizing Early Urbanization: Settlement Cores, Early States and Agro-Pastoral Strategies in the Fertile Crescent during the Fourth and Third Millennia BC. *Journal of World Prehistory* 27/1, 43–109.

Yener, K. A. (2000) *The Domestication of Metals. The Rise of Complex Metal Industries in Anatolia.* Culture and History of the Ancient Near East 4. Leiden *et. al.,* Brill.

Chapter 15

The Role of Metallurgy in Different Types of Early Hierarchical Society in Mesopotamia and Eastern Anatolia

Marcella Frangipane

In the course of the 4th millennium BCE, the production of metal objects, the beginnings of which date back in the Near East as far as Pre-Pottery Neolithic, became a socially, technologically and symbolically significant activity in connection with the emergence of early hierarchical societies. This phenomenon has often been linked to a new trade impetus presumed to result from the demand driven by the new elites and made possible by their capacity for managing and controlling craft activities and trading routes. Some scholars have also linked the occurrence of population movements and the development of wide-ranging interregional relations to this assumed new need, which is therefore considered to have had major transformation potential for the societies involved in the interaction. World systems models have been applied to various regions and various periods, usually assigning a central role to the circulation of metals (Algaze 1993; Sherratt 1993; Wengrow 2011). As for the Near East, the most hotly debated case has been that of the so-called Uruk expansion phenomenon with the foundation of 'colonies' in Northern Mesopotamia and the assumed formation of a 'world system' in the second half of the 4th millennium BCE involving the entire 'Greater Mesopotamia' and beyond (Algaze 1993; 2001; 2008; Kohl 1989; 2011; Stein 1999; Frangipane *et al.* 2001; 2009). In these 'world system' explanatory models the search for metal and other raw materials has again been an essential factor stimulating socio-political innovation and its dissemination.

The interpretation of centre-periphery relationships has been variously reconsidered and conceptualised by distinguished scholars, sometimes rejecting the too rigid view of systemic relations of dependence between urban and non-urban, nuclear and marginal societies (Sherratt 1993;

Wengrow 2011). However, I believe that the relationship between technological innovation, exchange systems and social and political change in pre- and proto-historic communities was far more complex and variable than these models imply, and that the processes of developing, transmitting and appropriating metallurgy varied much more widely depending upon when and where they occurred, the types of society involved and their different social needs and interests.

The first point to note is that extensive interregional relations characterised Mesopotamian and peri-Mesopotamian societies from the Neolithic, well before the spread of metallurgy. The second point is that, besides the presence of small tools (awls, small chisels) – some of which were already produced in the Neolithic – and a few axes and weapons, the metal objects produced in the 4th and early 3rd millennia BCE were essentially 'prestigious' items not designed for daily use and personal ornaments that had become more numerous and highly sophisticated. The weapons themselves, in the most remarkable cases, seem also to have had an ostentatious function and were used in elite environments. The majority of metals were indeed commonly found in elite areas, either public and ceremonial or funerary, depending on the type of society concerned, the ways metals were used and the nature of the 'value' attached to them.

Besides the emergent use of precious metals, such as gold and silver and certain 'alloys' with an extremely high arsenic content (used perhaps for brilliance and colour), the main metal used was copper, often with a significant proportion of 'minor' elements, such as arsenic, frequently combined with antimony, nickel or lead (Yener 2000). These compositions, which enhanced workability, finish and product quality, were most likely produced by the choice – intentional, in my opinion – of polymetallic ores rich in arsenic and other elements (Palmieri *et al.* 1993). While metal deposits were certainly not to be found everywhere, it is a fact that, within these deposits, copper ores, very often in the form of polymetallic ores, were widely available in the Near East, from the vast mountainous zones of northern and eastern Anatolia, the Caucasus and Transcaucasus, the Zagros mountain chains and the Iranian Highlands to the areas around the Persian Gulf, the southern Negev and Sinai and the eastern Egyptian desert. Apart from the presence of special objects in highly urbanised Lower Mesopotamia, metallurgy in the 4th and the early 3rd millennia reached the zenith of its development in areas that were not far from these deposits. It will suffice to recall the sophisticated metallurgical artefacts of the Late Chalcolithic and Early Bronze I cultures of eastern Anatolia and southern Caucasus or those of the Ghassulian communities in the southern Levant. Metal must certainly have circulated over a pretty vast area and have been a tradable commodity in the 4th millennium, but in my opinion, the forms of supply must have been rather simple, basically following the traditional exchange circuits, possibly extended to new routes linked to the intensified circulation of other commodities, such as precious stones or the secondary products of pastoralism, without necessarily upsetting the already existing systems of interregional interaction.

Considerable technological sophistication and specialisation were required to produce the metal artefacts of the 4th and early 3rd millennia BCE, some of which were highly advanced in their design, but they remained rather limited in number, mainly because they were essentially intended for elite circles.

In this connection, one interesting fact has been observed at Arslantepe, on the Turkish upper Euphrates, indicating a significant change over time in the ways and places in which metal was worked. In the 4th and early 3rd millennia BCE (periods VII, VIA and VIB2) minerals – including a large block of chalcopyrite – together with slag and crucibles for smelting (Fig. 15.1a–c) were found in the settlement, while an area set aside for smelting purposes was unearthed in a courtyard of the VIB2 village (EBI, about 2800 BCE) (Frangipane and Palmieri 1994–1995). Conversely, in one of the settlement levels from the end of the 3rd millennium (Early Bronze IIIB, Arslantepe, period VID), a small room was unearthed containing selected equipment for melting and moulding metal. This equipment consisted of various moulds and small crucibles with a spout for pouring the molten metal into the moulds (Palmieri 1973) (Fig. 15.1d). These findings suggest that, at least until 2800–2700 BCE, blocks of ore were transported to the sites to produce copper on the spot by smelting the minerals in crucibles. And this implies that the artefacts were only produced in limited quantities.[1] After the mid-3rd millennium, the likely increase

in demand for metal artefacts may have conversely driven the artisans to perform smelting in the mines and bring the metal ingots to the sites to be transformed into artefacts there. Such a trend, though not as clearly documented as at Arslantepe, has also been confirmed at other sites, as indicated by the types of crucibles attested and some evidence of metalworking close to the mines.

Bearing all this in mind, I believe that at least until 2600–2500 BCE metallurgy will not have been a fully-fledged, integral part of the production systems of Near Eastern societies and that the 'value' attributed to metal was mostly not 'use value' in economic terms but rather ostentatious and symbolic value, with a notable social function. The social/symbolic importance of these objects is suggested by the large-scale development of metal artefact production in conjunction with the establishment of the first genuine social and political hierarchies. But this very probably did not imply any substantial direct control over metallurgical activity and the metal trade by the emerging state institutions or the status of metal and metalworking as an important part of their political economy. This hypothesis is further supported by the fact that even in the case of the mature Mesopotamian states (Akkadian and Assyrian) later on in history, there is no proof whatsoever that artisans and traders were particularly dependent on the state (Yoffee 2011, 304).

Another possible confirmation of the limited importance of metals for the daily life of the population is the fact that in the 4th and early 3rd millennia BCE the lithic industry was still important everywhere, producing tools used for most of the activities performed. It was only from the second half of the 3rd millennium BCE that the manufacture of these tools began significantly to decline, finally dwindling down to specialisation in certain classes of tool only (for instance, arrow heads).

One of the major factors fuelling the increase in metal production in the course of the 3rd millennium was probably its effectiveness in the production of weapons, which gained increasing significance with the development of warfare, military classes and organised combat. The other factor I consider essential was the introduction of tin. It changed working conditions and improved quality assurance for metal products and, given its rarity in nature, generated a need to guarantee regular supplies by establishing well-defined political and commercial relations that were 'international' in scope. It is really only after 2600–2500 BCE that we can begin to talk about specific metal-trading routes recognisable as such, including even the establishment, at the very beginning of the 2nd millennium BCE, of trading expeditions and merchants' colonies. Once plentiful and regular supplies could be guaranteed in safety and the urban organisation of labour was fully accomplished, production could cover the whole range of artefacts, including common tools and objects in daily use,

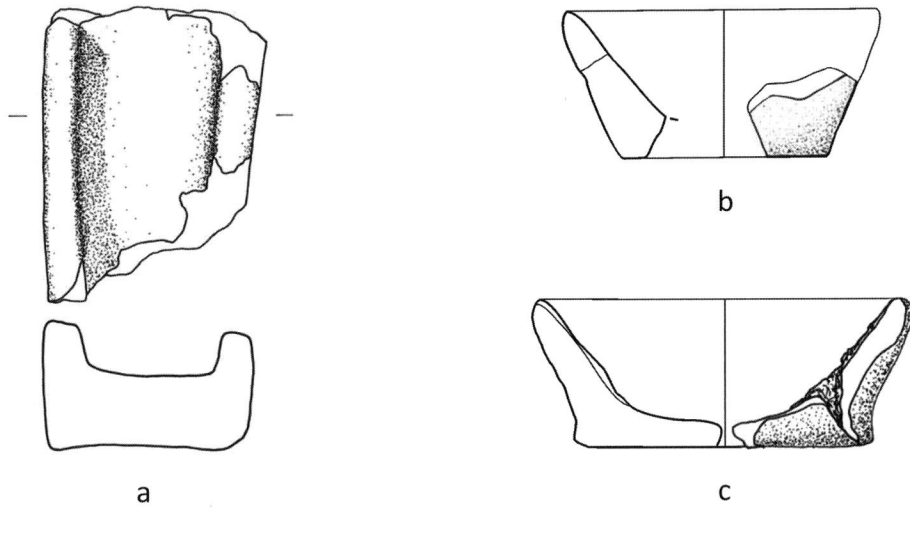

Figure 15.1: Arslantepe. Crucibles from various periods. (a–c) Smelting crucibles from period VII (a, c) and VIA (b), scale 1:3; (d) Melting crucibles from period VID.

thus becoming an industry in every sense of the term and an economically important productive activity. Metal trading and production had started to play an important role in the economy of the Near Eastern societies.

Although the two steps described above seem to have been the main stages in the general development of metallurgical technology in the Near East during the 4th

and the 3rd millennia BCE, wide-ranging differences emerged in the way metal was used and in the social and economic roles played by metal in different types of society. And it is this diversity that I would like to deal with here, stressing the contrasting patterns regulating the production and use of metals in different societies, patterns that have also had a variety of impacts on the potential for spreading, transmitting and disseminating this technology at different times and under different socio-political conditions.

I shall focus exclusively on the initial stages in the development of a 'socially important metallurgy' (4th and early 3rd millennia BCE), taking place at a time when metals and metalworking appeared to be mainly intended for purposes of prestige, display and elitism. I shall try to analyse the differences in the types of object produced and, most of all, in the ways in which metals displayed the prestige of high-status groups in two different types of early hierarchical society in the Mesopotamian and peri-Mesopotamian world. The societies analysed and compared here are (a) the emergent Uruk and Uruk-influenced state societies at the end of the 4th millennium, and (b) the clanic-tribal types of society that emerged, or re-emerged, from the collapse of those early state societies along the middle-to-upper Euphrates and the upper Tigris valleys at the beginning of the Early Bronze Age. In both cases I shall refer particularly to the findings from Arslantepe, where the two models of society clearly succeeded each other and with whose data I am directly familiar.

Metals in early centralised societies of the 4th millennium BCE

In terms of their social function and symbolism, the features and use of metallurgy in the various early centralised societies of the extensive 'Mesopotamian world' in the 4th millennium BCE may be considered similar, though it is likely that there were differences in scale and in production and supply systems between the large cities of Mesopotamia and the non-urban peripheral sites closer to the sources of metal, one of them being the palatial centre at Arslantepe. In the first case, the high degree of labour specialisation and the production and internal circulation systems inside the cities must have stimulated the development of sturdy local craftsmanship, which may have managed the supply of raw materials from far away, probably following the interregional communication routes that were already well established. In the case of northern non-urbanized centres such as Arslantepe (Frangipane 2010; 2012), metals, and perhaps even finished metal products, may have been obtained thanks to the intense structural relations established with some pastoralist communities moving around in the central-northern and north-eastern regions of Anatolia and the southern Caucasus, all rich in metal ores (Frangipane 2017). Between the end of the 4th and the beginning of the 3rd millennium BCE, these mobile groups reached the upper Euphrates valley, probably attracted by centres such as Arslantepe and the prospects of access to the urbanised world of Mesopotamia that these centres appeared to offer them. Seen from this perspective, they may have made the production of metals one of their specialised activities, bringing the metal with them

to the centres on the upper Euphrates, as evidenced by the similarities between the technical and morphological features of the various items found in the regions affected by the movements of these groups (the Kura-Araxes areas including Georgia, Armenia and north-eastern Turkey, the southern Black Sea coast and north-central Anatolia and the upper Euphrates valley). The remarkable continuity found in the metals used at Arslantepe in the Late Chalcolithic 5 palatial period (period VIA) and in the later post-palatial developments of Early Bronze I (period VIB1), which was characterized by the presence on the site of groups linked to the Kura-Araxes cultures, is further evidence in favour of metal production and circulation in the region as set out above (Frangipane 2017).

Regardless of the production systems involved, the objects from centralised 4th millennium societies exhibit a very high degree of technological sophistication and highly specialised craftsmanship. As we have seen, however, the main aim was to produce sumptuary or elite items, most of them found in public and ceremonial areas. In the large centres of the Late Uruk period, sophisticated artefacts made of various metals, including precious metals, mainly come from 'temples' or ceremonial places. One need only recall the luxurious nature of the numerous objects found in the sacred area of the Eanna at Uruk-Warka, including the two small spear or javelin heads, one them in silver, found in the *Riemchengebaude*, and from the slightly later Jemdet Nasr period, the series of opulent metal objects in the well-known *Sammelfund* from Uruk III (Heinrich 1936). Copper, silver and gold sheet cladding furniture and animal figurines, copper pins with caprine figurines, a silver vessel and the golden spout of another vessel were certainly artefacts worthy of exhibition.

Lead pots and human statuettes of copper were found in the other large southern centre of Susa (Le Breton 1957; Steve and Gasche 1971), while the famous altar with the panel covered with a sheet of gold and affixed with 42 gold-headed silver nails stands out in the Eye Temple at Tell Brak (Mallowan 1947; Oates and Oates 1992).

Arslantepe does not display the same wealth with respect to the use of precious metals, but various significant and important metal artefacts have been found in the late 4th millennium palatial complex, including the well-known group of arsenic copper weapons comprising 12 spearheads and 9 swords associated with a quadruple spiral plaque (Fig. 15.2f–g). The swords were an exceptional discovery in themselves because they were the first-ever example of the use of the sword as a weapon, far earlier than any other known example. Although not found elsewhere in the 4th millennium BCE, the spears were of a type that in the first centuries of the 3rd millennium was subsequently to spread throughout the whole middle and upper Euphrates valley, the upper Tigris and in many other sites in Transcaucasia and the Black Sea regions. The Arslantepe weapons were

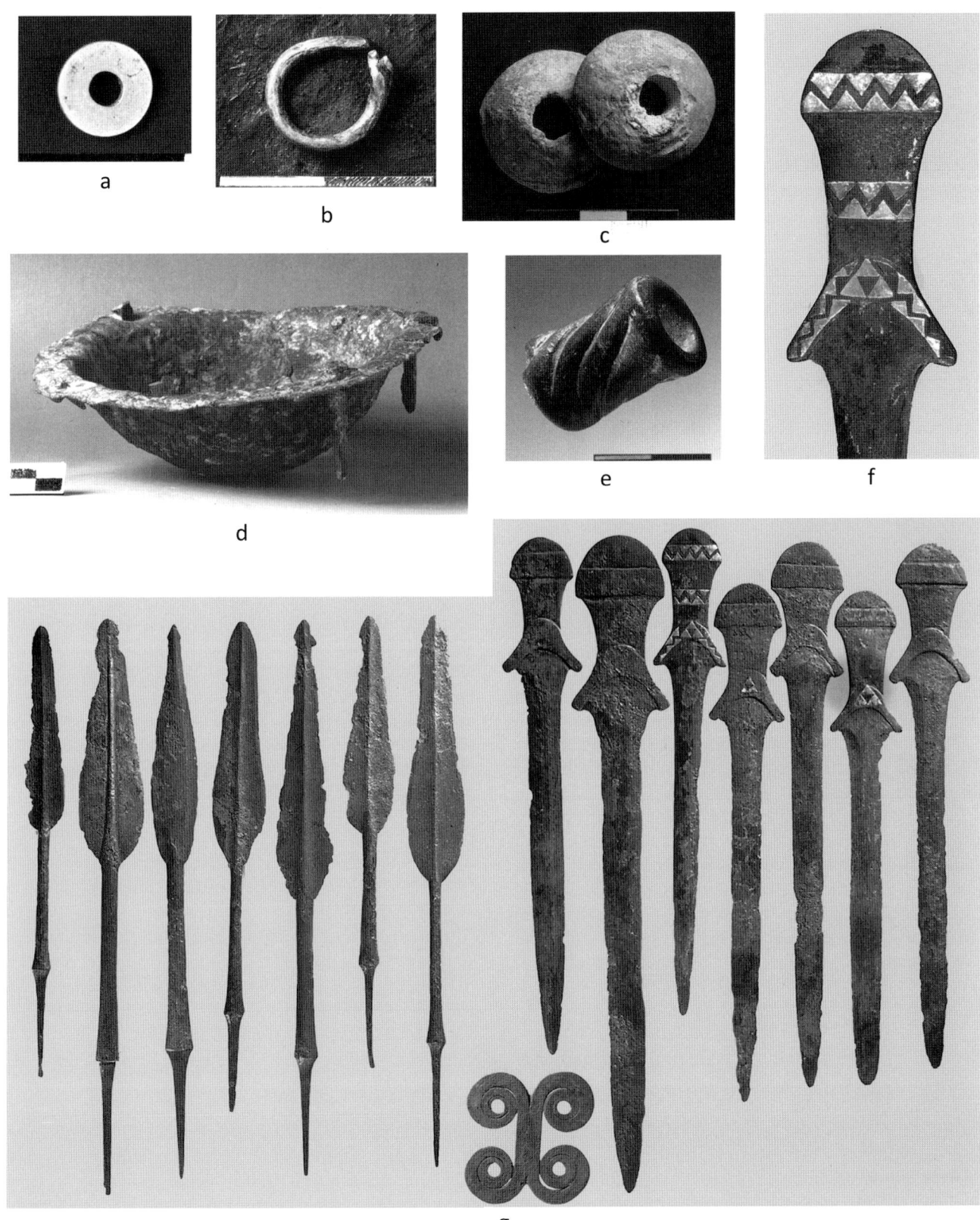

Figure 15.2: Metal objects from Arslantepe Period VIA. (a) Golden bead from Temple B; (b) Ag-Cu ring from the sealing dump A206; (c) Pb spindles/beads from corridor A796; (d) Cu-As door-socket from room A830; (e) Copper cylinder from a filling layer north of Temple A; (f–g) Cu-As weapons with silver inlay decoration from Building III (A113).

certainly used, as we can see from the care with which they were fashioned and the way the edge of the blade was sharpened. But at least as far as the swords were concerned, the completely flat hilts and the decorations on some of them (silver inlay and incisions) would suggest that these specific objects may have been parade weapons for display purposes. The position in which they were discovered, grouped together between two collapsed walls, also suggests that they may have been hanging on the back wall of the room with the most highly decorated sword on show (it was the last one to be found under the others). It is therefore probably safe to infer that (a) the weapons were exhibited as a token of prowess and a symbol of the power held by the élites of the palace and (b) they were reproductions of objects that were certainly in use.

It is also interesting to note that the presence of the swords together with the spears suggests the existence of organised combat involving different figures and functions: weapons to be thrown, perhaps for mass assaults, and weapons for hand to hand combat, perhaps reserved for high-ranking individuals. The first-time emergence of the sword may suggest the initial warrior role of the chiefs, probably in relation to some specific threat of instability that seems to have occurred at Arslantepe at the end of the palace period. It may have been the result of some internal weakness caused by the lack of urbanisation and any genuine social stratification. It may also have been prompted by encroaching pressure from the nomadic pastoralists, who after initial inclusion and integration into the centralised system may gradually have become less amenable to control.

This connotation of power is bound up with the process of early secularisation, expressed in the construction of the palace complex. It is unique in the 4th millennium BCE, in which the role of the temples, which were smaller and reserved for a few people, was no longer central. Confirmation of this has come from the discovery, in the most recent 2014 campaign, of an imposing building at the heart of the palace linking the residences of the elites to the public area. It was not used for activities of a ceremonial or religious character but to receive the public in a kind of 'audience hall' in line with a highly codified but 'secular' ceremony. In this context, the weapons would have been a symbol appropriate to the type of authority manifested at Arslantepe, which was different from the authority expressed in the large sacred areas of the Mesopotamian cities. In those areas, force and violence were depicted in iconographies rather than being displayed in real exhibits.

In addition to the weapons themselves and the sophisticated silver inlay on some of them, other minor metal objects indicate the ability to use different metals and complex alloys. These include a silver pin and an earring or ring, two small lead weights or ornaments, a small golden disc, probably from a dress, a cylindrical object or handle decorated with arsenical copper, a pendant made of

a very unusual copper–lead–arsenic alloy (Fig. 15.2a–c, e). Another object that stands out for its significance is an arsenic copper cup that had been nailed to a wooden beam at the monumental entrance to one of the palace buildings and was used as a door socket (Fig. 15.2d). Once again, the very fact that metal was used rather than the customary stone for the hinge on an entrance gate of an important building highlights the representative and symbolic role of metal in this context.

Despite the differences registered in the sacred and urbanised contexts of Mesopotamia and in the non-urban political-administrative centre Arslantepe, the metal objects found in all the early state societies of the 4th millennium appear to have had a prevailing symbolic value and to have been exhibited in places of power. They were not, however, used in funerary contexts. From the Uruk period almost no regular burials have been found except for rare examples with simple grave gifts and rituals. In fact, the funerary sphere seems to have been completely ignored or considered virtually insignificant. Metal was therefore something to be exhibited in the world of the living. Potentially susceptible of reuse, it was not removed from circulation. It seems fair to hypothesise that, besides the high symbolic value attaching to metal, there may also have been an incipient perception of its economic value in these early centralised societies stemming from the increasing inclusion of this activity in the productive system controlled by the state elites. Accordingly, there may have been an initial attribution of value as a non-perishable accumulative good and hence as a source of 'wealth'. This was subsequently to become much more pronounced in mature state societies and in those societies in which the elites based their power on real 'wealth finance'.

The beginning of the 3rd millennium BCE in eastern Anatolia: a new use of metals in a new society

At the very end of the 4th millennium BCE, the first early state societies, especially in the north, became embroiled in various ways in a crisis that disrupted the cultural and socio-political homogeneity created throughout the north by the dissemination of the Uruk culture and the intensive relations existing between all the communities of the Mesopotamian and peri-Mesopotamian world. But the different levels of urbanisation achieved in the 4th millennium BCE in different regions of the north led to very different outcomes from this crisis. In the Khabour and the heartland of Jezira, where as in Lower Mesopotamia urbanisation had developed considerably, the socio-political system did not have to cope with an all-out crisis. Indeed, albeit with a number of changes, the larger urban centres and the centralised management of the territories grew and flourished. In the middle and upper Euphrates valleys and in the Turkish Upper Tigris, on the other hand, where urban growth had not reached an equivalent level, the existing political and economic centralisation systems suffered a radical collapse at the beginning of the 3rd millennium. Although the outward forms and the intensity of the crisis in these areas were not the same everywhere, there was a general overthrow of the system generating new elites that were probably 'clanic/ aristocratic' in character. Though they were no longer able to exercise direct control over the staple economy of the people and the labour force and no longer focused on the management of ritual, they were still capable of exercising political leadership expressed through the ostentatious display of power and wealth. The outward and visible signs of the status of these new chiefs and their authority were no longer temples, palaces, mass-produced bowls and cretulae, but fortification walls, rich burials and the widespread use of metal mainly in the form of weapons (Fig. 15.3). The new elites were warriors and their symbols were put on display at the time of their death.

It is precisely in this period that we find a new focus on funerary rituals, with the dissemination of cist graves and rich metal burial gifts, both of which were wholly alien to the local Chalcolithic traditions and were perhaps adopted

a   b   c

Figure 15.3: Arslantepe. (a–b) Sword and spear from the 'Royal Tomb' (2950–2900 BC); (c) The upper town fortification wall (mud-brick elevation and stone foundations) from period VIB2 (2900–2800 BC).

from the customs of the Transcaucasian communities. As already mentioned, these communities were moving down into the south-west and the south-east, coming into contact with the valley dwellers there and instituting widely varying forms of penetration and hybridisation. These were highly invasive and clearly recognisable in the upper Euphrates but not very visible in the more southern areas. Their indirect influence manifested itself in the widespread transmission of certain selected customs and cultural features that were evidently easily adaptable to the new demands of the clanic and tribal societies and their elites emerging after the disintegration of the Uruk world (Cooper 2007; Frangipane 2007).

Metals very probably originally derived from, or inspired by, the Kura-Araxes traditions were definitely among the new features adopted widely in each of the zones in question as a new symbol of wealth. It was on this basis that a fully-fledged metallurgical tradition took shape in Early Bronze I that permeated the whole of the upper and middle Euphrates valley and – as we now know– also the upper Tigris. The metal objects produced were mainly weapons –spearheads of two kinds, butted and poker spears, daggers – axes, elaborate pins and other ornaments as well as ostentatious items such as decorated cylinders with animal statuettes on top of them (Fig. 15.4). Except for simpler pins and a few chisels, some of which have been found in the settlement, all these artefacts were almost exclusively discovered in graves, particularly cist graves, as funerary gifts. The effort required to build stone cist tombs was in itself a token of the new importance attached to the funerary sphere. Burials were elite tombs, graves set apart in some way in residential areas and cemeteries.

At Arslantepe, groups of transhumant pastoralists with links to the world of Kura-Araxes settled on the ruins of the 4th-millennium BCE palace that had been destroyed by a raging fire marking the ultimate collapse of the whole system. Previously, these groups had probably frequented the Malatya plain (Palumbi 2010; 2012). Now they occupied the tell, repeatedly staking their claim with seasonal settlements made of a few wattle and daub huts and large fenced areas for the livestock and slowly taking over the whole site, building a huge mud-brick building for public or communal use on the top of the mound (Frangipane 2014). This building, comprising a large hall almost certainly for receptions and two store rooms full of pots, was the only one where various metal objects were found, including two butted spearheads (Fig. 15.4a) identical in type to the spears found in the 4th millennium palace. A few copper rings of sheet metal with a variable number of nails may have been used to fix the spearheads to the wooden shafts. Other objects were a lead medallion with a decorated face, a double spiral pin and a few small awls/chisels (Frangipane 2014, fig. 10). The latter were the only items to have occasionally been found in the huts as well.

It is likely that an extraordinary stone cist tomb, the so-called 'Royal Tomb' of Arslantepe, dates from the end of this phase (Arslantepe VIB1) or the beginning of the phase immediately following it, which was marked by the construction of an imposing fortification wall on the peak of the mound (initial period VIB2) (Fig. 15.3c). The stone cist was built at the bottom of a large pit and bears witness to a complex ritual involving five individuals (Frangipane *et al.* 2001). The cist was the grave of an adult male. The very rich funerary gifts inside the cist included pots of various types and 64 metal objects. Four adolescents (three girls and one boy) had probably been sacrificed on the tomb and only the two of them who had been placed directly on the cist cover were wearing personal ornaments, comprising in each case diadems made of a copper-silver alloy, hair spirals of the same alloy and two copper pins. The metal gifts for the individual inside the cist encompassed a large variety of different objects (Fig. 15.5): precious silver and gold ornaments close to the body, seven arsenical-copper spearheads of the usual type symbolically stuck into the ground along the perimeter of the tomb all around the deceased's head and a variety of other objects piled behind his back in a sort of hoard. Here there were arsenical copper weapons (including daggers and swords), axes, tools (gouges, chisels, a knife) and two vessels, plus a decorated diadem and numerous arm-rings and hair spirals made of an unusual alloy of copper and silver.

The spearheads, diadems, spirals and almost all the objects in the tomb were closely related to similar items from the Kura-Araxes regions, as was the cist tomb and some aspects of the burial ritual (Kushnareva 1997). But the position of the tomb, which was built in isolation in a settlement and stood prominent on the mound, the human sacrifice and the extraordinary abundance of the metal artefacts, as well as the use of precious metals and the widespread presence of the highly unusual copper–silver alloy are special features of the Arslantepe tomb, making it a unique and very ancient example of this kind of burial for a high-ranking individual, perhaps a chief. Furthermore, the pottery found in the tomb was clearly locally produced, albeit with very close links to the Kura-Araxes counterparts, mixing typical examples of the Arslantepe VIB1 repertoire with wheel-made wares of the Uruk tradition inherited by the EBI cultures of the Turkish Euphrates valley.

This discovery thus represents a perfect example of the way in which new cultural traits were adopted by a profoundly transformed society, with the widespread use of metal items that were inspired by, or even imported from, the Transcaucasian world and intended as new and crucially important symbols of the prestige and power of the new leaders with powerful warlike connotations.

This phenomenon rapidly spread throughout the Turkish middle Euphrates valley during the course of Early Bronze

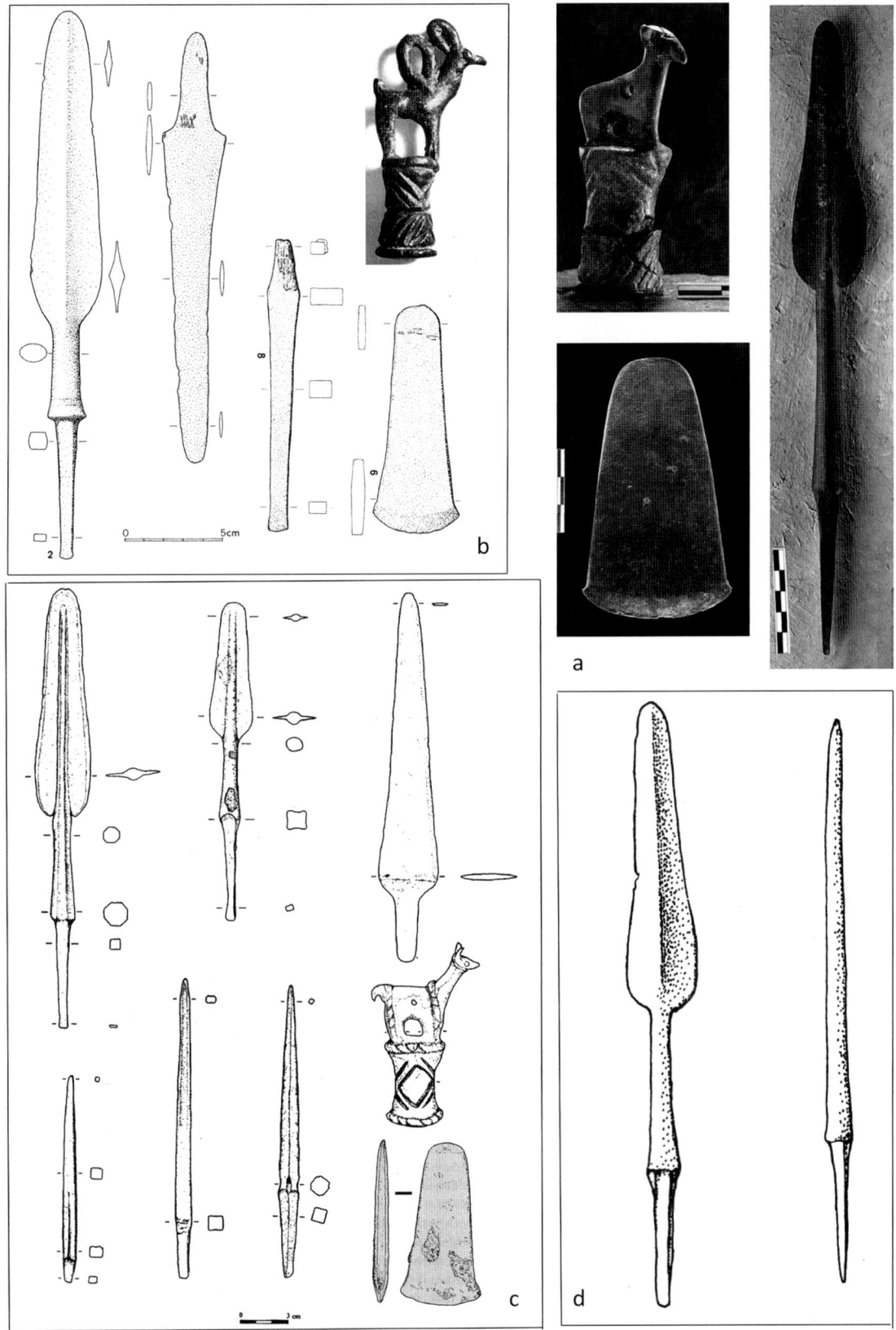

Figure 15.4: Early Bronze I metals form the Middle and Upper Euphrates region. (a) Arslantepe, periods VIB1 and VIB2; (b) Hassek Höyük, cist grave G12 (from Behm-Blancke 1984, fig. 8, pl. 12); (c) Birecik Cemetery (from Squadrone 2007, figs 13.5, 13.6); (d) Carchemish burials (redrawn from Woolley 1914).

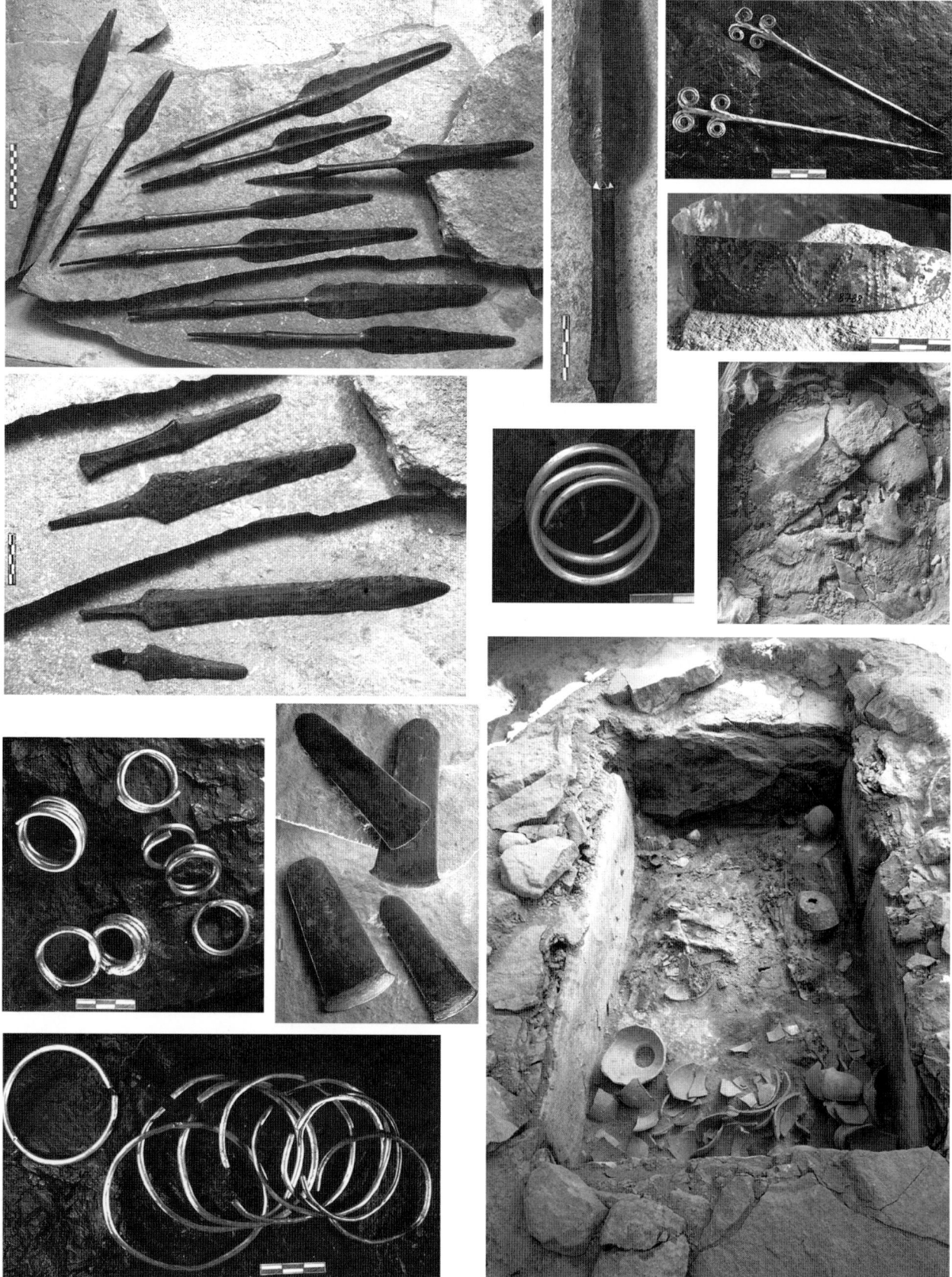

Figure 15.5: The so-called 'Royal Tomb' from Arslantepe with part of its rich metal gifts (see Frangipane et al. *2001).*

I, as evidenced by the two cist graves and a few special burials at Hassek Höyük (Behm-Blancke 1984) as well as the numerous cists in the Birecik necropolis, further to the south (Sertok and Ergeç 1999; Squadrone 2007). All are full of metal gifts similar to those found at Arslantepe, though less rich.

Lastly, a number of Early Bronze I cist graves excavated on top of Late Uruk levels have been recently brought to light at Başur Höyük (Sağlamtimur and Ozan 2014) in the Turkish sector of the Tigris region, revealing extraordinary rich metal gifts partly similar to those from the Arslantepe tomb and other Euphrates sites, but with a number of items of an even more surprising sophistication and complexity. An exceptional feature of this site is that it lies on the margins of the Ninevite 5 area, indeed to a certain extent forming a part of it as is evidenced by the painted Ninevite pottery found as burial gifts in these tombs, but no rich metal items have been found in the Ninevite 5 burials in the urbanised Jezira, nor has any cist grave been documented there, both features equally absent from the earlier urban and proto-state societies of the Uruk period. Accordingly, it is possible that the populations living on the margins of the northern urban areas underwent a pastoralist-oriented reconversion at the very beginning of the 3rd millennium BCE, entering into contact with the Kura-Araxes groups and acquiring from them technologies and symbols that in this case were appropriate for representing the new social order and the new needs for symbolic expression of communities sharing structural likenesses with their neighbours in the northern and western Taurus mountain range.

The use of metal mainly for funerary purposes and its resultant removal from circulation as a source of wealth seem to have been the prerogative of non-urban ranked societies, probably with warrior chiefs.

We may assume that in this case the symbolic power of metal objects as the expression of wealth and prestige was even greater than in early state societies in the Late Uruk periods, given that greater efforts were needed to produce these objects in such large quantities and that this wealth was then largely withdrawn from circulation.

Contrasting patterns in the use of metals in two different types of society: concluding remarks

In both the centralised Late Uruk societies of the 4th millennium BCE and the fragmented tribal and rank societies at the beginning of the 3rd millennium in eastern Anatolia and north-western Syria, the technology and the working commitment required to meet a need that was not theoretically economic in nature reveals the social importance of expressing prestige and its high potential for social reproduction. But the different way in which the

metal was used in these two environments suggests that its social and political role was profoundly different in each case. The most striking difference is that in the first case the metal was displayed in public places and at public events and was therefore somehow symbolically shared with the population, albeit as an expression of the might of the emergent state institutions and leaders. It was to some extent invested in the political performances of those leaders. In the second case, as Wengrow has brilliantly suggested in a recent paper analysing other contexts, the metal was hoarded and hidden, 'sacrificing it in spectacular performances' (Wengrow 2011, 137) Its withdrawal from circulation stressed the value of its exclusive possession by the chiefs (Wengrow 2011, 136). In the first case, the leaders symbolically shared their prerogatives with the people (religious events, redistribution practices and administration are forms of extended participation in politically and economically centralised performances) and the metal, though exclusively or almost exclusively destined for the elites, was reconverted and enjoyed by other people at the level of visual and symbolic sharing. In this case, as Wengrow says in his paper, 'we should expect to see a much smaller proportion of metalwork preserved,' whereas in the case of funerary use of metals, 'we should expect [...] metalwork in copious and impressive quantities' (2011, 137). And, in fact, this hypothesis is confirmed when we compare the relative rarity of metal products in the early state societies of the Mesopotamian world with their abundance in the cist graves of high-ranking persons in the chiefdom societies of the upper and middle Euphrates and Tigris valleys in the early 3rd millennium BCE.

When the initially very restricted use of metal for manufacturing weapons started to spread, this indicated a progressive change in the nature of political power in these eastern Anatolian societies.

Wherever real power of control, not only political but also economic, was exercised over the population, the 'prestige' objects were mainly used to emphasise and confirm that power. By contrast, when the authority of the new chiefs had to be established in the absence of effective instruments for economic control and coercion, the sumptuary objects probably had the function of 'creating' prestige, and hence status, in addition to merely representing it. The production of metal weapons and their possession by the new warrior leaders even after death allied the material instruments of warfare themselves with the symbolic display of authority.

In neither of the two systems did the acquisition of new or more advanced technological knowledge, in both cases related to elitist uses, have the effect of transforming society. On the contrary, it was society that transformed the function and use of the products of these new technologies, adapting them to suit its purposes. It was only when a new technology entailed a real change in the daily life of the population (as was the case, for instance, with the dissemination of agriculture) that, in my opinion, technological transfer exercised a transformative power on society.

Note

1   Limited production would only make it reasonable bringing to the site the heavy ore pieces from which a few metal objects could be produced.

References

Algaze, G. (1993) *The Uruk World System*. Chicago, IL, University of Chicago Press.

Algaze, G. (2001) The Prehistory of Imperialism: The Case of Uruk Period in Mesopotamia. In M. S. Rothman (ed.), *Uruk Mesopotamia and Its Neighbors: Cross-Cultural Interactions in the Era of State Formation*, 27–83. Santa Fe, NM, School of American Research (SAR) Press.

Algaze, G. (2008) *Ancient Mesopotamia at the Dawn of Civilization. The Evolution of Urban Landscape*. Chicago, IL, University of Chicago Press.

Behm-Blancke, M. (ed.) (1984) Hassek Höyük. *Istanbuler Mitteilungen* 34, 31–150.

Cooper, L. (2007) Early Bronze Age Burial Types and Social-Cultural Identity within the Northern Euphrates Valley. In E. J. Peltenburg (ed.), *Euphrates River Valley Settlement: The Carchemish Sector in the Third Millennium BC*. Levant Supplementary Series 5, 55–70. Oxford, Oxbow Books.

Frangipane, M. (2007) The Establishment of a Middle/Upper Euphrates EB I Culture from the Fragmentation of the Uruk World. New Data from Zeytinli Bahçe Höyük (Urfa, Turkey). In E. Peltenburg (ed.), *Euphrates River Valley Settlement. The Carchemish Sector in the Third Millennium BC*. Levant Supplementary Series 5, 122–141. Oxford, Oxbow Books.

Frangipane, M. (2009) Rise and Collapse of the Late Uruk Centres in Upper Mesopotamia and Eastern Anatolia. *Scienze dell'Antichità* 15, 15–31.

Frangipane, M. (2010) The Political Economy of the Early Central Institutions at Arslantepe. Concluding Remarks. In M. Frangipane (ed.), *Economic Centralisation in Formative States. The Archaeological Reconstruction of the Economic System in 4th Millennium Arslantepe*. Studi di Preistoria Orientale 3, 289–307. Rome, Sapienza University.

Frangipane, M. (2012) Fourth Millennium Arslantepe: The Development of a Centralised Society without Urbanisation. *Origini* 34, 19–40.

Frangipane, M. (2014) After Collapse: Continuity and Disruption in the Settlement by Kura-Araxes-Linked Pastoral Groups at Arslantepe-Malatya (Turkey). New Data. *Paléorient* 40/2, 169–182.

Frangipane, M. (2017) The Role of Metal Procurement in the Wide Interregional Connections of Arslantepe during the Late 4th–Early 3rd Millennia BC. In C. Maner, Horowitz, M. T. and Gilbert, A. S. (eds), *Overturning Certainties in Near Eastern Archaeology. A Festschrift in Honor of K. Aslıhan Yener*. Culture and History of the Ancient Near East 90. Leiden and Boston, Brill.

Frangipane, M. and Palmieri, A. M. (1994–1995) Un modello di ricostruzione dello sviluppo della metallurgia antica: il sito di Arslantepe. *Scienze dell'Antichità* 8–9, 59–78.

Frangipane, M., Di Nocera, G. M., Hauptmann, A., Morbidelli, P., Palmieri, A. M., Sadori, L., Schultz, M. and Schmidt-Schultz, T. (2001) New Symbols of a New Power in a 'Royal' Tomb from 3000 BC Arslantepe, Malatya (Turkey). *Paléorient* 27/2, 105–139.

Heinrich, E. (1936) *Kleinfunde aus den archaischen Templeschichten in Uruk*. Ausgrabungen der Deutschen Forschungsgemeinschaft in Uruk-Warka 1. Leipzig, Harrassowitz.

Kohl, P. L. (1989) The Use and Abuse of World System Theory: The Case of 'Pristine' West Asian State. In C. C. Lamberg-Karlovsky (ed.), *Archaeological Thought in America*, 218–240. Cambridge, Cambridge University Press.

Kohl, P. L. (2011) World-Systems and Modelling Macro-Historical Processes in Later Prehistory: An Examination of Old and a Search for New Perspectives. In T. C. Wilkinson, S. Sherratt and J. Bennet (eds), *Interweaving Worlds*, 77–86. Oxford, Oxbow Books.

Kushnareva, K. K. (1997) *The Southern Caucasus in Prehistory: Stages of Cultural and Socioeconomic Development from the Eighth to the Second Millennium B.C.* Translated by H. N. Michael. Philadelphia, PA, University of Pennsylvania Museum.

Le Breton, L. (1957) The Early Periods at Susa, Mesopotamian Relations. *Iraq* 19/2, 79–124.

Mallowan, M. E. L. (1947) Excavations at Brak and Chagar Bazar. *Iraq* 9, 1–259.

Oates, D. and Oates, J. (1992), Excavations at Tell Brak, 1990–91. *Iraq* 53, 127–145.

Palmieri, A. (1973) Scavi nell'area sud-occidentale di Arslantepe. *Origini* 7, 55–179.

Palmieri, A. M., Sertok, K. and Chernykh, E. (1993) From Arslantepe Metalwork to Arsenical Copper Technology in Eastern Anatolia. In M. Frangipane, H. Hauptmann, M. Liverani, P. Matthiae and M. Mellink (eds), *Between the Rivers and Over the Mountains: Archaeologica Anatolica et Mesopotamica Alba Palmieri Dedicate*, 573–599. Rome, Sapienza University.

Palumbi, G. (2010) Pastoral Models and Centralised Animal Husbandry. The Case of Arslantepe. In M. Frangipane (ed.), *Economic Centralisation in Formative States: The Archaeological Reconstruction of the Economic System in 4th Millennium Arslantepe*. Studi di Preistoria Orientale 3, 149–163. Rome, Sapienza University.

Palumbi, G. (2012) Bridging the Frontiers. Pastoral Groups in the Upper Euphrates Region in the Early Third Millennium BCE. *Origini* 34, 261–278.

Sağlamtimur, H. and Ozan, A. (2014) Siirt-Başur Höyük 2012 yılı kazı çalışmaları, 35. *Kazı Sonuçları Toplantısı* 3, 514–529. Ankara, Kültür ve Turizm Bakanlığı.

Sertok, K. and Ergeç, R. (1999) A New Early Bronze Age Cemetery: Excavations Near the Birecik Dam, Southeastern

Turkey. Preliminary Report (1997–1998). *Anatolica* 25, 87–107.

Squadrone, F. F. (2007) Regional Culture and Metal Objects in the Area of Carchemish during the Early Bronze Age. In E. J. Peltenburg (ed.), *Euphrates River Valley Settlement: The Carchemish Sector in the Third Millennium BC*. Levant Supplementary Series 5, 198–213. Oxford, Oxbow Books.

Sherratt, A. G. (1993) What Would a Bronze Age World System Look Like? Relations between Temperate Europe and the Mediterranean in Later Prehistory. *Journal of European Archaeology* 1, 1–57.

Stein, G. (1999) *Rethinking World Systems: Diasporas, Colonies, and Interaction in Uruk Mesopotamia*. Tucson, AZ, University of Arizona Press.

Steve, M. J. and Gasche, H. (1971) *L'Acropole de Suse*. Leiden, Brill.

Wengrow, D. (2011) 'Archival' and 'Sacrificial' Economies in Bronze Age Eurasia: An Interactionist Approach to the Hoarding of Metals. In T. C. Wilkinson, S. Sherratt and J. Bennet (eds.), *Interweaving Worlds*, 135–144. Oxford, Oxbow Books.

Woolley, C. L. (1914) Hittite Burial Customs. *Liverpool Annals of Archaeology and Anthropology* 6, 87–98.

Yener, K. A. (2000) *The Domestication of Metals: The Rise of Complex Metal Industries in Anatolia*. Leiden, Brill.

Yoffee, N. (2011) Unbounded Structures, Cultural Permeabilities and the Calyx of Change: Mesopotamia and its World. In T. C. Wilkinson, S. Sherratt and J. Bennet (eds.), *Interweaving Worlds*, 303–308. Oxford, Oxbow Books.

Chapter 16

The Use of Bronze Objects in the 3rd Millennium BC: A Survey between Atlantic and Indus

Lorenz Rahmstorf[1]

The spread of tin bronze technology is considered to have been 'a unified process that accompanied the transformation of society from a simple to a higher degree of organisation […] The spread did not occur by chance – now here, now there – but in a clear pattern from a relatively large region of origin' (Pernicka 1998, 140–141, translated from the German). The origins were to be sought in the Near East, with the technology subsequently spreading to the west through Anatolia in the Aegean Sea and finally to the rest of Europe as well as eastwards to the Indian subcontinent and finally to China. The whole distribution pattern followed the assumed dissemination routes of copper technology from south-west Asia to the west and east (Roberts *et al.* 2009; Mei 2009; Vandkilde 2016, fig. 1). These diffusion models, as they are called, have been contested, so it is worth reconsidering the material their assumptions are based on. The present article provides a survey on the appearance of bronze, an intentional alloy of copper and tin, before the end of the 3rd millennium BC, or more precisely before 2200/2100 BC. More than 140 sites have been assembled between the Atlantic in the west and the Indus or north-west India in the east. The analysis was originally intended as a detailed overview of the chronology and geographical distribution of early bronze objects and their contextual associations. Another aim was to investigate the kind of objects that were made of bronze when this material was still new and in most cases difficult to come by. No more than a preliminary outline of all these interesting aspects can be given here. Nevertheless, the data may hopefully provide a basis for others to investigate these questions further on a broader material basis than has been possible so far.

Tin is obtained from *cassiterite* (SnO_2), which occurs in primary deposits as mountainous tin, in secondary deposits, and as stream tin in rivers and coastal areas 'where weathering processes have destroyed the host rock and have deposited the chemically and physically stable, dense (SG 6.5–7) and hard (Mohs hardness 6–7) cassiterite mineral in a natural physical concentrating process' (Charles 1975, 20–21). The latter type of tin was probably often recovered with fluvial placer gold. Archaeological remains (slag) from the extraction of tin from tin oxide cassiterite cannot really be expected because alluvial tin is very pure and smelting produces virtually no slag (Wang *et al.* 2016, 88). The use of a copper-tin ore, *stannite* (Cu_2FeSnS_4), is also possible and will be discussed below. Compared to other important metals such as copper and iron, tin deposits are very rare, but Europe is relatively well supplied with them, notably on the western Iberian Peninsula, in Cornwall, in Brittany, in the Massif Central and the Ore Mountains, Vogtland and the Fichtelgebirge (Muhly 1985, fig. 1; Cierny *et al.* 2005, fig. 1; Hauptmann 2007, 567–568, fig. 109). By contrast, no significant deposits are known in the eastern Mediterranean and in the Near East (but *cf.* below). However, the oldest regular tin bronzes have been found in this region, argueably far away from substantial tin deposits. Considerable distances had to be overcome to obtain the raw materials, most probably necessitating exchange and trade networks. The enormous distances from potential sources, for example in the case of the Aegean, was the reason why some scholars (Renfrew 1967, 13) have considered it problematic that such distant places should have provided the tin used in the Early Bronze Age Aegean. The 'tin problem' has long been a feature of Bronze Age research. V. G. Childe (1928, 157) remarked nearly 90 years ago: 'The Sumerians drew supplies of copper from Oman, from the Iranian Plateau, and even from Anatolia, but the source of their tin remains unknown' (cited by C. C. Lamberg-Karlovsky in Weeks 2003, VIII). In recent decades, the 'tin problem' has been frequently discussed in the literature (*e.g.* Muhly 1985; Stech and Pigott 1986; Weeks 2003; Kaniuth 2007; Helwing 2009; Pigott 2011; Lehner 2014; Steinkeller 2016, 133–137). There are different perceptions about when it is legitimate to speak of a deliberate alloy of copper and tin to make bronze. For some, 1% tin is sufficient to blur the edges between unalloyed copper and tin bronze. To further reduce the possibility of accidentally produced tin bronze, others contend that tin bronzes are those that contain at least 2% tin (*cf.* Pernicka *et al.* 1990, 272). In some cases, we unfortunately lack precise information on the composition of the object. These sites have nevertheless been included in the distribution maps here. Older analyses are especially open to criticism for their accuracy. In heavily corroded samples, the tin content may now be higher than it was originally, since corrosion reduces copper more than it does tin (Pernicka and Schleiter 1997, 220).

Tin bronzes before the 3rd millennium BC

There are few tin bronzes that we know of from the 4th, let alone the 5th millennium BC, and in many cases their dating is questionable. A number of pins from Susa in south-west Iran allegedly date back to the 4th millennium BC (Susa A–B), but either they could not be relocated in the storerooms or their dating is simply open to debate (Müller-Karpe 1989, 184). The earliest graves in the cemetery of Kalleh Nissar in the Pusht-i Kuh-region in Zagros, western Iran, which also contained tin bronzes, seem to date back to around 3000 BC, the Jemdet Nasr phase (Fleming *et al.* 2005). A few sites in Armenia with tin bronzes (Thornton 2007, 129 with further references) may again date back to the late 4th millennium, but once more reliable data is missing. In the Balkans, there are allegedly tin bronzes from sites of the 5th and 4th millennia BC (Ottaway 1979; Glumac and Todd 1991), but the reliability of their dates/ stratigraphic positions has been questioned (Chapman 1991; Pernicka *et al.* 1993, 12). This discussion has recently been revived by the publication of tin bronze foil from the Vinča culture site of Pločnik in Serbia which is said to date back to *c.* 4650 BC (Radivojević *et al.* 2013). The authors discuss 15 other potentially very early tin bronzes from the 5th millennium BC in the Balkans, which they consider to have been produced from the smelting of the copper-tin ore *stannite*. This claim for much earlier use of tin bronze in the Balkans has proved controversial and the contextual date of the find has been rejected (Šljivar and Borić 2014). The debate remains open, but even if one accepts the results published by Radivojević and others (Radivojević *et al.* 2014), it is obvious that this potentially very early horizon of tin bronze did not last long. Similar claims for very early tin bronzes have also been made also for other

parts of Europe and western Asia. A high percentage of tin was found in the copper on the walls of a crucible from the settlement of Mandalo II (Papaefthymiou-Papanthimou and Pilali-Papasteriou 1997, 146; Zachos 2007, 171) in western Macedonia. This phase can be dated to the later stages of the 5th millennium BC (Maniatis and Kromer 1990, fig. 1). For typological reasons, a hammer axe of the Şiria type from Gopło in Poland must date back to the early 4th millennium BC, but in its material composition it remains an exceptional piece (Matuschik 1997, 87, 103, fig. 6; Krause 2003, 212, fig. 192; Rahmstorf 2011a, 105, fig. 9.1. 1). At Tel Tsaf in Israel, a copper awl with 6% tin was found in layers dating to the very late 6th and early 5th millennium BC (Garfinkel *et al.* 2014). At Aruchlo in Georgia, a copper bead with tin, iron and arsenic has even been dated to the middle of the 6th millennium BC (Hansen *et al.* 2012, 101, fig. 36), but the excavators do not rule out the eventuality that the small bead may have been moved by bioturbation. Once more, exceptionally tin-rich copper ore (*stannite*) was probably used. Hence these alloys were produced accidentally. The dating of tin bronzes (embossed sheets) allegedly from the late 6th millennium at Ghar-i-Mar in the Balkh province of Afghanistan is also questionable (Srivastava 1998, 176–177 with further references). It is important to note that, despite these early occurrences, there is no single site in Europe and Asia in the 5th or 4th millennium BC where regular and deliberate alloying of copper and tin can be documented. All these finds are exceptional and episodic. Fully deliberate production of tin bronzes and an understanding of tin as a specific metal cannot be proved to have existed in this period, which is therefore still correctly referred to as the Copper Age.

Tin bronzes in the 3rd millennium BC

In this section, I discuss the occurrence and frequency of tin bronzes in various regions between the Atlantic and north-western India in the 3rd millennium BC. A synopsis of the earliest tin bronzes in Europe has been published recently by C. Pare (2000b; for an earlier study, see Spindler 1971). The total distribution of tin-bronze objects between the European Atlantic coast and central Asia encompasses more than 140 sites (Fig. 16.1), most of them between the Aegean and western Iran (Fig. 16.2).

Tin bronzes in central Europe

As part of his study on early metallurgy and based mainly on the data from the Stuttgart project analysing the composition of early metal objects (SAM; Junghans *et al.* 1968; Junghans *et al.* 1974), R. Krause has recently dealt with early tin-bronze objects from Late and Final Neolithic contexts before the beginning of the Early Bronze Age, which in central Europe dates to around 2200 BC. Krause assembled eleven sites in central Europe that in his opinion are Final Neolithic

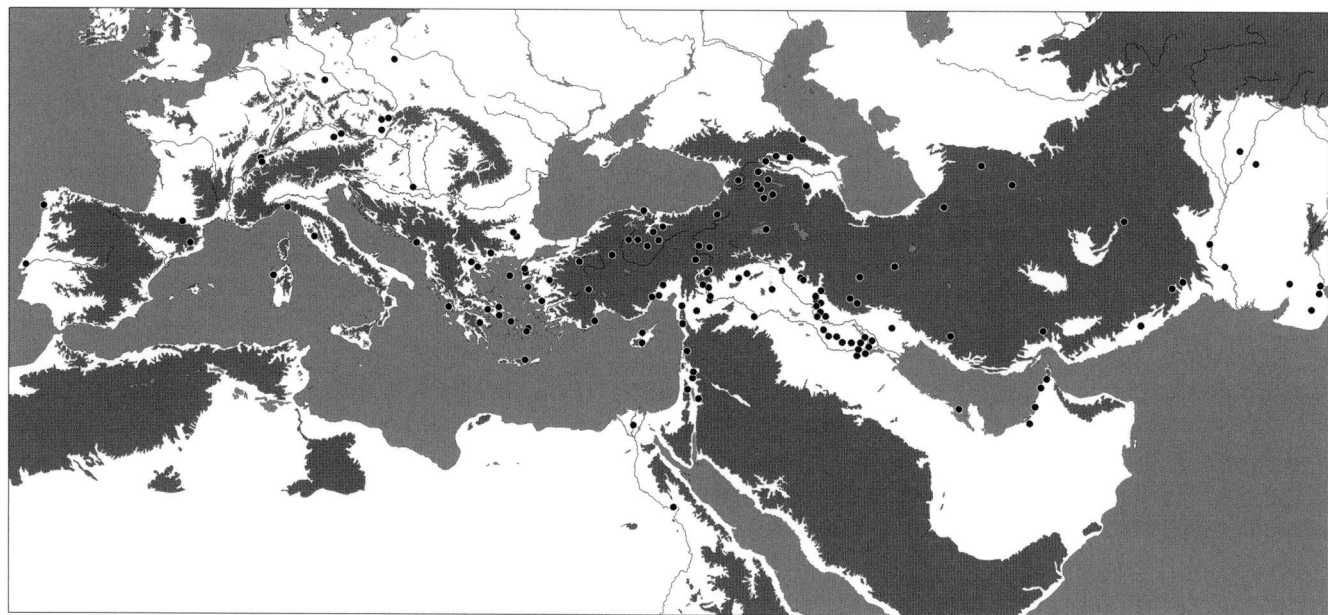

Figure 16.1: Distribution of tin-bronzes before c. *2200/2100 BC between the Atlantic and western India.*

(Krause 2003, 210–213; Schickler 1981 provides an older compilation). On closer inspection, however, some of these archaeological contexts can be dated to the transition to the Early Bronze Age. For example, an assumed Bell Beaker grave at Pfützthal near Salzmünde in Saxony-Anhalt contained a bulbous undecorated vessel and a needle-shaped bronze device with 10.5% tin content (Schlette 1948, 36; Krause 2003, 211, fig. 190). F. Schlette attributes the vessel to his Group E, asserting that this group was present in both the late phase of the Bell Beaker period and the early stages of Unetice culture (Schlette 1948, 56, 63, fig. 10.1). B. Zich also considers its cultural affiliation questionable (Zich 1996, 470 cat. no. E729). Similarly, the spiral ring from a grave at Niederkaina in Saxony with 2.2% tin and a metal strip as a single find from the same site (with 4.6% tin) remain doubtful in their precise dating. They have been dated to the Final Neolithic period due to a '*gehenkelter Schnurbecher*' (Schickler 1981, 436), but this feature also figures in the early part of the Early Bronze Age. For the mid- and late 3rd millennium there is a small cluster of sites with tin bronzes from the Bell Beaker period in the Czech Republic and Bavaria mirroring the slight concentration of central European locations with early gold and silver artefacts (Meller 2015, fig. 3). In this area, the advanced metallurgy of the 'Bell Beaker people' was not limited to precious metals, gold and silver, but included bronze, albeit in exceptional cases. Nevertheless, R. Krause is still right when he says that '*wir erst mit dem 23./22. Jahrhundert verlässlich mit Zinnlegierungen nördlich der Alpen rechnen können*' (Krause 2003, 213) [we cannot reliably expect tin alloys north of the Alps before the 23rd/22nd century BC] and not from the mid-3rd millennium BC. This is especially

true because the absolute number of metal analyses of samples from the early Metal Ages in central Europe runs into thousands, while in other regions samples have been taken unsystematically and only samples from specific sites have been examined. Similarly intensive sampling in other regions would probably result in a much higher density. As M. Primas assumed, metallic tin from the Early Bronze Age seems to have been negotiated in the form of beads or rings, as is indicated by beads from the Netherlands (Odoorn-Exloërmond), Bavaria (Buxheim), from Lake Zurich (Wädenswil-Vorder Au) and now also Tollense Valley in north-eastern Germany and Whitehorse Hill, Dartmoor, Devon in south-west England (Primas 2002, 311–312; Rahmstorf 2010, 687, fig. 7.1–3; Krüger *et al.* 2012; Jones *et al.* 2014). In central Europe, tin bronze is first used fairly frequently in the Bronze Age A1b (approximately from 2000 BC), but as C. Pare has demonstrated, only becomes widespread throughout central Europe a few centuries later with Bronze Age A2 (Pare 2000b, 16–20, figs. 1, 9–12, 14). The Únětice culture in central and eastern Europe seems to have been an innovation centre where halberds for example were produced in tin bronzes from approximately 2000 BC (Rassmann 2010). Addendum 2016: XRF analyses of copper-based metal objects from Bell Beaker Csepel Group sites in Hungary published in 2003 (Enrődi *et al.* 2003, tab. 1, nos. 4–5) revealed two bronze objects from the settlement at Albertfalva (not included in Figures 16.1 and 16.2).

Tin bronzes in western Europe

From western Europe, there are, to my knowledge, no tin bronzes before the start of the Early Bronze Age in the last

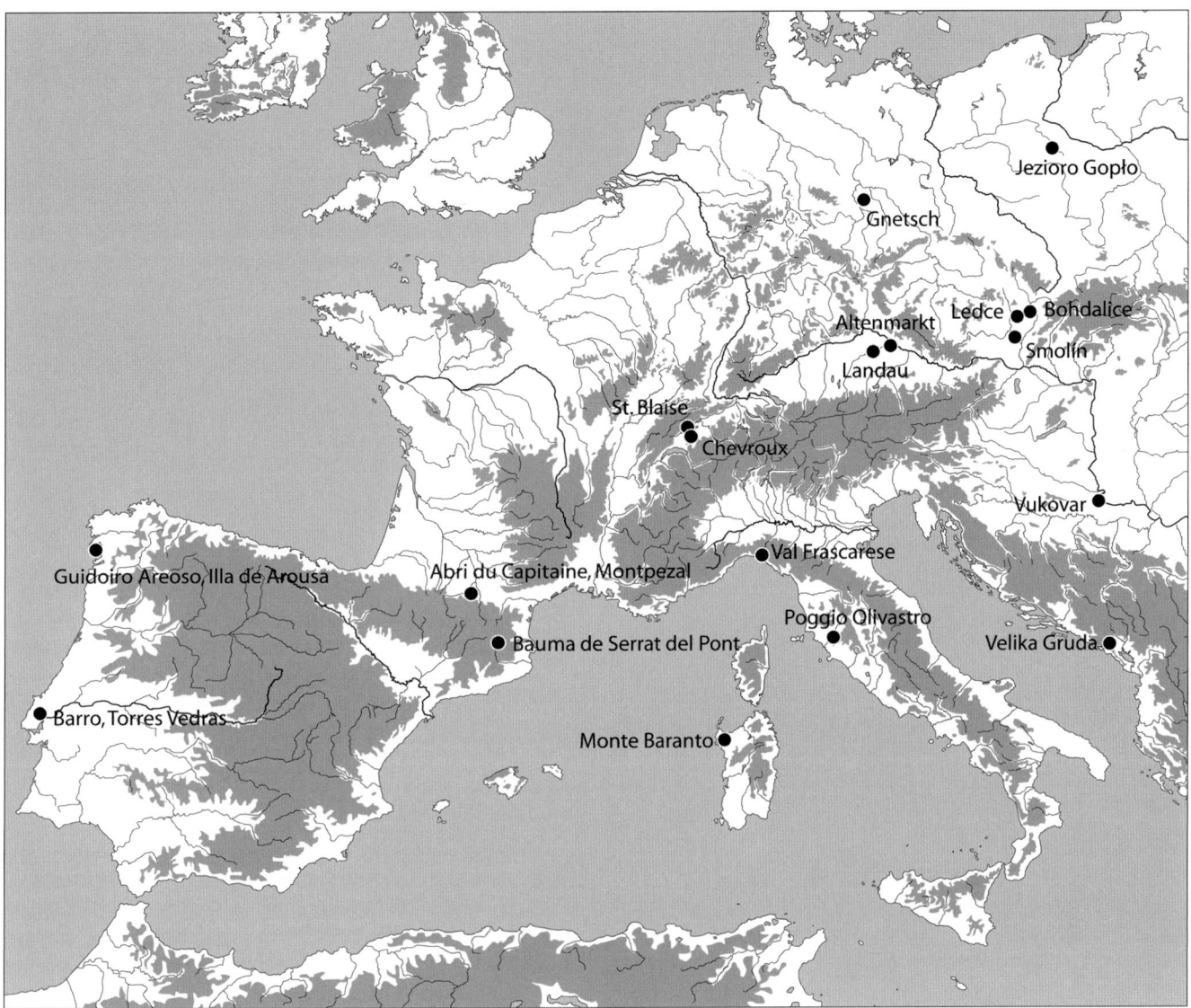

Figure 16.2: Distribution of tin bronzes in west, south and central Europe before c. *2200 BC.*

quarter of the 3rd millennium. From about 2200 BC, tin bronzes with a high tin content of 8–14% appear seemingly abruptly in the British Isles, while tin bronzes with only 1–5% tin are rare (Needham *et al.* 1989, 391–392, fig. 2; Bray 2012, 57). This coincides with S. Needham's 'fission horizon' where a new diversification in grave goods is observable (Needham 2005, 205, 209). Tin bronzes are then found from around 2000 BC in northern and western France, southern England and Wales. This suggests that, firstly, bronze metallurgy was accepted very quickly and that, secondly, from this time on, local, most likely alluvial deposits in Cornwall (Wang *et al.* 2016, 88), Brittany and the Massif Central were used. This is borne out by the mapping of tin content in copper alloys in parts of Europe by 2000 BC (Rassmann 2004, fig. 3). The assumption of an independent technological development is unconvincing as no evidence

of an experimental phase has been detected (Primas 2002, 304). It would be important to establish how, why, and from where this impulse came about (from metal prospectors?), unless one assumes an independent development in Britain. From the turn of the millennium, southern England may have supplied central Europe with tin (evidence summarised by Jockenhövel 2004). In comparison to other European countries, tin bronzes in the British Isles are generally liable to be found several hundred years earlier, *i.e.* before this technology asserted itself as standard in central Europe and other parts of Europe (Pare 2000b, 20–22, 27, fig. 1.14).

Tin bronzes in southern Europe

On the Iberian Peninsula, regular use of tin bronze for pendants and rings can only be detected from the earlier stages of the 2nd millennium (1800–1700 BC) with the

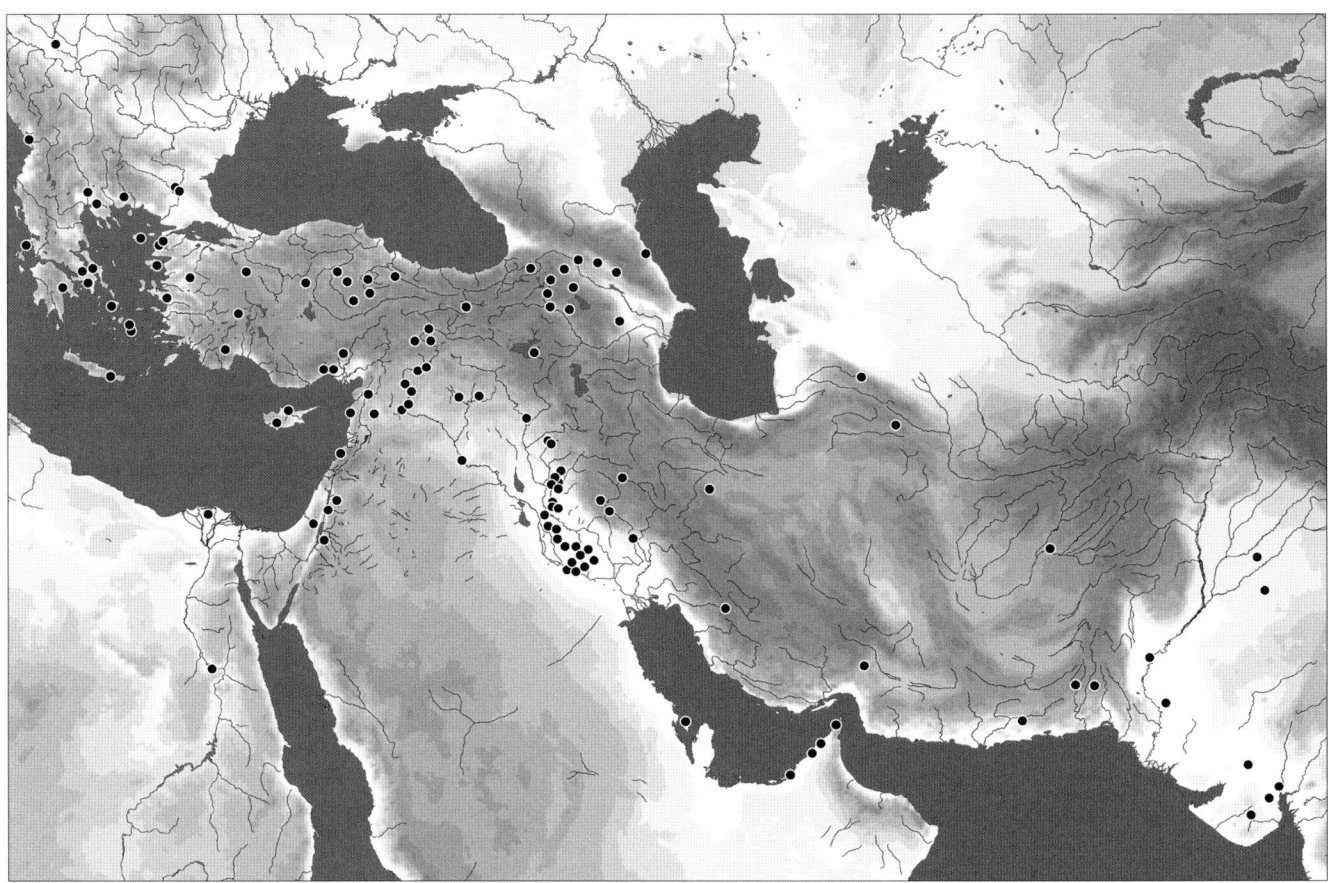

Figure 16.3: Distribution of tin bronzes between the Aegean and western India before c. *2200/2100 BC.*

El Argar culture (Primas 2002, 308–309, fig. 5; Lull *et al.* 2013, 599). Very few sites (Fig. 16.3) with tin bronzes can be dated before 2200 BC (*cf.* Fernandez-Miranda *et al.* 1995; Delibes *et al.* 1996). The hoard of Barro found next to the circular grave of Barro not far from Zambujal in central Portugal contained not only two axes and two daggers of arsenic bronze and a limestone cylinder but also a chisel of high-alloy tin bronze (Spindler 1981, 101–102, 243, fig. 41). Because of this chisel, K. Spindler has assigned the entire hoard to the beginning of the Early Bronze Age (VNSP III = culture of Vila Nova de São Pedro), although all other objects can be attributed to the late Copper Age/Bell Beaker period (*e.g.* the golden hair ornament, *cf.* Meller 2015, 617). In Spindler's opinion 'das Aufkommen der Zinnbronze auf der iberischen Halbinsel vor dem Einsetzen der Bronzezeit [ist] kaum denkbar' (Spindler 1981, 101) [the occurrence of tin bronze on the Iberian Peninsula prior to the Bronze Age is hardly conceivable]. But he cannot exclude the eventuality that the Barro hoard may have been first compiled in this composition by quarry workers. Nevertheless, there is more sparse evidence of the scattered use of tin bronzes during the later Copper Age, compiled by C. Pare (2000b, 22). Three further cases are early tin bronzes in the form of awls and daggers that can interestingly be related to the Bell Beaker

phenomenon of the late Copper Age, or more precisely the mid- and later 3rd millennium BC. These early tin bronzes are from Galicia (Guidoiro Areoso on the island of Arousa) and the foothills north and south of the Pyrenees (Abri du Capitaine; Bauma de Serrat del Pont) (Fig. 16.3).

On the Italian peninsula and in Sardinia, some initial, isolated instances of tin bronzes occurred at the same time. In Sardinia, a blade with a blunt tip (spatula) was found in the fortified settlement of Monte Baranta within the fortified horseshoe-shaped tower on the hilltop. It can be dated to the Monte Claro-Facies (mid-3rd millennium) and is contemporaneous with the Bell Beaker phenomenon, to which it displays similarities. No parallels for the object with a 4.2% tin content can be found in Sardinia or neighbouring regions (Lo Schiavo *et al.* 2005, 194, fig. 2; Atzeni *et al.* 2005, 124, tab. 5). Singular finds of early tin bronzes from the 3rd millennium BC came to light at Poggio Olivastro in Tuscany and in a grotto in the Val Frascarese in Genoa (Campana *et al.* 1996; Bulgarelli and Giumlia-Mair 2008; Dolfini 2010, 719). Progress to regular production of tin bronze did not take place before the transition from the 3rd to the 2nd millennium BC or the early 2nd millennium. Nevertheless, we still have no large-scale investigation of copper-based artefacts from the Early Bronze Age in Italy with a regionally differentiated measurement series (*cf.* Pare 2000b, 23–24; Dolfini 2010, 719, tab. 4).

Tin bronzes in south-east Europe

Very few sites in southeastern Europe (Fig. 16.4) have revealed tin bronzes from contexts of the 3rd millennium or before 2200 BC (*cf.* Appendix). The unusual double-edged knife from the tumulus tomb of Velika Gruda in Montenegro, which has been dated as early as the early 3rd millennium BC has a tin content of about 7.6% (Primas 1996, 97–98, 100, 104, fig. 7.5–6, tab. 7.1). The bronze has lead-isotope ratios that are very similar to those of early Aegean tin bronzes (Primas 2002, 304–305, fig. 2), suggesting that the bronzes arrived ready alloyed from wherever this may have been and that the material may have come from the same source. So far, lead isotopes relate to the copper in the objects, not to the tin (but see below). We cannot entirely rule out the use of small local tin deposits in the Balkans (Durman 1997; *cf.* Maran 2007, 12). Analyses of metal objects from the Maros culture in southern Hungary and the

Figure 16.4: Distribution of tin bronzes in southeast Europe, western Anatolia and Cyprus before c. *2200/2100 BC.*

Banat demonstrate that tin bronze was mainly used to make jewellery in the very late 3rd millennium BC (Primas 2002, 307–308, fig. 4). According to the findings from C. Pare's investigation, the transition to a regular use of tin bronze took place in the Balkans no earlier than the middle of the 2nd millennium BC (Pare 2000b, 12–16, figs 1.5–7.14).

Tin bronze in the Aegean

A dozen Early Bronze Age sites with tin bronzes can be listed for Greece (Fig. 16.4). From Sitagroi IV–Va in Macedonia come four or five objects with up to 8.1% Sn (Renfrew and Slater 2003, 300–307). According to J. Maran, these phases at Sitagroi can be parallelized with Early Helladic (EH) I and the beginning of the EH II, which implies a date in the first or an early 2nd quarter of the 3rd millennium BC (Maran 1998, 124–126; see also Maran 2007, 12 with n. 61). Finds of tin bronzes and even bronze slag from Axiochori/Vardarophtsa and Perivolaki/Saratse in Macedonia date to the end of the EBA, with the exception of a bronze fragment from Axiochori settlement 5 which is contemporaneous with the late EH II period on the southern Greek mainland (see Maran 1998, 268 with further references). There are also a few sites with tin bronzes on the southern Greek mainland and islands. Due to documentation errors we can no longer ascertain which objects from the R-graves of Steno-Nidri on Lefkas are tin bronzes (McGeehan-Liritzis and Taylor 1987, tab. 1; see Maran 1998, 267, n. 1109; Kilian-Dirlmeier 2005, 107, n. 244). A tweezer from Grave 23 (or found next to Grave 23) at Aghios Kosmas dates back to the transition from EH I to EH II, *i.e.* around 2800/2700 BC. Tin bronzes from Lithares (Tzavella-Evjen 1984, 140, cat. nos. 657, 663; Kayafa *et al.* 2000, 41–42, tab. 2.2) also indicate that the alloy was occasionally used during the earlier 3rd millennium BC not only in Macedonia, but also on the southern Greek mainland and Crete (see below). A bronze pin from the cemetery of Manika on Euboea can probably be dated to the later EH II (Sampson 1985, 177, 305–306; Maran 1998, 268). A loop pin was unearthed in Room 196 of horizon 8a (late EH II) in the lower citadel of Tiryns, which also contained 14 spool-shaped potential balance weights. K. Kilian (1982, 421, fig. 44.5) explicitly states that the pin is made of tin bronze even though there is no published analysis of the metal. Nevertheless, it must be emphasized that in the EH heartland tin bronzes are rare exceptions. Not even one tin bronze object was found in the sample of 25 copper-based metal objects from Lerna III (EH II) and IV (EH III) and seven objects from EBA Tsoungiza taken by M. Kafaya, S. Stos-Gale and N. Gale. Arsenic bronzes are common, but in the authors' opinion the variations in the ratios of arsenic to copper show that the alloy was probably not produced by a controlled process (Kafaya *et al.* 2000, 39–43, tabs 2.1, 2.3). It is also important to note that there are no changes in the popularity of arsenic in the copper-based artefacts from EH II to EH III in Lerna and Tsoungiza.

Because of the adoption of tin-bronze technology in the Cyclades along with new 'Anatolian' vessel forms, the Lefkandi 1-Kastri type has been understood by some authors as an indication of the advent and presence in the central Aegean islands of a foreign population from western Anatolia (see Maran 1998, 269, n. 1132 for further references). Until recently, this assessment was based only on the settlement of Kastri on Syros. All derive from the hoard found in room 11, which among other things also includes six spool-shaped potential balance weights and a slotted spearhead of tin bronze. The copper based metal objects from the hoard have tin ratios of 1.85/4.8 to 7.9% (Bossert 1967, 76, figs 1.1–12, 2.6). New data has now come from the contemporaneous settlement of Markiani on Amorgos. Twelve of the 17 copper-based metal objects have been analysed for their composition. Five of them are dated in Markiani III, 4 in Markiani IV and 3 remain undated. Only two of them are tin bronzes: a one-edged blade or knife (Birtacha 2006, 214–217, figs 8, 24.16, tab. 8.18: EE 639) derived from a layer (Marangou *et al.* 2006, 54) of late Markiani III, parallel with the earlier EBA II, *i.e.* 2800/2700–2500 BC (Marangou *et al.* 2006, 77–80, fig. 5.6), even though the ceramics have not yet been published. The second object is a pin with a hemispherical head (Birtacha 2006, 213, fig. 8.24.6, tab. 8.18: EE 241) from a layer in section 2 corresponding to Markiani IV, *i.e.* the younger EBA II (Lefkandi 1-Kastri). It has recently been argued that the shape of this pin shape is Anatolian in origin, and it is worth mentioning that the earliest bronzes in the Aegean (and Anatolia) are often pins of this kind with spherical heads (Rahmstorf 2015, 156–159, figs. 6–8). The other objects are made of pure copper (one object in Markiani III) or arsenic bronzes (three objects in Markiani III and IV, see Birtacha 2006, 216, 219, tab. 8.16, 8.18). It is interesting to note that in both cases the tin bronzes were found in direct relation or close proximity to the spool-shaped potential balance weights, but not in the same layers. The third site with EBA tin bronzes in the Cyclades is also situated on Amorgos. From a grave in the cemetery of Dokathismata we have two bronze bracelets that were bought with other inventory by the Archaeological Society of Athens before the excavations by C. Tsountas started in the late 19th century. One of the bracelets has a tin content of 13.52%. A closed find context is unfortunately missing, but the other finds from graves suggest a date in late EC (Early Cycladic) I or early EC II, although a much later, possibly even Middle Bronze Age dating cannot be entirely ruled out (Tsountas 1898, 187, pl. 8.2; Rambach 2000, 9, 13, pl. 3, 9–10). Finally, hardly any Early Bronze Age tin bronzes can be cited from Crete. Platanos in the Mesara plain in Crete, which was mentioned by Schillinger and Apakidze (Schillinger 1997; Apakadize 1999, 515, fig. 2), is not accepted here as an early site of tin bronze because numerous finds from the end of the 3rd millennium

(EM III) and Middle Bronze Age (Middle Minoan) have also been recovered from the site. However, two small daggers (Marinatos 1929, 131, 119, fig. 13, 30–31) from the Tholos of Krasi in north-east Crete with a tin content of 6% and 10%, can be dated to EM I–IIA or EM I. J. Muhly advocates a date in EM I (Muhly 2006, 170). In the north-eastern Aegean at Poliochni on Lemnos, a steady increase of tin bronzes is discernible from the Green settlement phase, datable to before the middle of the 3rd millennium BC. Tin bronzes are particularly common in the Yellow phase of the later 3rd millennium. But the ratio of arsenic bronzes also increases steadily from the Blue phase (early 3rd millennium BC) to the Yellow (Pernicka et al. 1990, 272, fig. 1; Pernicka et al. 2003, 163). Five of the six metal artefacts (with the exception of the pin) from the 'Potter's Pool' hoard found in Thermi IV (= late Troy I) consist of tin bronze (Begemann et al. 1992, tabs 1–2). In sum, I believe that we cannot reconstruct a time horizon from the point when tin bronzes suddenly occur in the Aegean. While there seems to be an accumulation of tin bronzes from the later phase of EBA II on the Greek mainland, on Euboea and the Cyclades, quite a few tin bronzes also date to early EBA II or even to EBA I (Sitagroi, Aghios Kosmas, Krasi). If metallic tin or alloyed tin bronze did indeed start coming into the Aegean from the east as of the early 3rd millennium BC, this eastern impact would have been paralleled by the adoption and local interpretation of the foreign innovations of sealing and weighing during this period (Rahmstorf 2016).

Of major significance are the results of recent investigations on copper-based artefacts found at several EBA Aegean sites: Katsambas, Dhaskalio Kavos and Çukuriçi Höyük. As mentioned earlier, Kayafa and others drew upon their analyses of samples from Lerna and Tsoungiza to argue that the intentional production of another copper-based alloy – arsenical copper, sometimes also called arsenic bronze – was not possible for Aegean metallurgists during the EBA. But crucible slags from Poros Katsambas on Crete indicate the 'purposeful admixing of arsenic-rich minerals to the copper melt' (Doonan et al. 2007, 112). This evidence is further underpinned by the recent find of a speiss fragment, the 'often accidental by-products of smelting arsenic or antimony-rich multimetallic ores' (Georgakopoulou 2013, 684–685, fig. 32.34) from Dhaskalio Kavos. This may have been used as an alloying agent to produce arsenical copper. In addition, a new find from Çukuriçi Höyük on the west Anatolian coast points in the same direction (see below). As arsenical copper has many of the advantages of real bronze, the Early Bronze Age metallurgists were apparently able to produce a substitute that was nearly as effective as tin bronze. This considerably undermines the importance that has been ascribed to this material and has (in modern research) given the whole period its name. It is a reasonable assumption

that it was especially the near-gold colour that made real bronze such a coveted material (see also below).

Tin bronzes in Anatolia

The earliest tin bronzes from western Anatolia (Fig. 16.4) belong to the Troy I period and come both from Troy itself and from nearby Beşiktepe. These bronzes contain up to 11.5% tin (Pare 2000b, 9; Begemann et al. 2003). From Tell Judaidah in the Amuq area on the Turkish-Syrian border, already part of the Mesopotamian catchment area, come numerous bronzes with tin contents of 5–37% that can be dated to Amuq phase G (c. 3000–2900 BC). Famous are the naked figurines (three female and three male: Braidwood and Braidwood 1960, 300–313, figs. 240–245, pls. 65–64) from this phase, though their stratigraphic (and stylistic) placement around 3000 BC has been doubted (Marchetti 2000). Nevertheless, a crucible from phase G shows that tin or bronze was processed (Yener 2009, 144–145 with further references). Apart from tin, arsenic was apparently used intentionally to produce arsenical copper. The compositional analysis of copper-based seals (Goldman 1956, fig. 392.13–14) of the Early Bronze (EB) 2 period at Gözlü Kule-Tarsus in Cilicia disclosed a 10% share of arsenic, which again can only be achieved through the deliberate smelting of copper and arsenic-rich ore material (Özbal et al. 2005). In addition, the recent findings of copper sulphides and iron arsenides in smelting debris from Çukuriçi Höyük on the delta of the Küçük Menderes River on the west Anatolian coast near ancient Ephesos indicate that in the Aegean and in Anatolia, intentionally produced arsenic-rich speiss was melted together with copper or copper ore to obtain arsenical copper (Mehofer 2014, 467–468) during the EBA. Although some 20 EBA locations with tin bronzes are known from Anatolia (see appendix), this is no proof of regular production at that stage. Nevertheless, they appeared in a developed phase of the EB 1 (early to mid-3rd millennium BC) and their number increased during EB 2 and EB 3. J. Yakar noted that tin bronzes occurred simultaneously with the earliest gold objects in Anatolia (Yakar 1984, 72). Now that the dating of the allegedly Early Bronze Age ring idols of gold and silver from Ikiztepe, Kalıkaya, Göller and other sites has been corrected to the 4th millennium (Lichter 2006, 528; Zimmermann 2007; Schoop 2011, 55, 59, fig. 3 – the latter for a Late Chalcolithic ring idol mould from Çamlıbel Tarlası), this assumed nexus is no longer tenable.

Tin bronzes in eastern Europe

To my knowledge, tin bronzes have not been recorded for the 3rd millennium BC in eastern Europe. In the North Pontic area metal objects were made of pure or arsenical copper until well into the 2nd millennium BC (Chernykh 1992; Pare 2000b, 13–16, fig. 1.6).

Tin bronzes in the Caucasus

The earliest tin bronzes from the Caucasus date from the turn of the 4th to the 3rd millennium and to the first half of the 3rd millennium BC, as in Telebi of the late Kura-Araxes culture (Kavtaradze 1999, 84) in eastern Georgia. A grave at mound 11 at Talin in Armenia contained a bronze curl ring with more than 10% tin (Meliksetian *et al.* 2003, 310 314; Avetisyan *et al.* 2010, 164, fig. 3.1). A single ^{14}C date from the grave of Talin has a calibrated margin of error (2σ) between 3330 and 2936 BC (Avetisyan *et al.* 2010, 163). Tin bronzes ('a curl ring and a standard') were also found in kurgans 3 and 5 in Martkopi (Kavtaradze 1999, 84) dating to the final stage (III) of the Kura-Araxes culture or the Bedeni culture, as it is called by the kurgans in the Bedeni Plateau. In absolute terms, this culture or phase dates to *c.* 2700–2200 BC (Batiuk and Rothman 2007, 9, fig. [without no.]; but *cf.* Bertram 2010, 255: '*die Chronologie [im Südkaukasus] hängt für weite Abschnitte des 3. Jts. v. Chr. bedauerlicherweise in der Luft*'). Kavtaradze also lists an alleged bronze pin and an awl from Delisi dating from the Middle Neolithic period, but there are 'certain doubts concerning the circumstances of their discovery' (Kavtaradze 1999, 71). Tin bronzes from the later EBA seem to be more common in the southern Caucasus. For example, an awl from Telebi contains 11.3% tin, two spearheads, a dagger and a thorn from Bakurcixe have tin ratios of 10 to 13.7% (Kavtaradze 1999, 84–86 with references). Other sites with tin bronzes in the southern Caucasus have been listed by C. P. Thornton (2007, 129) and A. Bobokhyan (2008, 67, map 15). On the other hand, of the 80 copper-based artefacts of Early Bronze Age sites in Armenia, only three are tin bronze while the rest are either arsenical copper or arsenical copper with a high lead content (Meliksetian *et al.* 2011, fig. 5 – no allocation of artefacts to sites). In the western and northern Caucasus, finds of tin bronzes from the second half of the 3rd millennium BC are unknown. The most northerly localition is Velikent in Daghestan (Fig. 16.5). The 195 sample analyses from the total of 1,500 metal artifacts from Hill 3, Grave 1 revealed 15 tin bronzes (8% of the total sample analysed) that are mainly rings and pendants with no tools or weapons (Kohl 2002, 174, fig. 12; Kohl 2003, 19, fig. 1, 7; Peterson 2003, tab. 2.1–2; *cf.* Rahmstorf 2010, 688, figs 7, 9a–b). In addition to these artefacts, there is also a slightly curved, approx. 7 cm-long object with a square-rounded cross-section designated by the excavators as bronze ingots, albeit with a question mark (Gadzhiev 1997, 191, fig. 8.3). P. Kohl therefore assumes that tin bronze arrived in Velikent in the shape of semi-finished products or ingots that were then further processed by local artisans (Kohl 2007, 108). Of particular interest are the lead-isotope analyses of tin bronzes from Velikent (L. Weeks in Kohl 2002, 178–183). For copper, they indicate similar lead-isotope ratios as in the sampled tin-bronze alloys from the EBA Aegean (Poliochni, Troy,

Thermi, Kastri) and from the United Arab Emirates (Al Sufouh, Shimal, Tell Abraq). This may suggest that all these places received their bronze (traded as an alloy) from the same deposit(s) (*cf.* Weeks 2003, 163). Addendum 2016: A few analyses of tin bronze artefacts from Early Bronze Age sites in Azerbaidjan have been published (Hasanova 2014, 66, tab. 1). They are not mapped in Figure 16.5.

Tin bronzes in Cyprus and the southern Levant

No tin bronzes are known from Late Chalcolithic Cyprus in the first half of the 3rd millennium. This was assumed (Coleman *et al.* 1996, 373) to also hold true for the second half of the millennium (Philia culture; Early Cypriote I–III), although the impact of Anatolian influences and of potential migration from the Anatolian coastal area to Cyprus has been established for this period in recent years and substantial cultural change has been emphasised (Webb and Frankel 2007; Peltenburg 2007; Kouka 2011). Recently, four rings with a tin content of 4.8–13.1% have been published from Grave 6 (Swiny *et al.* 2003, 111–114) at Sotira Kaminoudhia, which is assigned to the Philia culture (Swiny 2003, 376–377, 379, fig. 8.1 M13–M14, M21–M22; Giardino *et al.* 2003, 388–390, fig. 8.1.5, M13–M14, M21–M22). According to J. Webb and others, the tin bronzes (flat celt, a spearhead and a very early sword) from the richly endowed Kolokassides hoard, which may come from Vasilia in Northern Cyprus, are plausibly datable to the Philia culture (Webb *et al.* 2006, 268, tab. 2, fig. 1, 2. 4. 8).

In the southern Levant, tin bronzes from the Early Bronze Age are also quite a rare phenomenon (Fig. 16.4). This is demonstrated by the results of the analysis of Early Bronze Age metal finds. Nine copper-based metal objects from the EBA layers at Tell Abu al-Kharaz in the Jordan Valley have been analysed (Fischer 2008, 303–306, tab. 65). None of them contained amounts of tin that could be interpreted as resulting from a deliberate alloying process. So far, the only tin bronzes we know of from the southern Levant are individual objects, often daggers. In the chamber tombs cemetery of Tiwal Esh-Sharqi, a bronze dagger with four rivets was found with a tin content of 12%. The burial is dated to EB IV in the late 3rd millennium BC (Tubb 1990, 55, 58, 95–96, figs 39b, 40b SE1.6) Also from this period is a bronze dagger from Jericho (Prag 1974, 91, n. 63). Another bronze dagger from Bab edh-Dhra is slightly earlier, dating from EB III (Maddin *et al.* 2003, 513). The tin bronzes from Khirbet ez-Zeraqon (EB II–III) have not yet been published in detail (*cf.* Hauptmann 2000, 180).

Tin bronzes in Egypt

As is unfortunately often the case with ancient Egyptian artefacts, the exact origin and hence the precise dating has not been securely established for many metal objects from the First Dynasties and the Old Kingdom. R. M. Cowell has analysed 52 metal artefacts dated either by

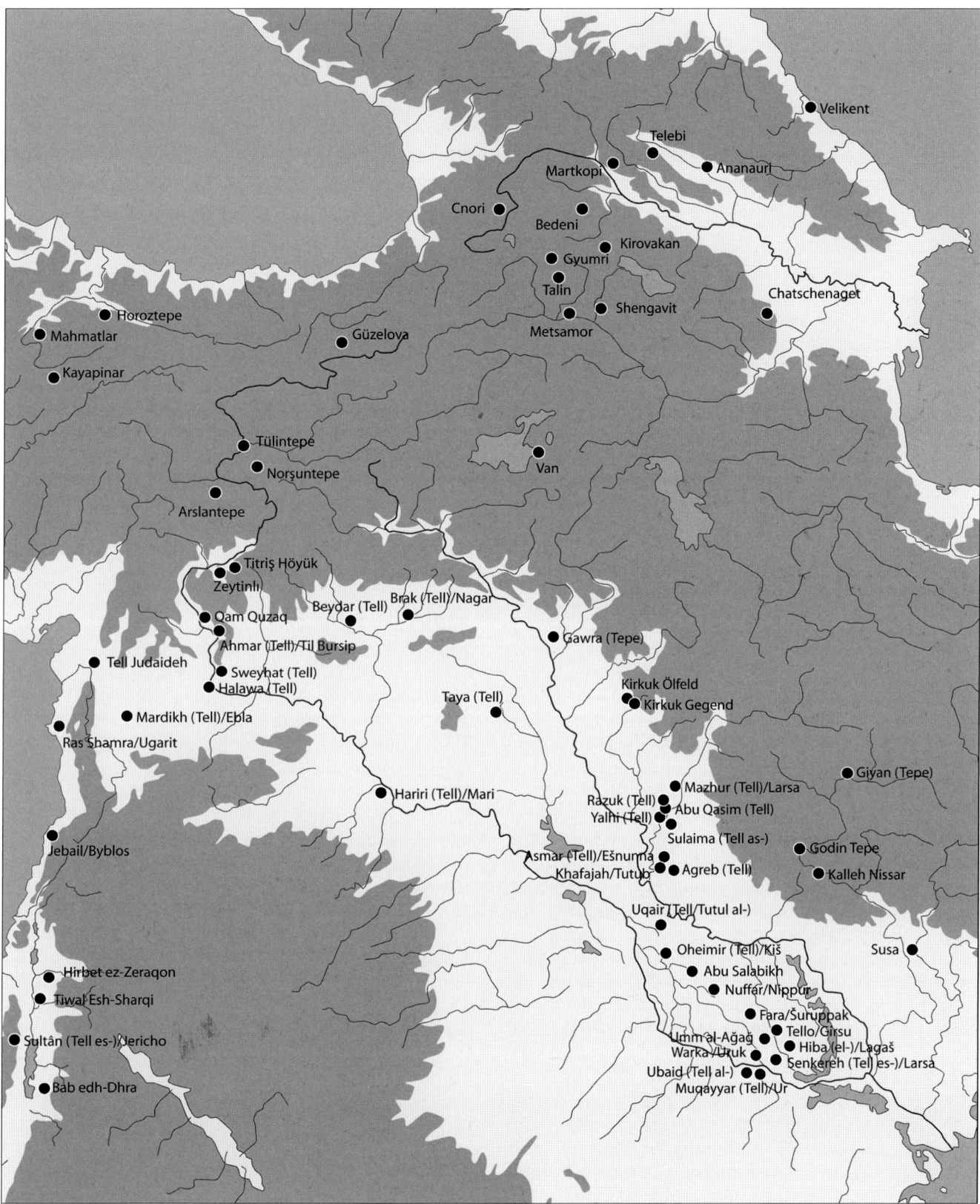

Figure 16.5: Distribution of tin bronzes in eastern Anatolia, the Levantine coast, the Caucasus, Mesopotamia and western Iran before 2200/2100 BC.

inscriptions, the archaeological context or ^{14}C datings. Only two of these are tin bronzes; both are bronze vessels from the grave of Khasekhemwy (2nd dyn.) at Umm el-Qa'ab at Abydos (Cowell 1987, BM 25571, BM 35572). These results are confirmed by the findings of modern settlement excavations in Egypt. Only one of the ten metal analyses of copper-based artefacts from prehistoric and early historic Buto in the Delta turned out to be tin bronze. This object, the function of which is unclear, was recovered from an early dynastic pit (Pernicka and Schleiter 1997, 220, tab. 2). Tin bronzes from 3rd millennium Egypt are therefore very rare (*cf.* Wertime 1978; Eaton and McKerrell 1976, fig. 9 [left], tab. 2; Muhly 1985, 283), which also holds true for the Middle Kingdom in the earlier part of the 2nd millennium BC, when metal objects made of this material were getting slightly more frequent. Only from the Second Intermediate Period and especially from the New Kingdom on do tin bronzes become common in Egypt (Garenne-Marot 1984, 113). The opening up of Egypt to (Levantine) Hyksos domination and the imperial power of New Kingdom Egypt in the eastern Mediterranean apparently led to new and fundamental links between Egypt and the international metal exchanges in the eastern Mediterranean and the Near East. Much greater Egyptian autarky with regard to innovations and resources can be observed for the Old and Middle Kingdoms. This fits in perfectly with the results of weight studies in Ancient Egypt. Only with the advent of the New Kingdom did the weight unit of *c.* 9.1–9.4g became common in Egypt (*cf.* Weigall 1908, IX). It had already been in use for more than 1000 years in many parts of the eastern Mediterranean, especially in Syria (nevertheless, this weight unit is still often referred to as the Egyptian *qedet*). Similarly, the typical east Mediterranean theriomorphic metal weights only occur in Egypt from the middle of the 2nd millennium in Egypt (already observed by Petrie 1926, 7). While other innovations that only arrived probably from northern Levant/Mesopotamia (*cf.* Genz 2013, 101–102) in the country during the Hyksos period are connected with warfare (chariots/horses, composite bows, *etc.*), tin bronze and especially weights are related to material and commercial improvements. In the Old and Middle Kingdom, the country was apparently supplied almost exclusively by metal deposits from the Egyptian eastern desert and Nubia (Sudan). Theoretically, tin could also have been mined, and there are deposits of alluvial *cassiterite* in the eastern desert as well (Wertime 1978, 42; Garenne-Marot 1984, figs 2–3; Cierny *et al.* 2005, fig. 1; Nezafati *et al.* 2011, tab. 2 with further references), but given the few tin bronzes dating from this period, this is unlikely. Accordingly, the relative absence of tin bronzes is another indication of the country's marked autonomy over and against the exchange networks in the eastern Mediterranean and western Asia in the mid- and late 3rd millennium BC. Of course, it is also conceivable that cultural and ideological reasons may have led to reservations

about tin bronze in Egypt. But Egypt's trade links differed at the time, and the extreme shortage of tin in the 3rd and early 2nd millennium BC is probably more convincingly explained by the fact that Egypt was not really a part of common trade networks during this period. Nevertheless, the potential correlations between object function, composition and colour would repay closer investigation in the Egyptian case as well (Ogden 2000, 154 with examples).

Tin bronzes in Syria and Mesopotamia

The project systematically analysing metal objects from the 3rd millennium BC headed by H. Hauptmann and E. Pernicka was the first to produce a large database of metal analyses from Bronze Age sites in Syro-Mesopotamia (for older series of analyses, see Potts 1994, 154, n. 86 with references). Nearly three dozen locations with tin bronzes, notably in southern Mesopotamia, are worthy of mention (Fig. 16.5). Few sites with tin bronzes are known from the earlier 3rd millennium in Syro-Mesopotamia and none at all from the 4th millennium. Only from Tepe Gawra VII do we have a bronze pin with a tin content of 5.6% (Speiser 1935, 102; Yakar 1984, 70). The dating problems besetting Tepe Gawra are well known (*cf.* Butterlin 2002). For this reason, Tepe Gawra should not be considered evidence for the presence of tin bronze in Mesopotamia in the 4th millennium BC. Examples of early tin bronzes are the vessels from Kiš in southern Mesopotamia dating back to the Early Dynastic (ED) I period at the beginning of the 3rd millennium BC (Hauptmann and Pernicka 2004, 116, 118, pls 30.458, 35.580; Rahmstorf 2011a, 105, fig. 9.1, 3) and from Tell Qara Quzaq (Montero Fenollós 1999, tab. 1) on the Syrian Euphrates. During this period (ED I) the word 'bronze' (*Zabar*) is mentioned for the first time in connection with a vessel in cuneiform (Reiter 1997, 292, 465). But it is only with the arrival of ED III that we observe a tremendous increase in the number of sites with tin bronzes (Tab. 16.1). 'Tin-bronze is not attested by analyses as a consistent feature of Mesopotamian and Susian metallurgy until Early Dynastic III' (Potts 1994, 154), which started around 2600 BC.

However, the ratio of tin bronzes at the individual sites is still very unevenly distributed. In Tello/Girsu there are almost no tin bronzes among the metal objects analysed, which are made of more or less pure copper or arsenical copper. By contrast, the ratio of tin bronzes in the large sample of metal objects from Ur (especially from the Royal Cemetery) is generally very high, but here again other copper-based alloys are present (Hauptmann and

Table 16.1: Chronological distribution of tin bronzes from 28 sites in Mesopotamia between the Uruk and Akkad period (n = 157) (from Müller-Karpe 2004 and Helwing and Müller 2004)

Uruk	Ǧemdet Nasr	EDI	EDII	EDIII	Akkad
60	2	3	8	65	79

Pernicka 2004, 122–141). This finding corresponds well with the written documentation, where different concepts for specific metals appear to reflect the different chemical composition of the various types of copper and copper-based alloys (Waetzoldt and Bachmann 1984). In short, there may be many reasons for the irregular frequency of tin bronzes in Early Bronze Age Mesopotamia. First, the difficulties involved in getting at the remote ore deposits and organising the tin trade meant that demand was not consistently covered and needs could frequently not be satisfied. Second, in terms of the hardness, viscosity or colour of the final product, it is fair to assume that the different properties of the materials were intentionally and deliberately exploited. Since with the exception of colour, arsenical copper and tin bronze do not differ greatly in their material properties, it would have been easy for metal-workers to switch to arsenical copper. Often, arsenical copper, tin bronzes and bronzes containing both arsenic and tin occur simultaneously at individual sites in Mesopotamia (*cf.* Fleming *et al.* 2005). Finally, it is conceivable that certain types of object were made preferably of a specific copper-based metal. An initial, not very sophisticated analysis of the published material does not point to any specific clusters. Vessels are often made of tin bronze, as are pins and weapons (Tab. 16.2). Differentiation may be possible by closer examination of individual sites or the chronological sequence, but this topic cannot be discussed here in any detail.

The sheer amounts of tin referred to in the texts can be quite considerable, *e.g.* over 30 kg tin (66.66 mines) in a text from Early Bronze Age Ebla. At this juncture, tin is no longer a rarity in the economic records (Reiter 1997, 268). The price of tin was also very high at Ebla, and the ratio to silver was 1:1.5. From the Akkadian period a ratio of 1:6 is attested (we have no other textual documentation from the Early Dynastic and Akkadian periods), but during the Ur III period at the very end of the 3rd millennium BC, the value ratios are closer to 1:12–1:30, while during the Old Babylonian period in the earlier stages of the 2nd millennium, the value relations leveling out at 1:10 (Reiter 1997, 273–276, appendix IV). This may imply a much more regular supply of tin to Mesopotamian cities in the late 3rd millennium than had been possible a few centuries before. The very high price in Ebla may also indicate a price increase for the material in trading from southern Mesopotamia to Syria. The depreciation from the middle of the 3rd to the earlier 2nd millennium BC is also noticeable for bronze. During the Early Dynastic and Akkadian periods, the value relation of silver to bronze was

1:12.8 and 1:9, while during the Old Babylonian period it was 1:120 (Reiter 1997, 334–336, appendix IV). The cities of Upper Mesopotamia are generally considered to have played a major role in the supply and trading of tin bronze during the second half of the 3rd millennium BC, especially between 2500 and 2300 BC (Lebeau 2006, 2; for the tin trade, especially in subsequent eras, see Mari 1985 and Schmidt 2005).

Tin bronzes in Iran

The *vase à la cachette* hoard from Susa contained three vessels and a flat celt made of tin bronze (more than 7% tin). The hoard is dated to the ED IIIb period (Amiet 1986, 126; Benoit 2003, 304–305, n. 7 with further references). From this time on, tin bronzes appear regularly at Susa, but the ratio between these and all copper-based artefacts remains below 10%–15% in the second half of the 3rd millennium (Weeks 2003, 175 with further references). Apart from Susa, there are few other EBA sites with tin bronzes in Iran (Fig. 16.6). Among them are Giyan Tepe, Tepe Godin and Tal-i Malyan (see appendix). At the two latter sites, the tin bronzes originate from layers of the Kaftari phase, which covers not only the final centuries of the 3rd but also encompasses the first centuries of the 2nd millennium. Tepe Godin and Tal-i Malyan may therefore fall well out of the time range (until 2200/2100 BC) under consideration here. The point on the map for Tepe Hissar may perhaps distort the general picture because of 203 artefacts analysed, only two objects contained more than 1% tin (Pigott 1989, 32). At this site, arsenical copper was clearly the rule. Tin bronzes in Iran are comparatively rare. C. C. Lamberg-Karlovsky (in Weeks 2003, VIII) remarks: 'There is a paucity, indeed a very great poverty, of contemporary tin-bronzes on the Iranian Plateau'. It is only with the arrival of the Iron Age that tin bronze becomes a common alloy in this region (*cf.* Helwing 2009, 213–214).

Tin bronzes on the Persian Gulf and the Gulf of Oman

We owe our basic understanding of bronze metallurgy in the Persian Gulf to the fundamental and pioneering work of L. Weeks, which has brought the 'tin problem' much closer to a solution. Weeks analysed the composition of copper-based artifacts from several archaeological sites (Al Sufouh, Shimal, Tell Abraq) of the Umm al-Nar culture in the United Arab Emirates (Fig. 16.6). These bronzes often bear a high tin content (15–20%; Weeks 2003, 94–96, tab. 4.1–23, figs 4, 16–17). One individual sample even contained a significantly higher tin content (up to 52%), but the objects

Table 16.2: Number of tin bronzes from 28 sites in Syro-Mesopotamia representing specific object groups between the Uruk and Akkad period (n = 205) (from Müller-Karpe 2004 and Helwing and Müller 2004)

Vessel	Pin	Dagger	Axe	Spear/Bident	Celt	Chisel	Ring	Toilet set	Figure	Other
60	45	25	23	17	10	7	5	5	3	5

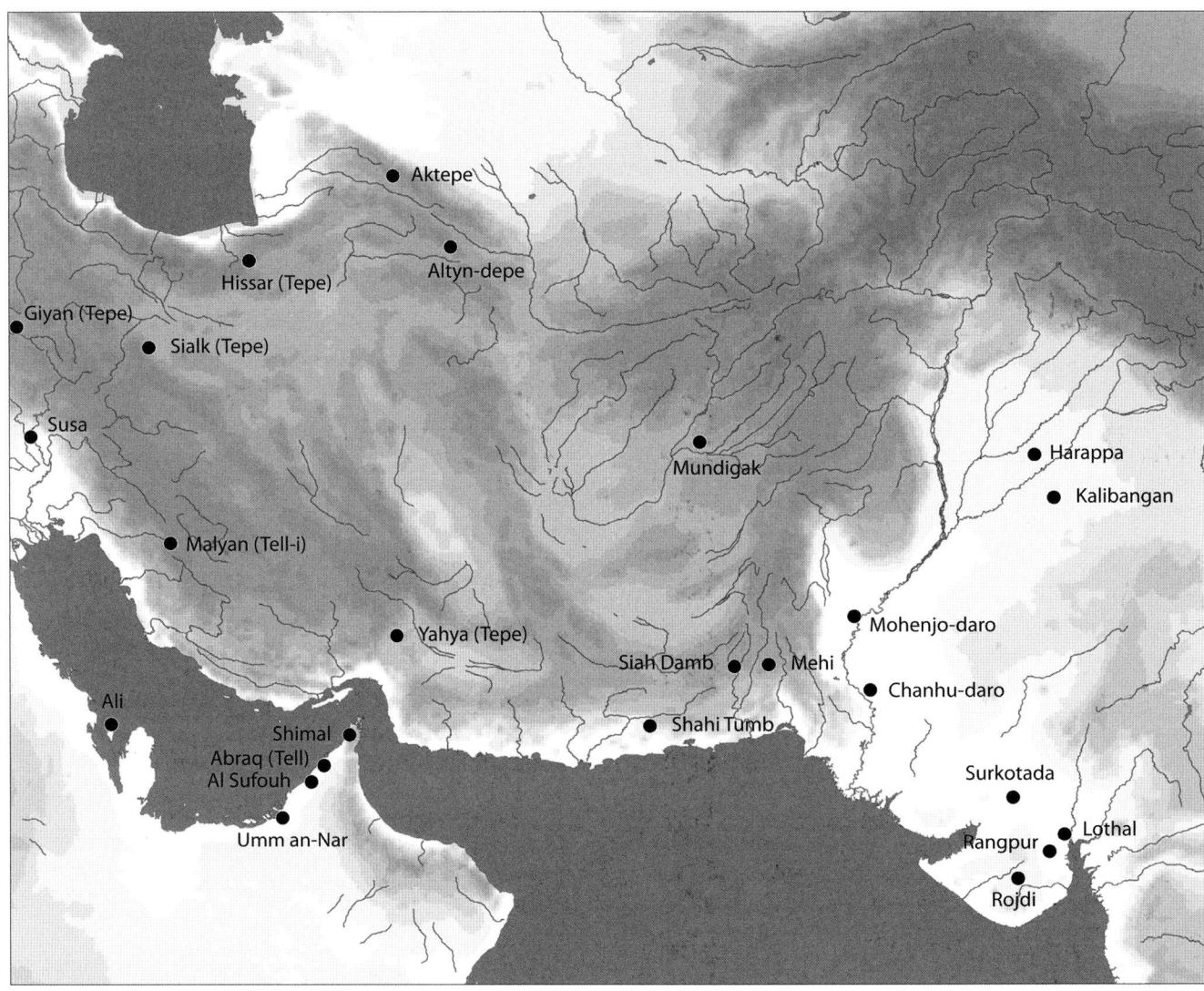

Figure 16.6: Distribution of tin bronzes in the Persian Gulf, Iran, Central Asia, Pakistan and western India in the 3rd millennium BC.

in question were subject to progressive corrosion, which increased the percentage of tin (Weeks 2003, 95–96). The objects are primarily rings or fragments of rings. The high-percentage tin bronzes from Shimal and Tell Abraq in the Emirates are particularly striking, since in the central Gulf and Oman there are no, or only minute, indications for tin bronzes dating from the 3rd millennium BC. In addition, Weeks carried out lead-isotope analyses that yielded very similar results to the tin bronzes from the Aegean and Velikent in the North Caucasus. This may indicate that the tin in the bronzes originates from the same deposit(s) (Weeks in Kohl 2002, 178–183; Weeks 2003, 163; L.).

Tin bronzes in Baluchistan, the Indus Valley and north-west India

It is still difficult to judge the role tin bronze really played in the Harappan culture. One reason is that only relatively few copper-based artefacts from different sites have been analysed. As tin bronze seems not to be very common, this might indicate the irregular use of tin bronze in the Harappan culture, which would by no means be unusual in comparison to some of the other regions already discussed. After all, there are eight Harappan sites that have yielded up bronzes with (in some cases) a high tin content. Of the sampled objects, 30% had a tin content of more than 1% (Agrawal 1984, fig. 6; Lahiri 1995, 123, tab. 2; Anthony 2007, 419). For example, three of the six analysed metal artifacts from Rangpur dating to the second half of the 3rd millennium (Rangpur period IIA) are tin bronzes (a celt, a hollow bangle and a pin with 4% and nearly 7% tin, respectively). Three other objects from period IIA only have trace amounts of tin (Rao 1962–63, 149–153 [nos 141, 169, 663], fig. 55, 1. 6). In Mohenjodaro and Harappa, almost one third of the objects analysed have a tin content of more than 1% (Agrawal 1984, 164; Weeks 2003, 177).

Tin bronzes in central Asia

Before the late 3rd millennium, *i.e.* before Namazga V, tin bronzes from central Asia are very rare (Fig. 16.6). This is a sound assertion given that some 1800 copper-based artefacts from the Bronze Age of central Asia (Turkmenistan, Uzbekistan, Tajikistan – but not Kyrgyzstan) have been analysed. The beginning of Namazga V is usually dated to *c.* 2300 or 2200 BC. A bronze rod with 7% tin from Aktepe at the lower Kopet Dag in southern Turkmenistan can be dated prior to this period (Ruzanov 1999, 104). From the Middle Bronze Age (Namazga V) about 10% of the copper-based artefacts are tin bronzes. From the BMAC (Bactria-Margiana Archaeological Complex), early Namazga VI and the beginnings of the Sapalli culture (in local terminology the Late Bronze Age!), only about 10% of the copper-based objects are tin bronzes, for example those in Dašly 3 (Anthony 2007, 425 with references) or at Altyn-Depe, where arsenical copper is typical and not tin bronze (Salvatori *et al.* 2002, 86–87). It was not until the Late Bronze Age II from *c.* 1700 BC onwards that the ratio increases to 30–45% (see Kaniuth 2007, 29–31, tabs 1–2 and Ruzanov 1999, 104). But from the middle of the 2nd millennium BC and the Andronovo culture, the ratio of tin bronzes nearly reaches 100% in the Eurasian steppe region (Kaniuth 2006, 169, n. 921 with further references). Information on Afghanistan is unfortunately absent due to lack of excavations and metal analyses. However, from a late stage of settlement phase III (early 3rd millennium BC) at Mundigak, we have an axe with 5% tin (Casal 1961, 244, 246–247, 249, fig. 139, 10, pl. XXXIX, B; Rahmstorf 2011a, 105, fig. 9.1 4), and some objects from Mundigak I in the late 4th millennium BC already contain at least about 1% tin (intentional alloys?). Not enough metal objects have been analysed from the 3rd millennium BC of Shortugai (I–II) (Francfort 1989, 75, 208) for us to assess whether and to what extent tin bronze was used at this site.

Tin bronzes, tin deposits and trade in tin during the 3rd millennium BC: some results

Promising new results suggest that in future it may be possible to determine the origin of tin by means of tin isotopes. For example, preliminary results have shown that the tin of the Nebra Sky Disc cannot come from the Erzgebirge or the Vogtland (Haustein and Pernicka 2008, 413–414). As with lead-isotope studies, the source cannot be determined beyond doubt, as only similiarites and dissimilarites in the isotope ratios can be observed. The development of tin-isotope analysis is still in its infancy, and it is not yet clear whether all potential tin resources are chemically distinct. But more and more research teams are

starting to test the potential of this new method (Haustein *et al.* 2010; Yamazaki *et al.* 2013; Molofsky *et al.* 2014; Nessel *et al.* 2014; Mason *et al.* 2016). In the meantime, all we can do is to use the results produced by lead-isotope analysis. The isotopic composition of the lead in an ore deposit is more or less constant and cannot be changed by chemical reactions caused by melting or corrosion (Nezafati *et al.* 2011, 223). The lead-isotope ratios of a deposit can therefore be regarded as a kind of fingerprint compared with those of finished products. Nevertheless, it may be that not all deposits are measurably different from each other (Begemann and Schmitt-Strecker 2008, 126–128). Re-melting and mixing of copper-based artefacts can distort the results considerably. Similarly, good interpretive results are only to be expected if the metal was not melted together from different deposits. Primarily, the lead will have come into the tin bronze with the copper, so lead-isotope analysis provides evidence of whether local copper was used or not. Accordingly, the local deposits have to be determined for their lead-isotope ratios. If this does not coincide with the signature of the sampled artifacts, one can be sure that the copper was brought in from elsewhere. Hence a very large database with the signatures of all European and Asian deposits is needed (and in part has already been built up) to verify potential sources of origin in the elimination procedure. Without results from the promising tin-isotope method not yet available, I shall give a brief summary here of the evidence for the possible exchange and trade of tin in the 3rd millennium BC. Lead-isotope analysis provides some clues. There has long been considerable controversy about whether there were tin deposits in the Aegean and Anatolia worth mining and whether they were identifiable by the ancient prospectors and metal-workers of the Early Bronze Age. As we have seen, the earliest tin bronzes occur in the Aegean at an early stage just after the beginning of the 3rd millennium, which has been taken by some as evidence of an indigenous development in bronze metallurgy. However, although numerous intensive archaeometallurgical surveys have been undertaken in recent decades in Greece and Turkey, not one undisputed deposit has been identified (Pernicka *et al.* 2003, 146).

The significance of the deposits at Kestel in the Taurus mountains of Turkey remains highly controversial. This was first presented in 1989 (Yener *et al.* 1989) as a kind of solution for the 'tin problem' (*cf.* Weeks 2003, 167–169, 180–181). The deposits only have minor concentrations of *cassiterite* (0.1–1%). By modern, but perhaps not by prehistoric standards, mining tin would not have been worthwhile or would at least have been very difficult (Muhly 1993, 246). While the ancient shafts and galleries that have been preserved do indicate ore mining both prehistoric and historical, it remains unclear which metal was in fact being mined at the site. At the nearby EBA settlement of Göltepe, tools for the processing of ores and powdered ore have been found (Earl and Özbal 1996, 289, 298; Yener 2000, 73–74). Concentrations of tin oxide were traceable in crucibles, but the individual results of analysis are contradictory (*cf.* the comments by Weeks 2003, 169 with references). Projections of the tin that could have been gained amount to only a few hundred kilograms, and this would have been obtained over a period of several hundred years (Bachmann 2004, 257). E. Pernicka points out that the glazed interiors of the crucibles with traces of tin oxide may have been caused by the fusion of loose gold grains. Accordingly, he feels that Kestel is more likely to have been a gold mine. In addition, *cassiterite* occurs at Kestel in mineralization with hematite and could therefore not have been detected by the ancient metallurgists (Pernicka 1998, 143). There are other arguments against the claim that during the Early Bronze Age much of the east Mediterranean and Near Eastern tin originated from Kestel and Göltepe. Lead-isotope studies by E. Pernicka, F. Begemann and others demonstrate that the copper of many EBA artefacts cannot have come from the Aegean or Anatolia because of its high precambrian age, which does not occur geologically in this region (Pernicka *et al.* 1984; Pernicka *et al.* 1990; Begemann *et al.* 1992; Pernicka 1998, fig. 2). Such copper is in fact typical of central Asia and Afghanistan. Thus, the conclusion for Kestel must be that '*die Lagerstätte [...] in der Zinnversorgung der Bronzezeit, wenn überhaupt, höchstens lokale Bedeutung gehabt haben kann*' (in terms of Bronze Age tin supply, its significance can only have been local at best: Hauptmann 2007, 568). In addition, cuneiform sources (literary sources, lexical lists, royal inscriptions and administrative texts) from 3rd-millennium Mesopotamia consistently point to the east when tin is mentioned, even if the geographic terms used sometimes make it difficult to pinpoint the locations that are being referred to (*cf.* Potts 1994, 9–36, fig. 2; Reiter 1997, 4, 7, 156; Steinkeller 2016, 133). In particular, the regions Dilmun (probably Bahrain/eastern Arabia), Magan (probably United Arabian Emirates and Oman) and Meluhha (probably Baluchistan and the Indus region) referred to in the texts can be sufficiently well localised (Reiter 1997, 212–258; Weeks 2003, 178–180 with further references). Nothing in the written sources indicates that tin may have come from the west. Nevertheless, most recent fieldwork indicates that tin may have been mined at Hisarcık near Kültepe in central Antolia in the later stages of the 3rd millennium BC (Lehner 2014, 31–34; Yener *et al.* 2015). This is especially challenging evidence in the light of the cuneiform texts from Karum Kaniš/Kültepe documenting extensive importation of tin from Mesopotamia to central Anatolia during the 19th century BC.

Investigations of metal objects from different locations in the United Arab Emirates on the Persian Gulf and the overall evaluation by L. Weeks have yielded interesting new results. Firstly, the lead-isotope ratios observed are similar to those from the EBA sites in the Aegean (Kastri, Thermi, Poliochni, Troy), Montenegro (Velika Gruda) and Daghestan in North Caucasus (Velikent). The lead-isotope ratios present in some Aegean tin bronzes indicate that the copper cannot have come from the Aegean-Anatolian region. This suggests that the copper or the bronze was imported, probably already as an alloy, and that this material may have originated from the same deposit(s). Since tin bronzes with high tin content from the United Arabian Emirates in the Persian Gulf are primarily small rings or the fragments left of them, it would be worth considering whether bronze or tin may not have been imported in the form of bronze rings/ingots. I have pointed out elsewhere (Rahmstorf 2010, 686–688, fig. 7) that tin bronzes from the eastern Mediterranean to the Indus region are frequently encountered in the shape of rings. Another clue is given by the written evidence, which suggests that metal was stored and transported in a ring form of some kind (Powell 1978, 213; Reiter 1997, 87–88, 334). Finally, we need to consider the shape of the earliest tin objects, for example, the famous tin bangle from Thermi on the east Aegean island of Lesbos or the oldest tin objects from central Europe, which are sectioned rings or beads (Primas 2002, 311–312). It would be worthwhile examining some questions in further detail, for example whether and to what extent rings were used as ingots during the 3rd millennium BC as in the Harappan culture, since they are remarkably common as tin bronzes in the Persian Gulf. Both L. Weeks and T. Potts assume that the tin or bronze reached the Persian Gulf during the Bronze Age via the Arabian Sea from Baluchistan and the Indus, *i.e.* the geographical sphere of the Harappan culture (Potts 1994, 281; Weeks 2003, 186, 200) and that the 'Harappans' controlled the tin trade. To clarify this question, we would need a concerted research programme involving the chemical examination of large series of copper-based artefacts of the Harappan culture plus lead-isotope analyses. Research programmes with lead-isotope studies have only recently begun in the Indus (Law and Burton 2006; Hoffman and Miller 2014) and there are fundamental obstacles: 'There is no comparative database of geological isotopic values from many potential sources surrounding the Indus, and there are potential logistical difficulties in obtaining the samples from several of these possible source areas' (Hoffman and Miller 2014, 703). Without these investigations, it can only be assumed that it was most likely the 'Harappans' who organised the tin/bronze trade westward across the Persian Gulf. The proximity and thus the possible control of the tin deposits in Afghanistan could have been a great advantage for traders of the Harappan culture. Locations such as Shortugai acted as Harappan trading bases in Afghanistan from which the trade of tin, gold and lapis lazuli to the Indus Valley was organised, although admittedly all these materials have hardly been found in Shortugai itself (however, we should bear in mind that hardly any tin or silver dating to the early 2nd millennium BC has been excavated at the well-known trading center of Karum Kaneš in central Anatolia although we know that dozens of tons of tin are referred to in the cuneiform texts from this site). In southern Afghanistan, the deposits of Mesgaran (with up to 6.6% tin content in the ore), in Herat and in the Iranian–Afghan–Pakistani border area would have been convenient sources of tin for the Harappan culture (Kaniuth 2006, 170; Nezafati *et al.* 2011, tab. 2 with references). Without fieldwork, which in this region will hardly be possible in the foreseeable future, such claims must unfortunately remain conjectures for the moment.

The central Asian region and, further north, the Eurasian steppe region cannot be regarded as the origin of early tin bronzes in the Near East because there are not many tin bronzes known from this region at all before 2200 BC. In the Eurasian steppes, the use of tin bronze does not occur before the Okunevo culture, where it has been detected in bronzes of the Sejma-Turbino type (Parzinger 2003, 160–164, 175, figs 2–3). The central Asian tin deposits sometimes displaying traces of mining at Karnab, Lapas and Čangali in Uzbekistan, at Mušiston (a site that gave its name to the copper-tin ore *mushistonite*, which may be up to 34% tin, see Nezafati *et al.* 2011, tab. 2) and Takfon in Tajikistan and now also in Kalai Topkan, Novaya Schulba and Askaraly in Kazakhstan (Stöllner *et al.* 2011) cannot help in solving the 'tin problem' for the early and middle 3rd millennium BC because so far tin mining has only been documented via field work relating to the turn of the 3rd to the 2nd millennium BC at the earliest (Parzinger and Boroffka 2003; Cierny *et al.* 2005). In my view, this evidence only leaves the Afghan area as a possible origin of tin demand in the Near East and the eastern Mediterranean during the earlier and middle 3rd millennium BC. Recently, the relatively small Deh Hosein deposit in the central eastern part of the Zagros Mountains in western Iran has been referred to as a possible origin of this early tin. In some cases, the lead-isotope analyses show a good correlation with analysed objects from Luristan (2nd and 1st millennium, partly also 3rd millennium BC). Some objects from the Persian Gulf, Mesopotamia and sometimes even the Aegean have similar isotopic ratios. But there is still no proof that Deh Hosein might be the area of origin for tin in the 3rd millennium, because once again the mining remains from Deh Hosein have as yet only been dated to the 2nd millennium BC, not before (Nezafati *et al.* 2006; Nezafati *et al.* 2011, 225).

For the 3rd millennium BC, we can rule out tin deposits in Europe having any significance for tin supply in the Near East and the East Mediterranean, although this has been suggested (*e.g.* Gerloff 1993, 86). As we have seen, tin bronzes are very rare at this time in Europe (outside the Aegean area) and this is a very serious argument against the region (as in the case of central Asia referred to earlier). However, it is certainly no coincidence that almost all early tin-bronze sites in central and southern Europe can be connected to the Bell Beaker phenomenon.

The particular affinity of the 'Bell Beaker people' to metals such as copper, and more importantly to gold and silver (Meller 2014, 616–620, figs 3–4), is further indicated by the numerous burials of metal craftsmen (*e.g.* Fitzpatrick 2009, 180–181). Also, the mobility of at least some Bell Beaker individuals was very high (Chenery and Evans 2011). They may have brought initial knowledge of the bronze alloy to central, western and southern Europe. The case of Velikent, though pre-Bell Beaker in date, indicates that assumptions about initial impulses from the Aegean and the eastern Mediterranean world should not be excluded outright as an explanatory hypothesis.

This survey of the use of tin bronzes before *c.* 2200 BC in the large area extending from the Atlantic to the Indus region can only afford a glimpse of the complicated issues posed by the early occurrence of tin bronzes. If we consulted all the recently published site reports, most likely dozens of sites could be added. This survey covers the evidence known to me until 2011.[2] It also falls short in terms of an in-depth analysis of the data. For example, more precise dating and chronological mapping of the sites with tin bronzes would be possible. A detailed contextual analysis of the find spots (where were they, what else was found?) at the various sites could also be rewarding. Another question of interest is the kind of objects produced using this material. Early tin bronzes were objects of adornment, weapons or tools. No clear preference is detectable, but maybe regional differences will emerge. Finally, tin bronze was often only one option among the different copper-based alloys metal-workers could intentionally produce during the 3rd millennium BC. It has been emphasised that 'at least originally, the preference of tin bronze over arsenic bronze was an esthetic and cultural choice, and not a utility consideration' (Steinkeller 2016, 136). It would be especially interesting to differentiate (regionally, chronologically, contextually and in object classes) between copper, arsenical copper and tin bronze (*cf.* Chernykh 1992 and Wilkinson 2014). It seems that the significance of tin bronze has been exaggerated since the beginnings of Bronze Age archaeology, in part quite simply because the material gave the whole period its name. In addition, as T. C. Wilkinson put it recently, 'the simple view of a unilinear technological evolution in which the discovery of the functional advantages of arsenic-bronze, then tin-bronze, facilitate the replacement of one material by another, does not account for the complex patterns of metal consumption' (Wilkinson 2014, 189).

Appendix: Tin bronze objects before *c.* 2200 BC (*cf.* also Jablonka 2014, fig. 10)

1. Jezioro Gopło, Poland (Matuschik 1997, 87, 103, fig. 6; Krause 2003, 212, fig. 192)
2. Altenmarkt, Germany (Krause 2003, 211, 213)
3. Gnetsch, Germany (Krause 2003, 211–212)

4. Landau, Germany (Krause 2003, 211, 213)
5. Chevroux, Switzerland (Krause 2003, 211–212)
6. St. Blaise, Switzerland (Krause 2003, 211–212)
7. Bohdalice, Czech Republic (Junghans *et al.* 1968, cat. no. 3248)
8. Bylany, Czech Republic (Junghans *et al.* 1968, cat. no. 3238)
9. Ledce, Czech Republic (Junghans *et al.* 1968, cat. no. 19935)
10. Smolín, Czech Republic (Novotný 1958, fig. 5, 13; Junghans *et al.* 1974, cat. no. 19936)
11. Vukovar, Croatia (Schmidt 1945, 71, pl. 28, 7; Junghans *et al.* 1968, cat. no. 2004)
12. Abri du Capitaine, Montpezal, France (Pare 2000b, 22)
13. Bauma del Serrat del Pont (Gerona), Spain (Alcalde *et al.* 1998; Pare 2000b, 22)
14. Guidoiro Areoso, Illa de Arousa, Spain (Comendador Rey 1999; Pare 2000b, 22)
15. Barro/Torres Vedras, Portugal (Spindler 1981, 102–103, fig. 41, 4)
16. Val Frascarese, Italy (Campana *et al.* 1996)
17. Poggio Olivastro, Italz (Bulgarelli and Giumlia-Mair 2008)
18. Ovcharitsa, Bulgaria (Leshtakov 1996, 250)
19. Mudrets (Tell), Bulgaria (Leshtakov 1996, 250)
20. Vasilia (?), Bulgaria [source missing, could not be verified]
21. Velika Gruda, Montenegro (Primas 1996, 104–105)
22. Aghios Kosmas, Greece (Mylonas 1959, 101, fig. 163, 13; McGeehan-Liritzis and Taylor 1987, 294, tab. 1)
23. Axiochori-Vardarophtsa, Greece (Maran 1998, 268 with further reference)
24. Dokathismata/Amorgos, Greece (Maran 1998, 268; Rambach 2000, 13, pl. 3, 9–10)
25. Kastri/Syros, Greece (Bossert 1967, 63, 76; Maran 1998, 269, n. 1128)
26. Krasi/Kreta, Greece (Marinatos 1929, 119, 131, fig. 13, 30–31; Muhly 2006, 170)
27. Lithares, Greece (Tzavella-Evjen 1984, 140, n. 657, 663; Kayafa *et al.* 2000, 41–42, tab. 2.2.)
28. Manika, Greece (Sampson 1985, 177, 305–306; Maran 1998, 268)
29. Markiani, Greece (Andreopoulou-Mangou 2006, 218–219, tab. 8.18)
30. Perivolaki-Saratse (?), Greece (Maran 1998, 268 with further references)
31. Poliochni/Lemnos, Greece (Pernicka *et al.* 1990; Maran 1998, 269, n. 1330 with further references)
32. Sitagroi, Greece (McGeehan-Liritzis and Taylor 1987, 293, tab. 1; Maran 1998, 267, n. 1106–1007 with further references)
33. Steno/Lefkas, Greece (McGeehan-Liritzis 1996, 186–187; Maran 1998, 267–268, n. 1009; Kilian-Dirlmeier 2005, 107)
34. Thermi/Lesbos, Greece (Begemann *et al.* 1992, tab. 1–2; Tin ring: Lamb 1936, 165, 171–172, fig. 50, 30.24; Pernicka 2001, 370, fig. 410)
35. Tiryns, Greece (Kilian 1982, 421, fig. 44, 5)
36. Ahlatlıbel, Turkey (Earl and Özbal 1996)
37. Alaca Höyük, Turkey (Kaptan 1990, 76 with references)
38. Ališar Höyük, Turkey (von der Osten 1937, 92–93)
39. Aphrodisias, Turkey (Joukowsky 1986, 511, 519, fig. 368, 372, 374, 15–16) [not securely analysed]
40. Arslantepe, Turkey (Bobokhyan 2008, 67 with references)
41. Beşiktepe, Turkey (Pare 2000b, 9; Begemann *et al.* 2003)
42. Bakla Tepe, Turkey (Şahoğlu 2005, 247–249, n. 9)
43. Demircihüyük, Turkey (Bachmann *et al.* 1987, 23) [traces of tin in mould]; Demircihüyük-Sarıket, Turkey (Pernicka 2000)
44. Göltepe/Kestel, Turkey (Earl and Özbal 1996, tab. 6)
45. Güzelova, Turkey (Bobokhyan 2008, 67)
46. Horoztepe, Turkey (Earl and Özbal 1996)
47. Judaida (Tell), Turkey (Yener 2009, 144–145 with further references)
48. Karataş-Semayük, Turkey (Warner 1994, 207)
49. Kayapınar, Turkey (Reeves 2003, Appendix 2 – generally refered to as bronze, but apparently no analyses performed)
50. Kusura, Turkey (Lamb 1936, 214–215)
51. Mahmatlar, Turkey (Earl and Özbal 1996)
52. Mersin, Turkey (Giardino *et al.* 2003, 389)
53. Norşuntepe, Turkey (Bobokhyan 2008, 628, pl. 15)
54. Resuloğlu, Turkey (Zimmermann 2007, fig. 6)
55. Tarsus, Turkey (Kuruçayırlı and Özbal 2005 with further references)
56. Titriš Höyük, Turkey (Palmieri and di Nocera 2004, 380, fig. 2, tab. 1)
57. Troia, Turkey (Pernicka *et al.* 1989, 575, 578, tabs 3–4)
58. Tülintepe, Turkey (Yalçın and Yalçın 2008)
59. Van, Turkey (Bobokhyan 2008, 628, pl. 15)
60. Yortan, Turkey (Pernicka *et al.* 1984 575, 579, tab. 3–4)
61. Zeytinli Bahçe, Turkey (Palmieri and di Nocera 2004, 380, fig. 3)
62. Velikent, Russia (Gadzhiev *et al.* 1997, fig. 8, 3; Kohl 2002, 174, fig. 12; Kohl 2003, 19, fig. 1, 7; Peterson 2003, tab. 2.1–2)
63. Ananauri, Georgia (Apakidze 1999, fig. 2)
64. Bakurcixe, Georgia (Kavtaradze 1999, 86, 97; Apakidze 2002, 762)
65. Bedeni, Georgia (Bobokhyan 2008, 67)
66. Tsnori/Cnori, Georgia (Apakidze 1999, fig. 2)
67. Martkopi, Georgia (Kavtaradze 1999, 84, 97; Apakidze 2002, 762)

68. Telebi, Georgia (Kavtaradze 1999, 84–86, 96; Apakidze 2002, 761–762)
69. Talin, Armenia (Meliksetian *et al.* 2003, 310, 314)
70. Gyumri, Armenia (Apakidze 1999, fig. 2)
71. Kirovakan, Armenien (Bobokhyan 2008, 67 with further references)
72. Metsamor, Armenien (Bobokhyan 2008, 628, pl. 15)
73. Shengavit, Armenien (Bobokhyan 2008, 67 with further references)
74. Chatschenaget, Aserbaidschan (Bobokhyan 2008, 67)
75. (Vasilia?) Kolokassides Hort, Zypern (Webb *et al.* 2006, 268, tab. 2, figs 1, 2. 4. 8)
76. Sotira Kaminoudhia, Zypern (Swiny 2003, 376–379, fig. 8.1, M13, M14, M21, M22; Giardino *et al.* 2003, 388–390, pl. 8.1.1)
77. Byblos, Libanon (Montero Fenollós 2001, 276 with reference)
78. Sultân (Tell es-)/Jericho, Palestine (Prag 1974, 91, n. 63)
79. Bab edh-Dhra, Jordania (Maddin *et al.* 2003, 513)
80. Hirbet ez-Zeraqon, Jordania (Hauptmann 2000, 180)
81. Abydos, Egypt (Pernicka and Schleiter 1997, 221, tab. 2)
82. Buto, Egypt (Cowell 1987, 96–118)
83. Ahmar (Tell)/Til Barsip, Syria (Lutz 2004, 148)
84. Beydar (Tell), Syria (Tonussi 2008, 210–211, 218, 221–222, pl. 11, 8503-M-2; pl. 13, 2640-M-3; pl. 14, 6138-M-3. 6138-M-5)
85. Brak (Tell), Syria (Oates *et al.* 2001)
86. Halawa (Tell), Syria (Lutz 2004, 148)
87. Hariri (Tell)/Mari, Syria (Lutz 2004, 149)
88. Mardikh (Tell)/Ebla, Syria (Peyronel 2012, fig. 3)
89. Qara Quzaq/Qarah Qazak, Syria (Montero Fenollós 2001, 275)
90. Ras Shamra/Ugarit, Syria (Montero Fenollós 2001, 276 with reference)
91. Sweyhat (Tell), Syria (Montero Fenollós 2001, 276 with reference)
92. Abu Qasim (Tell), Iraq (Lutz 2004, 110)
93. Abu Salabikh (Tell), Iraq (Lutz 2004, 110)
94. Agreb (Tell), Iraq (Lutz 2004, 110)
95. Asmar (Tell)/Eshunna, Iraq (Lutz 2004, 110–111)
96. Basmusian (Tell), Iraq (Lutz 2004, 112)
97. Fara/Šuruppak, Iraq (Lutz 2004, 113)
98. Gawra (Tepe), Iraq (Lutz 2004, 113–114)
99. Hiba (Tell al-)/Lagaš, Iraq (Lutz 2004, 114–115)
100. Khafadje/Tubub, Iraq (Lutz 2004, 115)
101. Kirkuk Ölfeld, Iraq (Lutz 2004, 116)
102. Kirkuk Gegend, Iraq (Lutz 2004, 116)
103. Mazhur (Tell), Iraq (Lutz 2004, 120)
104. Muqayyar (Tell al-)/Ur, Iraq (Lutz 2004, 123–141).
105. Nuffar/Nippur, Iraq (Lutz 2004, 120)
106. Oheimir (Tell el-)/Kiš, Iraq (Lutz 2004, 116–119)
107. Razuk (Tell), Iraq (Lutz 2004, 120)
108. Senkereh (Tell-es)/Larsa, Iraq (Lutz 2004, 120)
109. Sulaima (Tell as-), Iraq (Lutz 2004, 121–122)
110. Taya (Tell), Iraq (Lutz 2004, 122)
111. Tello/Girsu, Iraq (Lutz 2004, 122, 144–148)
112. Ubaid (Tell al-), Iraq (Lutz 2004, 122–123)
113. Umm al-Ağağ, Iraq (Lutz 2004, 123)
114. Uqair (Tell al-), Iraq (Lutz 2004, 123)
115. Warka/Uruk, Iraq (Lutz 2004, 141–143)
116. Yalhi (Tell), Iraq (Lutz 2004, 143)
117. Giyan (Tepe), Iran (Weeks 2003, 175 with reference)
118. Godin (Tepe), Iran (Weeks 2003, 175 with reference)
119. Kalleh Nissar, Iran (Fleming *et al.* 2005, 36–38, tab. 1)
120. Malyan (Tal-i), Iran (Pigott 2003, tab. 14.1)
121. Susa, Iran (Weeks 2003, 175 with reference; Benoit 2003, 304–305, n. 7)
122. Sialk, Iran (Ghirshman 1938)
123. Tepe Hissar, Iran (Pigott 1989, 32)
124. Tepe Yahya, Iran (Heskel and Lamberg-Karlovsky 1980)
125. Ali, Bahrain (Weeks 2003, 177–178 with reference)
126. Umm an-Nar, United Arabian Emirates (Weeks 2003, 96 with reference)
127. Al Sufouh, United Arabian Emirates (Weeks 2003, 72, tab. 4.1)
128. Tell Abraq, United Arabian Emirates (Weeks 2003, 75, tab. 4.4)
129. Shimal, United Arabian Emirates (Weeks 2003, 73, tab. 4.2–3)
130. Chanhu-daro, Pakistan (Lahiri 1995, tab. 2)
131. Harappa, Pakistan (Agrawal 1984, 164)
132. Mehi, Pakistan (Marshall 1931, 488; Asthana 1993, 277)
133. Mohenjo-daro, Pakistan (Agrawal 1984, 164)
134. Segak, Pakistan (Marshall 1931, 488; Asthana 1993, 277)
135. Shahi Tump (Nundara), Pakistan (Marshall 1931, 488; Asthana 1993, 277)
136. Shiah Damb, Pakistan (Marshall 1931, 488; Asthana 1993, 277)
137. Kalibangan, India (Lal *et al.* 2003, 265–266)
138. Lothal, India (Lahiri 1995, tab. 2)
139. Rangpur, India (Agrawal 1984, 164)
140. Rojdi, India (Lahiri 1995, tab. 2)
141. Surkotada, India (Lahiri 1995, tab. 2)
142. Mundigak, Afghanistan (Casal 1961, 244, 246–247, 249, fig. 139, 10a, pl. XXXIX, B)
143. Aktepe, Turkmenistan (Ruzanov 1999, 104)
144. Altyn-Depe, Turkmenistan (Salvatori *et al.* 2002, 86–87; Masioli *et al.* 2006)

Notes

1   I would like to thank Joseph Maran and Philipp W. Stockhammer for the invitation to the conference and the anonymous referee and the editors for suggestions for improvement. This paper is not the written version of the talk I gave in Heidelberg 'And

Childe was right after all? Vere Gordon Childe's thoughts on immigrant craftsmen, prospectors and the dissemination of key economic innovations during the 3rd millennium BC in the light of recent scholarship'. Unfortunately, due to time restraints it was not possible for me to write this paper at the time, though some archaeological evidence presented in my talk can be found in a recently published paper (Rahmstorf 2015). This contribution is a slightly modified chapter from my *Habiliation* thesis 'Studien zu Gewichtsmetrologie und Kulturkontakt im 3. Jahrtausend v. Chr.', which was submitted at the University of Mainz in 2012 and which is currently being prepared for publication. The chapter will not be included in this forthcoming monograph, but this almost exclusively material survey may still prove useful, especially in the proceedings of a conference that is focussing on tin bronze as an important innovation.

2	Very recently, P. Jablonka presented a map with early bronzes that is similar in its geographical range (Jablonka 2014, fig. 10), but it only includes 116 sites. However, his list also contains 17 sites not included in the distribution maps presented here. They were either unknown to me or I believe that either their dates (*e.g.* finds from Middle Bronze Age Emenska Pest, *cf.* Pare 2000b, 13 or 'Final Neolithic' tin bronzes in central Europe, *cf.* above) or their analytical results (*e.g.* Ezerovo II, *cf.* Pare 2000b, 13) suggest that their relevance is questionable. We may confidently hope that future work will create web-based databases and distribution maps of such material evidence that will permit continuous addition to what we have already and enable specialized researchers to engage in critical reflection on the material.

References

Agrawal, D. P. (1984) Metal Technology of the Harappans. In B. B. Lal and S. P. Gupta (eds), *Frontiers of the Indus Civilization*, 163–167. Janakpuri, Books & Books.

Alcalde, G., Molist, M., Montero, I., Planagumà, L., Saña, M. and Toledo, M. (1998) Producciones metalúrgicas en el nordeste de la Península Ibérica durante el III milenio cal. AC: Et taller de la Bauma del Serrat de Pont (Tortellà, Girona). *Trabajos de Prehistoria* 55, 81–100.

Andreopoulou-Mangou, H. (2006) Note on the Chemical Analysis of the Markiani Metal Objects. In L. Marangou, C. Renfrew, C. Doumas and G. Gavalas (eds), *MAPKIANH AMOPΓOY - Markiani, Amorgos. An Early Bronze Age Fortified Settlement. Overview of the 1985–1991 Investigations*. British School at Athens, Supplementary Volume 40, 218–219. London, British School at Athens.

Amiet, P. (1986) *L'âge des échanges inter-iraniens 3500–1700 avant J.-C.* Notes et Documents des Musées de France 11. Paris, Éd. de la Réunion des Musées Nationaux.

Anthony, D. W. (2007) *The Horse, the Wheel and Language. How Bronze-Age Riders from the Eurasian Steppes Shaped the Modern World*. Princeton, NJ, Princeton University Press.

Antoniadou, S. and Pace, A. (eds) (2007) *Mediterranean Crossroads*. Athens, Pierides Foundation.

Apakidze, J. (1999) Lapislazuli-Funde des 3. und 2. Jahrtausends v. Chr. in der Kaukasusregion – Ein Beitrag zur Herkunft des Lapislazulin in Troia. *Studia Troica* 9, 511–525.

Apakidze, J. (2002) Zinnbronze der Kolchis-Kultur: ein Beitrag zur Herkunft und Verbreitung der Zinnbronze im östlichen Schwarzmeergebiet im 2. und Anfang des 1. Jahrtausends v. Chr. In R. Aslan, S. Blum, G. Kastl, F. Schweizer and D. Thumm (eds), *Mauerschau. Festschrift für Manfred Korfmann*, 759–780. Remshalden-Grunbach, Greiner.

Aruz, J. and Wallenfels, R. (eds) (2003) *Art of the First Cities. The Third Millennium B.C. from the Mediterranean to the Indus*. New York, Yale University Press.

Asthana, S. (1993) Harappan Trade in Metals and Minerals: A Regional Approach. In G. L. Posehl (ed.), *Harappan Civilization. A Recent Perspective*. 2nd ed., 271–285. New Delhi, American Institute of Indian Studies and Oxford & IBH Publication.

Atzeni, C., Massidda, L. and Sanna, U. (2005) Archaeometric Data. Investigations and Results. In F. Lo Schiavo, A. Giumlia-Mair, U. Sanna and R. Valera (eds), *Archaeometallurgy in Sardinia from the Origin to the Early Iron Age*. Monographies Instrumentum 30, 115–183. Montagnac, Mergoil.

Avetisyan, P., Muradyan, F. and Sargsyan, G. (2010) Early Bronze Age Burial Mounds at Talin. In S. Hansen, A. Hauptmann, I. Mötzenbäcker and E. Pernicka (eds), *Von Majkop bis Trialeti. Gewinnung und Verbreitung von Metallen und Obsidian in Kaukasien im 4.–2. Jts. v. Chr.* Kolloquien zur Vor- und Frühgeschichte 13, 161–165. Bonn, Habelt.

Bachmann, H.-G. (2004) Zinn, das wichtigste Legierungsmetall der Bronzezeit in Mitteleuropa. In G. Weisgerber and G. Goldenberg (eds), *Alpenkupfer – Rame delle Alpi*. Der Anschnitt, Beiheft 17, 255–260. Bochum, Deutsches Bergbau-Museum.

Bachmann, H.-G., Otto, H. and Prunnbauer, F. (1987) Materialanalysen, Analyse von Metallfunden. In M. Korfmann (ed), *Demircihüyük. Die Ergebnisse der Ausgrabungen 1975–1978, 2: Naturwissenschaftliche Untersuchungen*, 21–24. Mainz, Philipp von Zabern.

Batiuk, S. and Rothman, M. S. (2007) Early Transcausian Cultures and their Neighbors. *Expedition* 49, 7–17.

Begemann, F. and Schmitt-Strecker, S. (2008) Bleiisotopie und die Provenienz von Metallen. In Ü. Yalçın (ed.), *Anatolian Metal IV*. Der Anschnitt, Beiheft 21, 125–134. Bochum, Deutsches Bergbau-Museum.

Begemann, F., Schmitt-Strecker, S. and Pernicka, E. (1992) The Metal Finds from Thermi III–V: A Chemical and Lead-Isotope Study. *Studia Troica* 2, 219–239.

Begemann, F., Schmitt-Strecker, S. and Pernicka, E. (2003) On the Composition and Provenance of Metal Finds from Beşiktepe (Troia). In G. A. Wagner, E. Pernicka and H.-P. Uerpmann (eds), *Troia and the Troad*, 173–201. Berlin, Springer.

Benoit, A. (2003) Vase à la cachette, 202a–h. In J. Aruz and R. Wallenfels (eds), *Art of the First Cities. The Third Millennium B.C. from the Mediterranean to the Indus*, 303–305. New York, Yale University Press.

Bertram, J.-K. (2010) Zum Martqopi-Bedeni-Horizont im Südkaukasusgebiet. In S. Hansen, A. Hauptmann, I. Mötzenbäcker and E. Pernicka (eds), *Von Majkop bis Trialeti. Gewinnung und Verbreitung von Metallen und Obsidian in Kaukasien im 4.–2. Jts. v. Chr.* Kolloquien zur Vor- und Frühgeschichte 13, 253–261. Bonn, Habelt.

Birtacha, K. (2006) The Metal Objects. In L. Marangou, C. Renfrew, C. Doumas and G. Gavalas (eds), *MAPKIANH*

ΑΜΟΡΓΟΥ – Markiani, Amorgos. An Early Bronze Age Fortified Settlement. Overview of the 1985–1991 Investigations. British School at Athens, Supplementary Volume 40, 211–217. London, British School at Athens.

Bobokhyan, A. (2008) *Kommunikation und Austausch im Hochland zwischen Kaukasus und Taurus, ca. 2500–1500 v. Chr.* British Archaeological Reports, International Series 1853. Oxford, BAR.

Bossert, E.-M. (1967) Kastri auf Syros: Vorbericht über die Untersuchung der prähistorischen Siedlung. *Archaiologikon Deltion* 22/1, 53–76.

Braidwood, R. J. and Braidwood, L. S. (1960) *Excavations in the Plain of Antioch 1. The Earlier Assemblages Phases A–J.* The University of Chicago Oriental Institute Publications 61. Chicago, IL, Chicago University Press.

Bray, P. (2012) Before [29]Cu Became Copper: Tracing the Recognition and Invention of Metallurgy in Britain and Ireland During the Third Millennium bc. In M. J. Allen, J. Gardiner, and J. A. Sheridan (eds), *Is there a British Chalcolithic: people, place and polity in the later third millennium*, 56–70. Oxford, Oxbow.

Bulgarelli, M. G. and Giumlia-Mair, A. (2008) Un anellino metallico dal sito neo-Eneolitico di Poggio Olivastro. In P. Petitti and F. Rossi (eds), *Aes: Metalli preistorici dalla Tuscia*, 12–13. Valentano, Museo della Preistoria della Tuscia e della Rocca Farnese.

Butterlin, P. (2002) Réflexions sur les pròblemes de continuité stratigraphique et culturelle à Tepe Gawra. *Syria* 79, 7–51.

Campana, N., Franceschi, E., Maggi, R. and Stos-Gale, Z. (1996) Grotticella sepolcrale di Val Frascarese (Genova): Nuove analisi dei reperti metallici. In D. Cocchi Genick (ed.), *L'antica età del Bronzo in Italia*, 556–557. Florence, Octavo Cantini.

Casal, J.-M. (1961) *Fouilles de Mundigak.* Mémoires de la Délégation Archéologique Française en Afghanistan 17. Paris, Klincksieck.

Chapman, R. (1991) *The Vinča-Culture of South-East Europe: Studies in Chronology, Economy and Society.* British Archaeological Reports, International Series 117. Oxford, BAR.

Charles, J. A. (1975) Where is the tin? *Antiquity* 49, 19–24.

Chenery, C. A. and Evans, J. A. (2011) A Summary of the Strontium and Oxygen Isotope Evidence for the Origins of Bell Beaker Individuals Found Near Stonehenge. In A. P. Fitzpatrick (ed.), *The Amesbury Archer and the Boscombe Bowmen. Bell Beaker Burials on Boscombe Down, Amesbury, Wiltshire.* Wessex Archaeology Report 27, 185–190. Salisbury, Wessex Archaeology.

Chernykh, E. N. (1992) *Ancient Metallurgy in the USSR: The Early Metal Age.* Cambridge, Cambridge University Press.

Childe, G. V. (1928) *The Most Ancient East. The Oriental Prelude to European Prehistory.* London, Kegan Paul, Trench, Trubner.

Cierny, J., Stöllner, T. and Weisgerber, G. (2005) Zinn in und aus Mittelasien. In Ü. Yalçın, C. Pulak and R. Slotta (eds), *Das Schiff von Uluburun. Welthandel vor 3000 Jahren. Katalog der Ausstellung im Deutschen Bergbau-Museum Bochum vom 15. Juli 2005 bis 16. Juli 2006.* Veröffentlichungen aus dem Deutschen Bergbau-Museum Bochum 138, 431–448. Bochum, Deutsches Bergbau-Museum.

Coleman, J. E., Barlow, J. A., Mogelonsky, M. K. and Schaar, K. W. (1996) *Alambra: A Middle Bronze Age Settlement in Cyprus: Archaeological Investigations by Cornell University 1974–1985.* Studies in Mediterranean Archaeology 118. Jonsered, Åström.

Comendador Rey, B. (1999) The Early Development of Metallurgy in the North-West of the Iberian Peninsula. In S. M. M. Young, A. M. Pollard, P. Budd and R. A. Ixer (eds), *Metals in Antiquity.* British Archaeological Reports, International Series 792, 63–67. Oxford, Archaeopress.

Cowell, R. M. (1987) Scientific Appendix I, Chemical Analysis. In W. V. Davies (ed.), *Catalogue of Egyptian Antiquities in the British Museum 7. Tools and Weapons 1, Axes*, 96–118. London, British Museum.

Delibes, G., Montero, I. and Rovira, S. (1996) The First Use of Metals in the Iberian Peninsula. In B. Bagolini, and F. Lo Schiavo (eds), *The Copper Age in the Near East and Europe. XIII International Congress of Prehistoric and Protohistoric Sciences Forlì, Italia, 8–14 September 1996. Colloquium XIX: Metallurgy: Origins and Technology, Section 10*, 19–34. Forlì, ABACO. Edizioni.

Dolfini, A. (2010) The Origins of Metallurgy in Central Italy: New Radiometric Evidence. *Antiquity* 84, 707–723.

Doonan, R. C. P., Day, P. M. and Dimopoulou-Rethemiotaki, N. (2007) Lame Excuses for Emerging Complexity in Early Bronze Age Crete: The Metallurgical Finds from Poros Katsambas and Their Context. In P. M. Day and R. C. P. Doonan (eds), *Metallurgy in the Early Bronze Age Aegean.* Sheffield Studies in Aegean Archaeology 7, 98–122. Oxford, Oxbow Books.

Durman, A. (1997) Tin in Southeastern Europe? *Opuscula Archaeologica (Zagreb)* 21/1, 7–14.

Earl, B. and Özbal, H. (1996) Early Bronze Age Tin Processing at Kestel/Göltepe, Anatolia. *Archaeometry* 38, 289–303.

Eaton, E. R. and McKerrell, H. (1976) Near Eastern Alloying and Some Textual Evidence for the Early Use of Arsenical Copper. *World Archaeology* 8, 169–191.

Enrődi, A., Reményi, L., Baradács E., Uzonyi, I., Kiss, A. Z., Montero, I., Rovira, S. (2003) Technological study of Bell Beaker metallurgy in Hungary. In *Archaeometallurgy in Europe (24–26 september 2003).* Proceedings Vol. 2, 29–38. Milan, Associazione Italiana di Metallurgia.

Fernandez-Miranda, M., Montero, I. and Rovira, S. (1995) Los primeros objetos de bronce en el occidente de Europa. *Trabajos de Prehistoria* 52, 57–69.

Fitzpatrick, A. (2009) In His Hands and in His Head: The Amesbury Archer as a Metal Worker. In P. Clark (ed.) *Bronze Age connections. Cultural Contact in Prehistoric Europe*, 176–188. Oxford, Oxbow Books.

Fischer, P. M. (2008) *Tell Abu Al-Kharaz in the Jordan Valley 1: The Early Bronze Age.* Österreichische Akademie der Wissenschaften, Denkschrift der Gesamtakademie 48. Contributions to the Chronology of the Eastern Mediterranean 16. Vienna, Verlag der Österreichischen Akadademie der Wissenschaften.

Fleming, S. J., Pigott, V. S., Swann, C. P. and Nash, S. K. (2005) Bronze in Luristan: Preliminary Analytical Evidence from Copper/Bronze Artifacts Excavated by the Belgian Mission in Iran. *Iranica Antiqua* 40, 35–64.

Francfort, H.-P. (1989) *Fouilles de Shortugaï. Recherches sur l'Asie centrale protohistorique.* Mémoires de la Mission Archéologique Française en Asie Centrale 2. Paris, Boccard.

Gadzhiev, M. G., Kohl, P. L., Magomedov, R. G. and Stronach, D. (1997) The 1995 Daghestan-American Velikent Expedition. *Eurasia Antiqua* 3, 181–222.

Garenne-Marot, L. (1984) Le cuivre en Égypte pharaonique: sources et métallurgie. *Paléorient* 10, 97–126.

Garfinkel, Y., Klimscha, F., Shalev, S. and Rosenberg, D. (2014) The Beginning of Metallurgy in the Southern Levant: A Late 6th Millennium CalBC Copper Awl from Tel Tsaf, Israel. *PLOS ONE*. DOI: 10.1371/journal.pone.0092591.

Genz, H. (2013) The Introduction of the Light, Horse-Drawn Chariot and the Role of Archery and the Near East at the Transition from the Middle to the Late Bronze Ages: Is there a Connection? In A. J. Veldmeijer and S. Ikram (eds), *Chasing Chariots. Proceedings of the First International Chariot Conference*, 95–106. Leiden, Sidestone Press.

Georgakopoulou, M. (2013) Metal Artefacts and Metallurgy. In C. Renfrew, O. Philaniotou, N. Brodie, G. Gavalas and M. J. Boyd (eds), *The Settlement at Dhaskalio. The Sanctuary on Keros and the Origins of Aegean Ritual Practice. The Excavations of 2006–2008*, 1, 667–692. Cambridge, McDonald Institute for Archaeological Research.

Gerloff, S. (1993) Zu Fragen mittelmeerländischer Kontakte und absoluter Chronologie der Frühbronzezeit in Mittel- und Westeuropa. *Prähistorische Zeitschrift* 68, 58–102.

Ghirshman, R. (1938) *Fouilles de Sialk, prés de Kashan, 1933, 1934, 1937*, 1. Paris, Geuthner.

Giardino, C., Gigante, G. E. and Ridolfi, S. (2003) Archaeometallurgical Studies. In S. Swiny, G. Rapp and E. Herrscher (eds), *Sotira Kaminoudhia. An Early Bronze Age Site in Cyprus*. American Schools of Oriental Research 8. Cyprus American Archaeological Research Institute, Monograph Series 4, 385–396. Boston, MA, American Schools of Oriental Research.

Glumac, P. D. and Todd, J. A. (1991) Early Metallurgy in Southeast Europe: The Evidence for Production. In P. D. Glumac (ed.), *Recent Trends in Archaeometallurgical Research*. MASCA Research Papers in Science and Archaeology 8/1. Philadelphia, PA, The University Museum of Archaeology and Anthropology and University of Pennsylvania.

Goldman, H. (1956) *Excavations at Gözlü Kule, Tarsu 2. From the Neolithic through the Bronze Age*. Princeton, NJ, Princeton University Press.

Hansen, S., Hauptmann, A., Mötzenbäcker, I. and Pernicka, E. (eds) (2010) *Von Majkop bis Trialeti. Gewinnung und Verbreitung von Metallen und Obsidian in Kaukasien im 4.–2. Jts. v. Chr.* Kolloquien zur Vor- und Frühgeschichte 13. Bonn, Habelt.

Hansen, S., Mirtskhulava, G., Bastert-Lamprichs, K. and Ullrich, M (2012) Aruchlo – eine neolithische Siedlung im Südkaukasus. *Altertum* 57, 81–106.

Hasanova, A. (2014) Influence of technological Anatolian traditions smelting ancient tin bronze on the development metallurgy III millennium BC on the territory of Azerbaidjan [sic]. In M. Kvachadze, M. Puturidze and N. Shanshashvili (eds.), *Problems of Early Metal Age archaeology of Caucasus and Anatoli*. Tbilisi, http://museum.ge/files/G%20Gamyrelidze/PDF/Problems_of_Early_Metal_Age_Archaeology_of_Caucasus_&_Anatolia.pdf (last accessed 13.07.2017)

Hauptmann, A. (2000) *Zur frühen Metallurgie des Kupfers in Fenan/Jordanien*. Der Anschnitt, Beiheft 11. Bochum, Deutsches Bergbau-Museum Bochum.

Hauptmann, A. (2007) Zinn. In *Reallexikon der Germanischen Altertumskunde* 34. 2nd ed., 566–572. Berlin and New York, De Gruyter.

Hauptmann, H. and Pernicka, E. (eds) (2004) *Die Metallindustrie Mesopotamiens von den Anfängen bis zum 2. Jahrtausend v. Chr.* Orient-Archäologie 3. Rahden/Westf., Verlag Marie Leidorf.

Haustein, M. and Pernicka, E. (2008) Die Verfolgung der bronzezeitlichen Zinnquellen Europas durch Zinnisotopie – Eine neue Methode zur Beantwortung einer alten Frage. *Jahresschrift zur Mitteldeutschen Vorgeschichte* 92, 387–415.

Haustein, M., Gillis, C. and Pernicka, E. (2010) Tin isotopy: a new method for solving old questions. *Archaeometry* 52, 816–832.

Helwing, B. (2009) Rethinking the Tin Mountains: Patterns of Usage and Circulation of Tin in Greater Iran from the 4th to the 1st Millennium BC. *TÜBA-AR. Turkish Academy of Sciences Journal of Archaeology* 12, 209–221.

Helwing, B. and Müller, U. (2004) Katalog II: Untersuchte Metallobjekte aus Mesopotamien und Syrien. In H. Hauptmann and E. Pernicka (eds), *Die Metallindustrie Mesopotamiens von den Anfängen bis zum 2. Jahrtausend v. Chr.* Orient-Archäologie 3, 91–103. Rahden/Westf., Verlag Marie Leidorf.

Heskel, D. and Lamberg-Karlovsky, C. C. (1980) An Alternative Sequence for the Development of Metallurgy: Tepe Yahya, Iran. In T. A. Wertime and J. D. Muhly (eds), *The Coming of the Age of Iron*, 229–266. New Haven, CT, Yale University Press.

Hoffman, B. C. and Miller, H. M.-L. (2014) Production and Consumption of Copper-Base Metals in the Indus Civilization. In B. W. Roberts and C. Thornton (eds), *Archaeometallurgy in Global Perspective. Methods and Syntheses*, 697–727. New York, Springer.

Jockenhövel, A. (2004) Von West nach Ost? Zur Genese der Frühbronzezeit Mitteleuropas. In H. Roche, E. Grogan, J. Bradley, J. Coles and B. Raftery (eds), *From Megaliths to Metal. Essays in Honour of George Eogan*, 155–167. Oxford, Oxbow Books.

Jones, A. M., Marchand, J., Sheridan, A. and Straker, V. (2014) Redeemed from the Peat: An Extraordinary Bronze Age Grave on Whitehorse Hill. *British Archaeology* 139, 16–23.

Joukowsky, M. (1986) *Prehistoric Aphrodisias. An Account of the Excavations and Artifact Studies*. Archaeologia Transatlantica 3. Providence, RI and Louvain-la-Neuve, Brown University, Center for Old World Archaeology and Art.

Junghans, S., Sangmeister, E. and Schröder, M. (1968) *Kupfer und Bronze in der frühen Metallzeit Europas* 3. *Katalog der Analysen Nr. 985–10040*. Studien zu den Anfängen der Metallurgie 2. Berlin, Mann.

Junghans, S., Sangmeister, E. and Schröder, M. (1974) *Kupfer und Bronze in der frühen Metallzeit Europas* 4. *Katalog der Analysen Nr. 10041–22000 (mit Nachuntersuchungen der Analysen Nr. 1–10040)*. Studien zu den Anfängen der Metallurgie 2. Berlin, Mann.

Kaniuth, K. (2006) *Metallobjekte der Bronzezeit aus Nordbaktrien*. Archäologie in Iran und Turan 6. Mainz, Philipp von Zabern.

Kaniuth, K. (2007) The Metallurgy of the Late Bronze Age Sapalli Culture (Southern Uzbekistan) and Its Implications for the Tin Question. *Iranica Antiqua* 42, 23–40.

Kaptan, E. (1990) Findings Related to the History of Mining in Turkey. *Bulletin of the Mineral Research and Exploration* 111, 75–84.

Kavtaradze, G. L. (1999) The Importance of Metallurgical Data for the Formation of a Central Transcaucasian Chronology. In A. Hauptmann, E. Pernicka, T. Rehren and Ü. Yalçin (eds), *The Beginnings of Metallurgy. Proceedings of the International Conference, Bochum 1985*. Der Anschnitt, Beiheft 9, 67–101. Bochum, Deutsches Bergbau-Museum.

Kayafa, K., Stos-Gale, S. and Gale, N. (2000) The Circulation of Copper in the Early Bronze Age in Mainland Greece: The Lead Isotope Evidence from Lerna, Lithares and Tsoungiza. In C. F. E. Pare (ed.), *Metals Make the World Go Round. The Supply and Circulation of Metals in Bronze Age Europe. Proceedings of a Conference Held at the University of Birmingham in June 1997*, 39–55. Oxford, Oxbow Books.

Kilian, K. (1982) Ausgrabungen in Tiryns 1980. *Archäologischer Anzeiger*, 393–430.

Kilian-Dirlmeier, I. (2005) *Die bronzezeitlichen Gräber bei Nidri auf Leukas. Ausgrabungen von W. Dörpfeld 1903–1913*. Römisch-Germanisches Zentralmuseum Monographien 62. Mainz, Römisch-Germanisches Zentralmuseum.

Kohl, P. L. (2002) Bronze Production and Utilization in Southeastern Daghestan, Russia: c. 3600–1900 BC. In M. Bartelheim, E. Pernicka and R. Krause (eds), *Die Anfänge der Metallurgie in der Alten Welt. The Beginnings of Metallurgy in the Old World*. Forschungen zur Archäometrie und Altertumswissenschaften 1, 161–183. Rahden/Westf., Verlag Marie Leidorf.

Kohl, P. L. (2003) Integrated Interaction at the Beginning of the Bronze Age: New Evidence from the Northeastern Caucasus and the Advent of Tin Bronzes in the Third Millennium BC. In A. T. Smith and K. S. Rubinson (eds), *Archaeology in the Borderlands. Investigations in Caucasia and beyond*. Cotsen Institute of Archaeology, Monograph 47, 9–21. Los Angeles, CA, Cotsen Institute of Archaeology at UCLA.

Kohl, P. L. (2007) *The Making of Bronze Age Eurasia*. Cambridge, Cambridge University Press.

Korfmann, M. (1987) (ed) *Demircihüyük. Die Ergebnisse der Ausgrabungen 1975–1978, 2: Naturwissenschaftliche Untersuchungen*. Mainz, Philipp von Zabern.

Kouka, O. (2011) Symbolism, Ritual Feasting and Ethnicity in Early Bronze Age Cyprus and Anatolia. In V. Karageoghis and O. Kouka (eds), *On Cooking Pots, Drinking Cups, Loomweights and Ethnicity in Bronze Age Cyprus and Neighbouring Regions. An International Archaeological Symposium Held in Nicosia, 6–7 November 2010*, 43–56. Nicosia, A. G. Leventis Foundation.

Krause, R. (2003) *Studien zur kupfer- und frühbronzezeitlichen Metallurgie zwischen Karpatenbecken und Ostsee*. Vorgeschichtliche Forschungen 24. Rahden/Westf., Verlag Marie Leidorf.

Krüger, J., Nagel, F., Nagel, S., Jantzen, D., Lampe, R., Dräger, J., Lidke, G., Mecking, O., Schüler, T. and Terberger, T. (2012) Bronze Age Tin Rings from the Tollense Valley in Northeastern Germany. *Prähistorische Zeitschrift* 87, 29–43.

Kuruçayırlı, E. and Özbal, H. (2005) New Metal Analysis from Tarsus-Gözlükule. In A. Özyar (ed.), *Field Seasons 2001–2003 of the Tarsus-Gözlükule Interdisciplinary Research Project*, 177–195. Istanbul, Ege Yayınları.

Lahiri, N. (1995) Indian Metal and Metal-Related Artefacts as Cultural Signifiers: An Ethnographic Perspective. *World Archaeology* 27, 116–132.

Lal, B. B., Joshi, P. J., Thapar, B. K. and Bala, M. (2003) *Excavations at Kalibangan. The Early Harappans (1960–1969)*. Memoirs of the Archaeological Survey of India 98. New Delhi, Archaeological Survey of India.

Lamb, W. (1936) *Excavations at Thermi in Lesbos*. Cambridge, Cambridge University Press.

Lamb, W. (1937) Excavations at Kusura Near Afyon Karahisar 2. *Archaeologia* 87, 217–273.

Law, R. W. and Burton, J. H. (2006) Non-Destructive Pb Isotope Analysis of Harappan Lead Artifacts Using Ethylenediaminetetraacetic Acid and ICP-MS. In J. Pérez-Arantegui (ed.), *Proceeding of the 34th International Symposium on Archaeometry, Zaragoza, Spain, 3–7 May 2004*, 181–185. Zaragoza, Sociedad Cooperativa, Librería General.

Lebeau, M. (2006) Nabada (Tell Beydar), an Early Bronze Age City in the Syrian Jezirah. Lecture Presented in Tübingen. http://www.beydar.com/pdf/nabada-conf-en.pdf (last accessed 20 October 2015).

Lehner, M. (2014) The Tin Problem reconsidered: Recent archaeometallurgical research on the Anatolian Plateau. *Backdirt. Annual Review of the Cotsen Institute of Archaeology at UCLA*, 24–35.

Leshtakov, K. (1996) Trade Centres from Early Bronze Age III and Middle Bronze Age in Upper Trace. In L. Nikolova (ed.), *Early Bronze Age Settlements Patterns in the Balkans (ca. 3500–2000 BC, Calibrated Dates)* 2. Reports of Prehistoric Research Projects 1, 239–287. Sofia, Agatho Publ.

Lichter, C. (2006) Varan und Ikiztepe: Überlegungen zu transpontischen Kulturbeziehungen im 5. und 4. Jahrtausend. In A. Erkanal-Öktü, E. Özgen and S. Günel (eds), *Hayat Erkanal'a Armağan, Kültürlerin Yansıması. Studies in Honor of Hayat Erkanal. Cultural Reflections*. Arkeoloji ve eskiçağ tarihi 44, 526–534. Istanbul, Homer Kıtabevi.

Lo Schiavo, F., Giumlia-Mair, A., Sanna, U. and Valera, R. (eds) (2005) *Archaeometallurgy in Sardinia from the Origin to the Early Iron Age*. Monographies Instrumentum 30. Montagnac, Mergoil.

Lull, V., Micó, R., Rihuete Herrada, C. and Risch, R. (2013) Bronze Age Iberia. In H. Fokkens and A. Harding (eds.) *The Oxford Handbook of the European Bronze Age*, 594–615. Oxford, Oxford University Press.

Lutz, J. (2004) Röntgenfluoreszenzanalysen. In H. Hauptmann and E. Pernicka (eds), *Die Metallindustrie Mesopotamiens von den Anfängen bis zum 2. Jahrtausend v. Chr.* Orient-Archäologie 3, 109–149. Rahden/Westf., Verlag Marie Leidorf.

Maddin, R., Muhly, J. D. and Stech, T. (2003) Metallurgical Studies on Copper Artifacts from Bab edh-Dhra. In W. E. Rast and R. T. Schaub (eds), *Bab edh-Dhra: Excavations at the Town Site (1975–1981)*. Reports of the Expedition to Dead Sea Plain, Jordan 2, 513–521. Winona Lake, IN, Eisenbrauns.

Maniatis, Y. and Kromer, B. (1990) Radiocarbon Dating of the Neolithic Early Bronze Age Site of Mandalo, W Macedonia. *Radiocarbon* 32/2, 149–153.

Maran, J. (1998) *Kulturwandel auf dem griechischen Festland und den Kykladen im späten 3. Jahrtausend v. Chr. Studien zu den kulturellen Verhältnissen in Südosteuropa und dem zentralen sowie östlichen Mittelmeerraum in der späten Kupfer- und frühen Bronzezeit*. Universitätsforschungen zur Prähistorischen Archäologie 53. Bonn, Habelt.

Maran, J. (2007) Sea-Borne Contacts between the Aegean, the Balkans and the Central Mediterranean in the 3rd Millennium BC: The Unfolding of the Mediterranean World. In I. Galanaki, H. Thomas, Y. Galanakis and R. Laffineur (eds), *Between the Aegean and Baltic Seas. Prehistory Across Borders. Proceedings of the International Conference: 'Bronze and Early Iron Age Interconnections and Contemporary Developments between the Aegean and the Regions of the Balkan Peninsula, Central and Northern Europe', University of Zagreb, 11–14 April 2005.* Aegaeum 27, 3–21. Liège, Université de Liége, Histoire de l'Art et Archéologie de la Grèce Antique.

Marangou, L., Renfrew, C., Doumas, C. and Gavalas, G. (eds) (2006) *MAPKIANH AMOPΓOY – Markiani, Amorgos. An Early Bronze Age Fortified Settlement. Overview of the 1985–1991 Investigations.* British School at Athens, Supplementary Volume 40. London, British School at Athens.

Marchetti, N. (2000) A Middle Bronze I Ritual Deposit from the `Amuq Plain: Note on the Dating and Significance of the Metal Anthropomorphic Figurines from Tell Judaidah. *Vicino Oriente* 12, 117–132.

Mari, A. (1985) Der Handel zwischen Syrien und Babylonien im achtzehnten Jahrhundert vor Christus. Unpublished Thesis, Julius-Maximilian University of Würzburg.

Marinatos, S. (1929) Πρωτομινωϊκός θολωτός τάφος παρά το χωρίον Κράσι Πεδιάδος. *Archaiologikon Deltion* 12, 102–141.

Marshall, J. (1931) *Mohenjo-Daro and the Indus Civilization: Being an Official Account of the Archaeological Excavations at Mohenjo-Daro Carried Out by the Government of India between the Years 1922 and 1927.* London, Probsthain.

Masioli, E., Artioli, D., Bianchetti, P., Pilato, S. di, Guida, G., Salvatori, S., Sidoti, G. and Vidal, M. (2006) Copper-Melting Crucibles from the Surface of Altyn-Depe, Turkmenistan (*ca.* 2500–2000 BC). *Paléorient* 32/2, 157–174.

Mason, A. H., Powell, W. G., Bankoff, H. A., Mathur, R., Bulatović, A., Filipović, V. and Ruiz, J. (2016) Tin isotope characterization of bronze artifacts of the Central Balkans. *Journal of Archaeological Science* 69, 110–117.

Matuschik, I. (1997) Eine donauländische Axt vom Typ Şiria aus Überlingen am Bodensee. Ein Beitrag zur Kenntnis des frühesten kupferführenden Horizontes im zentralen Nordalpengebiet. *Prähistorische Zeitschrift* 72, 81–105.

McGeehan-Liritzis, V. and Taylor, J. W. (1987) Yugoslavian Tin Deposits and the Early Bronze Age Industries of the Aegean Region. *Oxford Journal Archaeology* 6, 287–300.

Mehofer, M. (2014) Metallurgy during the Chalcolithic and the Beginning of the Early Bronze Age in Western Anatolia. In B. Horejs and M. Mehofer (eds), *Western Anatolia before Troy. Proto-Urbanisation in the 4th Millennium BC? Proceedings of the International Symposium Held at the Kunsthistorisches Museum Wien, Vienna, Austria, 21–24 November 2012.* Oriental and European Archaeology 1, 463–487. Vienna, Austrian Academy of Sciences Press.

Mei, J. (2009) Early Metallurgy in China: Some Challenging Issues in Current Studies. In J. Mej and T. Rehren (eds), *Metallurgy and Civilization: Eurasia and Beyond, Proceedings of the 6th International Conference on the Beginnings of the Use of Metals and Alloys,* 9–16. London, Archetype.

Meliksetian, K., Pernicka, E., Badalyan, R. and Avetisyan, P. (2003) Geochemical Characterisation of Armenian Early Bronze Age Metal Artefacts and Their Relation to Copper Ores. In *Archaeometallurgy in Europe: International Conference, Proceedings, Milan, 24–26 September 2003,* 597–606. Milan, Associazione Italiana Metallurgia.

Meliksetian, K., Kraus, S., Pernicka, E., Avetissyan, P., Devejian, S. and Petrosyan, L. (2011) Metallurgy of Prehistoric Armenia. In Ü. Yalçin (ed.), *Anatolian Metal V.* Der Anschnitt, Beiheft 24, 201–210. Bochum, Deutsches Bergbau-Museum.

Meller, H. (2014) Die neolithischen und bronzezeitlichen Goldfunde Mitteleuropas – eine Übersicht. In H. Meller, R. Risch and E. Pernicka (eds.) *Metalle der Macht – Frühes Gold und Silber. 6. Mitteldeutscher Archäologentag vom 17. bis 19. Oktober 2013 in Halle (Saale).* Tagungen des Landesmuseums für Vorgeschichte Halle 11/II, 611–716. Halle (Saale), Landesamt für Denkmalpflege und Archäologie Sachsen-Anhalt.

Molofsky, L. J. Killick, D., Ducea, M. N., Macovei, M., Chesley, J. T., Ruiz, J., Thibodeau, A. and Popescu, G. C. (2014) A Novel Approach to Lead Isotope Provenance Studies of Tin and Bronze: Applications to South African, Botswanan and Romanian Artifacts. *Journal of Archaeological Science* 50, 440–450.

Montero Fenollós, J.-L. (1999) Metallurgy in the Valley of the Syrian Upper Euphrates during the Early and Middle Bronze Ages. In G. del Olmo Lete and J.-L. Montero Fenollós (eds), *Archaeology of the Upper Syrian Euphrates in the Tishrin Dam Area. Proceedings of the International Symposium Held at Barcelona, January 28–30 1998.* Aula Orientalis, Supplementa 15, 443–465. Sabadell, Editorial Ausa.

Montero Fenollós, J.-L. (2001) Metalistería de la Edad del Bronce en Tell Qara Qūzaq. Estudio tipilógico y tecnológico. In G. del Olmo Lete, J.-L. Montero Fenollós and C. Valdés Pereiro (eds), *Tell Qara Quzaq II. Campanãs IV–VI (1992–1994),* 255–303. Barcelona, Editorial Ausa.

Müller-Karpe, M. (1989) Neue Forschungen zur frühen Metallverarbeitung in Mesopotamien. *Jahrbuch des Römisch-Germanischen Zentralmuseums* 36, 179–192.

Müller-Karpe, M. (2004) Katalog I: Untersuchte Metallobjekte aus Mesopotamien. In H. Hauptmann and E. Pernicka (eds), *Die Metallindustrie Mesopotamiens von den Anfängen bis zum 2. Jahrtausend v. Chr.* Orient-Archäologie 3, 1–89. Rahden/Westf., Verlag Marie Leidorf.

Muhly, J. D. (1985) Sources of Tin and the Beginnings of Bronze Metallurgy. *American Journal of Archaeology* 9, 275–291.

Muhly, J. D. (1993) Early Bronze Age Tin and the Taurus. *American Journal of Archaeology* 97, 239–253.

Muhly, J. D. (2006) Chrysokamino in the History of Early Metallurgy. In P. P. Betancourt (ed.), *The Chrysokamino Metallurgy Workshop and Its Territory.* Hesperia, Supplement 36, 155–177. Athens, American School of Classical Studies at Athens.

Mylonas, G. (1959) *Aghios Kosmas. An Early Bronze Age Settlement and Cemetery in Attica.* Princeton, NJ, Princeton University Press.

Needham, S. (2005) Transforming Beaker Culture in North-West Europa: Processes of Fusion and Fission. *Proceedings of the Prehistoric Society* 71, 171–217.

Needham, S. P., Leese, M. N., Hook, D. R. and Hughes, M. J. (1989) Developments in the Early Bronze Age Metallurgy of Southern Britain. *World Archaeology* 20, 383–402.

Nessel, B., Brügmann, G. and Pernicka, E. (2014) Discovering Bronze Age Tin: The Potential of Tin Isotopes. In J. M. Mata Perelló (ed.) *XV Congreso Internacional sobre Patrimonio Geológico y Minero. XIC Sesión Científica de la Sedpgym*, 13–14. Logrosán: Universidad Politécnica de Madrid.

Nezafati, N., Pernicka, E. and Momenzadeh, M. (2006) Ancient Tin: Old Question and a New Answer. *Antiquity* 80 (Project Gallery). http://www.antiquity.ac.uk/projgall/nezafati308/ (last accessed 26 October 2015).

Nezafati, N., Pernicka, E. and Momenzadeh, M. (2011) Early Tin-Copper Ore from Iran, a Possible Clue for the Enigma of Bronze Age Tin. In Ü. Yalçin (ed.), *Anatolian Metal V.* Der Anschnitt, Beiheft 24, 211–230. Bochum, Deutsches Bergbau-Museum.

Novotný, B. (1958) Hroby kultury zvoncovitých pohárů u Smolína na Moravě/Die Gräber der Glockenbecherkultur in Smolín (Mähren). *Památky Archeologické* 49, 297–311.

Oates, D., Oates, J. and McDonald, H. (2001*) Excavations at Tell Brak. Vol.2: Nagar in the third millennium BC.* Cambridge, McDonald Institute for Archaeological Research.

Ogden, J. (2000) Metals. In P. T. Nicholson and I. Shaw (eds.) *Ancient Egyptian Materials and Technology*, 148–176. Cambridge, Cambridge University Press.

del Olmo Lete, G., Montero Fenollós, J.-L. and Valdés Pereiro, C. (eds.) (2001) *Tell Qara Quzaq II. Campanãs IV–VI (1992–1994).* Barcelona, Editorial Ausa.

Özbal, H., Kuruçayırlı, E. and Mısırlı, Z. (2005) Two EB II Metal Stamp Seals from Tarsus-Gözlükule. In A. Özyar (ed.), *Field Seasons 2001–2003 of the Tarsus-Gözlükule Interdisciplinary Research Project*, 197–206. Istanbul, Ege Yayınları.

Ottaway, B. (1979) Analysis of the Earliest Metal Finds from Gomolava. *Otizak iz rada Vojvodanski Muzeja* 25 (Novisad), 53–59.

Palmieri, A. M. and di Nocera, G. M. (2004) Early Bronze Copper Circulation and Technology in Middle Euphrates Regions. In C. Nicolle (ed.), *Nomades et sédentaires dans le Proche-Orient ancien. XLVIe Rencontre Assyriologique Internationale. Paris, 10–13 juillet 2000.* Amurru 3, 253–265. Paris, Ed. Recherche sur les Civilisations.

Papaefthymiou-Papanthimou, A. and Pilali-Papasteriou, A. (1997) Οι προϊστορικοί οικισμοί στο Μάνδαλο και Αρχοντικό Πέλλας. *To Archaeologiko Ergo sti Makedonia kai ti Thraki* 10, 143–158.

Pare, C. F. E. (ed.) (2000a) *Metals Make the World Go Round. The Supply and Circulation of Metals in Bronze Age Europe. Proceedings of a Conference Held at the University of Birmingham in June 1997.* Oxford, Oxbow Books.

Pare, C. F. E. (2000b) Bronze and the Bronze Age. In C. F. E. Pare (ed.), *Metals Make the World Go Round. The Supply and Circulation of Metals in Bronze Age Europe. Proceedings of a Conference Held at the University of Birmingham in June 1997*, 1–38. Oxford, Oxbow Books.

Parzinger, H. (2003) Zinn in der Bronzezeit Europas. In H. Parzinger and N. Boroffka (eds), *Das Zinn der Bronzezeit in Mittelasien 1. Die siedlungsarchäologischen Forschungen im Umfeld der Zinnlagerstätten.* Archäologie in Iran und Turan 5, 287–296. Mainz, Philipp von Zabern.

Parzinger, H. and Boroffka, N. (2003) *Das Zinn der Bronzezeit in Mittelasien 1. Die siedlungsarchäologischen Forschungen im Umfeld der Zinnlagerstätten.* Archäologie in Iran und Turan 5. Mainz, Philipp von Zabern.

Peltenburg, E. (2007) East Mediterranean Interactions in the 3rd Millennium BC. In S. Antoniadou and A. Pace (eds), *Mediterranean Crossroads*, 141–161. Athens, Pierides Foundation.

Pernicka, E. (1998) Die Ausbreitung der Zinnbronze im 3. Jahrtausend. In B. Hänsel (ed.), *Mensch und Umwelt in der Bronzezeit Europas. Abschlußtagung der Kampagne des Europarates: Die Bronzezeit: Das Erste Goldene Zeitalter Europas, an der Freien Universität Berlin, 17.–19. März 1997. Beiträge und Ergebnisse*, 135–147. Kiel, Oetker-Voges.

Pernicka, E. (2000) Zusammenfassung der früh- und mittelbronzezeitlichen Metallfunde aus der Nekropole von Demircihüyük-Sarıket. In J. Seeher (ed.), *Die bronzezeitliche Nekropole von Demircihüyük-Sarıket.* Istanbuler Forschungen 44, 232–237. Tübingen, Wasmuth.

Pernicka, E. (2001) Metalle machen Epoche. Bronze, Eisen und Silber. In *Troia. Traum und Wirklichkeit. Ausstellung Stuttgart-Braunschweig-Bonn 17.3.2001–17.2.2002*, 369–372. Stuttgart, Theiss.

Pernicka, E. and Schleiter, M. (1997) Untersuchung der Metallproben vom Tell el-Fara'în (Buto). In T. von der Way (ed.), *Tell el-Fara'în – Buto* 1. *Ergebnisse zum frühen Kontext. Die Kampagnen der Jahre 1983–1989.* Archäologische Veröffentlichungen Deutsches Archäologisches Institut Kairo 83, 219–223. Mainz, Philipp von Zabern.

Pernicka, E., Seelinger, T. C., Wagner, G. A., Wagner, G. A., Begemann, F., Schmitt-Strecker, S., Eibner, C., Öztunali, Ö. and Baranyi, I. (1984) Archäometallurgische Untersuchungen in Nordwestanatolien. *Jahrbuch des Römisch-Germanischen Zentralmuseums* 31, 533–599.

Pernicka, E., Begemann, F., Schmitt-Strecker, S. and Grimanis, A. P. (1990) On the Composition and Provenance of Metal Artefacts from Poliochni on Lemnos. *Oxford Journal of Archaeology* 9, 263–298.

Pernicka, E., Begemann, F., Schmitt-Strecker, S. and Wagner, G. A. (1993) Eneolithic and Early Bronze Age Copper Artefacts from the Balkans and Their Relation to Serbian Copper Ores. *Prähistorische Zeitschrift* 68, 1–54.

Pernicka, P., Eibner, C., Öztunalı, Ö. and Wagner, G. A. (2003) Early Bronze Age Metallurgy in the North-East Aegean. In G. A. Wagner, E. Pernicka and H.-P. Uerpmann (eds), *Troia and the Troad*, 143–172. Berlin, Springer.

Peterson, D. L. (2003) Ancient Metallurgy in the Mountain Kingdom: The Technology and Value of Early Bronze Age Metalwork from Velikent, Dagestan. In A. T. Smith and K. S. Rubinson (eds), *Archaeology in the Borderlands. Investigations in Caucasia and beyond.* Cotsen Institute of Archaeology, Monograph 47, 22–37. Los Angeles, CA, Cotsen Institute of Archaeology at UCLA.

Petrie, W. M. F. (1926) *Glass Stamps and Weights. Ancient Weights and Measures Illustrated by the Egyptian Collection in University College, London.* London, Department of Egyptology, University College.

Peyronel, L. (2012) Resources Exploitation and Handicraft Activities in at Tell Mardikh/Ebla (Syria) during the Early and Middle Bronze Ages. In R. Matthews and J. Curtis (eds), *Proceedings of the Seventh International Congress on the Archaeology of the Ancient Near East, 12–16 April, the British Museum and UCL, London 2. Ancient and Modern Issues in*

Cultural Heritage. Colour and Light in Architecture, Art and Material Culture. Islamic Archaeology, 475–496. Wiesbaden, Harrassowitz.

Pigott, V. C. (1989) Archaeo-Metallurgical Investigations at Bronze Age Tappeh Hesar, 1976. In R. H. Dyson and S. M. Howard (eds), *Tappeh Hesār. Reports of the Restudy Project 1976*. Monografie di Mesopotamia 2, 28–33. Florence, Casa ED. Le Lettere.

Pigott, V. C. (2003) Archaeometallurgical investigations at Malyan. The evidence for tin-bronze in the Kaftari Phase. In N. F. Miller and K. Abdi (Hrsg.), *Yeki bud, yeki nabud. Essays on the archaeology of Iran in honor of William M. Sumner.* Cotsen Institute of Archaeology, Monograph 48, 161–176. Los Angeles, CA Cotsen Institute of Archaeology at UCLA.

Pigott, V. C. (2011) Sources of Tin and the Tin Trade in Southwest Asia: Recent Research and Its Relevance to Current Understanding. In P. P. Betancourt and S. Ferrence (eds), *Metallurgy: Understanding How, Learning Why. Studies in Honor of James D. Muhly.* Prehistory Monographs 29, 273–291. Philadelphia, PA, INSTAP Academic Press.

Potts, T. (1994) *Mesopotamia and the East: An Archaeological and Historical Study of Foreign Relations, ca. 3400–2000 BC.* Oxford, Oxbow Books.

Powell, M. A. (1978) A Contribution to the History of Money in Mesopotamia Prior to the Invention of Coinage. In B. Hruška, and G. Komoróczy (eds), *Festschrift Lubor Matouš* 2, 211–243. Budapest, ELTE.

Prag, K. (1974) The Intermediate Early Bronze–Middle Bronze Age: An Interpretation of the Evidence from Transjordan, Syria and Lebanon. *Levant* 6, 69–116.

Primas, M. (1996) *Velika Gruda 1. Hügelgräber des frühen 3. Jahrtausends v. Chr. im Adriagebiet – Velika Gruda, Mala Gruda und ihr Kontext.* Universitätsforschungen zur Prähistorischen Archäologie 32. Bonn, Habelt.

Primas, M. (2002) Early Tin Bronze in Central and Southern Europe. In M. Bartelheim, E. Pernicka and R. Krause (eds), *Die Anfänge der Metallurgie in der Alten Welt.* Forschungen zur Archäometrie und Altertumswissenschaften 1, 303–314. Rahden/Westf., Verlag Marie Leidorf.

Radivojević, M., Rehren, T., Kuzmanović-Cvetković, J., Jovanović, M. J. and Northover, P. (2013) Tainted Ores and the Rise of Tin Bronzes in Eurasia, c. 6500 Years Ago. *Antiquity* 87, 1030–1045.

Radivojević, M., Rehren, T., Kuzmanović-Cvetković, J. and Jovanović, M. J. (2014) Context is Everything Indeed: A Response to Šljivar and Borić. *Antiquity* 88, 1315–1319.

Rahmstorf, L. (2010) Die Nutzung von Booten und Schiffe in der bronzezeitlichen Ägäis und die Fernkontakte der Frühbronzezeit. In H. Meller and F. Bertemes (eds), *Der Griff nach den Sternen. Wie Europas Eliten zu Macht und Reichtum kamen. Internationales Symposium in Halle (Saale) 16.–21. Februar 2005.* Tagungen des Landesmuseums für Vorgeschichte Halle 5, 675–697. Halle, Landesamt für Denkmalpflege und Archäologie Sachsen-Anhalt und Landesmuseum für Vorgeschichte.

Rahmstorf, L. (2011a) Re-Integrating 'Diffusion': The Spread of Innovations among the Neolithic and Bronze Age Societies of Europe and the Near East. In T. C. Wilkinson, S. Sherratt and J. Bennet (eds), *Interweaving Worlds: Systemic Interactions in Eurasia, 7th to 1st Millennia BC. Papers from a Conference in Memory of Professor Andrew Sherratt*, 100–119. Oxford, Oxbow Books.

Rahmstorf, L. (2011b) Maß für Maß. Indikatoren für Kulturkontakte im 3. Jahrtausend. In Badisches Landesmuseum Karlsruhe (ed.), *Kykladen. Lebenswelt einer frühgriechischen Kultur. Ausstellung im Badischen Landesmuseum Karlsruhe 17.12.2011–22.04.2012*, 144–153. Karlsruhe, Badisches Landesmuseum.

Rahmstorf, L. (2015) The Aegean before and after *ca.* 2200 BC between Europe and Asia: Trade as a Prime Mover of Cultural Change. In H. Meller, H. W. Arz, R. Jung and R. Risch (eds), *2200 BC. Ein Klimasturz als Ursache für den Zerfall der Alten Welt? 7. Mitteldeutscher Archäologentag vom 23.–26. Oktober 2014 in Halle (Saale).* Tagungen des Landesmuseums für Vorgeschichte Halle 12/1, 149–180. Halle, Landesamt für Denkmalpflege und Archäologie Sachsen-Anhalt.

Rahmstorf, L. (2016) Emerging Economic Complexity in the Aegean and Western Anatolia During the Earlier Third Millennium BC. In B. Molloy (ed.), *Of Odysseys and Oddities: Scales and Modes of Interaction Between Prehistoric Aegean Societies and Their Neighbors.* Sheffield Studies in Aegean Archaeology 10, 225–276. Oxford and Philadephia, PA, Oxbow Books.

Rambach, J. (2000) *Kykladen 1. Die frühe Bronzezeit. Grab- und Siedlungsbefunde.* Beiträge zur ur- und frühgeschichtlichen Archäologie des Mittelmeerraumes 33. Bonn, Habelt.

Rao, S. R. (1962–63) Excavation at Rangpur and other Explorations in Gujarat. *Ancient India* 18–19, 5–207.

Rassmann, K. (2004) Chronologie – Die zeitliche Ordnung der Dinge als Schlüssel zur Geschichte der Bronzezeit. In *Mythos und Magie. Archäologische Schätze der Bronzezeit aus Mecklenburg-Vorpommern. Ausstellungskatalog Schwerin.* Archäologie in Mecklenburg-Vorpommern 3, 38–49. Schwerin, Archäologisches Landesmuseum and Landesamt für Bodendenkmalpflege Mecklenburg-Vorpommern.

Rassmann, K. (2010) Die frühbronzezeitlichen Stabdolche Ostmitteleuropas – Anmerkungen zu Chronologie, Typologie, Technik und Archäometallurgie. In: H. Meller and F. Bertemes (eds), *Der Griff nach den Sternen. Wie Europas Eliten zu Macht und Reichtum kamen. Internationales Symposium in Halle (Saale) 16.–21. Februar 2005.* Tagungen des Landesmuseums für Vorgeschichte Halle 5, 807–821. Halle, Landesamt für Denkmalpflege und Archäologie Sachsen-Anhalt and Landesmuseum für Vorgeschichte.

Reeves, L. C. (2003) Aegean and Anatolian Bronze Age Metal Vessels: A Social Perspective, 2 vols. Doctoral thesis, University of London. http://discovery.ucl.ac.uk/1383530/ (last accessed 13.07.2017)

Reiter, K. (1997) *Die Metalle im Alten Orient.* Alter Orient und Altes Testament 249. Münster, Ugarit-Verlag.

Renfrew, C. (1967) Cycladic Metallurgy and the Aegean Early Bronze Age. *American Journal of Archaeology* 71, 2–26.

Renfrew, C. and Slater, E. A. (2003) Metal Artifacts and Metallurgy, with Appendix 'The Finds: Metal Objects'. In E. S. Elster and C. Renfrew (eds), *Prehistoric Sitagroi: Excavations in Northeast Greece, 1968–1970*, 2: *The Final Report.* Monumenta Archaeologica 20, 301–322. Los Angeles, CA, Cotsen Institute of Archaeology, University of California.

Roberts, B., Thornton, C. P. and Pigott, V. C. (2009) Development of Metallurgy in Eurasia. *Antiquity* 83, 1012–1022.

Ruzanov, V. (1999) Zum frühen Auftreten der Zinnbronze in Mittelasien. In A. Hauptmann, E. Pernicka, T. Rehren and Ü. Yalçin (eds), *The Beginnings of Metallurgy. Proceedings of the International Conference 'The Beginnings of Metallurgy', Bochum 1995.* Der Anschnitt, Beiheft 9, 103–105. Bochum, Deutsches Bergbau-Museum.

Şahoğlu, V. (2005) The Anatolian Trade Network and the Izmir Region during the Early Bronze Age. *Oxford Journal of Archaeology* 24, 339–361.

Salvatori, S., Vidale, M., Guida, G. and Gigante, G. (2002) A Glimpse on Copper and Lead Metalworking at Altyn-Depe (Turkmenistan) in the 3rd Millennium BC. *Ancient Civilizations from Scythia to Siberia* 8/1–2, 69–101.

Sampson, A. (1985) Μάνικα. Μία προτοελλαδική πόλη στη Χαλκίδα 1. Manika 1. An Early Helladic Town in Chalcis. Athens, Society for Euboean Studies.

Schickler, H. (1981) 'Neolithische' Zinnbronzen. In H. Lorenz (ed.), *Studien zur Bronzezeit. Festschrift für Wilhelm Albert v. Brunn*, 419–445. Mainz, Philipp von Zabern.

Schillinger, A. (1997) Die früheste Zinnbronze im Schwarzmeergebiet. Unpublished thesis, University of Tübingen.

Schlette, F. (1948) Die neuen Funde der Glockenbecherkultur im Lande Sachsen-Anhalt. In K. Schwarz (ed.), *Strena Praehistorica. Festgabe zum 60. Geburtstag von Martin Jahn*, 29–77. Halle, Niemeyer.

Schmidt, C. (2005) Überregionale Austauschsysteme und Fernhandelswaren in der Ur III-Zeit. *Bagdader Mitteilungen*, 7–135.

Schmidt, R. R. (1945) *Die Burg Vučedol.* Zagreb, Arheološki muzej u Zagrebu.

Schoop, U.-D. (2011) Çamlıbel Tarlası, ein metallverarbeitender Fundplatz des vierten Jahrtausends v. Chr. im nördlichen Zentralanatolien. In Ü. Yalçin (ed.), *Anatolian Metal V.* Der Anschnitt, Beiheft 24, 53–68. Bochum, Deutsches Bergbau-Museum.

Šljivar, D. and Borić, D. (2014) Context is Everything: Comments on Radivojević *et al.* 2013. *Antiquity* 88, 1310–1315.

Speiser, E. A. (1935) *Excavations at Tepe Gawra 1: Levels I–VIII.* Publications of the Baghdad School, Excavations 1. Philadelphia, PA, University of Pennsylvania Press.

Spindler, K. (1971) Zur Herstellung der Zinnbronze in der frühen Metallurgie Europas. *Acta Praehistoria et Archaeologia* 2, 199–253.

Spindler, K. (1981) *Cova da Moura. Die Besiedlung des Atlantischen Küstengebietes Mittelportugals vom Neolithikum bis an das Ende der Bronzezeit.* Madrider Beiträge 7. Mainz, Philipp von Zabern.

Srivastava, V. C. (1998) Bronze Tools and Technology in Protohistoric Afghanistan: A Review. In V. Tripathi (ed.), *Archaeometallurgy in India*, 176–183. Delhi, Sharada Publishing House.

Stech, T. and Pigott, V. C. (1986) The Metal Trade in Southwest Asia in the Third Millennium B.C. *Iraq* 48, 39–64.

Steinkeller, P. (2016) The Role of Iran in the Inter-Regional Exchange of Metals: Tin, Copper, Silver and Gold in the Second Half of the Third Millennium BC. In K. Maekawa (ed.), *Ancient Iran New Perspectives from Archaeology and Cuneiform Studies. Proceedings of the International Colloquium held at the Center for Eurasian Cultural Studies, Kyoto University, December 6–7, 2014.* Ancient Text Studies in the National Museum, Vol. 2, 127–150. Kyoto, Nakanishi Printing Company.

Stöllner, T., Samaschev, Z., Berdenov, S., Cierny, J., Doll, M., Garner, J., Gontsharov, A., Gorelik, A., Hauptmann, A., Herd, R., Kusch, G. A., Merz, V., Riese, T., Sikorski, B. and Zickgraf, B. (2011) Tin from Kazakhstan – Steppe Tin for the West? In Ü. Yalçin (ed.), *Anatolian Metal V.* Der Anschnitt, Beiheft 24, 231–251. Bochum, Deutsches Bergbau-Museum.

Swiny, S. (2003) The Metal. In S. Swiny, G. Rapp and E. Herrscher (ed.), *Sotira Kaminoudhia. An Early Bronze Age Site in Cyprus.* American Schools of Oriental Research 8. Cyprus American Archaeological Research Institute, Monograph Series 4, 369–384. Boston, MA, American Schools of Oriental Research.

Swiny, S., Rapp, G. and Herrscher, E. (ed.) (2003) *Sotira Kaminoudhia. An Early Bronze Age Site in Cyprus.* American Schools of Oriental Research 8. Cyprus American Archaeological Research Institute, Monograph Series 4. Boston, American Schools of Oriental Research.

Thornton, C. P. (2007) Of Brass and Bronze in Prehistoric Southwest Asia. In S. La Niece, D. Hook and P. Craddock (eds), *Metals and Mines. Studies in Archaeometallurgy. Selected Papers from the Conference 'Metallurgy: A Touchstone for Cross-Cultural Interaction' Held at the British Museum 28–30 April 2005 to Celebrate the Career of Paul Craddock during his 40 Years at the British Museum*, 123–135. London, Archetype.

Tonussi, M. (2008) Metal workshops and metal finds from third millennium Tell Beydar/Nabada (1992–2005 seasons). In M. Lebeau and A. Suleiman, *Beydar Studies 1.* Subartu 21, 195–257. Turnhout, Brepols.

Tsountas, C. (1898) Κυκλαδικά 1. *Archaiologike Ephemeris*, 137–212.

Tubb, J. N. (1990) *Excavations at the Early Bronze Age Cemetery of Twal Esh-Sharqi.* London, British Museum Publications.

Tzavella-Evjen, H. (1984) *Λιθαρές.* Δημοσιεύματα του Αρχαιολογικού Δελτίου 32. Athens, T.A.P.A.

von der Osten, H. H. (1937) *The Alishar Hüyük. Seasons of 1930–32, 2.* The University of Chicago Oriental Institute Publications 29. Researches in Anatolia 8. Chicago, IL, University of Chicago Press.

Vandkilde, H. (2016) Bronzization: The Bronze Age as Pre-Modern Globalization. *Prähistorische Zeitschrift* 91, 103–123.

Waetzoldt, H. and Bachmann, H. G. (1984) Zinn- und Arsenbronzen in den Texten aus Ebla und aus dem Mesopotamien des 3. Jahrtausends (mit einem Beitrag von E. Pernicka). *Oriens Antiquus* 23, 1–18.

Wang, Q., Strekopytov, S., Roberts, B. W., Wilkin, N. (2016), Tin ingots from a probable Bronze Age shipwreck off the coast of Salcombe, Devon: Composition and microstructure. *Journal of Archaeological Science* 67, 80–92.

Warner, J. L. (1994) *Elmali-Karataş. The Early Bronze Age Village of Karataş.* Bryn Mawr, Bryn Mawr College.

Webb, J. M., Frankel, D., Stos, Z. A. and Gale, N. (2006) Early Bronze Age Metal Trade in the Eastern Mediterranean. New Compositional and Lead Isotope Evidence from Cyprus. *Oxford Journal of Archaeology* 25/3, 261–288.

Webb, J. and Frankel, D. (2007) Identifying Population Movements by Everyday Practice: The Case of 3rd Millennium Cyprus. In S. Antoniadou and A. Pace (eds), *Mediterranean Crossroads*, 189–216. Athens, Pierides Foundation.

Weeks, L. R. (2003) *Early Metallurgy of the Persian Gulf. Technology, Trade, and the Bronze Age World.* Boston, MA, Leiden.

Weigall, A. E. P. (1908) *Catalogue général des antiquités égyptiennes du Musée du Caire. No. 31271–31670. Weights and balances.* Cairo, Impr. de l'Institut Français d'Archéologie Orientale.

Wertime, T. A. (1978) Tin and the Egyptian Bronze Age. In D. Schmandt-Besserat (ed.), *Immortal Egypt. Invited lectures on the Middle East at the University of Texas at Austin*, 37–42. Malibu, Undena Press.

Wilkinson, T. C. (2014) *Tying the Threads of Eurasia. Trans-Regional Routes and Material Flows in Transcaucasia, Eastern Anatolia and Western Central Asia, c. 3000–1500 BC.* Leiden, Sidestone Press.

Yalçın, Ü. and Yalçın, H. G. (2008) Der Hortfund von Tülintepe, Ostanatolien. In: Ü. Yalçın (ed.), *Anatolian Metal IV.* Der Anschnitt, Beiheft 21, 101–123. Bochum, Deutsches Bergbau-Museum.

Yakar, J. (1984) Regional and Local Schools of Metalwork in Early Bronze Age Anatolia 1. *Anatolian Studies* 34, 59–86.

Yamazaki, E. Nakai, S., Sahoo, Y., Yokoyama, T., Mifune, H., Saito, T., Chen, J., Takagi, N. Hokanishi, N. and Yasuda, A. (2013) Feasibility studies of Sn isotope composition for provenancing ancient bronzes. *Journal of Archaeological Science* 52, 458–467.

Yener, K. A. (2000) *The Domestication of Metals. The Rise of Complex Metal Industries in Anatolia.* Culture and History of the Ancient Near East 4. Leiden, Brill.

Yener, K. A. (2009) Strategic Industries and Tin in the Ancient Near East: Anatolia Updated. *TÜBA-AR* 12, 143–154.

Yener, K. A., Özbal, H., Kaptan, E., Pehlivan, A. N. and Goodway, M. (1989) Kestel: An Early Bronze Age Source of Tin Ore in the Taurus Mountains. *Science* 244, 200–203.

Yener, K. A., Kulakoğlu, F., Yazgan, E., Kontani, R., Hayakawa, Y. S., Lehner, J. W., Dardeniz, G., Öztürk, G., Johnson, M., Kaptan, E., and Hacar, A. (2015) New Tin Mines and Production Sites near Kültepe in Turkey: A Third-Millennium BC Highland Production Model. *Antiquity* 89, 596–612.

Zachos, K. (2007) The Neolithic Background: A Reassessment. In P. M. Day and R. C. P. Doonan (eds) *Metallurgy in the Early Bronze Age Aegean.* Sheffield Studies in Aegean Archeology 7, 168–206. Oxford, Oxbow Books.

Zich, B. (1996) *Studien zur regionalen und chronologischen Gliederung der nördlichen Aunjetitzer Kultur.* Vorgeschichtliche Forschungen 20. Berlin and New York, de Guyter.

Zimmermann, T. (2007) Anatolia and the Balkans, Once Again – Ring-Shaped Idols from Western Asia and a Critical Reassessment of Some 'Early Bronze Age' Items from İkiztepe, Turkey. *Oxford Journal of Archaeology* 26, 25–33.

Chapter 17

Appropriation of Tin-Bronze Technology: A Regional Study of the History of Metallurgy in Early Bronze Age Southern Mesopotamia

Ulrike Wischnewski

Introduction

This paper presents the results of an ongoing research project designed to achieve greater insight into the entanglement of society and technology by collecting and analysing archaeological and textual evidence concerning tin-bronze casting in southern Mesopotamia.

In the middle of the 3rd millennium BC, tin-bronze casting became common in southern Mesopotamia. The adoption and integration of this innovation into the society of the day is reflected both by an increasing number of tin-bronzes in burials and by other depositions. Besides the archaeological sources, contemporary administrative, cultic and literary texts all underline the social impact of the new technology. Appropriation of innovation always needs some kind of transfer of knowledge. It calls for contact with other regions, with craftsmen and traders or with objects themselves. But how was it possible to arouse interest in a new substance and a new technology in a region such as Mesopotamia, with no metal resources at its disposal? Even if the reason for appropriating an innovation is not traceable, there are clearly many factors involved and the intermesh between technology, material culture and society is one of them. Accordingly, both Pfaffenberger's (1992) approach to the interlock between new techniques, material culture and the socio-technological system and the Latourian *Actor-Network Theory* (ANT) (Latour 1991; Akrich and Latour 1992) are reliable methodological resources in this research. Latour states that not only people, but also objects have to be seen as actors. Whether it was the greater durability, the superior casting properties or the golden shimmer of the material that Mesopotamian societies found attractive, it was surely contact with an object made of tin-bronze that aroused interest in the innovation that tin-bronze casting represented. After awakening interest in a new substance, the technology has to be adjusted to the given regional conditions. What kind of fuel is available and how much? What kind of kiln will obtain the best results and save the most fuel? What resources can be used for moulds, crucibles and bellows? Were there existing crafts that could help to develop the most suitable tin-bronze casting technology for southern Mesopotamia? Via cross-craft interaction, pottery manufacture as an established pyrotechnical craft may have served as a source of know-how (Brysbaert 2011). Potters may have supplied tools for casting, like moulds or crucibles, they may also have imparted information on heating and the control of high temperatures (Miller 2007, 240). But tin-bronze did not in fact introduce the technology of metalworking to Mesopotamia. At least part of the technology was already known, as copper has been established as figuring there previously. Tin-bronze casting could therefore be integrated into the existing *chaîne opératoire* (Leroi-Gourhan 1943; 1945) of copper manufacture. The experience gathered in the course of alloying copper and arsenic could be translated and transformed to develop the new alloy. Before tin-bronze was introduced, arsenic-bronze and ceramics or stone had been used to make tools, weapons and vessels. Here, the question arises whether traditional and local forms of these articles were still produced with the new substance, or whether the new substance introduced new forms as well.

After, or even during, the appropriation and integration of an innovation, it will surely have some kind of impact on society. Whether and to what degree tin-bronze casting exerted an influence on social change is one of the questions that I am currently investigating, not in this paper, but in the broader context of my research, where tin-bronze casting is not only regarded as a technological but also (potentially) as a social innovation. This involves extending one's purview to the entire historical, social and geographical background. What were the social and geographical conditions obtaining in Early Bronze Age (EBA) Mesopotamia? Which resources were locally available and which were not? Metallurgical analyses undertaken and published *e.g.* by Hauptmann and Pernicka (2004) or Moorey and Schweizer (1972) are taken into account for the technological information they offer.

Geographical and historical background

Early Bronze Age[1] (EBA, *c.* 2900–2000 BC) Mesopotamia is characterised by a series of urbanisation processes. During the so-called 'Early Dynastic Period' (ED, *c.* 2900–2350 BC), many different competing city states had a formative influence on the political, social and economic landscape in southern Mesopotamia. The influence of each city state was defined or even limited by its economic wealth. Each local ruler tried to exercise control over resources like land, craftsmen and importation (Kuhrt 1995, 26). A very sophisticated system of administration was established, placed entirely in the hands of the palace and the temple. These elite-run households were in charge of trade, the storage of goods and local production and agriculture. As southern Mesopotamia had very limited resources – notably barley agriculture, cattle for milk and leather and sheep providing wool for textiles – an elaborate system of irrigation canals and long-distance trade routes were essential if the local elites were to retain control and their economic power (see *e.g.* Kuhrt 1995; Gülçur 2002, 29–30.) Accordingly, most conflicts between the city states centred around water and the irrigation of the fields surrounding the cities.

During the Akkadian period (*c.* 2350–2200 BC), military rivalry actually increased, spurred on as it was by the desire to conquer foreign territory. In contrast to their predecessors, the kings of Agade ruled over a united empire with only one capital and royal seat – Agade. They undertook military campaigns against the surrounding lands, sometimes going even further afield to expand their power. Control over trade, local resources and newly conquered foreign regions was exercised via Agade (see *e.g.* Kuhrt 1995; Yoffee 1995; Edzard 2004).

Agricultural productivity, the formation of the first city states and the establishment of a very complex administrative system were fruitful soil for the emergence of a high level of craft specialisation. Surplus production and storage led to a degree of continuity that guaranteed the subsistence of craftsmen working in the workshops, as they and their

families were paid in rations of barley, wool or textiles, or else with land itself (Yoffee 1995, 1398).

The unique situation southern Mesopotamia found itself in was partly bound up with its geographical location (Fig. 17.1), slotted in between different regions richly endowed with much-needed resources like metals, stone and cedar-wood. Ease of access to overland and overseas trade via the Euphrates, the Tigris and the Persian Gulf facilitated the importation of these goods. The major trade asset was textiles produced in skilled workshops. In exchange for fine garments, metals and other luxury goods like lapis lazuli, building stone and cedar-wood found their way to Sumer. All these foreign products required skilled craftsmen either educated for the purpose or arriving with the resources or deported from foreign regions. This was the time when the specialised manufacture of copper, tin-bronze and arsenic-bronze weapons, tools and vessels first reared its head in southern Mesopotamian society. Texts stored in archives help to create a larger picture of the formative effect tin-bronze technology will have had on EBA Mesopotamia.

Metals in Mesopotamia

The knowledge of copper or arsenic-bronze manufacture was already widespread in southern Mesopotamia in the 4th millennium BC, when metal became the most important substance for a variety of artefacts in the material culture of Mesopotamia. Figurines, vessels – either for ritual purposes or everyday usage – weapons, tools, jewellery and cultic or royal symbols were made of metal (Müller-Karpe 2002, 137). The first objects fashioned in native copper were manufactured by hammering and subsequent annealing at high temperatures. During this heating process, metal craftsmen may have discovered the melting effect high temperature has on metal. Smelting technology developed (De Ryck *et al.* 2005, 261) and paved the ground for casting technology, brought to perfection in the second half of the 3rd millennium BC. A statue of Lahmu (found near Bassetki in northern Iraq) made of pure copper is discussed and described in detail by Müller-Karpe (2002). It dates to the reign of Naramsîn of Agade (*c.* 23rd century BC) and shows the high degree of perfection casting technology had reached in EBA Mesopotamia.

Arsenic alloys with up to 5% arsenic concentrations became quite common around the turn of the 4th to the 3rd millennium BC (De Ryck *et al.* 2005, 261), and until the 2nd millennium BC arsenic-bronze remained the most popular copper alloy (Moorey 1985, 15–16). Although tin-bronze was introduced to Mesopotamia in the middle of the 3rd millennium BC (ED III, *c.* 2600–2350 BC), it took until around 1500 BC for tin-bronze to completely replace the arsenic-copper alloy (De Ryck *et al.* 2005, 261–262). Analyses undertaken and published by different scholars, *e.g.* Hauptmann and Pernicka (2004), Moorey and Schweizer

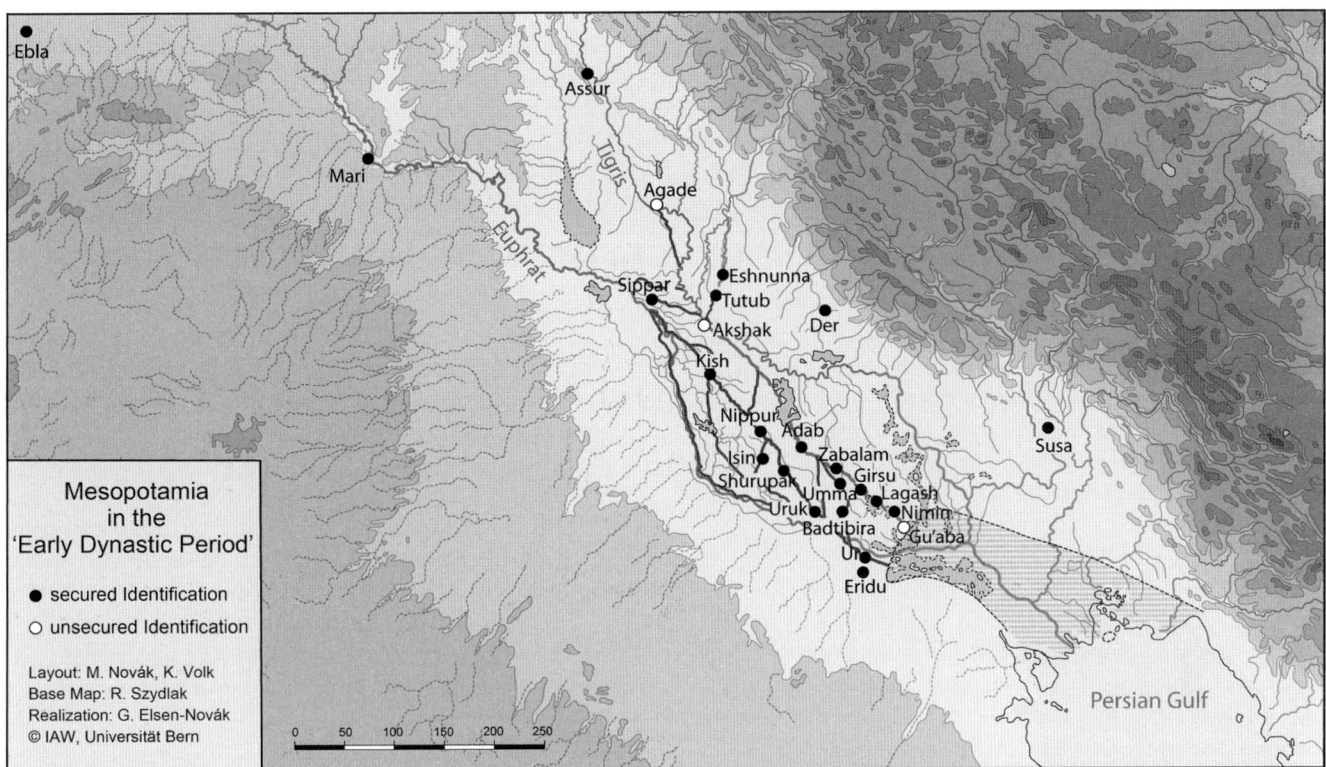

Figure 17.1: Map of Southern Mesopotamia (base map Richard Szydlak, version by U. Wischnewski, © IAW, University of Bern).

(1972), or De Ryck *et al.* (2005), demonstrate the different concentrations of arsenic or tin alloys in Mesopotamia throughout the EBA. Tin concentration in ED III could vary from low (2%) to high (>10%) (De Ryck *et al.* 2005, 267). The majority of tin-bronze objects from the royal cemetery in Ur, for example, had a middling concentration of about 8% (De Ryck *et al.* 2005, 265). Approximately 40% of the copper-based objects from the ED IIIa (*c.* 2600–2450) graves in Ur were made of tin-bronze (Helwing 2007, 249).

The origin of the metals used for manufacture in Mesopotamia still remains uncertain. Anatolia, Iran and the Gulf region may have been suppliers of copper (Moorey 1985, 12) and tin could have reached Mesopotamia from Iran, Afghanistan (Reiter 1997, 209) or the Kestel mines in the Taurus mountains of Anatolia with their nearby production site Göltepe (Yener 2000; De Ryck *et al.* 2005, 263; Lehner and Yener 2014). The high degree of recycling and repair still makes it difficult to identify the major sources for copper and tin. Different regions are mentioned in Sumerian texts. For example, Dilmun (today's Bahrein), Magan (modern Oman) and Meluḫḫa (today's Indus Valley) are referred to as sources of copper and tin in 3rd millennium BC administrative texts (Limet 1960, 86–87; Potts 1997, 174; Moorey 1985, 128; Reiter 1997, 155–159, 212). However; recent geological research in Bahrein and the coastal area of Saudi Arabia as well as the Indus Valley rule these regions out as direct suppliers,

at least in the case of tin (Nezagati *et al.* 2011, 214–215). They were more important as intermediate trading stations. Third millennium BC texts already indicate Dilmun's role as a distributor of goods from regions in the southeast (Reiter 1997, 155). Besides trading, warfare could also be a source for metals, as an Akkadian text from the reign of Rimuš (*c.* 2278–2270 BC) demonstrates (Frayne 1993). After the conquest of Elam and Paraḫšum, copper from these regions was brought to the temple of Enlil in Nippur (Frayne 1993, RIME 2.1.2.6, 131–144; Moorey 1985, 11; Reiter 1997, 159). Analyses published by Nezagati, Pernicka and Momenzadeh (2011) put the focus on Iran as a copper and tin supplier for 3rd millennium Mesopotamia, as is demonstrated by one artefact from Kish featuring a composition compatible with a tin/copper source in Deh Hosein (Nezagati *et al.* 2011, 225).

The evidence

Richly furnished burial and hoard findings provide an enormous amount of archaeological evidence for EBA Mesopotamia. Vessels made of stone, ceramics or metals figure prominently in burial sites. Besides these vessels, tools, weapons and jewellery were also laid down in the graves together with the bodies. The metal artefacts from burials and hoards are not the only part of the material culture the present author has been investigating in her research. Contemporary images like seal impressions or reliefs are consulted both for the purpose of comparison and for information about the contexts these objects were used in.

Social reorganisation and the complicated bureaucratic system also promoted the development of cuneiform writing. With scrupulous accuracy, administrative listings describe how metals underwent the transformation from ingot or raw material to manufactured objects. Besides these more prosaic texts, literary sources can also provide information about the influence metal had on society (or vice versa) and the role it played in social contexts.

Hopefully, the combination of archaeological and textual evidence will help us to understand the impact tin-bronze may have had and what it was that prompted 3rd-millennium BC society in Mesopotamia to adopt and continuously improve tin-bronze casting.

Archaeological evidence

As we have seen, copper and arsenic-bronze were already firmly established in the 3rd millennium. So why introduce tin-bronze? Was the golden shimmer or the harder consistency of tin-bronze an argument for using it for tools and weapons? Did tin-bronze replace copper or arsenic-bronze or even other substances like ceramic or stone as the material of choice for certain kinds of objects? What changes did the use of bronze and the development of bronze casting technology bring about? These are some of the questions asked the archaeological evidence.

At present, the author of this article has started collecting data from different settlements in Southern Mesopotamia, like Ur and Kish, using published analytic data for the distinction of copper, arsenic- and tin-bronze. Most important are the publications by Moorey (1970; 1978), Moorey and Schweizer (1972), Müller-Karpe (1993) and Hauptmann and Pernicka (2004). The author also wishes to thank Federico Zaina for granting access to his Kish database. Most of the data comes from burial contexts dating to ED III.

In ED III the number of vessels made of arsenic-bronze decreased. At the same time, an increase both in tin-bronze and arsenic-bronze tools and weapons is observable (Fig. 17.2). Presumably, better control of the alloying process and new sources encouraged the slow replacement of arsenic-bronze by tin-bronze. Most of the vessels were made by embossing a probably roughly cast workpiece or plate, a technique designed to produce vessels with very thin walls (less than 3 mm). In the 3rd millennium, casting technology had only been able to achieve thicknesses of more than 3mm. Another argument for embossing rather than casting was the small amount of metal required: metal was an expensive resource in Mesopotamia (Müller-Karpe 1993, 267–277). The substance used for metal vessels seems to have been largely determined by the form of the respective vessel. All basins, cauldrons, handled bowls (*Griffschalen*)

and spherical bowls (*Kalottenschalen*) found in graves were without exception copper or bronze, with no counterparts in silver or gold. On the other hand, vessel-forms like spouted jugs and '*Knickwandschalen*' were made of gold or silver but also had their parallels in copper/bronze. These two types of vessel often turn up as a set in which both of them are always made of the same substance: gold, silver or copper/bronze (Müller-Karpe 1993, 271–272).

The next section is a case study of the spouted jug as an example for the use of metal for locally common, traditional types of vessels. The question is whether the innovation had any impact on existing tradition or not. In other words, did this new substance have any impact on the form, function or usage of the vessels?

The spouted jug – an example

The Type D spouted vessels in Mackay's classification (Fig. 17.3) (Mackay 1925–1929, 149 pl. LI 18–27) found among ED I (*c.* 2900–2750 BC) and ED II (*c.* 2750–2600 BC) pottery below Palace A in Kish were very common in the ceramic tradition of 4th millennium BC Mesopotamia (Moorey 1970, 99). Ceramic or stone versions

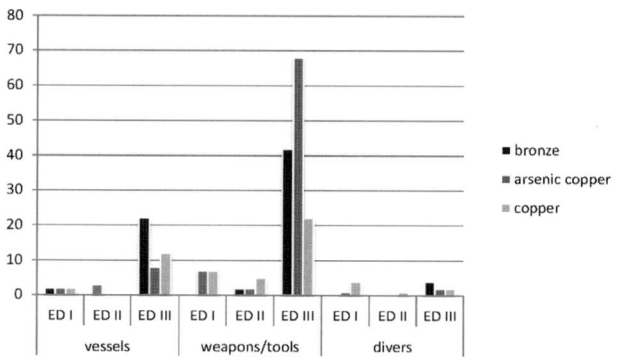

Figure 17.2: Diagram: Development of tin-bronze usage in the first half of the 3rd millennium.

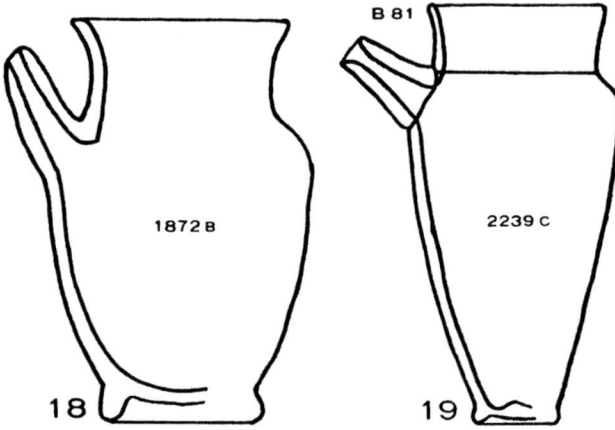

Figure 17.3: Type D ceramic spouted vessels (Mackay 1925, 149 pls LI 18, 19).

were already widely used during the Uruk period (*c.* 4000–3100/3000 BC), as depictions on the so-called 'Uruk Vase' demonstrate. They were in use at least until the onset of the Akkadian period (*c.* 2350–2200 BC). The spouted jug made of metal – more precisely copper or arsenic-bronze – first occurred at the end of the 4th millennium BC, and its incidence climaxed in the middle of the 3rd millennium BC. The earliest metal examples were ball-shaped and had a rather short spout of the kind also found in contemporary pottery (Moorey 1970, 99; Müller-Karpe 1993, 24–25). During the long period when they were in use, the form of these metallic spouted jugs changed slightly (Fig. 17.4.20). Unlike their ceramic counterparts, the spout of the metallic vessels lengthened, while the neck could vary from short to

long and slim. The foot either took the form of a standing ring (Fig. 17.4.23) or could also be rounded (Fig. 17.4.22), to mention only one or two variations.

In Early Dynastic III burials, the spouted jug is a relatively common item among the deposited objects. Some examples from the so-called 'Royal Cemetery' in Ur are an excellent instance of its use in the funerary context.

Burial PG 755 (Fig. 17.5) dates to the end of ED III. The site is a rectangular pit with a wooden coffin near its north-eastern wall. The body is that of a young man lying on his left side in a crouching position. Both inside and outside the coffin the grave is richly endowed with jewellery, vessels made of ceramic, metal and stone plus weapons and tools like daggers, arrow-heads or chisels (Woolley 1934, 155–159; Müller-Karpe 1993, 249–251). Among the metal vessels are the spouted jug (U.10035) and the bowl (*Knickwandschale*) referred to earlier, both made of silver. Additionally, a second spouted vessel made of bronze (U.10085) was deposited just next to the set (Woolley 1934, 159). All three vessels (Fig. 17.6) were laid down outside the coffin in the northern corner of the pit with no direct link to the corpse. Remarkably, there are two variations of the same vessel type lying side by side in

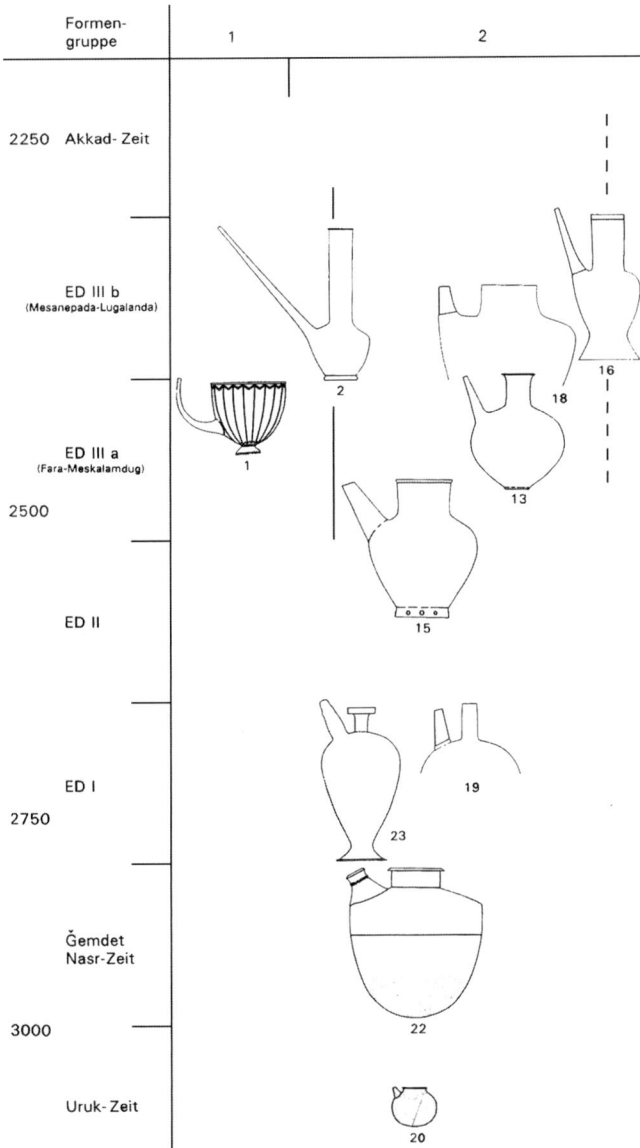

Figure 17.4: Typology of spouted vessels (Müller-Karpe 1993, pl. 172, left).

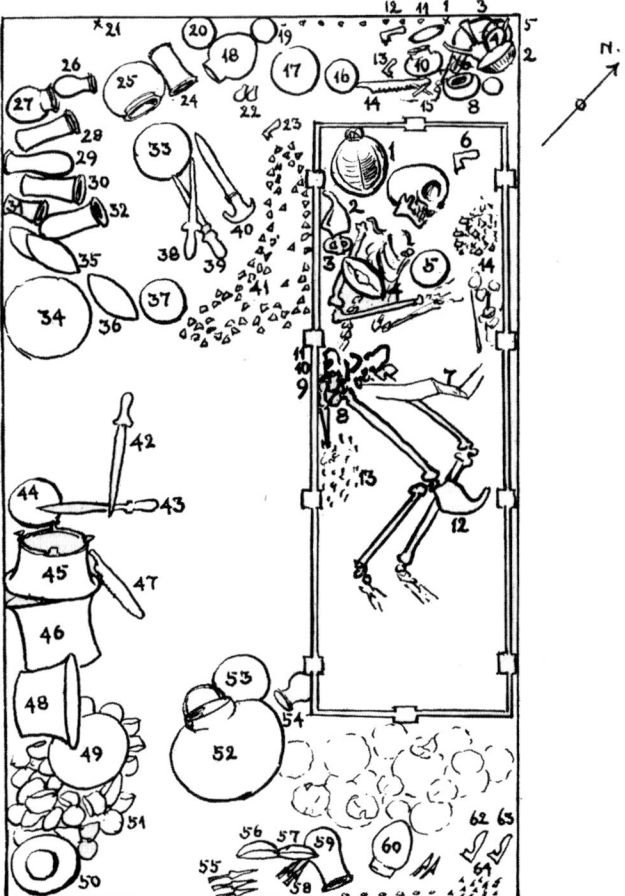

Figure 17.5: PG 755 (Woolley 1934, 157 fig. 35).

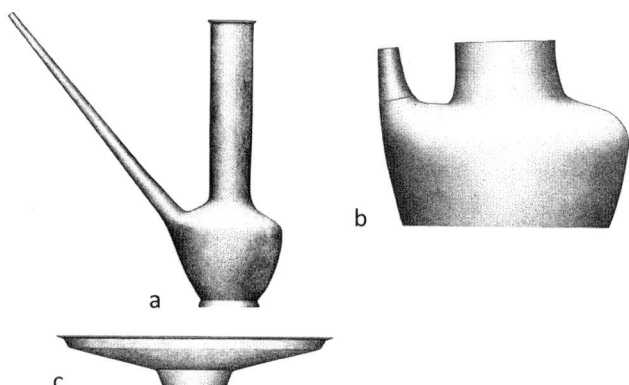

Figure 17.6: Spouted jugs U.10035 (a), U.10085 (b) and the bowl (c) from PG 755 (Müller-Karpe 1993, pl. 152.2, 18, 699).

Figure 17.7: Modern seal impression of an Early Dynastic cylinder seal from Ur (U. 9315) (Woolley 1934, pl. 200.98).

the same burial site. Spouted jug U.10035 (Müller-Karpes variant form 2 I 1) has a rather compact body and a long cylindrical neck constituting more than half of the whole vessel. A long, slim spout starts at the vessel's shoulder (Müller-Karpe 1993, 13, pl. 152.2). By contrast, U. 10085 (variant form 2 III b) has a shorter neck and spout (Müller-Karpe 1993, 16, pl. 152.18). The spout also seems to be more upright in its positioning.

A spouted jug (U. 11921) of variant form 2 I 1, in this case made of bronze and again forming a set with a *Knickwandschale* (U. 11922) of the same substance, was also found in burial PG 1054. It dates to early ED III (Woolley 1934, 107; Müller-Karpe 1993, 14, 249). Unlike PG 755, the two vessels were found in direct contact with the female skeleton's back, at pelvis level (Woolley 1934, 105, fig. 17; 107).

This cylinder seal (Fig. 17.7) depicts a banquet scene and shows the spouted vessel and bowl set 'in action'. In the second row of the scene, a naked man - presumably a servant - holds a spouted vessel by the neck in his right hand and a bowl in his left. The jug has a long neck and spout, fully comparable with the examples of variant form 2 I 1 from PG 755 and PG 1054. Obviously, these vessels were used as libation vessels in cultic or ceremonial contexts. Spouted jugs of variant form 2 I with a long neck and spout like U. 10035 (Fig. 17.6) and U. 1191 were held by the neck using one or two hands, as can be seen on the contemporary cylinder seal from Ur referred to earlier. By contrast, jugs with a shorter neck and longer foot were held by the narrowest section of the foot (Müller-Karpe 1993, 22–23).

Obviously, these innovative metal vessels did not immediately replace the other traditional substances like stone or clay outright. The same type of vessel was manufactured from different materials.

Philological evidence

In addition to the archaeological material, textual evidence (administrative, literary and cultic texts or letters) tell us

more about the origin of resources, the organisation and specialisation of crafts and the use of metal objects.

Social aspects

Here is a literary text that sheds some more light on the potential functions of artefacts deposited in burial sites. *Urnamma's Death* (Castellino 1957, Kramer 1967, see also more recently http://etcsl.orinst.ox.ac.uk/cgi-bin/etcsl.cgi?text=t.2.4.1.1#), a lament over the death of Urnamma, the first ruler (*c.* 2112–2095 BC) of the IIIrd dynasty of Ur (*c.* 2112–2004 BC), sets out a specific view of the afterlife. After dying, the king has to offer gifts to the seven main gods of the nether world. Furthermore, Urnamma also has to host a huge banquet (lines 76–87). Kramer (1967) reconstructed the whole text by filling in the fragmented parts of CBS 4560 col. ii–v (lines 42–202 of the text, see Castellino 1957) located in Pennsylvania with the help of other tablets. Pertinent to the lines cited here are two texts now located in Jena: HS 1428 and 1560 obv. ii (lines 75–83). A more recent translation of a version from Nibru has been published online in *The Electronic Text Corpus of Sumerian Literature (ETCSL).*

> Lines 76–87: He presented gifts to the seven chief porters of the nether world. As the famous kings who had died and the dead išib priests, lumaḫ priests, and nindiĝir priestesses, all chosen by extispicy, announced the king's coming to the people, a tumult arose in the nether world. As they announced Ur-Namma's coming to the people, a tumult arose in the nether world. The king slaughtered numerous bulls and sheep, Ur-Namma seated the people at a huge banquet. The food of the nether world is bitter, the water of the nether world is brackish. The trustworthy shepherd knew well the rites of the nether world, so the king presented the offerings of the nether world, Ur-Namma presented the offerings of the nether world: as many faultless bulls,

faultless kids, and fattened sheep as could be brought. (http://etcsl.orinst.ox.ac.uk/cgi-bin/etcsl.cgi?text=t.2.4.1.1#)

Were all these goods (and vessels) placed in the burial site perhaps intended either as gifts for the gods or as equipment needed to host a banquet like the one described above? Burial rituals might also be a reason for these libation vessels. Descendants of the deceased person may have used them for libation ceremonies provided for in the funeral arrangements. Given their wide range and sheer quantity, the diverse objects buried with the dead may have been both gifts for the gods and a token of the wealth of the dead person.

Another text from the 3rd millennium BC, the *Debate between Copper and Silver* published online in translation and transliteration by the *Cuneiform Digital Library Initiative (http://etcsl.orinst.ox.ac.uk/cgi-bin/etcsl.cgi?text=t.5.3.6#)* sheds light on the wide range of use and the social significances of the two substances copper and silver. Silver and copper are arguing about which of them is the more important to mankind. Silver sees its role in the house and at banquets: 'But my assigned task is in houses, in [...] and at banquets' (segment B, line 12). Copper argues that silver is always hidden in boxes or buried in the ground or inside graves:

> Segment D 18–23: Men caulk tiny, very strong boxes for you (silver), as they do a boat. They cover you over with their oldest rags, and someone digs a hole for you in the middle of the cattle-pen. Or they pour clay on top of you, as on a jar with a sealed mouth, and then, in the darkest place inside the house, someone buries you in the most obscure corner of a grave.

No one sees it or actually uses it and therefore it cannot be as valuable for people as copper itself, which is used for agriculture, chopping firewood and building houses. Copper is needed for the manufacture of different tools like hoes, adzes, axes, sickles and chisels:

> Segment D lines 24–37: When the time of wet ground has arrived for me, you do not supply the copper hoes that chop weeds [...]. When sowing time has arrived for me, you do not supply the copper adzes that make ploughs [...]. When winter time has arrived for me, you do not supply the copper axes that chop firewood [...]. When harvest time has arrived for me, you do not supply the copper sickles that reap grain [...]. For your harvest or winter, you do not supply the copper adzes and chisels which build houses [...].

Later, silver argues that copper is fragile and is often replaced by the 'old' traditional tools made of wood and stone, perhaps already a sign of the awareness that metal technology has to be advanced in order to produce more reliable tools.

> Segment D, lines 63–66: The copper hoe has its digging taken over by the wooden hoe in the harder ground. The copper

sickles need to have the hard weeds burned. The copper axes which chop trees, stripping and pulling out tamarisks and ash shrubs, have their blades dulled. The copper saws have to lie down for a rest beside the mountain trees.

Copper actually wins this dispute, which is conducted before Enlil, the main god.

Technological aspects

Administrative texts cast light not only on the organisation of crafts but also on technological processes like alloying, different stages of production, technical terms, material used by the smiths and the provenance of metals. According to administrative texts from the end of the 3rd millennium BC, for example, Meluḫḫa, today's Indus Valley, served as a source of tin and the so-called 'urudu me-luḫ-ḫa', *i.e.* Meluḫḫa copper (Limet 1960, 35, 87; Potts 1997, 174; Moorey 1999, 298). As we have seen, other sources were Dilmun and Magan. But literary texts also refer to other regions like Aratta as a source of both copper and tin (Reiter 1997, 160, 212).

At least at the end of the 3rd millennium BC, smiths used not only metal but also flour, sheep fat or oil required to facilitate the extraction of objects after they had been cast. The stone or clay moulds were greased or coated with these substances prior to the casting process (Reiter 1997, 459–461).

The term z a b a r (U D . K A . B A R = akk. *siparru(m)*) for bronze as an alloy of copper (u r u d u = akk. *werû(m)*, *erû*) and tin (A N . N A = a n - n a, n a g g a = akk. *annaku(m)* (this term probably sometimes stood for lead as well)) first appears in administrative texts from Ur dating back to ED I (Moorey 1999, 252). In these archaic texts, bronze and copper were named separately for the first time (Moorey 1985, 18). Limet (1960, 66) translated the term UD.KA.BAR as 'strong metal', an interesting choice if we assume that initially vessels in particular were made of bronze. According to the *electronic Pennsylvania Sumerian Dictionary* (ePSD), z a b a r also means '(to be) shiny' (akk. *namru(m)* 'bright, shiny'). An ED III administrative text from Lagaš sets out the precise ratio of tin to copper in alloys for vessels: 80 shekels (*c.* 669 g) of copper and 13.3 shekels (*c.* 111 g) of tin are delivered to the smiths. The result is a loss of 18.3 shekels (153 g) and a 17.73% bronze vessel weighing 75 shekels (627 g) (Moorey 1985, 18; Potts 1997, 180).

These administrative and literary texts also indicate that there were three different groups of smiths. First, there was the s i m u g (akk. *nappāḫu(m)* 'smith'), who cast all sorts of metal objects and is later defined more specifically by adding the metal - copper, bronze, gold and iron - he works with (s i m u g u r u d u / z a b a r / a n - b a r / k ù - G I). Then there was the t i b i r a (D U B / U R U D U . N A G A R, akk. *qurqurru(m)*), a 'metal/copper worker' whose work consists in hammering metal. He was originally connected to the n a g a r (*akk. nagaru(m)*), 'carpenter' or 'wood worker'. Thirdly (but not so relevant for this study) we have the 'gold-/silversmith', k ù - d í m (akk. *kuīmu(m)*, *kuttimmu(m)*), who worked on luxury metals such as gold and silver (Joannés 1993–1997, 100–101). These definitions give some indication of how the specialisation of crafts translated into language. While the eponymic aspect of the s i m u g is fire, the t i b i r a is related to the work of imposing a design on a sheet of metal by hammering, probably on a wooden form. Both professions often ran in families the specialist knowledge involved was handed down from father to son (Matthews 1995, 463; Renger 1996, 228).

Discussion

In conclusion, it seems fair to say that it was the long tradition of ceramic production plus specialisation processes in copper hammering and casting that paved the way for the appropriation of tin-bronze technology in EBA Mesopotamia. Transfer of knowledge is traceable at least within alloying technology. The knowledge that different substances could be alloyed to come up with a new substance with different properties had been discovered by alloying copper and arsenic. So tin-bronze seems to have tied in smoothly with an existing system. But it replaced neither copper/arsenic-bronze nor even older substances like ceramic or stone. Indeed, a certain parallelism is observable, as ceramic vessels like the spouted jug were still made of ceramics after the introduction of metal technology. Initially, the different sorts of metal – silver, copper, arsenic-bronze and tin-bronze – were used as additional substances in the creation of these vessels. Later, during the 3rd millennium, the types of metal were associated with different formal variations, but by no means did they completely supplant the ceramic versions of the spouted jug.

The use of stone or clay did not decline, quite the contrary, as now the production of moulds and crucibles was necessary for casting and 'cross-craft interaction potters' supported smiths with their skills. 'Cross-craft interaction' is also expressed by the term t i b i r a, which originates in woodworking and therefore describes the connection between these two crafts. Metal plates could be shaped by hammering them on a wooden form. The different terms for smiths – s i m u g and t i b i r a – appear to foreground technological aspects, not the metal itself. The appropriation of tin-bronze did not mean new specialisation. It was part and parcel of an existing system of casting and hammering.

Knowledge was presumably transferred by intermediate traders. The Sumerians did not explore mines themselves and traders represented a contact zone between the mining areas and the production centres in Mesopotamia. The sources for tin and copper used in Sumer have not yet been located, but Kestel in Anatolia and regions in Iran or Afghanistan can be taken into consideration as possible supply sources.

Note

1   The author is fully aware of the recent debates culminating in the establishment of the ARCANE project dedicated to a new chronology for the 3rd millennium BC. For more information, see *http://www.arcane.uni-tuebingen.de/*. All dates used refer to the Middle Chronology.

References

Akrich, M. and Latour, B. (1992) A Summary of Convenient Vocabulary for the Semiotics of Human and Non-Human Assemblies. In W. E. Bijker and J. Law (eds), *Shaping Technology, Building Society. Studies in Sociotechnical Change*, 259–264. Cambridge, MIT Press.

Brysbaert, A. (2011) Technologies of Reusing and Recycling in the Aegean and Beyond. In A. Brysbaert (ed.), *Tracing Prehistoric Social Networks through Technology*, 183–203. New York, Routledge.

Castellino, G. (1957) Urnammu Three Religious Texts. *Zeitschrift für Assyriologie und Vorderasiatische Archäologie* 52, 1–57.

De Ryck, I., Adriaens, A. and Adams, F. (2005) An Overview of Mesopotamian Bronze Metallurgy during the 3rd Millennium BC. *Journal of Cultural Heritage* 6, 261–268.

Edzard, D. O. (2004) *Geschichte Mesopotamiens. Von den Sumerern bis zu Alexander dem Großen*. Munich, Beck.

ePSD <http://psd.museum.upenn.edu/epsd1/nepsd-frame.html> accessed 15.08.2016.

ETCSLtranslation: t.2.4.1.1. Latest update 19.12.2006. Available at http://etcsl.orinst.ox.ac.uk/cgi-bin/etcsl.cgi?text=t.2.4.1.1#, last accessed 15.08.2016.

ETCSLtranslation: t.5.3.6. Latest update 19.12.2006. Available at http://etcsl.orinst.ox.ac.uk/cgi-bin/etcsl.cgi?text=t.5.3.6#, last accessed 26.11.2015.

Frayne, D.R. (1993) Sargonic and Gutian Periods (2334–2113 BC). The Royal Inscriptions of Mesopotamia Early Periods Volume 2, Toronto, University of Toronto Press.

Gülçur, S. (2002) Handelsbeziehungen des 4. und 3. Jahrtausends. v. Chr. im Vorderen Orient. In Ü. Yalçın (ed.), *Anatolian Metal II*. Der Anschnitt, Beiheft 15, 27–37. Bochum, Deutsches Bergbau-Museum.

Hauptmann, H. and Pernicka, E. (ed.) (2004) *Die Metallindustrie Mesopotamiens von den Anfängen bis zum 2. Jahrtausend v. Chr.* Orient-Archäologie 3. Rahden/Westfalen, Verlag Marie Leidorf.

Helwing, B. (DRAFTONLY, latest update 2007) Klassifikation, Typologie und Datierung der untersuchten Metallobjekte, to appear as 2 chapters in planned volume 2 of: Hauptmann, H.; Pernicka E. (ed.) (2004): *Die Metallindustrie Mesopotamiens von den Anfängen bis zum 2. Jahrstausend v. Chr.* Orient Archäologie 3, available at https://www.academia.edu/2184593/, last accessed 15.08.2016.

Joannés, F. (1993–1997) Metalle und Metallurgie: A. I. in Mesopotamien. In D. O. Edzard (ed.), *Reallexikon der Assyriologie und Vorderasiatischen Archäologie* 8, 96–112. Berlin, de Gruyter.

Kramer, S. N. (1967) The Death of Ur-Nammu and His Descent to the Netherworld. *Journal of Cuneiform Studies* 21, 104–122.

Kuhrt, A. (1995) *The Ancient Near East c. 3000–330 BC* 1. London, Routledge.

Latour, B. (1991) Technology is Society Made Durable. In J. Law (ed.), *A Sociology of Monsters? Essays on Power, Technology and Domination*, 103–131. London and New York, Routledge.

Lehner, J. W. and Yener, K. A. (2014) Organization and Specialization of Early Mining and Metal Technologies in Anatolia. In B. W. Roberts and C. P. Thornton (eds), *Archaeometallurgy in Global Perspective*, 529–557. New York, Springer Science and Business Media.

Leroi-Gourhan, A. (1943) L'homme et la matière. Paris, Michel.

Leroi-Gourhan, A. (1945) Milieu et technique. Paris, Michel.

Limet, H. (1960) *Le travail du métal au pays de Sumer au temps de la IIIe dynastie d'Ur*. Paris, Les Belles Lettres.

Mackay, E. J. (1925–1929) *Report on the Excavation of the 'A' Cemetery at Kish, Mesopotamia* 1–2. Chicago, IL, Field Museum Press.

Matthews, D. (1995) Artisans and Artists in Ancient Western Asia. In J. M. Sasson (ed.), *Civilizations of the Ancient Near East* 1, 455–468. New York, Scribner.

Miller, H. M.-L. (2007) *Archaeological Approaches to Technology*. Amsterdam, Elsevier.

Moorey, P. R. S. (1970) Cemetery A at Kish: Grave Groups and Chronology. *Iraq* 32/2, 86–128.

Moorey, P. R. S. (1978) *Kish Excavations 1923–1933. With a Microfiche Catalogue of the Objects in Oxford Excavated by the Oxford Field Museum, Chicago Expedition to Kish in Iraq, 1923–1933*. Oxford, Clarendon.

Moorey, P. R. S (1985) *Materials and Manufacture in Ancient Mesopotamia: The Evidence of Archaeology and Art; Metals and Metalwork, Glazed Materials and Glass*. British Archaeological Reports International Series 237. Oxford, BAR.

Moorey, P. R. S. (1999) *Ancient Mesopotamian Materials and Industries: The Archaeological Evidence*. Winona Lake, IN, Eisenbrauns.

Moorey, P. R. S. and Schweizer, F. (1972) Copper and Copper Alloys in Ancient Iraq, Syria and Palestine: Some New Analyses. *Archaeometry* 14/2, 177–198.

Müller-Karpe, M. (1993) *Metallgefäße im Iraq* I *(Von den Anfängen bis zur Akkad-Zeit)*. Stuttgart, Franz Steiner Verlag.

Müller-Karpe, M. (2002) Zur Metallverwendung im Mesopotamien des 4. und 3. Jt. In Ü. Yalçın (ed.), *Anatolian Metal II*. Der Anschnitt, Beiheft 15, 137–148. Bochum, Deutsches Bergbau-Museum.

Nezagati, N., Pernicka, E. and Momenzadeh, M. (2011) Early Tin-Copper Ore from Iran, a Possible Clue for the Enigma of Bronze Age Tin. In Ü. Yalçın (ed.), *Anatolian Metal V.* Der Anschnitt, Beiheft 24, 211–230. Bochum, Deutsches Bergbau-Museum.

Pfaffenberger, B. (1992) Social Anthropology of Technology. *Annual Review of Anthropology* 21, 491–516.

Potts, D. T. (1997) *Mesopotamian Civilization: The Material Foundations*. London, Athlone.

Reiter, K. (1997) *Metalle im Alten Orient unter besonderer Berücksichtigung altbabylonischer Quellen*. Alter Orient und Altes Testament 249. Münster, Ugarit-Verlag.

Renger, J. (1996) Handwerk und Handwerker im alten Mesopotamien. Eine Einleitung. *Altorientalische Forschungen* 23, 211–231.

Woolley, C. L. (1934) *The Royal Cemetery. A Report on the Predynastic and Sargonid Graves Excavated Between 1926 and 1931*. Ur Excavations 2, Texts. London, British Museum.

Yener, K. A. (2000) *The Domestication of Metals: The Rise of Complex Metal Industries in Anatolia*. Leiden, Brill.

Yoffee, N. (1995) The Economy of Ancient Western Asia. In J. M. Sasson (ed.), *Civilizations of the Ancient Near East* 3, 1387–1399. New York, Scribner.

Chapter 18

Gonur Depe (Turkmenistan) and its Role in the Middle Asian Interaction Sphere

Federica Lume Pereira

As Lauren Zych remarks in her article on Chalcolithic Turkmenistan, archaeological exploration of this region has had a tendency to focus on the interregional connections of Bronze Age Middle Asia. This often happens at the expense of a deeper understanding of the rich and varied local development of human settlement in the area (Zych 2006). However, in the context of the key terms of this conference – 'appropriation' and 'innovation' – it seems appropriate to discuss some further evidence for the very active links between sites in southern Turkmenistan – particularly the site of Gonur Depe – and other regions of Middle Asia in the late 3rd and early 2nd millennia BC.

In the Near and Middle East, a marked expansion of interregional communication networks is discernible during the late 4th millennium BC (most prominently within the Late-Uruk horizon). This expansion was mirrored to some extent during the Early Harappan phases on the Indian subcontinent (Algaze 1993; Potts 1993; Possehl 2002, 40; Kenoyer 2004, 42). However, by the middle of the 3rd millennium BC – after a long phase of consolidation – intensive contacts of a different nature had developed. Amongst other factors, the emergence of maritime routes is regarded as a driving force for this intensification. These networks now connected the domains of the Indus Valley, southern Central Asia, the Iranian plateau and coast, the Persian Gulf coast and Mesopotamia. While evidence of these contacts is often quite ephemeral and uncontextualised, the quality and wide distribution of certain object categories point to very dynamic and far-reaching exchange mechanisms (Potts 1994; Ratnagar 2004).

In this paper, I will present further evidence of interconnectedness from one of the regions of this dynamic network: southern Central Asia, specifically the site of Gonur Depe (Turkmenistan) that became the starting point of fieldwork undertaken in the course of my ongoing Ph.D. project.

The Region

The archaeological exploration of southern Turkmenistan began with Raphael W. Pumpelly, who first surveyed the Karakum desert and the adjacent Kopet Dagh-piedmont, bringing to light a dense pre- and protohistoric settlement pattern (Pumpelly 1908). Research in the following decades demonstrated that the Late Eneolithic and Bronze Age (*c.* 3100–1500 BC) settlement areas are to be found in the Kopet Dagh range, its piedmont area, and the central and north-eastern portion of the Murghab river delta (Marcolongo and Mozzi 1998, 4). The prevalent periodisation for Turkmen sites was initially derived from the long material sequence excavated at Namazga Depe in the central Kopet Dagh piedmont (Table 18.1).

Early Soviet surveys performed in the delta and the middle reaches of the Murghab river during the fifties, and more recent surveys carried out by Italian scholars during the early nineties, have identified almost a hundred prehistoric sites in the area varying greatly in size from five to over 10 hectares (Gubaev *et al.* 1998; Salvatori *et al.* 2008). A total of nine discernible micro-oasis site conglomerates were identified (Kelleli, Egri Boaz, Taip, Adji Kui, Gonur, Auchin, Adam Basan, Togolok, Takhirbai), each containing settlements of different phases of the *Margiana* chronology, starting in the Middle Bronze Age (*c.* 2500–2000 BC) (see Salvatori 2004, 93 for earlier finds in the area).

The largest settlement of the Gonur oasis – Gonur 1 – is located on one of the eastern branches of the delta. It is

Table 18.1: Archaeological periodisation of Southern Turkmenistan (after Vidale 2010 and Hiebert 1994)

Site	Period	Date BC
Djeitun	Neolithic	6500–4500
Namazga I	Eneolithic	4500–3800
Namazga II		3800–3100
Namazga III		3100–2900
Namazga IV	Early Bronze Age	2900–2500
Namazga V	Middle Bronze Age	2500–2000/1700
Namazga VI	Late Bronze Age	2000/1700–1500

the most prominent site of the so-called 'Oasis' or 'Oxus Civilisation' of Central Asia, also referred to as 'Bactria and Margiana Archaeological Complex' (BMAC) by the excavator of the site, Viktor I. Sarianidi (for a discussion of the term, see Lamberg-Karlovsky 2003; Salvatori *et al.* 2008, 75). Material related to this archaeological complex is mainly found in southern Uzbekistan, the Zeravshan Valley (Tajikistan), the upper Oxus/Amu Darya valley (Afghanistan) and the upper course of the river Murghab in southern Turkmenistan (Salvatori 2004, 92).

The settlement of Gonur Depe

The settlement of Gonur 1 or 'Gonur Depe' consists of two mounds: the larger 'Gonur North' and the smaller 'Gonur South'. Gonur North dates mainly to the (Late) Namazga V sequence of the piedmont chronology (second half of the 3rd millennium BC), which corresponds to the Central Asian Middle Bronze Age. The younger southern mound containing a fortified compound (*temenos*) dates to the early centuries of the 2nd millennium BC (correlated to Namazga VI *i.e.* Central Asian Late Bronze Age). Radiocarbon dating was undertaken by Hiebert (1994, 76) and Jungner (2007, appendix) and roughly corroborated these dates. Occupation at the site has been subdivided into three major periods: MBA occupation I and II and a third post-urban LBA occupation after the main buildings had been abandoned for some time (Hiebert 1994, 80).

The mound of Gonur North was a large urban settlement of roughly oval shape and partially enclosed by an outer wall. At its centre lies the 'Palace', a vast complex of storage facilities, specialised craft areas and representative rooms surrounded by a double mudbrick wall with defensive turrets. After the first structure was destroyed by fire, three reconstruction phases have been identified so far (Sarianidi 2005, 43 fig. 1). Furthermore, a large necropolis was located to the west of the settlement and stretched along a bend in one of the arms of the delta in a north–south axis. It consisted of more than 3000 graves which were mostly sacked in antiquity but nevertheless have provided a large quantity of objects illustrating the varied local material culture and its links with neighbouring regions. Correlation with finds from the settlement and comparison

with Bactrian finds enable us to date the Necropolis to the last third of the 3rd millennium BC, *i.e.* 2170–2050 BC/ Late Namazga V (Salvatori 2003).

Numerous graves of different types were also found within the site. Amongst these are a series of eleven graves (numbers 3200, 3205, 3206, 3210, 3220, 3225, 3230, 3235, 3240, 3245, 3250) of the 'hypogeum type' – large subterranean multi-roomed mudbrick chambers – at the south-eastern edge of the settlement which, although sacked in antiquity as well, still contained an impressive array of grave goods (Sarianidi 2005, 205 fig. 68). Three of them (nos 3200, 3225, 3240) contained four-wheeled carts along with camels and possibly an equid skeleton (Salvatori 2010). The northern mound itself has been excavated almost in its entirety and stretches over an area of 40 hectares, thus surpassing other settlements of the region by far. Around it, numerous smaller 'satellite settlements' can be seen in what is now a desert landscape.

Gonur and the MAIS

When Gregory Possehl coined the term 'Middle Asian Interaction Sphere' (MAIS) (Possehl 2002, 215), he was referring to the aforementioned intensification of long-range contacts during the second half of the 3rd millennium BC as evidenced by varied objects and materials distributed mainly across the Near East and Central and South Asia. Gonur Depe does not figure prominently amongst the Central Asian sites mentioned in his study, but its role within the MAIS became widely acknowledged after 2004. At that time, excavations in the so-called 'Northern Water Temple' yielded an Indus Stamp seal (Fig. 18.1) which was found in enclosure 19 of Area 9 (north of the Main South Basin; dimensions face h: 3.8 cm, w: 3.9 cm, t: 1 cm). The seal is an exquisite example of Harappan glyptic and can be neatly connected with the glyptic tradition at its major sites (Parpola 2009). It is also noteworthy that the seals' inscription has no direct parallels and the elephant motif is comparatively rare in the repertoire, as Parpola indicates in his detailed analysis (2009, 74).

5 cm

Figure 18.1: Indus stamp seal (courtesy Margiana Archaeological Expedition).

In the next section, further finds attesting to Gonur Depe's established place within the MAIS will be discussed. The opportunity to examine these presented itself when I took part in a new excavation project of the Margiana Archaeological Expedition. One of the aims of this project undertaken within the frames of an agreement between Bern University (Institut für Archäologische Wissenschaften – IAW), the Institute of Ethnology and Anthropology (Russian Academy of Science, Moscow) and the Ministry of Culture of Turkmenistan (Ashgabat) is to provide further stratigraphic data for the site as well as to establish a reliable ceramic sequence.

Beads on a string

Geographically, the most widespread object along these networks is the so-called 'etched' agate bead. Initially found in Southern Mesopotamia and the core area of the Indus Valley region, further examples have been identified in sites in Iran and the Gulf. The westernmost example of this particular bead (type C3 after Reade's typology, 1979) was found in 2000 as part of a hoard in Kolonna on the Aegean island of Aegina and was dated to the Early Helladic III (*c.* 2200–2000 BC) (Reinholdt 2008, 127, pls 3.032, 13.1).

The complex manufacture of these beads has been studied in detail (Beck 1933; Mackay 1933). For our present purposes, suffice it to say that through a combination of heat and soda the painted designs on the agate were permanently 'etched' on the surface of the stone. This technique was used to create white surfaces or linear patterns in black or white on (mostly) red agate (for an in-depth discussion of this material, see Law 2011).

Etched beads are attested in the Indus Valley at Mohejodaro, Harappa, Chanhudaro, Kalibangan, Lothal and Rojdi (Possehl 1996, 157) as well as further north at the Indus outpost of Shortugai (Francfort 1989, pls 72.3, 36e). References to etched bead-finds from Baluchistan (Moghul-ghundai, Tor-Dherai) and southern Afghanistan (Mundigak) on the Bolan Pass route exist. However, they could not be contextualised and probably date to later periods (During-Caspers 1971, 90). This is also the case for several etched beads from Bactria shown in the Venice catalogue of private collection items (Ligabue and Rossi-Osmida 2007, 204–205, 208–209). Their motifs do not fit the known repertoire of 3rd millennium beads and correspond mainly to the 'Übergangsphase' in Eger's typology of 1st millennium AD etched beads (specifically 3rd–5th century AD; Eger 2010, 250 fig. 23), which is well attested, for example at Taxila (Pakistan).

The etched beads found in northern Iran (Tepe Hissar, Shah Tepe), the Pusht-i Kuh area of Luristan (Kalleh Nissar) and Iraki Kurdistan (Assur, Bakr 'Awa) date to the end of the 3rd and the beginning of the 2nd millennia BC and were all found in the funerary context of comparably rich graves

(During-Caspers 1971, 88; Miglus 2011, pl. 4). Several etched beads are also known from Jalalabad (southern Fars), Susa (Khuzistan) and Tepe Yahya (Soghun Valley). Their designs correspond to types of the 3rd millennium, but their context is often unclear (During-Caspers 1971, 92; Chakrabarti and Moghadam 1977, 167; Tallon 1995, 59 figs. 23–24). A large number of etched beads has also emerged from recent excavations on the Arabian shores of the Gulf. They were mostly found in graves and cairns of the Umm an-Nar period (second half of the 3rd millennium BC) and other contemporary settlements along the coast. De Waele and Haerinck (2006) recently compiled some of the known etched beads from Medinat Hamad and Sar el-Jisr (Bahrain), Umm an-Nar and Hili/Hili North (Abu Dhabi), Al-Sufouh (Dubai), Mowaihat (Ajman), Shimal (Ras al-Khaimah) and at the large 3rd millennium port-settlement of Tel Abraq (Sharjah/Umm al-Qawain). So far, the largest corpus of etched beads has been found in Southern Mesopotamia at the sites of Tell Asmar (Diyala), Kish, Abu Salabikh, Nippur and most prominently Ur (Reade 1979). They were almost exclusively found in funerary contexts dated to the Early Dynastic III period (second half of the 3rd millennium BC).

In southern Central Asia examples of this bead type are comparatively rare. Kaniuth (2011, 8) mentions one possible example from the Bronze Age Necropolis of Rannij Tulkhar in south-western Tajikistan. At the major urban site of Altyn Depe on the Kopet Dagh piedmont two beads were found in burials dating to the Middle and Late Namazga V period (Masson and Beriozkin 2005, pls 56.9, 74.2). In the Murghab delta region, one etched bead from Adji Kui 9 is depicted amongst other beads found during recent excavations (Ligabue and Rossi-Osmida 2007, 161 fig. 8), while another unpublished bead comes from the neighbouring oasis of Togolok (Salvatori 2000, 133). The most numerous occurrence of etched beads in the area (all showing a white design on a red stone) is indeed found at Gonur Depe.

One lentoid bead – of type D3 according to Reade's typology – was found in burial 2710 of the Necropolis (Sarianidi 2007, 116 fig. 211). The burial was undisturbed and contained the remains of a middle-aged woman. Another etched bead is depicted amongst two fragments of long bicone beads (another noted category of Indus-related beads) in Sarianidi's volume (2007, 117 fig. 221). The etched design does not show up particularly well on the picture, but can be easily recognized on the bead itself, which is on display at the National Museum in Ashgabad, where a significant part of the Gonur Depe material is housed. A definite provenance is not given, but the parallel zig-zag line design on a cylindrical body corresponds to type A2 of Reade's typology of 3rd millennium beads (Reade 1979, 8). I was further able to identify two more etched beads on display at the National Museum. They were part of a long string of eclectic beads made from very diverse materials and in different forms. The string apparently contained finds from the disturbed grave 2647 of the Necropolis. One spherical bead probably corresponds to Reade's type C3, but the design is quite faded. The other one is an uneven cylindrical bead with an alternating line-and-circle motif (Fig. 18.2). A similar bead was collected by Sir M. Aurel Stein in Baluchistan (Beck 1933, pl. 68.1). However, the uncertainty of its contexts makes further correlations impossible.

A set of seven etched beads was found in the filling of the so-called 'dog mausoleum' (hypogeum-burial 1999) of the Necropolis named after a dog burial near the entrance. The inventory of this heavily disturbed grave was partly published but unfortunately the beads were not depicted (Rossi-Osmida 2002, 121). Kaniuth parallels two silver bowls found in it with finds at Altyn Depe dating to the Late Namazga V period – he also mentions another etched bead found by Rossi-Osmida in grave 011 (Kaniuth 2011, 8). However, the necklace in question does not include any etched beads. Finally, during my work at the Mary Museum that houses the most recent finds from Gonur Depe, I came upon another hitherto unpublished etched bead. It is a small, elongated lentoid bead that corresponds to Reade's type D9 (Fig. 18.3). It was found in 2003 in burial 2065 (Sektor 5) of the Necropolis (Mary Museum Number AK 3529/46). It is listed as the disturbed shaft burial of an adult female.

The most significant and surprising find however, was made during the most recent joint Swiss-Russian-Turkmen autumn campaign at Gonur Depe. In October 2015 two agate beads – one of them etched – were found *in situ* in what seems to be the working area of a building. During the first campaign – which began in September 2014 – a team from the IAW Bern had conducted excavations in three sectors (A – B – C) of Gonur Depe that were all located to the west of the Palace complex. The excavation unearthed part of a long ditch – a continuation of the northern city-wall – which at one time probably constituted the town's westernmost enclosure (Winkelmann 2014). In Sector A (10 × 20 m) – within area ('Raskop') 23 of the Margiana expedition's ground plan – several rooms and freestanding ovens were found that were tentatively attributed to occupation Period II. In 2015, the team from the IAW Bern partially expanded Sector A to the south to complete excavation of 'House A2'. Seven occupation phases (according to the preliminary stratification) were identified in this subsector alternating between occupation phases associated with architectural remains and phases of exposure. 'House A5' in phase 6 consists of a room of rectangular shape with an entrance to the south-west (Fig. 18.4). It contained several elaborate mudbrick installations. Two square niches in the northern wall were probably used as hearths. Soundings inside 'House 5' determined that it was built upon an area with open hearths (probably a courtyard or a similar working

Figure 18.2: Etched bead on a string. National Museum Ashgabat (drawing by J. Radosavljević after author's photograph).

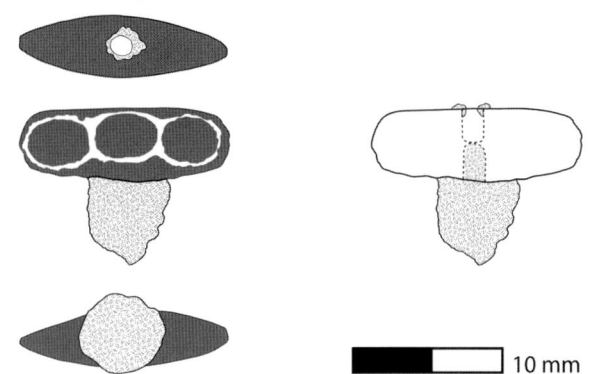

Figure 18.3: Etched bead. Mary Museum AK 3529/46 (drawing by author).

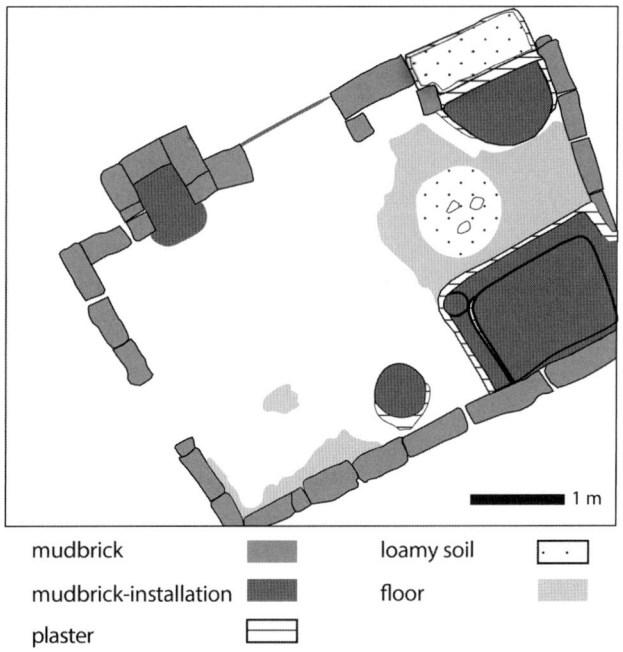

mudbrick ▮ loamy soil ⬚

mudbrick-installation ▮ floor ▨

plaster ▭

Figure 18.4: Gonur 15-campaign. Plan of Phase 6, House A5, phase b, Sectors A1/A6 (S. von Peschke).

area). The house itself contained traces of two distinct floors separated by a layer of sandy fill. The earliest floor continued to the area outside the entrance. One short rounded bead of translucent orange-coloured agate (G15-A055) and a

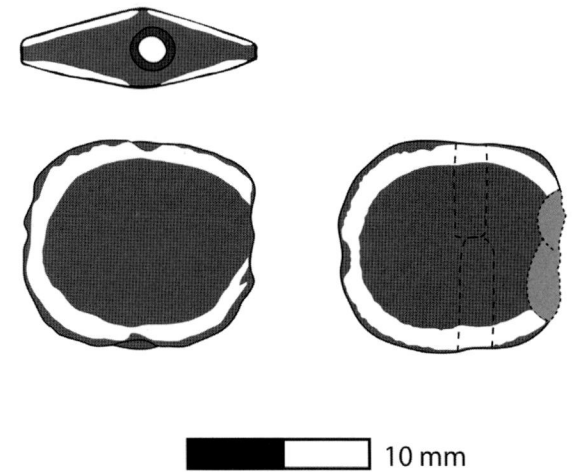

Figure 18.5: Etched bead. Field-Number G15-A054 (drawing by author).

lentoid bead of slightly darker colour with a well-preserved etched design (G15-A054) were found right outside the entrance on the floor, covered by the fill separating the two occupation levels of the house (Fig. 18.5). The etched bead neatly corresponds to type D1 of Reade's typology of early etched beads. The domestic context of the find makes it the more surprising. Hopefully, radiocarbon dating of samples taken from the older hearths will provide a basis for a clearer dating.

Sticks and bones

Several other objects found at Gonur Depe and neighbouring sites indicate contact towards south-eastern Iran, the Helmand Valley and the Indus Valley domains. Most of them have been treated exhaustively (Possehl 2002; Salvatori 2004; Kaniuth 2011). Accordingly, I will only enlarge on those groups for which new evidence can be presented.

One relevant group of objects includes the 'ivories' found at Gonur Depe. No analysis has been made to determine the exact nature or provenance of the material. However, the very well preserved examples housed at the Mary Museum were smooth and the cracks revealed horizontal layers of material filaments and no porosity, pointing to true ivory. This group includes plaques and discs with incised motifs that are often interpreted as gaming pieces (Cortesi *et al.* 2008, 24), so-called 'stick dice' or 'ivory sticks', and combs and pins (for a discussion of similar ivory combs found in Mesopotamia, the Indus Valley and the Gulf, see Mackay 1976, pl. 89.12; Ratnagar 2004, 161; Potts 2008). Furniture inlays, figurative ornaments and particularly fine arrowheads of different shapes were also made in ivory and bone, but they are rooted in the material and visual repertoire of southern Central Asia.

'Ivory sticks' are elongated rectangular or tapered pieces of bone/ivory. They are 10 to 15 cm long and have a square,

rectangular or triangular section. Their faces show incised designs with parallel lines, crosses and/or circle-and-dot motifs. A large group of these 'ivory sticks' or 'stick dice' was found at Mohenjodaro (Marshall 1931, pl. 132.22–26; Mackay 1938, pls 138.41–59, 141.32–40), Harappa (Vats 1940, pl. 119) and Chanhudaro (Mackay 1976, pl. 60.12.16). Similar pieces have been found in a burial and a hoard at Altyn Depe (Masson and Sarianidi 1972, 117 fig. 29; Kaniuth 2011, 6 fig. 3). Several of these – always square in section – were also reported from Gonur Depe. Two very long examples were found in hypogeum-type grave 3210 of the 'Royal Necropolis': one stick and fragments of three others were found inside the largest chamber of the complex (Sarianidi 2005, 250). Tomb 3220 – also part of the 'Royal Necropolis' – yielded several fragments of such sticks that were found in its smallest chamber (Sarianidi 2005, 250; 2008a, 195 fig. 107). One further stick was found in grave 575 of the Necropolis, and 'ivory bar' fragments are reported from grave 1898 (possibly 'rectangular dice'; Sarianidi 2007, 122). Furthermore, two sets of three sticks were retrieved during excavations of the 'Royal Sanctuary' north of the Palace entrance (N. Dubova, personal communication; partially depicted in Sarianidi 1998a, 55 fig. 21). Those examples that have been published with depictions of all four sides show an increasing number (1 to 3) of circle-and-dot incisions on three faces and a series of hatched lines and/or crosses on the fourth.

Another large group of ivory objects associated with gaming also includes discs and square plaques decorated with circle-and-dot incisions/drillings. The aforementioned hoard – of uncertain date – at Altyn Depe contained thirteen plaques and discs with varied circle-and-dot combinations and incisions. Three very similar discs (one square, two octagonal) with six circle-and-dot motifs placed in lines of three were found in two refuse pits at the Bactrian site of Dzharkutan located in southern Uzbekistan (Kaniuth 2011, 7). They were dated to the first centuries of the 2nd millennium BC based on the accompanying ceramic assemblage (Central Asian Late Bronze Age I). At Gonur Depe, tomb 3220 of the 'Royal Necropolis' contained a similar array of ivories. Besides the sticks already enumerated, it also contained five flat ivory discs. One was incised with a floral design that shows similarities with designs common in Bactria and the Indus area (Possehl 1996, 181 fig. 31) and another showed the incised image of a scorpion. Two other discs displayed several circle-and-dot motifs unevenly arranged across the surface, whilst two others seem to be blank. Several more fragments of discs and square plaques were amongst the ivory objects of that tomb. Four square plaques were found almost intact, two of them show traces of circles-and-dots arranged in rows (Sarianidi 2005, 231 fig. 92; 2008a, 195 fig. 107). In tomb 3155 from the 'Southern burial ground' next to the Palace Complex of Gonur (North) a similar group of ivory objects was found including pins and rectangular fragments. Amongst the objects retrieved from beneath the burial pit there were several fragments of square plaques (three of them mostly intact), discs and octagonal plaques. At least one of them shows rows of circle-and-dot motifs (Sarianidi 2005, 199 fig. 65). One further ivory disc with four circle-and-dot motifs and one square plaque with a checkerboard design with alternating circle-and-dots from Gonur were published without specifying the context (Sarianidi 1998a, 56 fig. 22.8–9; 2002, 153). The Mary Museum collection of Gonur material contains four other particularly well-preserved ivory pieces. Mary Museum number AK 1786 shows three parallel rows of three circle-and-dot motifs on one side, the other was left blank. It was found during the 1999 campaign in Room 48 near the western entrance of the Palace Complex (Fig. 18.6). A very similar piece – Mary Museum number AK 3516 – was found in 2002 in a room numbered 96 (the specific sector could not be verified). Like the aforementioned example, one side has three parallel rows of three incised/drilled circle-and-dot motifs. One of them was drilled around a perforation at one of its corners (Fig. 18.6). Mary Museum number AK 1456/601 (ø 5 cm, t: 0.6–0.7 cm) is a blank, very smooth disc found in 1992 in the upper filling of room 212 in Sector 1 of that year's campaign around the Palace Complex (Fig. 18.6). One small disc AK 1836/83 (ø 4 cm, t:0.4 cm) was found in 1999 with burial 340/66 of the Necropolis. It is perforated through the middle and shows no other incisions. Whether it was some sort of gaming piece, or it was used in another context, cannot be determined (Fig. 18.6). Plaques and discs of this kind thus occur more frequently in Central Asia and do not seem to

have been particularly popular in the Indus domains. The closest parallel in that area is a soft-stone stamp seal from Harappa (H-128 A; Joshi and Parpola 1987, 196) that shows the same three parallel lines of three circle-and-dot motifs each. However, another small ivory/bone disc (AK 3069, Fig. 18.6) may again represent closer ties with the south. The disc in question was found in 2001 within burial 2013/148 of the Necropolis. It is a small (ø 3.3 cm, t: 0.1–0.3 cm), convex disc with a central perforation made of more porous material than the previously mentioned pieces. Regularly spaced incisions along the edges give it the appearance of a six-petalled flower. Several similar examples are known from Mohenjodaro (Mackay 1937, pls 136.82–83, 138.18, 139.52.71). They may be button-like ornaments rather than gaming pieces, but their function remains unclear.

Cylinder Seals

After examining objects attesting to the Bactrian- and Indus-oriented connections of Gonur Depe, this section briefly addresses several objects that point towards the Iranian Plateau. So far, fifteen cylinder seals and six cylinder seal-impressions have been found in different sites of the Murghab delta (Sarianidi 1998b, 316–324). Many were collected from the surface, as is the case for Taip Tepe, and are often not well-preserved. Interestingly, almost all the sealings were found within the *temenos* area in Gonur South; this includes both sealed bullae and a pottery sherd sealed on the shoulder before firing (Olijdam 2008, 272 fig. 4). Further sealed shoulder-sherds were found on the surface at Taip Tepe (Sarianidi 1998b, 321 fig. 1774, 1776).

Cylinder seals clearly stand out amongst the locally predominant metal stamp seals. Five of these were recovered in Gonur Depe North. One cylinder seal was made of black veined stone and shows a scene with standing deities flanking a seated figure surrounded by a 'halo'. It was found in between the two circular structures dated to Phase I that constitute the so-called 'Complex of Sacrifice' in the immediate vicinity of the Palace's south-west fortification (Sarianidi 2008a, 101 fig. 25). Another cylinder seal with a seated deity above a snake came from the surface above burial 23 of the Necropolis (Sarianidi 2007, 105 fig. 180). A further seal – incised with a clustered scene of two symposiasts and several mythological creatures – was found within burial 1393 of the Necropolis near the midriff of a female aged over sixty (Sarianidi 2007, 106 fig. 181). According to Salvatori, these and other two cylinder seals found at the neighbouring site of Togolok 1 (Hiebert 1994, 20 fig. 9.16; seals: burial 10, Sarianidi 1998a, 65 fig. 28.6; surface, Sarianidi 1998b, 321 fig. 1767) are rooted in the glyptic tradition found at Tepe Yahya and in broader southeastern Iran (Salvatori 2000, 133). Isolated seals of South-Bactrian provenance also belong to this group (Winkelmann 1997, fig. 1a–c).

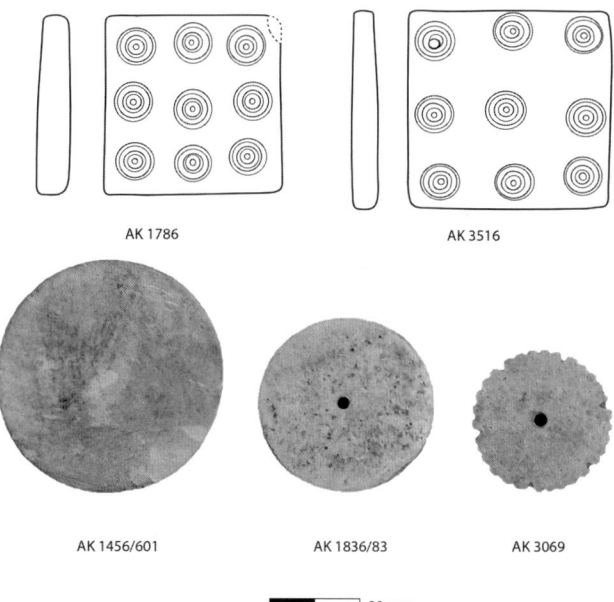

Figure 18.6: Ivory plaques and discs in the Mary Museum (drawings and photos by author).

Two rather different seals were found in burial 2550 of the Gonur Necropolis and on the surface of site no. 1220 of the 'Archaeological Map of the Murghab Delta' project (Salvatori 2008). The one from Gonur Depe bears an inscription in Sumerian characters. It is quite worn, but it has been interpreted as the name of a cup-bearer (Sarianidi 2002, 334). The carvings and shape suggest that it may have been a western import and can be placed within the Late Akkadian glyptic tradition, whereas the seal from site 1220 shares some of its stylistic features but has been interpreted as 'provincial Akkadian' (Salvatori 2008, 116; Kaniuth 2011, 14).

One further – albeit uncarved – example is housed in the Mary Collection (Fig. 18.7). It is a rough-out of a cylinder seal (Mary Museum number AK 3529/17) made out of a soft-stone (chlorite, according to the label). It was found during the 2000 campaign in the disturbed cist-grave of a young adolescent male (no. 1337). It is a rather short cylinder with a straight and wide perforation along the middle axis. The ends show traces of roughly perpendicular polishing streaks while the rounded surface is smooth and polished. This new addition gives yet further strength to the argument of 'regional re-elaboration' brought forward by Salvatori (2008, 116) and also underpins the simultaneous use of several administration/identification practices at Gonur.

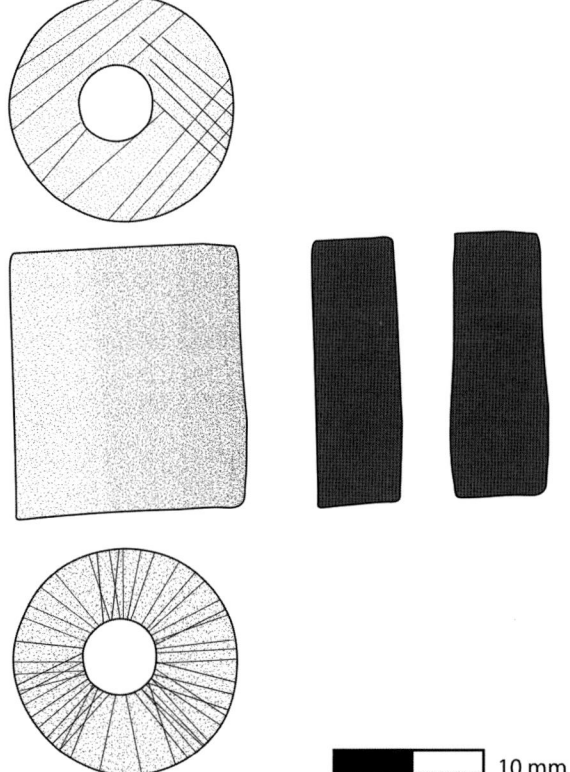

Figure 18.7: Unfinished cylinder seal. Mary Museum AK 3529/17 (drawing by author).

10 mm

Same but different

There are many other instances of objects excavated at Gonur and its vicinity that bear mention in this context. Links to the Gulf trade are clearly shown in various types of small soft-stone phials with geometrical incised decor. Several types seem to have been produced in different regions of Central Asia, south-eastern Iran and the Gulf, and show a wide distribution (Salvatori 2008, 84). Small soft-stone vessels of the *série récente* type provide a more direct link with the Gulf coast (Potts 2003, Ziolkowski 2006, 157). A very distinct series of hemispherical bowls and 'kidney-shaped' vessels have been found at Gonur Depe in the Palace area (rooms 1, 35, 147, 85) and in burials (1654, 1750) (Hiebert 1994, fig. 9.10; for other instances of 'heart'- or 'kidney-shaped' decor see Mackay 1937 pls 141.11–13, 142.34; Possehl 1996, 170). They represent types of the *série récente* typological groups A, C and D (Potts 1994, 262 after de Miroschedji 1973). Other examples come from tomb 5 of the Adji Kui necropolis (Ligabue and Rossi-Osmida 2007, 167 fig. 17) and Togolok (Mary Museum AK 1014; registered in 1977 without specifying the site-number). According to Hiebert they correspond to the local 'BMAC style' variant of these vessel types dated to Periods 1 and 2 of the Margiana chronology (Hiebert 1994, 148; for the hemispherical bowl found during the old excavations at Mohenjodaro see Potts 2008, 175 fig. 26). A quite different example comes from tomb 3245 of the Royal Necropolis of Gonur. The inventory of the tomb has been described by Salvatori (2010) and includes bronze vessels and a thick segment of an elephant tusk. It also contained a distinct two-compartment soft-stone vessel of the Umm an-Nar type that has been firmly linked to manufacturing centres on the Gulf shore (Potts 2008).

The list of objects found in Gonur Depe that can be related to external sources has become quite long over the years. It includes statues such as the fragment of a crouched figure from room 132 of the 'Royal Sanctuary' (Sarianidi 2004, 238 fig. 12; Sarianidi 2005, 122 fig. 30), a recumbent ram statuette from tomb 3220 of the Royal Necropolis (Sarianidi 2008a, 196 fig. 108) and a small duck-shaped weight from burial 1200 of the Necropolis similar to those in use in the Mesopotamian sphere (Sarianidi 2007, 119 fig. 226). A group of strainer-like ceramic tubes and tall jars of the Period 1 pottery assemblage has also been linked to vessels from the Indus Valley region *e.g.* 'pear-shaped jars' (Salvatori 2000, 33; Sarianidi 2007, 55 fig. 9–10; 2008b, 193 fig. 13.19). All these examples are mainly based on stylistic features and some – like the fragment of a crouched statuette – have already been refuted (Kaniuth 2011, 8 with reference to Winkelmann 1994).

In turn, objects of possibly Central Asian origin have been found at Shar-i Sokhta and Shahdad; in the Quetta Valley and at Mohenjo-daro; Susa; on the Islands of Failaka and Tarut, as well as Bahrain (for a detailed discussion see During-Caspers 1994; Possehl 2002, 231; Possehl 2006; Olijdam 2008; Potts 2008; Franke 2010).

Conclusion and final remarks

The dating of regular interactions between Gonur Depe, its immediate neighbours and the Indus Valley domains is somewhat debated. Salvatori pleads for an early date well within the Namazga V period (Salvatori 2003, 10), while Hiebert argues that no such finds come from Early/Mid-Namazga V contexts in Margiana (Hiebert 1994, 85). The late Umm an-Nar vessel and the Akkadian seals help to provide a chronological setting within the last few centuries of the 3rd millennium BC, while other finds like the ivory plaques seem to have more parallels with objects dated to the early 2nd millennium BC. The etched beads span a rather long period of time. However, the types identified at Gonur have clear parallels in late 3rd millennium contexts. And, as always, the long-lasting circulation of certain items has to be taken into account. Seals and other identifiers or 'prestige items' may have been deposited long after their manufacture. Whichever the preferred view, due to the limited chronological depth of the finds from Gonur a more precise dating is impossible at the moment.

Many models can be applied when discussing the nature of these interactions between urban centres in such diverse landscapes. Hiebert interprets the spread of BMAC-related material as a clear sign of people moving for the purpose of resource acquisition (Hiebert 1994, 141, 163). Similarly, Potts sees many of the aforementioned objects as some sort of 'peripheral items' and curiosities that circulated in the wake of a primary trade in raw materials and whose spread was inhibited only by 'cultural discontinuities' (1993, 394). For Possehl, the 'Middle Asian Interaction Sphere' (MAIS) was a unique confluence of 'economic, political and technological configurations' and the readiness for 'Harappan participation' was rooted in their special relationship to technology (a view that goes back to the emphasis scholarly tradition has placed on manufacturing processes and material development within Indus Valley crafts; *e.g.* Vidale 2000). Most of all, according to Possehl, a shared set of symbols and some kind of shared ideology was one of the main motors of the 'MAIS' (2002, 216). The 'technological revolution' and/or change in knowledge that he refers to is closely linked to the 'sudden rise' in maritime activity (*e.g.* 'Dilmun trade') thought to have taken place during the second half of the 3rd millennium BC (Potts 1994, 38; Vogt 1996, 107).

Taking this technological innovation as the starting point for what became the MAIS, we may seek inspiration in current sociological studies of innovation (see Schubert's contribution to this volume). According to Rogers' study of the sequential spread of innovation (1983, 247), the

initiators of this maritime link are likely to have been intrepid personalities on the margins of society. These 'venturesome innovators' would then have been followed in their enterprise by the more integrated and respectable 'early adopters', *e.g.* tradespeople, before the socially conformed centre, *i.e.* the ruling elite, took an interest in this exchange. Following this model would mean diverting from the prevalent paradigm of a (Mesopotamian) elite in need of prestige goods as providing much of the momentum for such an increase in interconnectedness (at least in the case of maritime trade). In fact, the premise that marginal groups display a higher permeability to innovation is based on the 'less to loose' hypothesis put forward by Rogers (1983, 252) and requires amongst other things the ability to better absorb losses (in financial and social capital). As Schubert points out in his contribution, this can apply to other marginal or subaltern groups as well. Those with an already marginal standing in their local social system may feel more inclined to adopt an innovation, as Rogers' example on 'Water Boiling in a Peruvian Village' clearly illustrates (Rogers 1983, 3).

However, to reduce the 'MAIS' to the maritime connection would be a misrepresentation, especially concerning the large continental networks evidenced by the material in southern Central Asia. In a somewhat simplistic analogy, let me point to yet another case in which 'the margins' become conduits of innovation: the Steppe, as seen from the urban settlements described above. Increasing archaeological evidence indicates that especially those often considered 'marginal' to the urban landscape played an instrumental role in spreading both resources and innovations (Frachetti 2012, 17). In the Murghab delta, these 'steppe connections' are evidenced by seasonal encampments in the vicinity of urban settlements (Hiebert and Moore 2004; Spengler *et al.* 2014) that significantly increase in number at the beginning of the 2nd millennium BC (Cattani 2008). The involvement of pastoralist communities of these 'key interstitial territories' (Frachetti and Rouse 2012, 690) is an integral part of the 'MAIS'.

Hard as it is to trace technological innovation and/or appropriation in an archaeological context, the objects presented above may provide a glimpse of the inner workings of these interaction networks. At this point the only way to shed some light on questions of circulation and interaction is through the careful contextualisation of the known objects. Many of the objects presented above lack such a clear context. Nevertheless, some information can be inferred, for instance from the case of the 'gaming pieces' or game-board inlays of ivory or bone. Recent studies show that gaming can be well localised within settlements (Rogersdotter 2011). Unfortunately, the examples from Gonur were largely found in graves, while the rest were found in and around the main representative building in areas without preserved room inventories ('Palace Complex'). Often gaming pieces are associated with divinatory practices (Rogersdotter 2011,

51). Do the Gonur finds thus point to the shared ideological and symbolic fabric that Possehl proposed as backdrop for the MAIS?

Another case is illustrated by the small etched bead found in grave 2065. The marked difference between deposition practices for jewellery in the Indus Valley domains as opposed to Mesopotamia and Iran has already been discussed (Kenoyer 2014). While this bead has clear typological parallels, its metal 'shafting' is puzzling. Etched beads with thin golden loops for suspension are shown in Woolley's publication of the Royal Cemetery of Ur (1934, 374 fig. 80, pl. 133). However, the present example is attached to a solid bronze stub that probably constitutes the remnant of a pin. Since pins were not a common feature of Indus ornament tradition (Kenoyer 2014), this find could represent an interesting case of appropriation of a foreign ornament into local traditions of dress. An equally interesting example is provided by the uncarved cylinder seal. Seals of this type that were found in southern Central Asia are usually seen in the context of cultural ties with south-eastern Iran, although the scarcity of material and chronological disparities make it difficult to ascertain where this extraneous element really came from (Salvatori 2000, 135). The presence of an uncarved seal in the grave of a young person can certainly be interpreted as a sign of foreign presence within the site. However, being an unfinished piece, it could evidence a local interest or need in imitating these objects, especially in the context of goods exchange.

Some of these examples point to an appropriation of certain habits and objects into the local repertoire. This was not an appropriation of images or visual narratives but rather of specific objects associated with administrative practices (cylinder seals) and entertainment or divination ('gaming pieces'). Furthermore, in view of the Indus seal (in the vicinity of the Palace Complex) and the recent example of an etched bead found within the settlement area, it is very tempting to conjecture that at some point individuals with close links to the Indus Valley were present in Gonur Depe.

Ultimately, the disentanglement of the MAIS into its different 'social worlds' (Strauss 1978) and their respective contact mechanisms (Potts 2015) will require much more research. Intense excavation endeavours in key regions have already filled many 'gaps' in the evidence (*e.g.* the Gulf coast of the eastern Arabian Peninsula), and modern stratigraphic methods can be expected to provide a better basis for correlation with old finds in many of the sites referred to here. The fact that the oases of Margiana had their own active share in the MAIS has been amply corroborated by a comparatively large pocket of material evidence. Further research in the area of the Kopet Dagh as well as the Margiana Oases would therefore close another large gap in this 'string' of interlocking exchange networks of the late 3rd millennium BC.

Acknowledgements

I am particularly indebted to Professor Nadezhda Dubova, Dr. Sylvia Winkelmann and Professor Mirko Novák of the Margiana Archaeological Expedition for their constant support of my research, as well as to Viktor Turik (Mary Museum antiquities department) and Ruslan Muradov (Ministry of Culture, Turkmenistan) for the immense amount of information they so graciously shared with me. I would also like to thank Professor Massimo Vidale and Dr Christoph Gerber for their invaluable comments and Martina Morello for first introducing Gonur to me. I am very grateful to the editors of this volume for encouraging this work by every means possible and to Tobi Reu for lending a sociologist's eye. Warmest thanks also go out to all my fellow team-members of the Gonur campaigns for letting me stray off the dig once in a while and supplying me with the all the latest documentation. Finally, I am eternally grateful to Natalie Golikov for helping me decipher handwritten object-labels in Russian.

References

Algaze, G. (1993) *The Uruk World System*. Chicago, IL, University of Chicago Press.

Beck, H. C. (1933) Etched Carnelian Beads. *The Antiquaries Journal* 13, 382–398.

Cattani, M. (2008) The Final Phase of the Bronze Age and the 'Andronovo Question' in Margiana. In S. Salvatori, M. Tosi and B. Cerasetti (eds), *The Bronze Age and Early Iron Age in the Margiana Lowlands: Facts and Methodological Proposals for a Redefinition of the Research Strategies*, 133–151, Oxford, Archaeopress.

Chakrabarti, D. K. and Moghadam, P. (1977) Some Unpublished Indus Beads from Iran. *Iran* 15, 166–168.

Cortesi, E., Tosi, M., Lazzari, A. and Vidale, M. (2008) Cultural Relationships beyond the Iranian Plateau: The Helmand Civilization, Baluchistan and the Indus Valley in the 3rd Millennium BCE. *Paléorient* 34/2, 5–35.

de Miroschedji, P. (1973) Vases et objets en stéatite susiens du Musée du Louvre. *Cahiers de la Délégation Archéologique Française en Iran* 3, 9–79.

De Waele, A. D. and Haerinck, E. (2006) Etched (Carnelian) Beads from Northeast and Southeast Arabia. *Arabian Archaeology and Epigraphy* 17/1, 31–40.

During-Caspers, E. C. L. (1971) Etched Cornelian Beads. *Bulletin of the Institute of Archaeology* 10, 83–98.

During-Caspers, E. C. L. (1994) Non-Indus Glyptics in a Harappan Context. *Iranica Antiqua* 29, 83–106.

Eger, C. (2010) Indisch, persisch oder Kaukasisch? Zu den Karneolperlen mit Ätzdekor der Gruppe C nach Beck und den östlichen Fernkontakten der Provinz Arabia. *Sonderdruck aus dem Jahrbuch des Römisch-Germanischen Zentralmuseums Mainz* 57, 221–278.

Frachetti, M. D. (2012) Multiregional Emergence of Mobile Pastoralism and Nonuniform Institutional Complexity across Eurasia. *Current Anthropology* 53/1, 2–38.

Frachetti, M. D. and Rouse, L. M. (2012) Central Asia, the Steppe, and the Near East, 2500–1500 BC. In D. T. Potts (ed.), *A Companion to the Archaeology of the Ancient Near East*, 687–705. Chichester *et al.*, Blackwell Publishing Ltd.

Francfort, H.-P. (1989) *Fouilles de Shortugai. Recherches sur L'Asie Centrale Protohistorique*. Paris, Diffusion de Boccard.

Franke, U. (2010) From the Oxus to the Indus: Two Compartmented Seals from Mohenjo-daro (Pakistan). In A. Parpola, B. M. Pande and P. Koskikallio (eds), *Corpus of Indus Seals and Inscriptions. Vol. 3: New material, untraced objects, and collections outside India and Pakistan. Memoirs of the Archaeological Survey of India 96*, xvii–xlii. Helsinki, Suomalainen Tiedeakatemia.

Gubaev, A., Koshelenko, G. and Tosi, M. (eds) (1998) *The Archaeological Map of the Murghab Delta. Preliminary Reports 1990–95*, Reports and Memoirs, Series Minor 3. Rome, IsIAO.

Hiebert, F. T. (1994) *Origins of the Bronze Age Oasis Civilization in Central Asia*. Cambridge, MA, Peabody Museum of Archaeology and Ethnology, Harvard University.

Hiebert, F. T. and Moore, K. M. (2004) A Small Steppe Site Near Gonur. In M. F. Kosarev, P. M. Kozhin and N. A. Dubova (eds), *Near the Sources of Civilizations. Issue in Honor of the 75-Anniversary of Victor Sarianidi*. Moscow, Staryi Sad.

Joshi, J. P. and Parpola, A. (1987) *Corpus of Indus Seals and Inscriptions 1. Collections in India. Memoirs of the Archaeological Survey of India 86*. Helsinki, Suomalainen Tiedeakatemia.

Jungner, H. (2007) Radiocarbon Dating of Samples from the Necropolis of Gonur in Turkmenistan. In V. Sarianidi (ed.), *Necropolis of Gonur, Appendix*. Athens, Kapon Editions.

Kaniuth, K. (2011) Long Distance Imports in the Bronze Age of Southern Central Asia. Recent Finds and Their Implications for Chronology and Trade. In A. V. G. Betts and F. Kidd (eds), *New Directions on Silk Road Archaeology. Proceedings of a Workshop Held at ICAANE V, Madrid, 2006*. Archäologische Mitteilungen aus Iran und Turan 42, 3–22. Berlin, Deutsches Archäologisches Institut and Dietrich Reimer.

Kenoyer, J. M. (2004) Chronology and Interrelations between Harappa and Central Asia. *Journal of the Japanese Society for West Asian Archaeology* 5, 38–45.

Kenoyer, J. M. (2014) Ornament Styles of the Indus Valley Tradition. In S. Kaul (ed.), *Cultural History of Early South Asia*. New Delhi, Orient Blackswan.

Lamberg-Karlovsky, C. C. (2003) Civilization, State or Tribe? Bactria and Margiana in the Bronze Age. *The Review of Archaeology* 24/1, 11–19.

Law, R. (2011) *Inter-Regional Interaction and Urbanism in the Ancient Indus Valley: A Geologic Provenience Study of Harappa's Rock and Mineral Assemblage*. New Delhi, Manohar Publications.

Ligabue, G. and Rossi-Osmida, G. (eds) (2007) *Sulla via delle oasi: Tesori dell'Oriente antico*. Trebaseleghe (Padova), Il Punto.

Mackay, E. J. H. (1933) Decorated Carnelian Beads. *Man 33, Royal Anthropological Institute of Great Britain and Ireland*, 143–146.

Mackay, E. J. H. (1937–1938) *Further Excavations at Mohenjodaro*. New Delhi, Manager of Publications.

Mackay, E. J. H. (1943, reprinted 1976) *Chanhu-Daro Excavations: 1935–36*. Varanasi, Bharatiya Publication House.

Marcolongo, B. and Mozzi, P. (1998) Outline of Recent Geological History of the Kopet-Dagh Mountains and the Southern Karakum. In A. Gubaev, G. Koshelenko and M. Tosi (eds), *The Archaeological Map of the Murghab Delta. Preliminary Reports 1990–95*. Reports and Memoirs, Series Minor 3, 1–14. Rome, IsIAO.

Marshall, J. (1931) *Mohenjo-daro and the Indus Civilization*. London, A. Probsthain.

Masson, V. and Sarianidi, V. (1972) *Central Asia – Turkmenia before the Achaemenids*. London, Thames and Hudson.

Masson, V. and Beriozkin Yu. E. (eds) (2005) *Хронология эПохи позднего энеолита – средней бронзы Средней Азии (погребения Алтын-депе)*. Российская Академия Наук, Институт Истории Материалной Культуры, Труды 16. St Petersburg.

Miglus, P. A. (2011) Ausgrabungen in Bakr Āwa. *Zeitschrift für Orient-Archäologie* 4, 136–194.

Olijdam, E. (2008) A Possible Central Asian Origin for the Seal-Impressed Jar from the 'Temple Tower' at Failaka. In E. Olijdam and R. H. Spoor (eds), *Intercultural Relations between South and Southwest Asia. Studies in Commemoration of E.C.L. During Caspers (1934–1996)*. British Archaeological Reports, International Series 1826, 268–287. Oxford, Archaeopress.

Parpola, A. (2009) A New Indus Seal Excavated at Gonur (Turkmenistan) in November 2004. In T. Osada (ed.), *Linguistics, Archaeology and Human Past in South Asia*, 71–76. New Delhi, Manohar Publ.

Possehl, G. L. (1996) Meluhha. In J. Reade (ed.), *The Indian Ocean in Antiquity*, 133–208. London, Kegan Paul International.

Possehl, G. L. (2002) *The Indus Civilization: A Contemporary Perspective*. Walnut Creek, CA, AltaMira Press.

Possehl, G. L. (2006) The Bactria-Margiana Archaeological Complex and the Greater Indus Valley. In V. Sarianidi *et al.* (eds), *Ancient Margiana – New Center of World Civilization. Proceedings of the International Scientific Conference, 14–16 November 2006 in Mary*. Ashgabat, Ministry of Culture of Turkmenistan.

Potts, D. T. (2003) A Soft Stone Genre from Southeastern Iran: 'Zig-Zag' Bowls from Magan to Margiana. In T. Potts (ed.), *Culture through Objects: Ancient Near Eastern Studies in Honour of P. R. S. Moorey*, 77–92. Oxford, Griffith Institute.

Potts, D. T. (2008) An Umm an-Nar-Type Compartmented Soft-Stone Vessel from Gonur Depe, Turkmenistan. *Arabian Archaeology and Epigraphy* 19, 168–181.

Potts, D. T. (2015) Et dona ferentes. Foreign Reception of Mesopotamian Objects. In R. Rollinger (ed.), *Mesopotamia in the Ancient World: Impact, Continuities, Parallels. Proceedings of the Seventh Symposium of the Melammu Project Held in Obergurgl, Austria, 4–8 November 2013*, 143–144. Münster, Ugarit-Verlag.

Potts, T. F. (1993) Patterns of Trade in Third-Millennium BC Mesopotamia and Iran. *World Archaeology* 24/3, 379–402.

Potts, T. F. (1994) *Mesopotamia and the East: An Archaeological and Historical Study of Foreign Relations, ca. 3400–2000 BC*. Oxford, Oxbow Books.

Pumpelly, R. (1908) *Explorations in Turkestan. Expedition of 1904*. Carnegie Institution of Washington Publication 73. Washington DC, Carnegie Institution.

Ratnagar, S. (2004) *Trading Encounters: From the Euphrates to the Indus in the Bronze Age*. New Delhi, Oxford University Press.

Reade, J. (1979) *Early Etched Beads and the Indus-Mesopotamia Trade*. British Museum Occasional Papers 2. London, British Museum.

Reinholdt, C. (2008) Der *frühbronzezeitliche Schmuckhortfund von Kap Kolonna : Ägina und die Ägäis im Goldzeitalter des 3. Jahrtausends v. Chr.* Vienna, Verlag der Österreichischen Akademie der Wissenschaften.

Rogers, E. M. (1983) *Diffusion of Innovations*. 3rd ed. New York, Free Press.

Rogersdotter, E. (2011) Gaming in Mohenjo-daro – An Archaeology of Unities. Unpublished thesis, University of Gothenburg, Gothenburg.

Rossi-Osmida, G. (ed.) (2002) *Margiana: Gonur-Depe Necropolis; 10 Years of Excavations by Ligabue Study and Research Centre*. Padova, Il Punto Edizioni.

Salvatori, S. (2000) Bactria and Margiana Seals: A Typological Survey and a New Assessment of their Chronological Position. *East and West* 50, 97–145.

Salvatori, S. (2003) Pots and Peoples: The 'Padora's Jar' of Central Asia Archaeological Research. On Two Recent Books on Gonur Graveyard Excavations (Margiana, Turkmenistan). *Rivista di Archeologia* 27, 5–20.

Salvatori, S. (2004) Oxus Civilization Cultural Variability in the Light of its Relations with Surrounding Regions: The Middle Bronze Age. In M. F. Kosarev, P. M. Kozhin and N. A. Dubova (eds), *Near the Sources of Civilizations. The Issue in Honor of the 75th Anniversary of Victor Sarianidi*, 92–101. Moscow, Margiana Archaeological Expedition.

Salvatori, S. (2008) A New Cylinder Seal from Ancient Margiana: Cultural Exchange and Syncretism in a 'World Wide Trade System' at the End of the 3rd Millennium BC. In S. Salvatori, M. Tosi and B. Cerasetti (eds), *The Bronze Age and Early Iron Age in the Margiana Lowlands: Facts and Methodological Proposals for a Redefinition of the Research Strategies*, 111–118, Oxford, Archaeopress.

Salvatori, S. (2010) Thinking Around Grave 3245 in the 'Royal Graveyard' of Gonur (Murghab Delta, Turkmenistan). In P. M. Kozhin, M. F. Kosarev and N. Dubova (eds), *On the Track of Uncovering a Civilization. A Volume in Honor of the 80th Anniversary of Victor Sarianidi*. Transactions of the Margiana Archaeological Expedition 3, 244–257. Moscow and St Petersburg, Aletheia.

Salvatori, S., Tosi, M. and Cerasetti, B (eds) (2008) *The Bronze Age and Early Iron Age in the Margiana Lowlands: Facts and Methodological Proposals for a Redefinition of the Research Strategies*. Oxford, Archaeopress.

Sarianidi, V. (1998a) *Margiana and Protozoroastrism*. Athens, Kapon Editions.

Sarianidi, V. (1998b) Myths of Ancient Bactria and Margiana on its Seals and Amulets. Moscow, Pentagraphic Publishing House.

Sarianidi, V. (2002) *Margush – Ancient Oriental Kingdom in the Old Delta of the Murghab River*. Ashgabat, Türkmendöwlethabarlary.

Sarianidi, V. (2004) *Дворцово-Культововый Ансамбль Северного Гонура*. In M. F. Kosarev, P. M. Kozhin and N. A. Dubova (eds), *Near the Sources of Civilizations. The Issue in Honor of the 75th Anniversary of Victor Sarianidi*, 229–253. Moscow, Margiana Archaeological Expedition.

Sarianidi, V. (2005) *Gonur Depe. City of Kings and Gods*. Ashgabad, Miras.

Sarianidi, V. (2007) *Necropolis of Gonur*. Athens, Kapon Editions.

Sarianidi, V. (2008a) *Margush – Mystery and Truth of the Great Culture*. Ashgabat, Türkmendöwlethabarlary.

Sarianidi, V. (2008b) 'Hypogaeum' Type Burials in the Gonur Necropolis. In E. M. Raven (ed.), *South Asian Archaeology 1999. Proceedings of the fifteenth International Conference of the European Association of South Asian Archaeologists, Held at the Universiteit Leiden, 5–9 July 1999*, 190–195. Groningen, Forsten.

Spengler III, R. N., Cerasetti, B., Tengberg, M., Cattani, M. and Rouse, L. M. (2014) Agriculturalists and pastoralists: Bronze Age economy of the Murghab alluvial fan, southern Central Asia. *Vegetation History and Archaeobotany* 23, 805–820.

Strauss, A. L. (1978) A Social World Perspective. In N. Denzin (ed.), *Studies in Symbolic Interaction* 1, 119–128. Greenwich, Jai Press.

Tallon, F. (ed.) (1995) Les pierres précieuses de l'Orient ancien des Sumériens aux Sassanides. Paris, Réunion des Musées Nationaux.

Vats, M. S. (1940) *Excavations at Harappa*. Delhi, Government of India.

Vogt, B. (1996) Bronze Age maritime trade in the Indian Ocean: Harappan traits on the Oman peninsula. In J. Reade (ed.), *The Indian Ocean in Antiquity*, 107–132. London, Kegan Paul International.

Vidale, M. (2000) *The Archaeology of Indus Crafts*. Rome, IsIAO.

Vidale M. (2010) *A oriente di Sumer. Archeologia dei primi stati euroasiatici, 4000–2000 a.C.* Rome, Carocci editore.

Winkelmann, S. (1994) Intercultural Relations between Iran, Central Asia and Northwestern India in the Light of Squatting Stone Sculptures from Mohenjo-daro. In A. Parpola and P. Koskikallio (eds), *South Asian Archaeology 1993. Proceedings of the Twelfth International Conference of the European Association of South Asian Archaeologists Held in Helsinki University 5–9 July 1993*, 815–832. Helsinki, Suomalainen Tiedeakatemia.

Winkelmann, S. (1997) Ein neues Trans-Elamisches Siegel. *Archäologische Mitteilungen aus Iran und Turan* 29, 135–146.

Winkelmann, S. (2014) Gonur Depe, Turkmenistan. Bericht über die 1. russisch-schweizerische Kampagne 2014, *Jahresbericht Schweizerisch-Liechtensteinische Stiftung für archäologische Forschung im Ausland*, 135–166.

Woolley, C. L. (1934) *The Royal Cemetery – Ur Excavations*. London, British Museum.

Ziolkowski, M. and Al-Sharqi, A. (2006) Dot-in-Circle: An Ethnoarchaeological Approach to Soft-Stone Vessel Decoration. *Arabian Archaeology and Epigraphy* 17, 152–162.

Zych, L. (2006) Perspectives on Chalcolithic Turkmenistan: The Global and the Local. In D. L. Peterson (ed.), *Beyond the Steppe and the Sown. Proceedings of the 2002 University of Chicago Conference on Eurasian Archaeology*, 112–121. Leiden, Brill.

Chapter 19

The Appropriation of Early Bronze Technology in China

Jianjun Mei, Yongbin Yu, Kunlong Chen, Lu Wang[1]

Introduction

Over the past two decades, there has been increasing archaeological evidence from north-west China suggesting a strong link between the beginnings of bronze metallurgy in East Asia and the eastward spread of bronzes across the Eurasian steppe (Fitzgerald-Huber 1995; Mei 2003a; Li, S. 2005; Mei *et al.* 2012). It has been observed that in terms of object types and manufacturing technologies bronze finds of the late 3rd and early 2nd millennia BC from north-west and northern China contrast sharply with those found in the Central Plains of China, indicating significant distinctions in the appropriation of early bronze technology in different regions (Mei *et al.* 2015, 223). The appearance of ritual bronzes and the rise of piece-mould casting technology at the Erlitou site in Henan province marked a breakthrough in the early development of bronze metallurgy in the Central Plains. Both regional interaction and local innovations provided further impetus for the development and spread of bronze metallurgy, which has been widely considered a highly significant technological and economic factor in the rise of Chinese civilisation in the Central Plains (Mei 2009a, 14; Fang 2010; Liu and Chen 2012, 271). This paper explores the appropriation of early bronze technology in the Central Plains and asks what it was that paved the way for innovations in casting technology, notably piece-mould casting. In an attempt to understand the driving forces behind these innovations, it highlights the role of local cultural and ritual traditions in shaping the trajectories of early bronze technology in this region. Finally, it focuses on the development of bronze technology in regions peripheral to the Central Plains and explores the crucial role played by socio-cultural factors in the evolution of some local bronze traditions.

The beginnings of metallurgy in China

Over the past decades, various theories have been put forward concerning the beginnings of copper and bronze metallurgy in China. During the 1940s–50s it was widely believed that copper and bronze metallurgy was introduced into China from elsewhere (Bishop 1942, 14; Loehr 1949; 1956). From the 1960s to the 1980s, the view that metallurgy developed independently in China gradually gained favour because more and more archaeological discoveries in the Gansu-Qinghai region in northwest China provided a growing body of evidence for the use of copper and bronze long before the Shang dynasty (16th–11th centuries BC), suggesting the existence of a primitive stage in the development of copper and bronze metallurgy (Barnard 1961, 108; 1983; Chang 1963, 139–141; Barnard and Sato 1975, 1–16; Chêng 1974; Ho 1975, 177–221; Sun and Han 1981). Since the 1990s, however, archaeological evidence suggesting strong cultural links between the Eurasian steppe and northwest China has prompted a number of scholars to again argue in favour of the major role of outside influence in the beginnings of copper and bronze metallurgy in China (An 1993; Fitzgerald-Huber 1995; Mei, 2003a; Li, S. 2005; Mei *et al.* 2012).

Though the idea that copper and bronze metallurgy was introduced into China from elsewhere now seems to be widely accepted among scholars in China and abroad, many issues remain unresolved, such as when, how, and where the introduction process actually took place (Linduff *et al.* 2000; Linduff 2004). The argument for cultural links between the Eurasian steppe and northwest China is largely based on typological similarities of copper and bronze objects from these two regions, notably back-curved knives, socketed axes and spearheads, bone-mounted awls, and earrings with flared or trumpet-shaped ends (Fitzgerald-Huber 1995; Mei 2003a; 2003b). In north-west China, archaeological cultures with which early copper and bronze objects are associated include Xiaohe and Tianshanbeilu in eastern Xinjiang, Siba in the Hexi Corridor in Gansu, and Qijia in eastern Gansu and Qinghai (Fig. 19.1.1–6), while in the Eurasian steppe, early copper and bronze finds are related to the Okunev culture in Minusinsk and Altai and to the eastern part of the Seima-Turbino complex at Rostovka in the upper Ob River Valley (Fig. 19.1). To date, there are two major hypotheses concerning the transmission route of early metallurgical technology from the Eurasian steppe to the Central Plains of China. One suggests a route starting in Altai, then moving southward to eastern Xinjiang, running eastward along the Hexi Corridor to eastern Gansu, and finally reaching the Central Plains (Li, S. 2005). The other favours a route running southward from the Minusinsk basin to the northern border areas of China via Mongolia (Fig. 19.1; Fitzgerald-Huber 1995; Mei 2003a; Liu and Chen 2012, 332). For the time being, the paucity of archaeological evidence for the use of early copper and bronze objects in Mongolia makes the latter route unlikely, while recently discovered evidence for early metallurgical activities along the route via the Hexi Corridor suggests that this route is a more plausible candidate.

The geochemical analysis of sediments from Huoshiliang in the middle of the Hexi Corridor in northwestern Gansu has revealed significant occurrences of Cu and As for the period 2200–1700 BC, suggesting the existence of early metallurgical activities in the region (Fig. 19.1.4; Dodson *et al.* 2009; Li, X. *et al.* 2010). A copper smelting site was recently found at the so-called Heishuigou site (now also known as 'Xichengyi') near Zhangye, again in the middle of the Hexi Corridor. It has been dated to the late 3rd and early 2nd millennia BC (Fig. 19.1.5). According to a brief report by Wang and Chen (2012), more than 30 copper and bronze artefacts have been unearthed at this site, mainly small awls, knives, rings and buttons plus remains indicative of metallurgical activity, such as ores, furnace fragments, clay tuyeres, slag, and stone casting molds. Other significant finds from this settlement site include houses built with clay bricks, painted and plain pottery vessels, wheat, barley and millet, animal bones, and ornaments made of turquoise, quartz, and carnelian. Preliminary examination of slag samples indicates the practice of copper smelting at the site (Mei *et al.* 2012). Findings from the analysis of dozens of copper and bronze objects recovered at the Xichengyi site have recently been published, revealing the use of copper and its alloys, such as Cu-As, Cu-Sn, Cu-Sb, and Cu-As-Sb, but with unalloyed copper predominant (Chen *et al.* 2015). A recent study of metallurgical remains (slag and ore samples) from the Xichengyi site (Li, Y. *et al.* 2015) also indicates that both unalloyed copper and Cu-As were produced at this smelting site via the reduction of oxidized copper minerals, some of which may occasionally contain arsenic.

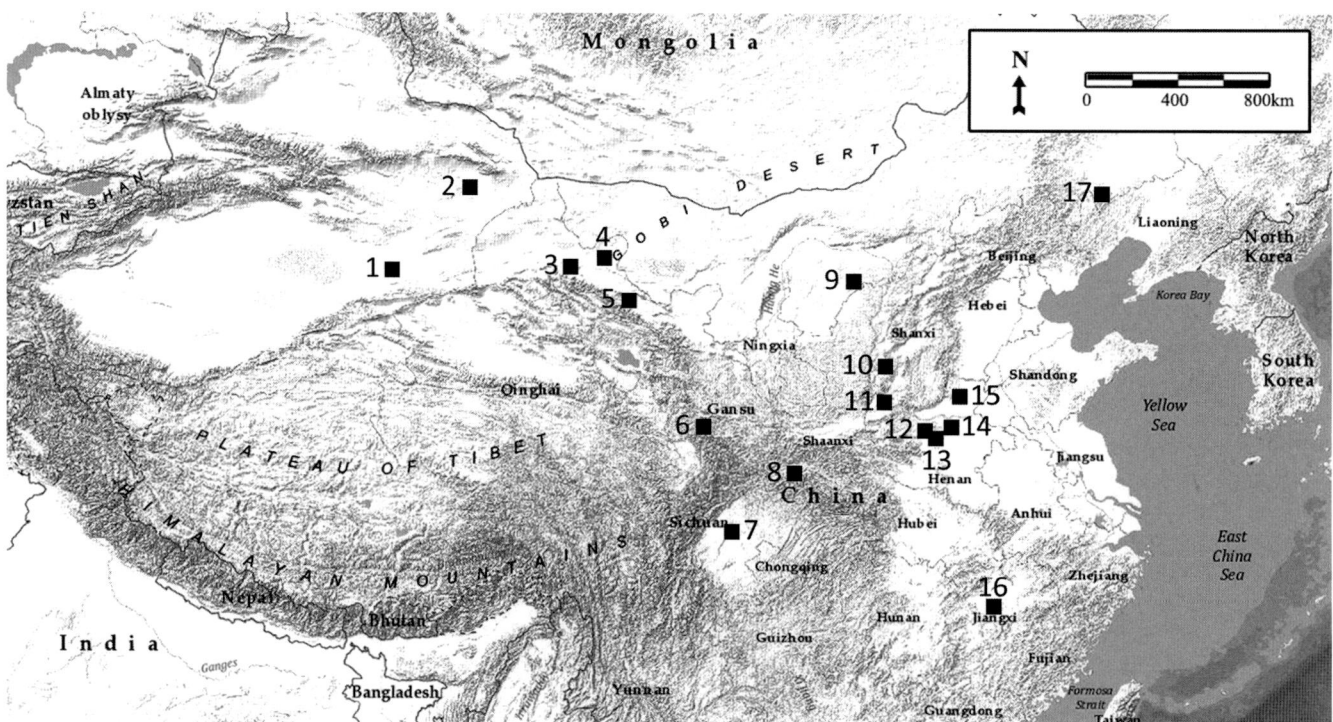

Figure 19.1: The map showing the major archaeological sites mentioned in the text: (1) Xiaohe; (2) Tianshanbeilu; (3) Huoshaogou; (4) Huoshiliang; (5) Heishuiguo (Xichengyi); (6) Mogou; (7) Sanxingdui; (8) Hanzhong; (9) Zhukaigou; (10) Shilou; (11) Taosi; (12) Erlitou; (13) Wangchenggang; (14) Erligang; (15) Yinxu; (16) Xin'gan; (17) Dadianzi.

This site is important in at least two respects. First, it clearly represents a significant juncture in the appropriation of metallurgical technology in the Hexi Corridor region. Second, finds such as clay bricks and wheat are crucial evidence for the existence of early cultural influences from the west. Although it is still far from clear how these western influences arrived in the Hexi Corridor region, the finds from the Heishuiguo (Xichengyi) site have revealed several interesting features characteristic of early metallurgical activities in the region: a) the mining was carried out in other places, b) mineral ores were transported to the settlement site, c) smelting operations took place on a small scale, d) tuyeres were used to achieve high temperatures, e) casting was carried out in stone moulds, f) copper and arsenical copper were produced, and g) metals were used to make small implements and ornaments. These features provide us with some general ideas about how the earliest metallurgy found so far in northwest China was organised and what role it may have played in a village-like community. On the basis of the evidence available to date, it seems quite possible that early metallurgical technology may have been transmitted as a package of skills and lore, including mining (ores), smelting (furnace and tuyeres or blowing pipes), casting (stone moulds), and usage (awls and knives). Of great interest is the fact that this transmission may have taken place together with other cultural elements, such as architectural material (clay bricks), the domestication of goats, cattle, wheat, and barley, and interest in ornaments made of precious stones and metals (Jaang 2015). In other words, the beginnings of copper and bronze metallurgy in northwest China were not merely isolated technological events, instead they were part of a longstanding and widespread process of cultural interaction between east and west via the Eurasian steppe in the course of the late 3rd millennium BC.

The problem of personal ornaments

More than thirty years ago, Ursula M. Franklin (1983, 95) raised the problem of personal ornaments in her discussion of the beginnings of metallurgy in China: 'When discussing archaeological inventory in China, serious attention has to be paid to the culturally determined absence of personal decoration made of metal. This fact has to be considered when comparing and contrasting the Chinese situation with that of other civilizations [...] For those who are looking for "proof" of the indigenous nature of Chinese metallurgy, this very characteristic idiosyncrasy should be another indication that there were no metal-using outsiders who brought copper and bronze work into China. The people to the west, north and south of the Chinese heartland used metal extensively for personal ornamentation and would have brought this application to them.' Although archaeological discoveries over the past three decades reveal a rather different picture

concerning the uses of personal ornaments made of metal in China, this acute observation remains both stimulating and significant.

As we saw earlier, over the past three or more decades, hundreds of copper and bronze objects have been found in northwest China and the northern border areas of China, such as those unearthed from the Xiaohe and Tianshanbeilu cemeteries in Eastern Xinjiang (Fig. 19.1.1–2), the Huoshaogou cemetery in the Hexi Corridor, Gansu (Fig. 19.1.3), the Mogou cemetery in Lintan, Gansu (Fig. 19.1.6), and the Dadianzi cemetery in Inner Mongolia (Fig. 19.1.17). These early copper and bronze objects are mostly either small implements and weapons or personal ornaments, such as earrings, finger rings, armbands, and bracelets (Fig. 19.2). Personal gold and silver ornaments, such as earrings and finger rings, also figure among the archaeological finds from the Xiaohe cemetery in Xinjiang and the Huoshaogou and Mogou cemeteries in Gansu. These new archaeological discoveries do not, however, refute Franklin's observation because we have yet to discover personal ornaments made of metal in the Central Plains of China.

The earliest undisputed evidence for the use of copper and its alloys in the Central Plains of China comes from the Taosi site in Xiangfen, southern Shanxi (Fig. 19.1.11), which has been ascribed to the Longshan culture and dated to the late third millennium BC (Bai 2002, 31). We now know that a small copper bell excavated at the Taosi site is in fact made of copper with a small amount of lead (Cu 97.86%, Pb 1.54%) (Li, M. *et al.* 1984). Further metal finds from this site include two rings (one with 29 protrusions) and a fragment. While the small ring is made of unalloyed copper, the ring with protrusions and the fragment are both Cu-As, the earliest instance found so far in the Central Plains of China (Gao and He 2014). Though some scholars suggest that the use of Cu and Cu-As at the Taosi site may have resulted from cultural influence from northwest China (Liu and Li 2014), the whole picture remains obscure because a clear chronological framework has yet to materialise. It is worth noting that while the two rings may be considered 'personal ornaments,' the bell and the fragment display clear links with the later metal finds from the Erlitou site in Yanshi, Henan province.

Among the copper and bronze objects found at the Erlitou site (Fig. 19.1.12), the earliest site providing direct evidence for metallurgical production found in the Central Plains, ritual vessels, knives, awls, dagger-axes, arrowheads, and plaques are the most common types encountered, while personal ornaments such as earrings and bracelets are conspicuous by their absence (Fig. 19.3). The manufacturing technology in the Central Plains is characterised by the predominance of piece-mould casting, while the bronze technology employed in north-west and northern China was limited to simple casting and forging.

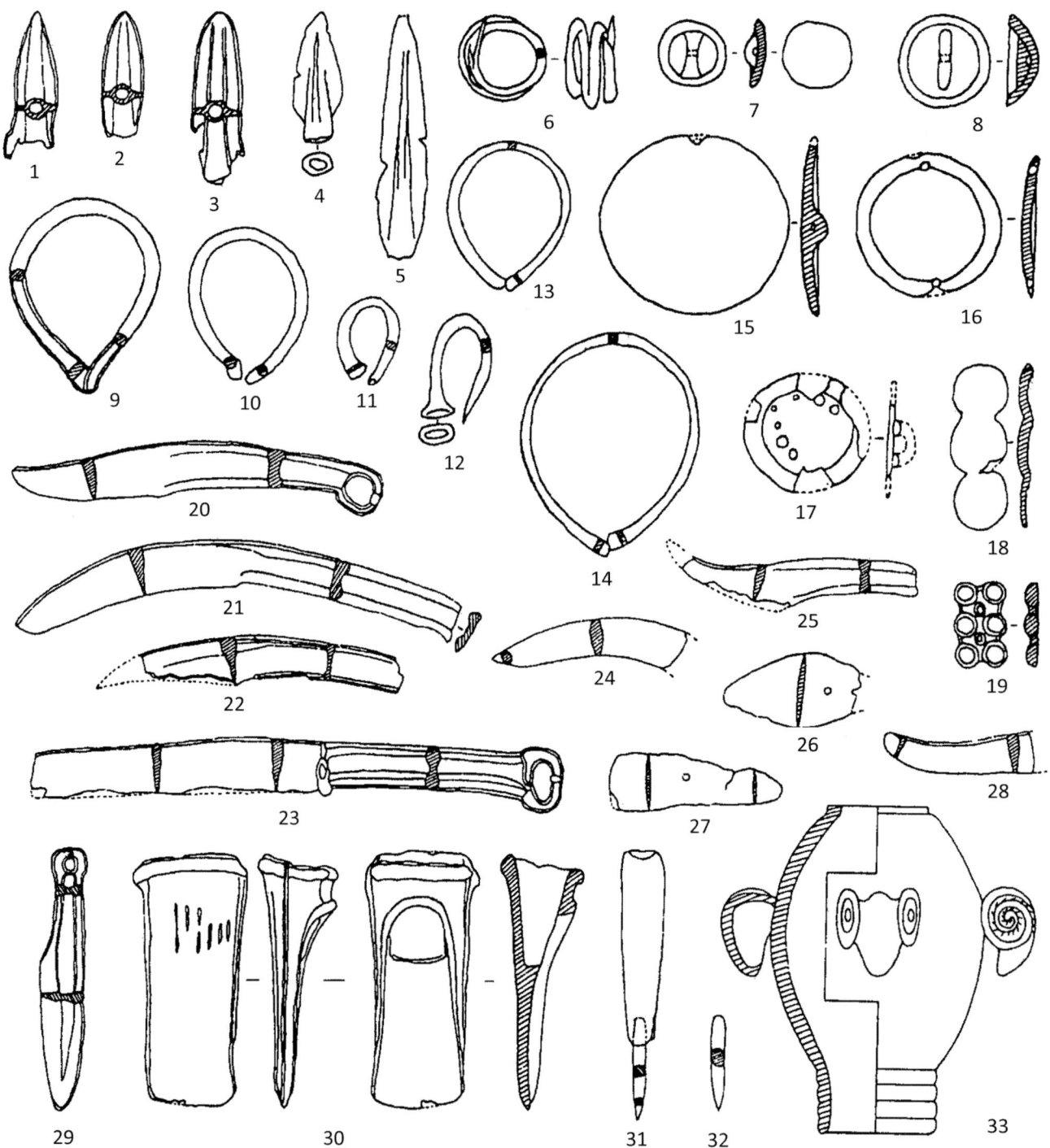

Figure 19.2: Copper and bronze artefacts of the Siba culture in Gansu: (1–5) arrowheads; (6) ring; (7–8) buttons; (9–14) earrings and rings; (15–19) ornaments; (20–29) knives; (30) socketed axe; (31–32) awls; (33) macehead (after Bai 2002, 29 fig. 3).

In other words, the initial stage of metallurgy in the Central Plains seems to present an entirely different picture, one that contrasts strongly with what we know about early metal-using cultures in north-west China or the northern border areas of China.

Should we, then, chime in with Franklin's suggestion and see this absence of personal metal ornaments as proof of the indigenous nature of Chinese metallurgy? To answer this question, we need to examine the technological and socio-cultural backgrounds of the rise of ritual

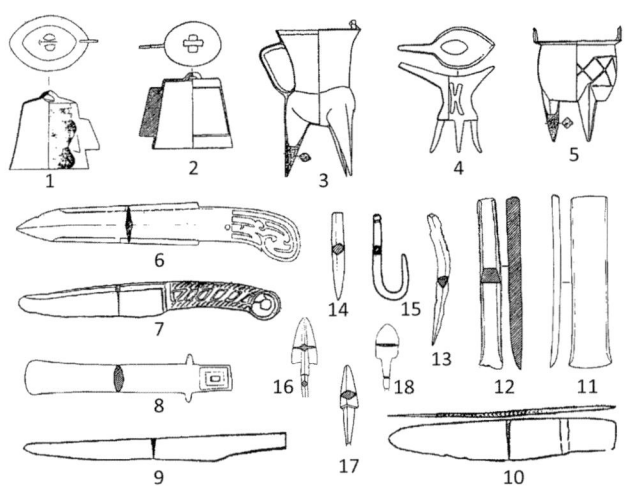

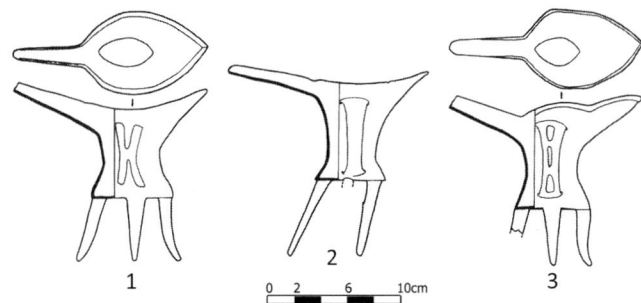

Figure 19.4: Three bronze jue *vessels of the Phase III unearthed at the Erlitou site (after IA-CASS 1999, 252 fig. 164).*

Figure 19.3: Copper and bronze artifacts of the Erlitou culture found in Yanshi, Henan, Central Plains of China: (1–2) bells; (3–5) jia, jue *and* ding *vessels; (6) dagger-axe; (7, 9, 10) knives; (8) axe; (11–15) small implements; (16–18) arrowheads (after Bai 2002, 33–36 figs. 7–9, 11).*

bronze vessels and piece-mould casting in the Central Plains region.

The rise of ritual bronze vessels and piece-mould casting in the Central Plains

As we have seen, the earliest copper and bronze objects found so far in the Central Plains come from Taosi (Fig. 19.1.11), a Neolithic site dated to 2600–2000 BC. They include a bell, a ring, a gear-shaped object, and a fragment. Some scholars suggest that at this time metal was not a part of the regular inventory of prestige goods found at elite burial sites (Liu and Chen 2012, 222–225). Others think differently and believe that metal production and consumption may have already been under the control of the elite (Fang 2010, 75–76). It has also been suggested that the making of the bell implied a casting technology for assembling the inner and outer moulds. In addition, the fragment may have been part of a bronze vessel and as such represent further evidence for the use of piece-mould casting (Gao and He 2014). Another Neolithic site, Wangchenggang (Fig. 19.1.13), has yielded a copper fragment dated to around 1900 BC and thought to come from a broken vessel (Bai 2002, 31). The earliest concrete evidence for the appearance of bronze vessels in the Central Plains comes from the Erlitou site, where several *jue* vessels have been unearthed from the Phase III remains and are dated to *c.* 17th century BC (Fig. 19.4; IA-CASS 1999, 252). Lian *et al.* (2011, 566) have examined 7 *jue* vessels excavated from the Erlitou site and conclude that they were all made by using piece-mould casting technology (Fig. 19.5). This conclusion is strongly supported by the finds of more than 20 fragments of clay moulds at the site (IA-CASS 1999, 270). It should be noted

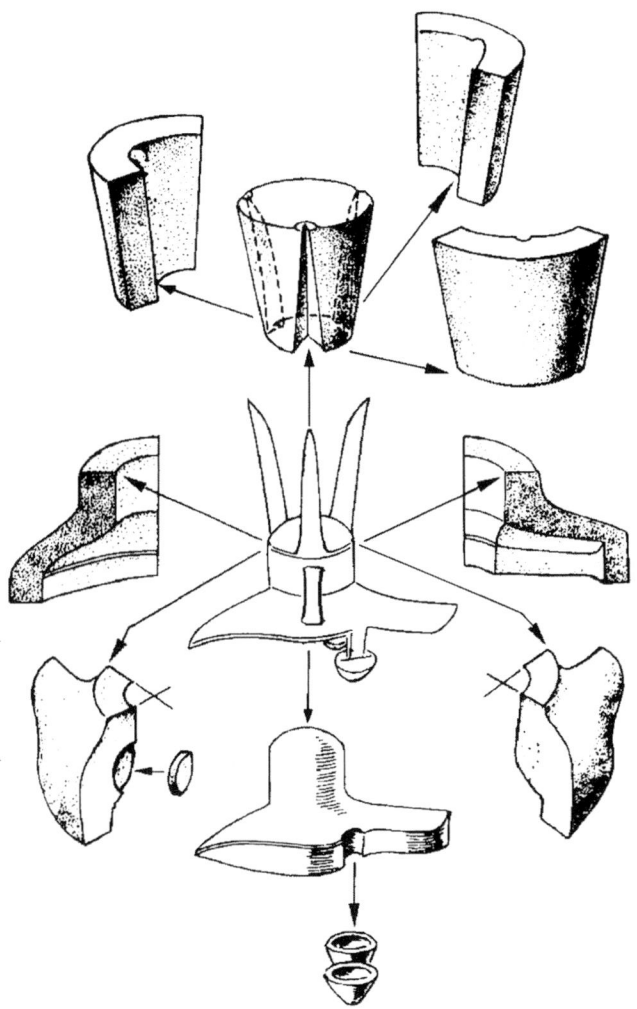

Figure 19.5: Wan Jiabao's reconstruction of casting a jue *vessel by using piece-moulds technology during the Yinxu period (after Lian* et al. *2011, 567 fig. 2).*

that small knives already appeared during Phase I at Erlitou, while Phase II featured a clear increase in object types such as awl, bell, and plaque, though no vessels (Chen 2008, 152).

Two major hypotheses have been advanced to explain the rise of ritual bronze vessels and piece-mould casting in

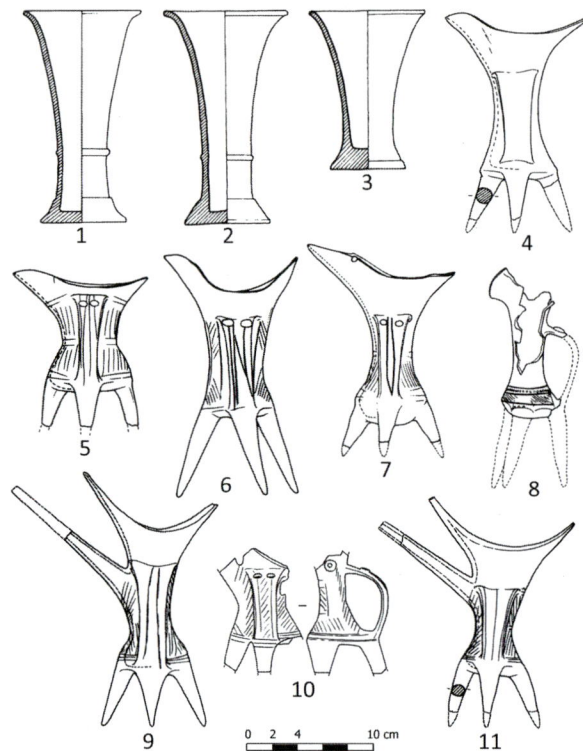

Figure 19.6: Ceramic vessels of the Phase II unearthed at the Erlitou site (after IA-CASS 1999, 136 fig. 81).

Figure 19.7: A gui vessel found in Northern Shaanxi showing a local decorating taste (after Cao 2009, 88).

Figure 19.8: A bronze figure head recovered at the Sanxingdui site, Guanghan, Sichuan (after EC 1994, 7).

272; Mei 2009b, 226). The other draws upon typological comparisons and observations of so-called 'hammering' features on some Erlitou ceramic vessels to argue in favour of the introduction of bronze vessel types from the Eurasian steppe (Hwang 2014, 35). Bagley (2014, 39) has long argued for the presence of vessels hammered from sheet metal in north China around 2000 BC, but so far there has been no archaeological find to bear this out, except for a few imitations in pottery and cast bronze. Accordingly, the appearance of hammered vessels prior to that of cast vessels must for the moment remain an unresolved issue in the study of early Chinese metallurgy. Considering the fact that the earliest bronze vessels, such as *jue*, *jia* and *he*, have their ceramic counterparts in an earlier context at the Erlitou site, it would seem more likely that these bronze vessels were made to imitate their ceramic prototypes (Fig. 19.6). In other words, both the existing ritual system and the desire to integrate newly obtained metal materials into the system may have acted as a major driving force in the rise of ritual bronze vessels.

From a technological point of view, it was the social and ritual need for bronze vessels that stimulated the invention of piece-mould bronze casting, a technology that assembles

the Central Plains. One suggests that bronze technology was most likely adapted to existing ritual practices characterized by the use of ceramic vessels for 'ritual feasting as a part of the ancestral cult ceremony' (Liu and Chen 2012, multiple inner and outer moulds for casting purposes (Liu and Chen 2012, 271–272). Most scholars agree that the invention of piece-mould casting technology marked a major breakthrough in the development of Chinese metallurgy. Franklin (1983, 96) goes so far as to say that piece-mould technology 'has had profound consequences on the mode of bronze production, on the division of labor, and on the organization and coordination of bronze manufacturing.'

Why did such a major technological innovation take place at Erlitou? A long time ago, Barnard pointed out a possible link between pottery kilns and bronze-casting furnaces (Barnard and Sato 1975, 27). Franklin (1983, 96) also feels that 'the intimate connection between ceramic and metal technology has resulted in the evolution of a uniquely Chinese system of casting bronze into piece molds.' On the basis of very recent archaeological discoveries, Zhang and Chen (2013) point out that in the Central Plains significant technological knowledge in mining, pottery kiln operation, and mould-making had accumulated since the Neolithic period, representing a crucial technological premise for the appearance of bronze metallurgy in the region. The crucial role of pottery-making technology in the emergence of piece-mould casting is widely recognized among scholars, though in terms of current archaeological evidence, the emergence of ritual bronze vessels and piece-mould casting technology at the Erlitou site admittedly still seems rather sudden, and as Bagley (1999, 139–142) suggests, there is insufficient evidence to trace a process of transition from small-scale metallurgy to large-scale metallurgy.

The emergence of local bronze traditions during the Erligang and Yinxu periods

During the Erlitou period, the production of ritual bronze vessels appears to have been restricted to the Erlitou site, with no evidence of such production at other sites such as Dongxiafeng, suggesting 'an increased state monopoly in producing high status symbols' (Liu and Chen 2012, 268). The subsequent Erligang (c. 1600–1400 BC) and Yinxu periods (1250–1046 BC) witnessed a flourishing in the production and use of ritual bronze vessels in the Central Plains and some of its surrounding regions. While major technological influence from such centres as Erligang (Fig. 19.1.14) and Yinxu (Fig. 19.1.15) can be seen in many bronze discoveries in the surrounding regions, local or regional bronze traditions also emerged and developed, exhibiting a degree of exoticism in taste and choice. Some scholars have argued that it was the so-called Erligang expansion that resulted in the spread of the piece-mould casting technology from the Central Plains centres to the surrounding regions and the emergence of local bronze traditions in those regions (McNeal 2014; Steinke 2014). Wang Haicheng (2014) proposes that ideological motives plus resource procurement and trade may have played a

major role in the large-scale dissemination of a distinctive material culture such as ritual bronze vessels as observed in the Erligang expansion.

Bronze objects found in the region to the north of the metropolitan region are clearly local in taste, examples being the objects found in Shilou, Shanxi province (Fig. 19.1. 10). First, they are mostly weapons and implements such as daggers and knives typically decorated with a terminal in the shape of an animal head, as well as axes with tubular sockets (Di Cosmo 1999, 894). In fact, this is a local bronze tradition associated with the Zhukaigou and Lower Xiajiadian cultures that can be traced back to the early Bronze Age (Fig. 19.1.9, 19.1.17). A tradition like this has a strong connection with the spread of early bronze technology across the eastern Eurasian steppe. Of particular interest is the region between the Central Plains and the eastern Eurasian steppe, where some bronze vessels with local characteristics have been found that seem to reflect the great efforts made by locals to cast vessels in imitation of the Central Plains prototypes, albeit not so successfully (Fig. 19.7). This example demonstrates once again the crucial role played by local socio-cultural settings. Without the support of a dominant and appropriate ritual or ideological system in which to situate the vessels, the production of such imitations based on 'purely' technical considerations, such as type or casting method, would not have been sustainable in the long term.

Bronze objects found in Sichuan, southwest China are another example demonstrating the significant role of socio-cultural factors in defining technological choice. Hundreds of extraordinary sculptures and masks cast in bronze, mostly human heads (Fig. 19.8), have been excavated at the Sanxingdui site in Sichuan (Fig. 19.1.7) and demonstrate that 'the Sichuan basin was the home of a sophisticated bronze-using culture at a time far earlier than hitherto assumed' (Bagley 1988, 78). The Sanxingdui finds are so unexpected that few scholars would consider them to be the result of a secondary development rooted in the bronze technology of the Central Plains of China. Yet, together with these exotic sculptures and masks a few ritual bronze vessels, *zun* and *lei*, were found, indicating some sort of connection between Sanxingdui and Hunan in the central Yangtze river region (Bagley 1990). Although cultural and technological influence from the Central Plains region of the Shang Kingdom may have existed indirectly in Sanxingdui, the general characteristics clearly derive from a local bronze tradition. We know very little about that tradition, but its existence is beyond doubt. The casting technology for the manufacturing of these bronze objects may have come originally from the Central Plains, but in terms of artistic and ritual expression they are entirely indigenous, and they most likely performed a role that was completely different from the part they played in the Central Plains region (Liu and Chen 2012, 390).

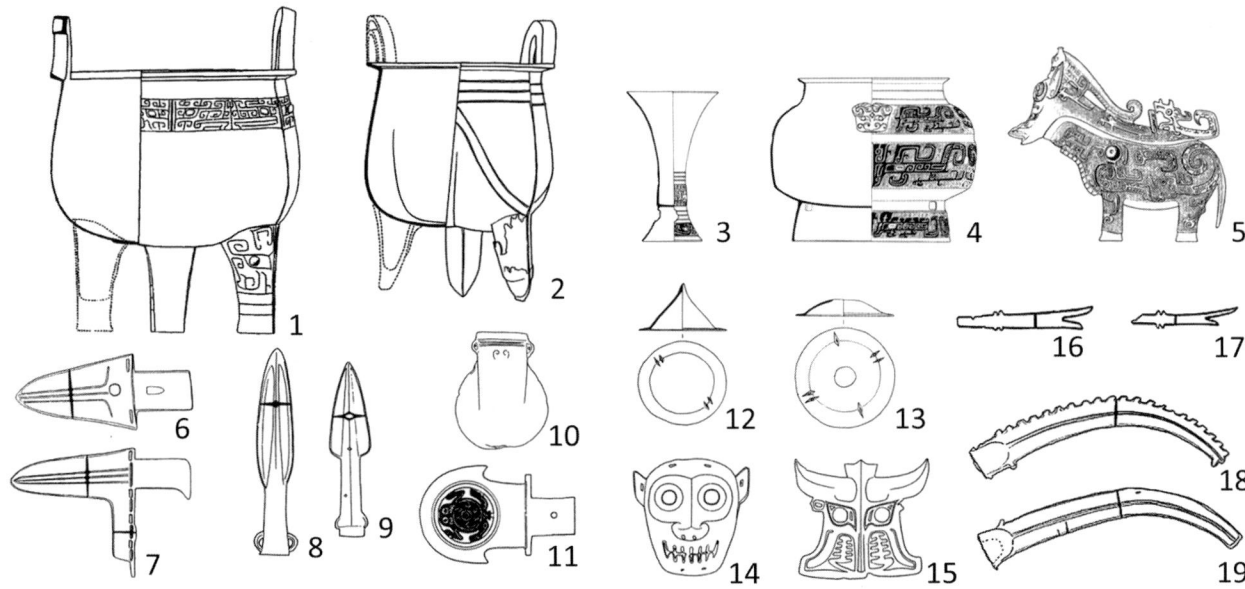

Figure 19.9: Drawing of different types of objects from Hanzhong, Shaanxi (after Cao 2006 and Zhao 2006; in different scales).

Hundreds of bronze objects recovered in the Hanzhong region, located roughly between Sichuan and the Central Plains region, are also of great interest. Relatively speaking, these objects show much stronger connections with the bronze products in the Central Plains, as is evidenced by the existence of dozens of ritual bronze vessels, such as *zun*, *ding*, and *li*. However, types of a distinctly local character also abound, such as sickle-shaped objects, ornaments, and masks (Fig. 19.9; Chen *et al.* 2009). It is difficult to determine what ritual practices were current during the late Shang period in the Hanzhong region, but they were most likely different from those in the Central Plains region.

Bronze objects recovered in southern China, such as those from Xin'gan in Jiangxi Province (Fig. 19.1.16), remind us of the expansion of the Central Plains influence to the Yangtze River valley. Steinke (2014, 170) points out Xin'gan's connection with Erligang and interprets it as evidence for Erligang expansion. Yet he also emphasises the contribution of southern cultures, stating that 'the Erligang expansion brought northern ideas into contact with southern cultures that were fully prepared to exploit them.' While the repertory of ritual vessels recovered at the Xin'gan site is generally similar to that of the Central Plains, distinctive local additions can be easily identified in many bronze vessels, such as the bird and tiger motifs. Such additions are clearly reflections of local artistic preferences, and this has prompted some scholars to argue in favour of the existence in the Yangtze region of civilized centres or city-states independent of the Shang power at Anyang at the beginning of the Anyang period (Bagley 1993, 35–36). Here we have a further example of the selective adaptation of a ritual system to a local setting with some deliberate changes made to accommodate local ideological or artistic characteristics.

In summary, the existence of diverse local bronze traditions in peripheral regions around the Central Plains provides us with examples demonstrating the crucial role of various ideological or ritual systems in shaping the development of bronze technology in Bronze Age China. On the one hand, it was the ritual system that decided on the direction taken by the development of bronze technology; on the other hand, bronze technology and its products, like piece-mould casting and vessels, became an integral part of that ritual system. It is this kind of interaction between technology and its socio-cultural context that has proved a longstanding force driving the development of Bronze Age civilisation in China.

Conclusions

On the basis of our discussion, four conclusions suggest themselves. 1) There has been increasing archaeological evidence indicating that copper and bronze metallurgy may have been introduced into the Gansu-Qinghai region, north-west China from the Eurasian steppe during the 3rd millennium BC, though the nature and the mechanisms of this process remain extremely obscure. 2) The appropriation of early bronze technology in the Central Plains of China was crucially shaped by the existing ideological system, which was characterised by the use of ritual vessels. 3) The social and ritual need for producing bronze vessels and the highly developed ceramic manufacturing system already in place may have led to the invention of piece-mould casting technology as well as the subsequent innovations in Bronze Age China. 4) The metallurgical diversity in Bronze Age China suggests that during the Erligang and Yinxu periods, local cultural and ritual traditions must have played a major role in the appropriation of bronze technology in the areas surrounding the Central Plains of China.

Acknowledgements

We are most grateful to Prof. Dr Joseph Maran and Prof. Dr Philipp W. Stockhammer for their kind encouragement and patience. We would also like to thank Mr John Moffett for kindly revising this paper. Our work was supported by grants awarded by the National Natural Science Foundation of China (Nos. 51074026; 51474029) and the Administration of the Cultural Heritage of China under the 'Excellent Young Scholars Research Project' (No. 2014220).

Note

1   Jianjun Mei: The Institute of Historical Metallurgy and Materials, University of Science and Technology Beijing, 100083, China and The Needham Research Institute, Cambridge, United Kingdom, jjm1006@cam.ac.uk; Yongbin Yu, Kunlong Chen, Lu Wang: The Institute of Historical Metallurgy and Materials, University of Science and Technology Beijing.

References

An, Z. (1993) On Early Copper and Bronze Objects in China. *Kaogu* (Archaeology) 12, 1110–1119 (in Chinese).

Bagley, R. W. (1988) Sacrificial Pits of the Shang Period at Sanxingdui in Guanghan County, Sichuan Province. *Arts Asiatiques* 43, 78–86.

Bagley, R. W. (1990) A Shang City in Sichuan Province. *Orientations* 21/11, 52–67.

Bagley, R. W. (1993) An Early Bronze Age Tomb in Jiangxi Province. *Orientations* 24/7, 20–36.

Bagley, R. W. (1999) Shang Archaeology. In M. Loewe and E. L. Shaughnessy (eds), *The Cambridge History of Ancient China: From the Origins of Civilization to 221 B.C.*, 124–231. Cambridge, Cambridge University Press.

Bagley, R. W. (2014) Erligang Bronzes and the Discovery of the Erligang Culture. In K. Steinke (ed.), *Art and Archaeology of the Erligang Civilization*, 19–48. Princeton, NJ, Princeton University Press.

Barnard, N. (1961) *Bronze Casting and Bronze Alloys in Ancient China*. Canberra, Australian National University.

Barnard, N. (1983) Further Evidence to Support the Hypothesis of Indigenous Origins of Metallurgy in Ancient China. In D. N. Keightley (ed.), *The Origins of Chinese Civilization*. Berkeley, CA, University of California Press.

Barnard, N. and Sato, T. (1975) *Metallurgical Remains of Ancient China*. Tokyo, Nichiosha.

Bai, Y. (2002) The Origins of Early Chinese Copper and Bronze Objects. *Dongnan Wenhua* (South-east Culture) 7, 25–37 (in Chinese).

Bishop, C.W. (1942) *Origin of the Far Eastern Civilizations. A Brief Handbook*. Smithsonian Institute, Washington, DC.

Cao, W. (ed.) (2006) *Shang Dynasty Bronzes Unearthed from Hanzhong*1. Chengdu, Bashu Publishing House (in Chinese).

Cao, W. (2009) *Bronzes from Northern Shaanxi* 2. Chengdu, Bashu Publishing House (in Chinese).

Chang, K. C. (1963) *The Archaeology of Ancient China*. New Haven, CT and London, Yale University Press.

Chen, G. (2008) A Study of Copper Objects of the Erlitou Culture. In Institute of Archaeology, Chinese Academy of Social Sciences (ed.), *Early Bronze Cultures in China*, 124–274. Beijing, Kexue Chubanshe (in Chinese).

Chen, G., Li, Y., Qian, W. and Wang, H. (2015) Preliminary Studies of Copper Objects Excavated from the Xichengyi Site in Zhangyi. *Kaogu yu Wenwu* (Archaeology and Cultural Relics) 2, 105–118 (in Chinese).

Chen, K., Rehren, T., Mei, J. and Zhao, C. (2009) Special Alloys from Remote Frontiers of the Shang Kingdom: Scientific Study of the Hanzhong Bronzes from Southwestern Shaanxi, China. *Journal of Archaeological Science* 36, 2108–2018.

Chêng, T. K. (1974) Metallurgy in Shang China. *T'oungPao* 60/4–5, 209–229.

Di Cosmo, N. (1999) The Northern Frontier in Pre-Imperial China. In M. Loewe and E. L. Shaughnessy (eds), *The Cambridge History of Ancient China: From the Origins of Civilization to 221 B.C.*, 885–966. Cambridge, Cambridge University Press.

Dodson, J., Li, X., Ji, M., Zhao, K., Zhou, X. and Levchenko, V. (2009) Early Bronze in Two Holocene Archaeological Sites in Gansu, NW China. *Quaternary Research* 72, 309–314.

Editorial Committee for the Collection of Chinese Bronzes (1994) *The Collection of Chinese Bronzes* 13. Beijing, Cultural Relics Press (in Chinese).

Fang, H. (2010) On the Formation of the Ritual Bronze System during the Early States Period in China. *Wenshizhe* (Journal of Literature, History and Philosophy) 1, 73–79 (in Chinese).

Fitzgerald-Huber, L. G. (1995) Qijia and Erlitou: The Question of Contacts with Distant Cultures. *Early China* 20, 17–67.

Franklin, U. M. (1983) The Beginnings of Metallurgy in China: A Comparative Approach. In G. Kuwayama (ed.), *The Great Bronze Age of China: A Symposium*, 94–99. Los Angeles, CA, Los Angeles County Museum of Art.

Gao, J. and He, N. (2014) A Preliminary Investigation of Copper Objects Excavated from the Taosi site. *Nanfang Wenwu* (Cultural Relics from South) 1, 91–95 (in Chinese).

Ho, P. T. (1975) *The Cradle of the East: An Inquiry into the Indigenous Origins of Techniques and Ideas of Neolithic and Early Historic China, 5000–1000 BC*. Hong Kong, Chinese University of Hong Kong.

Hwang, M. C. (2014) Toward an Age of Monumental Bronze: Importation and Formation of Bronze-Making Technology in China. *Bulletin of the Institute of History and Philology Academia Sinica* 85/4, 575–678 (in Chinese).

Institute of Archaeology, Chinese Academy of Social Sciences (1999) *Yanshi Erlitou*. (The Erlitou Site in Yanshi). Beijing, The Encyclopaedia of China Publishing House (in Chinese).

Jaang, L. (2015) The Landscape of China's Participation in the Bronze Age Eurasian Network. *Journal of World Prehistory* 28, 179–213.

Li, M., Huang, S., and Ji, L. (1984) Compositional Report on a Copper Object Excavated from the Taosi Site in Xiangfen, Shanxi. *Kaogu* (Archaeology) 12, 1068, 1071 (in Chinese).

Li, S. (2005). The Regional Characteristics and Interaction of Early Metallurgical Industries in the North-west and the Central

Plains. *Kaogu Xuebao* (Acta Archaeologica Sinica) 3, 239–277 (in Chinese).

Li, X., Ji, M., Dodson, J., Zhou, X., Zhao, K., Sun, N. and Yang, Q. (2010) Records of Element Geochemistry on the Bronze Smelting in Hexi Corridor Since 4200 aBP. *Hubo Kexue* (J. Lake Sci.) 22/1, 103–109 (in Chinese).

Li, Y., Chen, K., Qian, W. and Wang, H. (2015) Studies of Metallurgical Remains from the Xichenyi site in Zhangyi. *Kaogu yu Wenwu* (Archaeology and Cultural Relics) 2, 119–127 (in Chinese).

Lian, H., Tan, D. and Zheng G. (2011) The Research and Exploration to the Bronze Casting Techniques of Erlitou Site. *Kaogu Xuebao* (Acta Archaeologica Sinica) 4, 561–575 (in Chinese).

Linduff, K. M., Han, R. and Sun, S. (eds), (2000) *The Beginnings of Metallurgy in China*. Lewiston, The Edwin Mellen Press.

Linduff, K.M. (ed.) (2004) *Metallurgy in Ancient Eastern Eurasia from the Urals to the Yellow River*. Lewiston, The Edwin Mellen Press.

Liu, L. and Chen, X. (2012) *The Archaeology of China: From the Late Paleolithic to the Early Bronze Age*. Cambridge, Cambridge University Press.

Liu, X. and Li, W. (2014) Prehistoric 'Bronze Road' and Central Plains Civilizition. *Xinjiang Shifan Daxue Xuebao* (Journal of Xinjiang Normal University) 2, 79–88 (in Chinese).

Loehr, M. (1949) Weapons and Tools from Anyang and Siberian Analogies. *American Journal of Archaeology* 53, 126–144.

Loehr, M. (1956) *Chinese Bronze Age Weapons*. Ann Arbor, MI, University of Michigan Press.

McNeal, R. (2014) Erligang Contacts South of Yangzi River: The Expansion of Interaction Networks in Early Bronze Age Hunan. In K. Steinke (ed.), *Art and Archaeology of the Erligang Civilization*, 173–187. Princeton, NJ, Princeton University Press.

Mei, J. (2003a) Qijia and Seima-Turbino: The Question of Early Contacts between Northwest China and the Eurasian Steppe. *Bulletin of the Museum of Far Eastern Antiquities* 75, 31–54.

Mei, J. (2003b) Cultural Interaction between China and Central Asia during the Bronze Age. *Proceedings of the British Academy* 121, 1–39.

Mei, J. (2009a) Early Metallurgy in China: Some Challenging Issues in Current Studies. In J. Mei and T. Rehren (eds), *Metallurgy and Civilisation: Eurasia and beyond*. London, Archetype Publications.

Mei, J. (2009b) Early Metallurgy and Socio-Cultural Complexity: Archaeological Discoveries in North-west China. In B. Hanks and K. Linduff (eds), *Monuments, Metals and Mobility: Trajectories of Social Complexity in the Late Prehistoric Eurasian Steppe*, 215–232. Cambridge, Cambridge University Press.

Mei, J., Xu, J., Chen, K., Shen, L. and Wang, H. (2012) Recent Research on Early Bronze Metallurgy in Northwest China. In P. Jett, B. McCarthy and J. G. Douglas (eds), *Scientific Research on Ancient Asian Metallurgy*. London, Archetype Publications.

Mei, J., Wang, P., Chen, K., Wang, L., Wang, Y. and Liu, Y. (2015) Archaeometallurgical Studies in China: Some Recent Developments and Challenging Issues. *Journal of Archaeological Science* 56, 221–232.

Steinke, K. (2014) Erligang and the Southern Bronze Industries. In K. Steinke (ed.), *Art and Archaeology of the Erligang Civilization*, 151–170. Princeton, NJ, Princeton University Press.

Sun, S. and Han, R. (1981) A Preliminary Study of Early Chinese Copper and Bronze Artefacts. *Kaogu Xuebao* (Acta Archaeologica Sinica) 3, 287–301 (in Chinese). English translation: Linduff *et al.* 2000, 129–151.

Wang, H. (2014) China's First Empire? Interpreting the Material Record of the Erligang Expansion. In K. Steinke (ed.), *Art and Archaeology of the Erligang Civilization*, 67–93. Princeton, NJ, Princeton University Press.

Wang, H. and Chen, G. (2012) The Xichengyi Site in Zhangye, Gansu. In State Administration of Cultural Heritage (ed.), *Major Archaeological Discoveries in China*, 20–23. Beijing, Cultural Relics Press (in Chinese).

Zhang, H. and Chen, J. (2013) The Relationship between Prehistoric Bronze Metallurgy and the State Formation in the Central Plains of China. *Zhongyuan Wenwu* (Cultural Relics of Central China) 1, 52–59 (in Chinese).

Zhao, C. (ed.) (2006) *Bronze Objects Unearthed from the Chenggu and Yangxian Counties*. Beijing, Science Press (in Chinese).

Chapter 20

Patterns of Transformation from the Final Neolithic to the Early Bronze Age: A Case Study from the Lech Valley South of Augsburg[1]

Ken Massy, Corina Knipper, Alissa Mittnik, Steffen Kraus, Ernst Pernicka, Fabian Wittenborn, Johannes Krause, Philipp W. Stockhammer

Introduction

The transition from the Final Neolithic[2] to the Early Bronze Age in Central Europe has long been a matter of major interest. Archaeologists have published comprehensively on the Corded Ware Complex (CWC), the Bell Beaker Complex (BBC), and the Early Bronze Age and compiled large corpora of material (*e.g.* Buchvaldek 1967; Hájek 1968; Matthias 1974; Geber 1978; Ruckdeschel 1978; Zich 1996; Bartelheim 1998; Heyd 2000; Furholt 2003; Dresely 2004). They have been the basis both for sophisticated typo-chronological sorting activity and for approaches to social and economic systems in the 3rd and early 2nd millennia BC. The major concern of archaeologists, however, has been to understand the dynamics involved by exploring the cultural similarities and differences of CWC, BBC, and the Early Bronze Age with a view to shedding light on the continuities and discontinuities manifesting themselves in that period.

Scientific analyses have repeatedly modified archaeological assumptions about this period, beginning with radiocarbon datings for the Early Bronze Age that indicated that it started earlier than formerly assumed, *i.e.* in the late 3rd millennium BC already (Becker *et al.* 1989). Especially in the last few years, scientific data have helped to understand the 3rd millennium BC as a crucial period of transformation for European societies. Palaeogenetic analyses have demonstrated that the ancestors of those individuals identified as representatives of the CWC may have migrated from the eastern Pontic Steppe regions to Central Europe, thus importing entirely new genetic components (Allentoft *et al.* 2015; Haak *et al.* 2015). Moreover, it has been shown that the migrants from the east brought bacteria (*Yersinia pestis*) with them that were likely to cause plague, and we have only just started exploring the potential impact diseases may have had in the 3rd millennium BC (Rasmussen *et al.* 2015; Andrades Valtueña *et al.* 2017). Furthermore, stable isotope analyses have contributed to our understanding of human dietary habits and mobility from the Final Neolithic to the Early Bronze Age. The BBC has been the subject of one of the earliest large-scale strontium-isotope studies, which have done much to elucidate the role of residential changes during the spread of this supra-regional archaeological phenomenon (Grupe *et al.* 1997; Price *et al.* 2004). Following earlier investigations on BBC (Heyd *et al.* 2002–2003; Bertemes and Heyd 2015), the CWC (Sjögren *et al.* 2016) and Únětice contexts (Knipper *et al.* 2016a) have now begun to yield up complex and regionally diverse residential patterns and dietary habits.

Despite such completely new insights, the supra-regional narratives are still not able to adequately explain local or regional patterns of transformation taking place in the 3rd and early 2nd millennia BC. The large-scale scientific studies have captured individuals from an extensive area but only a very small number of individuals per region. We pursue a micro-historical approach and focus on a micro-region with a unique density of archaeological sources for the period in question: the Lech Valley south of Augsburg.

Regional and archaeological background

The south-north-oriented Lech Valley is situated in southern Bavaria, some 50 km west of Munich. Two rivers run through it, the Wertach in the west and the Lech in the east. They converge in what is today the city of Augsburg (Fig. 20.1). A fertile loess ridge lies in the center of the valley, extending 45 km from north to south and at its widest point 4 km from west to east. On the eastern side of the loess ridge lies a large gravel plain, the lower part of which (close to the Lech) has frequently been flooded. The upper part is perfectly suited for building settlements and placing graveyards because it is protected from the river and at the same time the buildings do not cover the fertile loess terrace with its favourable arable conditions. Until now, we have had no evidence of settlements or burial grounds on the loess despite the large-scale excavations that have taken place there (Gairhos 2007, 86). The geographical situation suffices to explain the density of the archaeological evidence, as on the gravel part of the landscape there was only very limited space for building hamlets and burying the dead.

Recent large-scale excavations have revealed an astonishing number of graveyards from the 3rd and 2nd millennia BC and sometimes also associated hamlets (Fig. 20.2). We assume that a cemetery accompanied every hamlet and that the individuals buried there represent at least in part the inhabitants of each hamlet. The best evidence for this relation between hamlet and burial-site is found in Haunstetten 'Unterer Talweg 85' (Dumler and Wirth 2001)[3]. This Early Bronze Age hamlet consisted of a long-house (*Langhaus*) and a few smaller buildings around it and was built right next to the edge of the loess terrace. Five BBC graves were discovered to the north-east of the hamlet and two more to the south-west. Directly to the east of the hamlet three burials from the Early Bronze Age were found that were probably part of a larger cemetery. Similar patterns have been found at various intervals along the gravel plain, indicating that hamlet-based communities lived and died next to one another along the riverbanks.

Over the last thirty years, all these burials have been excavated in accordance with the latest archaeological standards. Bone preservation is very good, furnishing excellent preconditions for bio-archaeometric analyses. While only a few CWC burials exist, the number of graves from the BBC and EBA exceeds 400, with 341 of them belonging to the Early Bronze Age or the early Middle Bronze Age (Tabs 20.1–20.3). Up to the 1970s, only 15 Early Bronze Age burials were known and none from the BBC.[4] According to the archaeological evidence we have, the Lech Valley was one of the most densely populated areas in Central Europe during these periods. CWC burials were mostly found in small groups of only two or three inhumations, whereas in the Early Bronze Age the burial sites were larger, with anything up to 63 graves, as exemplified by the cemetery in Kleinaitingen. At several sites, post alignments with single cattle-teeth

placed below the posts ran east from the burials for up to 30 metres. The burial rites in the Lech Valley followed general trends in southern Germany from the Final Neolithic to the Middle Bronze Age. Typical of the CWC are inhumations in a crouched position oriented west-east with the deceased facing south. They were differentiated by sex, with men lying to the west and women to the east. Inhumation, body positioning and sex-specific segregation persisted during the BBC, but the orientation changed to north–south with the deceased now facing east and the position of the sexes reversed, so that the men now lay to the north and the women to the south. This practice remained unchanged until the end of the Early Bronze Age. In the Middle Bronze Age, by contrast, people started placing their deceased in a fully extended position without any sex-specific distinctions.

When we collate the information on the geological conditions of the Lech Valley with the archaeological evidence from the settlements and the burial sites, the results are astonishing. In the central part of the valley (*i.e.* to the west of the settlements), plants were most likely grown on the fertile loess ridge in order to guarantee permanent supplies for the inhabitants of the hamlets and communities strung out like pearls along the eastern edge of the loess terrace. The space allotted to the dead was situated to the east of each hamlet, and each hamlet had its own burial site. Like the post alignments, the faces of the deceased looked towards the east. Although settlement evidence and alignments have yet to be documented for the BBC, the orientation and other burial practices appear to be so similar to the subsequent Early Bronze Age that one can assume a degree of continuity in the respective practices, related ideas, and possibly also in the local population interred there.

The research issue

While the global perspective on the 3rd and early 2nd millennia BC suggests that this was a highly dynamic and transformative period, the archaeological evidence in the micro-region of the Lech Valley provides evidence rather for long-term continuity (in the sense of continuous change; *cf.* Stockhammer 2008, 1) among the local inhabitants, their practices and ideas. We are fully aware that to a significant extent the selection of a specific scale already predetermines the results of the subsequent study (Jiménez 2005). To integrate global and local perspectives and to understand the local impact of global transformations – especially the introduction of the new bronze-casting technology – we will first take a closer look at the Lech Valley itself.

Since 2012, we have pursued an interdisciplinary research approach for our studies on Final Neolithic and Early and Middle Bronze Age burials in this region. We have dovetailed sophisticated archaeological analysis based on a practically-oriented approach (*cf.* Stockhammer 2012) with a broad range of scientific analyses.

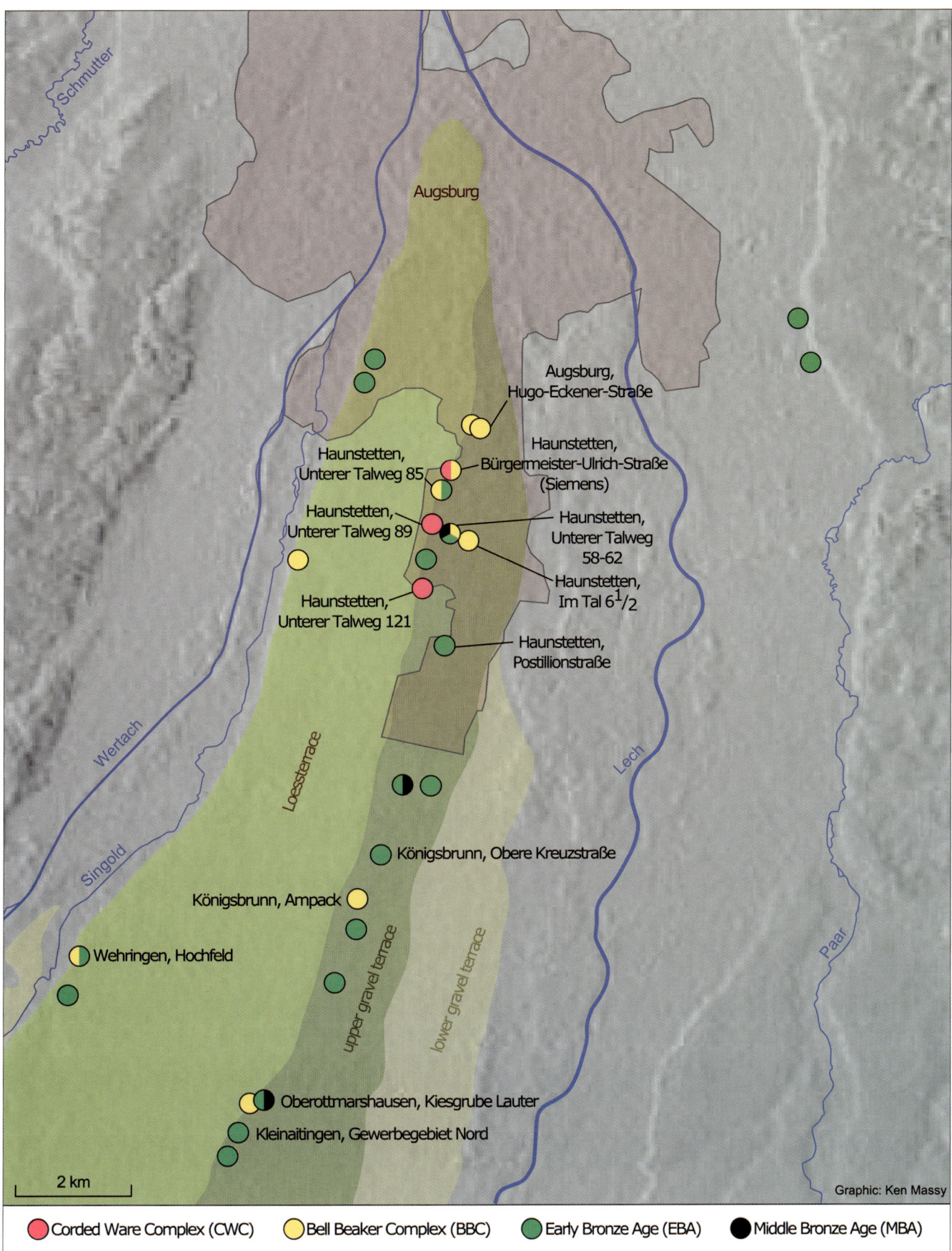

Corded Ware Complex (CWC)   Bell Beaker Complex (BBC)   Early Bronze Age (EBA)   Middle Bronze Age (MBA)

Figure 20.1: Map of cemeteries and single burial sites south of the city of Augsburg. The labelled sites provide samples included in this study.

Figure 20.2: Map of Final Neolithic and Early Bronze Age as well as Middle Bronze Age settlements and cemeteries at the 'Unterer Talweg' in Haunstetten, Augsburg.

The overall research project investigates the following questions: Was there a temporal sequence or a degree of overlap between the Neolithic and the Early Bronze Age as well as between the different archaeologically defined complexes?[5] What can the use times of the burial grounds tell us about the time in which the different hamlets existed? Did the hamlets figure in a temporal sequence with a pattern of shifting farmsteads, or were they contemporaneous? What can we say about interaction between the different hamlets at a local level? Can we trace changes or persistence in marital patterns, subsistence strategies, and food practices – bearing in mind that the hamlets lay close to each other and the inhabitants could communicate by shouting to each other without even leaving their hamlets? Can we trace the local impact of the mass migrations from eastern to central Europe in the 3rd millennium BC (*cf.* Allentoft *et al.* 2015; Haak *et al.* 2015)? Do we identify continuity in large-scale mobility after the CWC as well, and how is this mirrored by individual mobility during childhood? How did networks of interaction – relevant for the acquisition of raw materials (especially copper and tin) and objects – change over time and who was involved? Can we identify hamlet-specific networks? And how and to what extent can these continuities and changes be related to the introduction of the new bronze technology the introduction of which has always been regarded as an impetus for societal change? How can we understand the development of metalworking technology now that it has become increasingly obvious that

Table 20.1: Results of the radiocarbon dating of the individuals from H-UT 58–62

Grave no.	Grave no. after	Feat. no.	Lab. no.	C[14] age	Cal BC 1 sigma	Cal BC 2 sigma	N%	C%	C/N atom	δ^{13}C (‰ VPBD)	δ^{15}N (‰ AIR)
Haunstetten, Unterer Talweg 89 (H-UT 89)											
CWC											
		FK 231	MAMS 23729	4155 ± 23	2867−2678	2875−2635					
Haunstetten, Unterer Talweg 58-62 (H-UT 58-62)											
BBC											
7	Kociumaka	67	MAMS 18934	3840 ± 20	2340−2211	2455−2204	15.1	41.5	3.2	-20.71	8.61
8	Kociumaka	68 SK 1, Fznr. 33	MAMS 29074	3909 ± 29	2466−2346	2471−2300	15.2	42.3	3.3	-20.87	8.99
8	Kociumaka	68 SK 2, Fznr. 34	MAMS 29075	3870 ± 30	2453−2294	2464−2212	10.8	29.4	3.2	-20.17	9.52
EBA											
10	Massy	146	MAMS 18933	3570 ± 19	1940−1894	2009−1881	12.4	34.2	3.2	-20.66	9.46
9	Massy	147	MAMS 18937	3612 ± 25	2020−1938	2031−1900	16.4	45.0	3.2	-20.68	9.48
7	Massy	149	MAMS 18938	3597 ± 24	2011−1917	2023−1892	14.5	39.9	3.2	-20.76	9.19
6	Massy	150	MAMS 18939	3559 ± 24	1939−1885	2009−1780	16.5	45.2	3.2	-20.81	9.59
5	Massy	151	MAMS 18940(2) + 18941(2) Combined	3566 ± 11	1930−1894	1946−1886	13.7	37.7	3.2	-20.87	8.90
4	Massy	152	MAMS 18942	3558 ± 23	1939−1885	2007−1779	16.6	45.8	3.2	-21.26	8.79
3	Massy	153	MAMS 18943	3553 ± 24	1941−1836	1971−1776	16.7	45.9	3.2	-21.06	9.10

Table 20.2: MtDNA analyses from the sampled individuals from H-UT 58-62

Grave no.	Grave no. after	Feat. no.	Sex (archaeological determination)	Sex (aDNA analysis)	Age (anthropological determination)	Shotgun sequencing: % endogenous	Shotgun sequencing: average frag. length	mtDNA capture: average coverage of mtDNA	mtDNA capture: 5'-deamination (%)	mtDNA capture: final contamination estimate	mtDNA Haplogroup
Haunstetten, Unterer Talweg 58-62 (H-UT 58-62)											
BBC											
7	Kociumaka	67	male	XY	20–35	0.27	52.4	1521.5	0.42	1–3%	K1a
8	Kociumaka	68 SK 1	male	XY	> 21 yrs Middle adult	40.61	48.5	691.0	0.43	2–4%	J1c
8	Kociumaka	68 SK 2	female	XX	> 21 yrs Middle adult	4.36	49.1	641.4	0.42	1–3%	K1a
EBA											
10	Massy	146	female	n/a	15–25	0.46	57.8	852.2	0.41	1–3%	H1c
9	Massy	147	female	XX	35–55 ±	0.91	59.0	372.4	0.43	~2%	H2a1
8	Massy	148	female								
7	Massy	149	male	n/a	12 yrs ± 36 mths	0.15	68.0	767.1	0.43	1–3%	K1a1b1g
6	Massy	150	male	n/a	5 yrs ± 16 mths	0.16	59.4	1225.1	0.39	1–3%	U5b1c2
5	Massy	151	male	n/a	10 yrs ± 30 mths	0.21	59.3	64.2	0.38	1–3%	U5b1c2
4	Massy	152	female	n/a	9 yrs ± 30 mths	0.23	66.4	774.5	0.41	1–3%	U5b1c2
3	Massy	153	male	XY	12–21 yrs	0.37	60.6	267.6	0.43	2–4%	R1a1a

Table 20.3: Oxygen and strontium isotope data analyses from the sampled individuals from H-UT 58-62

Grave no.	Grave no. after	Feat. no.	Tooth	$\delta^{18}Op$ (‰ VSMOW)	1 sd	$^{87}Sr/^{86}Sr$	2 SD
Haunstetten, Unterer Talweg 89 (H-UT 89)							
CWC							
		FK 231	M 47	16.65	0.32	0.71279	0.00006
		FK 231	M 28			0.71200	0.00001
Haunstetten, Unterer Talweg 58-62 (H-UT 58-62)							
BBC							
7	Kociumaka	67	M 27	16.29	0.15	0.70820	0.00001
8	Kociumaka	68 SK 1	M 48			0.71505	0.00001
8	Kociumaka	68 SK 1	M 18			0.71384	0.00001
8	Kociumaka	68 SK 2	M 37	16.53	0.15	0.70898	0.00001
8	Kociumaka	68 SK 2	M 17	16.62	0.08	0.70899	0.00001
EBA							
10	Massy	146	M 48	15.92	0.02	0.71027	0.00001
9	Massy	147	M 17	16.04	0.10	0.71361	0.00002
9	Massy	147	M 28			0.71389	0.00002
8	Massy	148	M 46	18.15	0.29	0.71301	0.00001
7	Massy	149	M 17	15.92	0.02	0.70993	0.00001
6	Massy	150	M 26	16.85	0.09	0.70935	0.00001
5	Massy	151	M 16	15.87	0.15	0.70950	0.00001
4	Massy	152	M 16	16.56	0.03	0.70955	0.00001
3	Massy	153	M 26			0.70937	0.00001
3	Massy	153	M 48	15.49	0.06	0.71162	0.00001

linear perspectives need to be replaced by complex, non-linear alternatives and well-established insights from science and technology studies (*cf.* Schubert this volume) have now found their way into archaeology (*e.g.* Maran 2004a; 2004b; Kienlin 2008; 2010; Bernbeck *et al.* 2011; Burmeister 2011; Hofmann and Patzke 2012; Burmeister and Müller-Scheeßel 2013; Kienlin 2014; Stockhammer 2015)? Technological innovations have the potential for triggering social change, but they do not invariably do so (Braun-Thürmann 2005, 30–64; Hofmann 2012; Hofmann and Patzke 2012). So what is the relation between technological and social change in our micro-region?

To find an answer to all these questions, we sampled over 170 individuals from the CWC up to the beginning of the MBA.[6] All of these individuals were radiocarbon-dated and mostly determined by stable isotope ratios (C, N, O, Sr) in order to obtain insights on individual diet and mobility. Also, we extracted aDNA from the dentine of 87 individuals, sequenced their mitochondrial DNA, and produced genome-wide data for 60 of these individuals.[7] More than 500 copper and bronze artefacts from all the burials within our research area were analysed with X-ray fluorescent (XRF) scanning – including burials where the human remains were not selected for further scientific analyses. Based on the results of the XRF measurements, we chose 12 metal objects for subsequent lead-isotope and trace element analyses. With all these results available, the Lech Valley is now the micro-region with the highest incidence of interdisciplinary archaeological and scientific analysis worldwide. Due to these enormous amounts of data, we will not be able to

present all our findings here but will concentrate on a small area within the Lech Valley excavated in 2007, 2008, and 2011 by private excavation companies operating under the name of 'Augsburg-Haunstetten, Unterer Talweg 58–62 and 89' (H-UT 58–62 and H-UT 89 for short).

Chronology in the Lech Valley

Until recently, there has been hardly any radiocarbon dating of features from the Final Neolithic and Early Bronze Age in southern Bavaria. This has hindered adequate understanding of the sequence of cultural changes that took place in this region. It was unclear whether and to what extent Final Neolithic groups and Early Bronze Age communities coexisted and hence perhaps interacted. In response to this challenge, we first conducted a comprehensive radiocarbon dating of burials in the Augsburg region and arrived at some highly pertinent insights (Stockhammer *et al.* 2015a; 2015b). The CWC burials (n = 2) date back to the centuries between 2900 and 2600 BC (2-sigma ranges), whereas the 2-sigma ranges of the earliest BBC burials only start around 2560 BC (Fig. 20.3). Due to the very much larger number of radiocarbon-dated BBC (n = 28) and Early Bronze Age burials (n = 106), we can say with a degree of certainty that there was neither any significant overlap nor a hiatus between the two phases. The end of the Final Neolithic can be placed around 2150/2140 BC. The onset of the Bronze Age in the Augsburg valley around 2150/2140 BC corresponds well with the (slightly later) new radiocarbon dates for the Early Bronze Age cemetery of Singen[8].

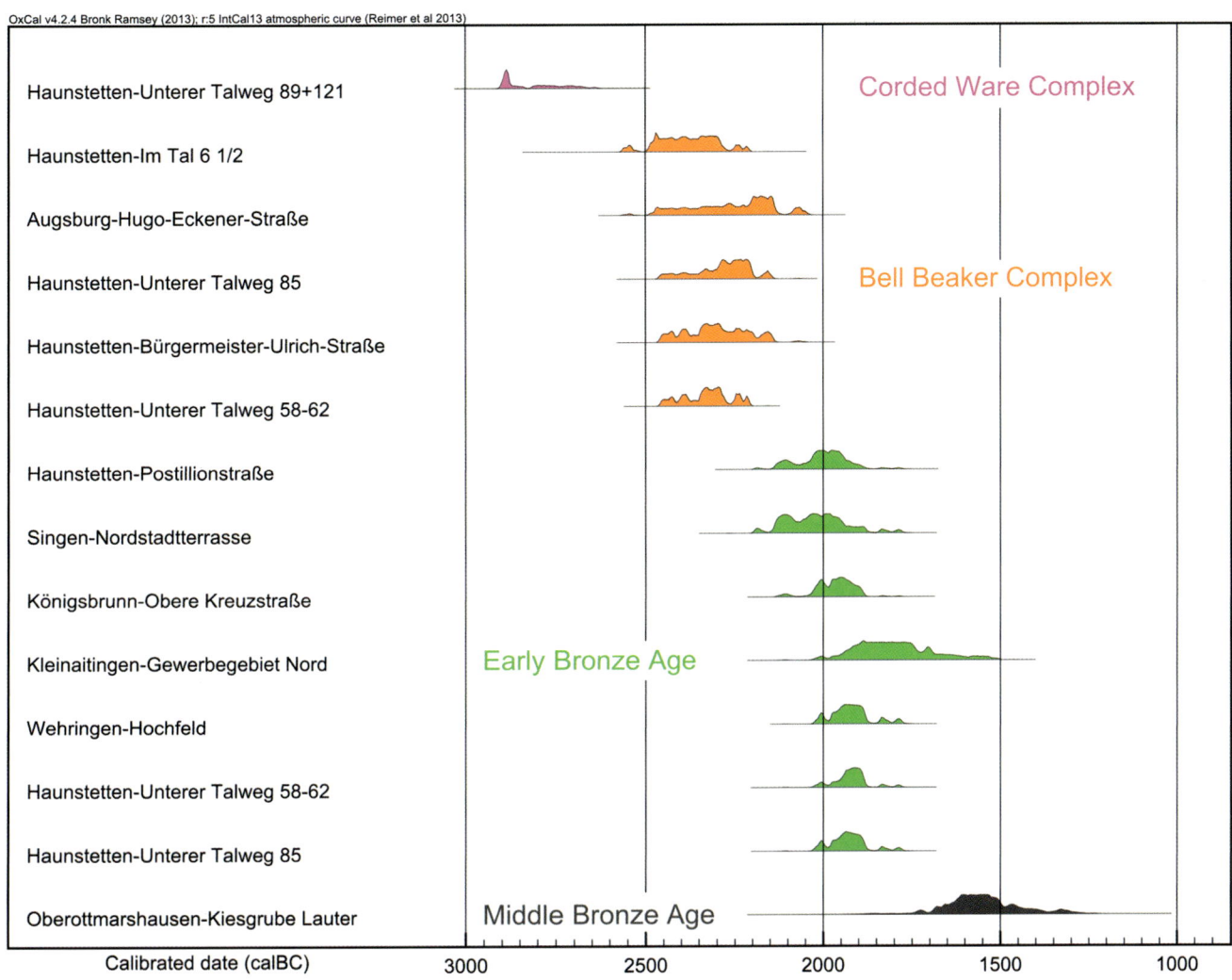

Figure 20.3: Sum calibrations of the radiocarbon dates of human skeletal remains from the Final Neolithic and Bronze Age sites conducted within the research project.

Accordingly, we can understand the Early Bronze Age as a continuation of the BBC – especially with regard to unchanged burial practices – with some slight transformations in the material culture (this at least is what the burial gifts suggest). Major changes can be assumed to have taken place in house construction, given that we so far have no traces of wooden-post structures in the Final Neolithic, whereas during the Early Bronze Age large buildings with wooden posts are known to have existed (Nadler 2001; Schefzik 2001; Bartelheim 2010).

The total of 106 radiocarbon-dated Early Bronze Age burials from the entire Lech Valley south of the city of Augsburg has also enabled us to contradict the still widely held assumption of a linear development of bronze-working technology from the simple hammering of objects (mostly on the basis of bronze sheets) to complex casting techniques (Stockhammer *et al.* 2015a; Stockhammer *et al.* 2015b). The validity of this traditionally assumed progression from

the simple to the more complex (*cf.* already Reinecke 1902; 1924) has already been queried in the recent past (Kienlin 2006a, 115; 2006b, 528–529; 2008; 2010; 2014, 453–454; Burmeister and Müller-Scheeßel 2013; Brumlich 2014) and can now be dismissed altogether and replaced with narratives based on new data (Stockhammer 2015). This is due to the large number of metal objects in the graves, which enable us to better understand the time span involved in the deposition (and hence probably also the use) of particular types of bronze objects. This new understanding was important mainly for the different types of pin, as the shapes of these pins have always played a major role in the chronological subdivision of the Early Bronze Age (Ruckdeschel 1978; David 1998). We were able to show that the rather simple 'Bz A1' pins were in use from the very beginning of the Early Bronze Age until *c.* 1700 BC, whereas the Early Bronze Age continues until the 16th century BC – albeit without any indicative Bz A1 pin types. Únětice-related 'Bz A2' pins appeared on top of the 'Bz A1'

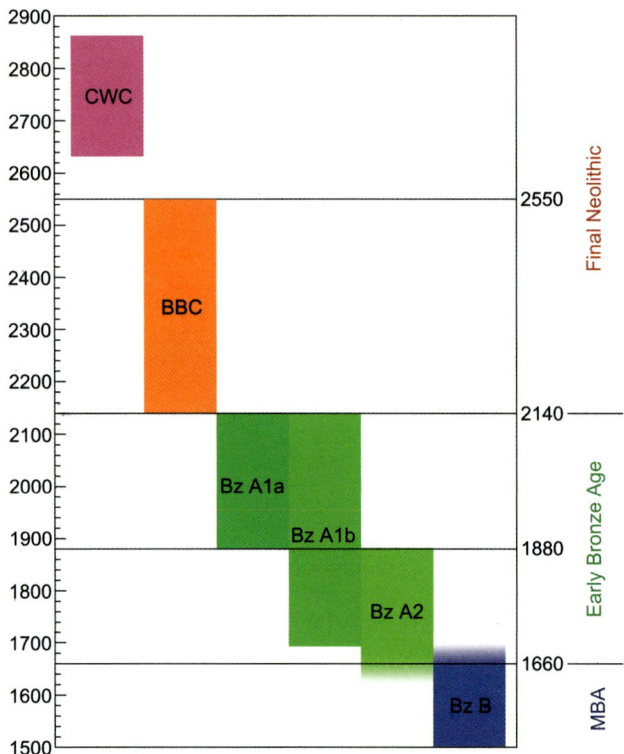

Figure 20.4: Chronological sketch of the Final Neolithic and Early and Middle Bronze Age based upon data from the project.

inventory from *c.* 1900 BC and continued until 1700 BC and disappeared together with the Bz A1 pin type (Stockhammer *et al.* 2015a; 2015b). None of the Early Bronze Age burials with respective pins has a 2-sigma range of the radiocarbon date that reaches the 17th century BC. If we correlate the end of the Early Bronze Age with the end of the relevant pin types, its end could already be placed around 1700 BC (Fig. 20.4) Therefore we can conclude that the traditional phases Bz A1 and Bz A2 do not represent a chronological sequence but reflect regional differences in the willingness to appropriate the new technology. Whereas the inhabitants of Central Germany (in the so-called Únětice culture) quickly appropriated and mastered complex casting techniques, the inhabitants of Southern Germany continued with the simple hammering of objects until the end of the Early Bronze Age.

Case study: Haunstetten 'Unterer Talweg 58–62 and 89'

Archaeological evidence and chronology

To demonstrate the complexity of the transformations taking place from the Neolithic to the Bronze Age, we now narrow our perspective within the Lech Valley further to a small area of a few hundred square metres, *i.e.* the area of Haunstetten 'Unterer Talweg 58–62' and 'Unterer Talweg 89' (H-UT 58–62 and H-UT 89 for short) (Fig. 20.5). This area is of

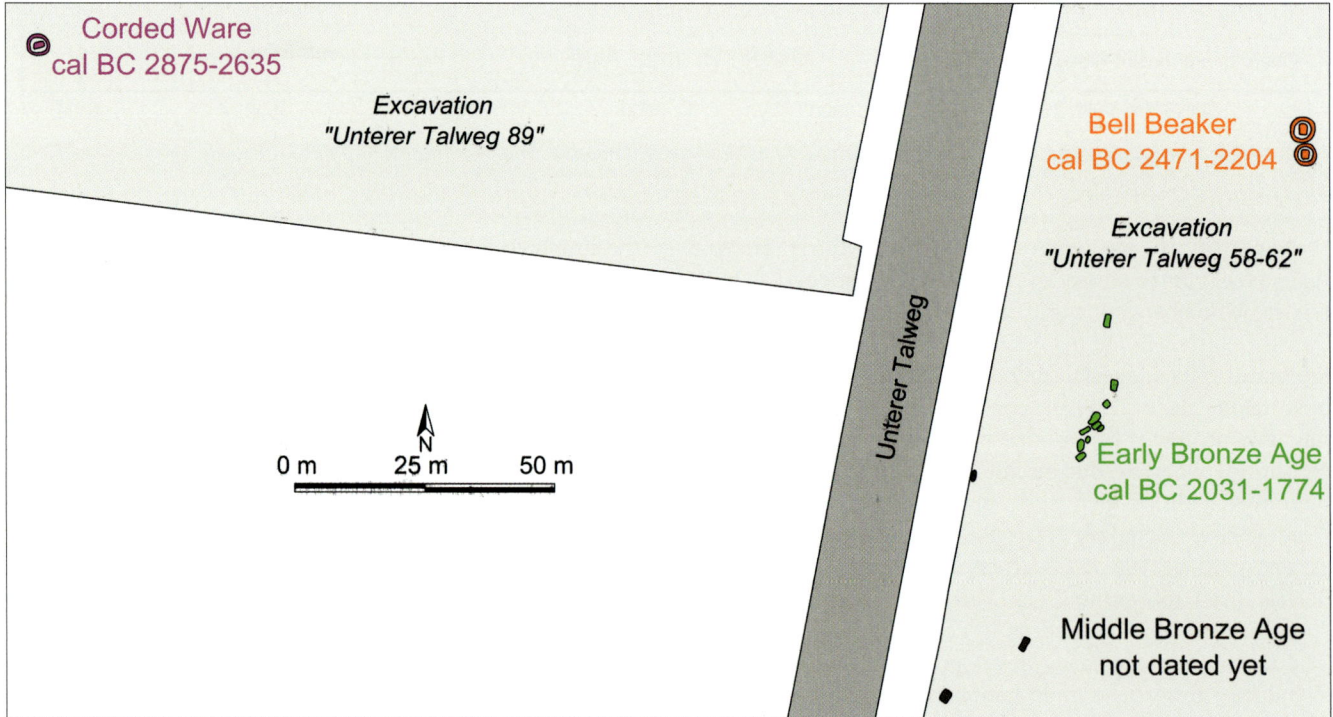

Figure 20.5: Close up of the case study site of H-UT 58-62 and H-UT 89. Features from other periods are not shown on the map.

special interest due to the existence of CWC, BBC, Early, and Middle Bronze Age burials in the direct vicinity. The westernmost evidence is a single CWC burial (H-UT 89 feat. 231) near the edge of the loess terrace with a small ring ditch. The burial pit was west–east oriented and contained an inhumation in a crouched position with the head to the west plus a stone axe. Only 200 metres to the east, two BBC graves (H-UT 58–62 feat. 67 and 68) were placed in a north–south orientation and again accompanied by ring ditches (Fig. 20.6). The southern one (feat. 67) contained a single male individual in a crouched position on the left side with the head to the north and furnished with a tanged dagger and a wrist guard made of sandstone. A few metres to the north, a BBC double burial (feat. 68) of a male (SK 1) and a female (SK 2) individual was excavated. The male individual was again equipped with a wrist guard. The two accompanying vessels could not be attributed to one or the other of the individuals.

A group of ten Early Bronze Age burials was situated 45 metres to the southwest (Fig. 20.7). Metal artefacts have been found in six of these graves. Grave 5 (feat. 151) contained a sub-adult male and is the richest burial with a hilted dagger, a small hammered pin ('*Ruderkopfnadel*'), a bracelet, and small spiral rolls (Fig. 20.8). The orientation of these Early Bronze Age burials is shifted slightly clockwise, most of them being aligned NNE–SSW. This orientation was preserved for the burials to the southwest of this group, which date to the beginning of the MBA. The bodies were placed in an extended position. One of the burials contained two pins of the '*Paarstadl*' type, another a pin of the '*Wetzleinsdorf*' type, both characteristic of the earliest stage of the Middle Bronze Age (Innerhofer 2000, 36, 40, 45).

In the following, our analysis of these burials is designed to shed light on the continuities and changes taking place between 2900 and 1500 BC. The places inhabited by the living have yet to be excavated, but we conclude from neighbouring excavations that each of the small cemeteries was associated with a single hamlet to its west. Only burial-pit orientation and body placement changed continuously over time, as described above. The same is true of the selection of burial goods. During the CWC, the axe was the typical attribute of the male burial (Ruckdeschel 1978, 219), replaced in the BBC by dagger and archery equipment (Bosch 2009, 135). BBC body placement traditions persisted during the Early Bronze Age. While pottery vessels were no longer selected as burial gifts during the Early Bronze Age, the selection criteria for other grave goods (daggers, archery equipment *etc.*) continued. Marked discontinuity is discernible at the transition to the Middle Bronze Age. Sex-specific differentiation and the crouched body posture of the deceased were relinquished, which raises important questions about novel or modified perspectives on gender roles in the communities both of the living and the dead.

While the archaeological evidence indicates the continuous existence of hamlets in the H-UT area, the radiocarbon dates suggest a shifting of the hamlets over the years. This cannot be explained by a lack of fertile farmland, as the neighbouring loess terrace provided extensive scope for cultivation. The CWC burial dates to 2875–2635 BC, the 2-sigma ranges and the BBC burials to 2471–2204 BC. If we can assume that the people of the CWC built stable hamlets, this indicates a gap of at least 150 years without a hamlet in the H-UT area. The same is true of the transition from the BBC to the Early Bronze Age. Again, a span of at least 170 years separates the latest BBC from the earliest EBA datings (2031–1776 BC). The placement of the deceased nevertheless suggests an awareness of older burials. We interpret both the positioning of the burials and the spaces between the burial groups as an expression of allegiance to the past combined at the same time with a desire to set oneself off from that past.

Metal analysis

The established use of the terms 'Stone Age' and 'Bronze Age' suggests in itself that an important shift in material culture took place in the transition between the two eras. However, it became clear several decades ago that the 'Bronze Age' in Southern Germany did not begin with 'real' bronzes (*i.e.* alloys of copper and tin) but with copper objects that contained hardly any tin at all. Back in the 1980s, Christian Strahm proposed a new term, the '*Metallikum*', as a replacement for this misleading terminology (Strahm 1982). Even before that it was noted that bronzes with relatively high concentrations of tin already appeared during the Final Neolithic (Schickler 1981). Nevertheless, it was still often taken for granted that Final Neolithic metal objects basically consisted of copper only.

In the course of our research, we have scanned all the metal objects from the H-UT area with the help of a portable XRF spectrometer. In addition, the metal dagger from the BBC burial feat. 67 was further analysed at the Curt-Engelhorn-Centre for Archaeometry in Mannheim for lead isotope ratios using multi-collector ICP mass spectrometry, while trace element analyses were accomplished with energy-dispersive X-ray fluorescence. These additional analyses not only revealed the true bulk composition of the alloy but also yielded information on the possible provenance of the copper. When corrosion sets in, tin tends to concentrate at the surface of alloyed metal objects. Since with portable XRF only the surface can be analysed, the values for tin are generally overestimated. But in most metal objects, the tin content was below the detection limit of about 0.1%.

The CWC burial contained no metal objects. The only BBC metal object is the dagger referred to earlier. Portable XRF already suggested that this object contained 19% tin at the surface of the object. Subsequent analysis of a drill sample from the interior came up with a value of 4.6%, which nevertheless suggests that the tin was intentionally added and the alloy can thus be classified as bronze.[9] Against

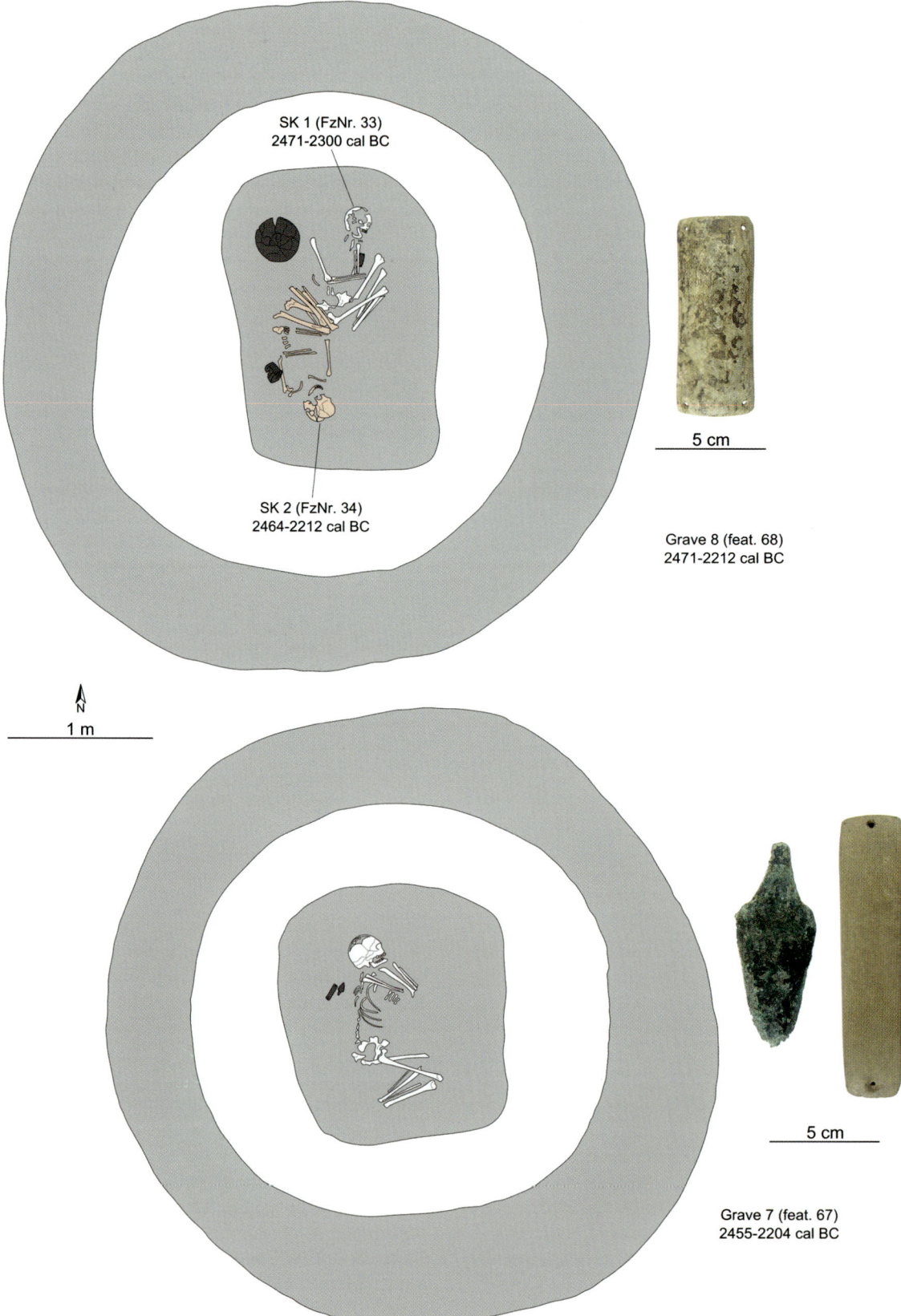

SK 1 (FzNr. 33)
2471-2300 cal BC

SK 2 (FzNr. 34)
2464-2212 cal BC

5 cm

Grave 8 (feat. 68)
2471-2212 cal BC

N
1 m

5 cm

Grave 7 (feat. 67)
2455-2204 cal BC

Figure 20.6: The two graves from H-UT 58-62 dating to the BBC. The two vessels from feat. 68 are not depicted.

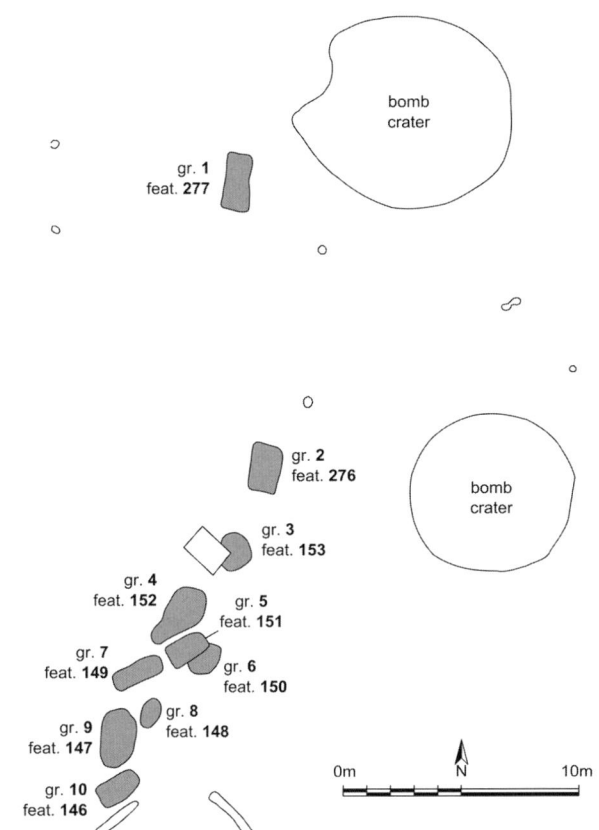

Figure 20.7: Map of the Early Bronze Age cemetery of H-UT 58-62.

all expectations, most of the 28 metal objects from the Early Bronze Age burials contained no tin at all (Fig. 20.9). While in the Final Neolithic the BBC inhabitants of H-UT were obviously able to obtain tin, this supply network seems to have collapsed at the onset of the Bronze Age. The evidence for this is supported by the 540 analyses with portable XRF and the 12 lead-isotope analyses plus trace-element analyses conducted on metal objects from the Lech Valley. 'Real' tin bronzes repeatedly occur in the BBC, while hardly any tin bronzes figure in the first centuries of the Early Bronze Age up to *c.* 1900 BC. This raises serious doubts about the pertinence of our material-based terminologies, as the Lech Valley seems to have shifted from a Neolithic CWC to a Bronze-Age BBC, then back to a Copper Age (also known as 'Early Bronze Age') and finally to an Early Bronze Age with real bronzes.

DNA analysis

The purpose of studying the genetic makeup of the individuals buried in the Lech Valley was to address the possibility of (dis-)continuity in the population at the transition between Final Neolithic and Early Bronze Age. This can be identified, for example, in a frequency change in mitochondrial DNA (mtDNA) lineages, which are inherited maternally. A stark shift in frequencies would normally be

interpreted as a sign of a population replacement, *e.g.* as observed in Europe at the onset of the Neolithic Revolution (*e.g.* Bramanti *et al.* 2009; Brandt *et al.* 2013). On the other hand, haplotype sharing (*i.e.* several individuals featuring an identical mtDNA genome) across different epochs would indicate close kinship and local continuity. The detection of direct kinship between individuals is interesting for the reconstruction both of marital/inheritance patterns and of burial customs.

All in all, 87 individuals from the Lech Valley were sampled for aDNA studies, among them the three individuals attributed to the BBC and seven Early Bronze Age individuals from H-UT 58–62 plus the CWC individual from H-UT 89. Where possible, second molars were chosen for sampling, as they were the primary target for subsequent isotope analyses. Complete mitochondrial genomes were sequenced for all the individuals from H-UT 58–62 by enriching the DNA libraries for mtDNA (Maricic *et al.* 2010). In addition, nuclear DNA was targeted by enriching for around 1.2 million informative single-nucleotide positions (Fu *et al.* 2013; Mathieson *et al.* 2015) in the three BBC and four EBA individuals at H-UT 58–62 and the CWC individual at H-UT 89.

Overall, a high diversity of maternal lineages could be detected, consistent with exogamous marriage practices (Haak *et al.* 2008). Several haplogroups appear only in the later EBA, but population continuity over the periods cannot be rejected and is strengthened by one case of haplotype sharing – an indication of direct kinship in the maternal lineage – that bridges the epochs (Knipper *et al.* forthcoming).

MtDNA haplotype sharing was observed in several other burials and will now be discussed in further detail here for the H-UT 58–62 site chosen for our case study. The BBC male from the single burial (feat. 67) and the female of the double burial (feat. 68 SK 2) carry the same haplotype, meaning they could be related as (grand)mother and (grand) child, as siblings, cousins, or other constellations via the female line. It seems fair to assume that the three burials represent a nuclear family: father, mother, and son.

In the small EBA cemetery we find similar structures based on family connections. The three children buried next to each other, a girl and two boys, share the same mtDNA haplotype. What makes these two male burials unusual is the fact that both of them have a copper bangle around the right wrist. It seems to be a grave good of some social significance, possibly an object of distinction in their lifetime and linked to family bonds. Furthermore, one of the two young males was given a dagger, a copper pin, and a necklace, a significant amount of material wealth that he cannot have amassed for himself during his short lifetime. We can safely assume that this wealth is the result of particular burial practices. The proximity of the graves also stands for close relation in death.

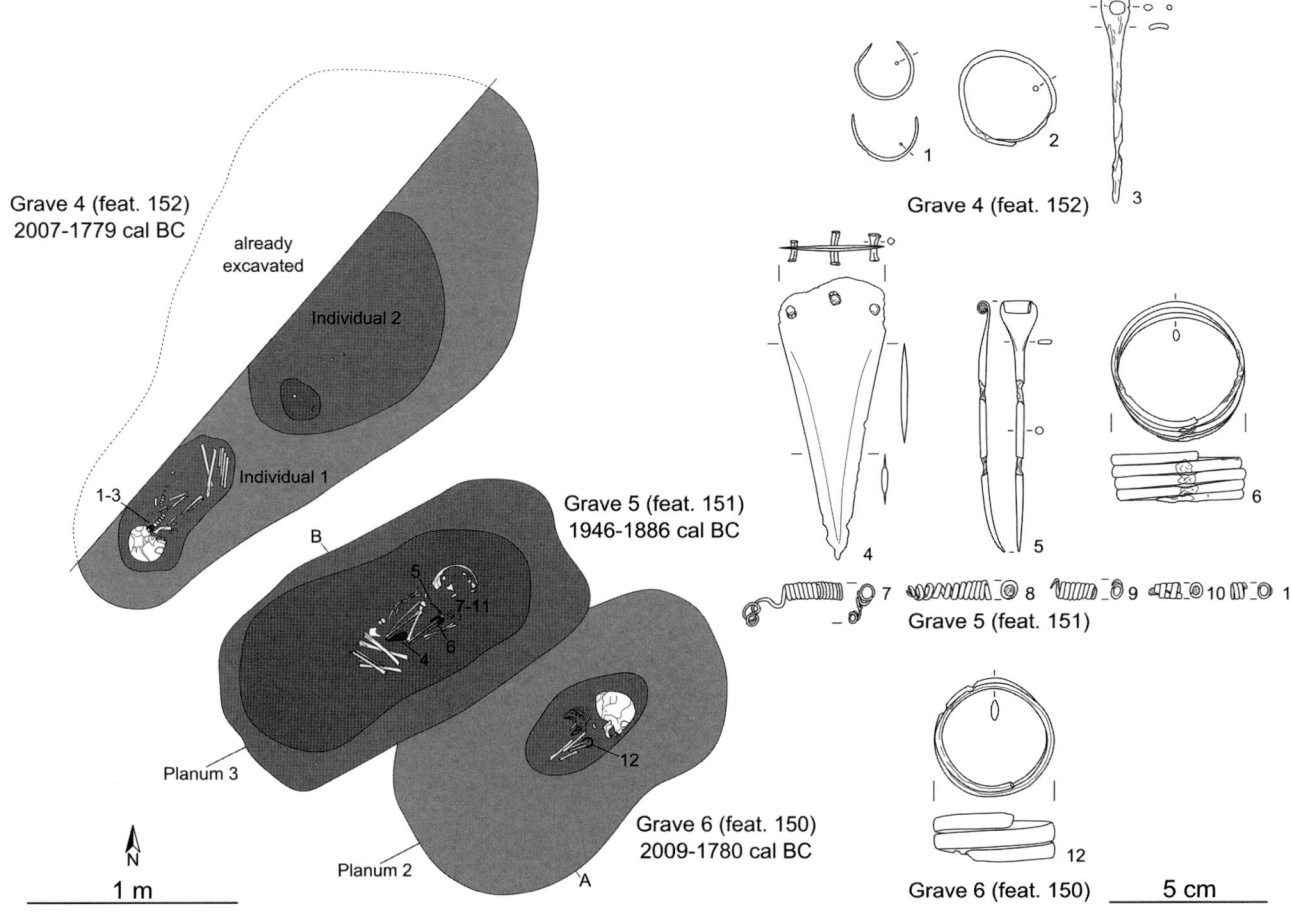

Figure 20.8: Detail of the three EBA graves from H-UT 58-62 with haplotype sharing.

All individuals amenable to genetic sexing – the CWC individual of H-UT 89, the three BBC individuals, and two of the EBA individuals from H-UT 58–62 – were found to conform with their gender-specific burial positions, again emphasising the continuity of this tradition throughout the epochs.

Isotope analysis ($^{87}Sr/^{86}Sr$, $\delta^{18}O$, $\delta^{13}C$, $\delta^{15}N$)

Methodological background and sampling strategy

To further explore continuity and change with regard to diet and mobility between the Final Neolithic and the Early Bronze Age, we performed a series of radiogenic and stable isotope analyses on the skeletal remains that also underwent radiocarbon dating and ancient DNA analysis. While the mtDNA data not only indicate direct kinship between several individuals but also large-scale population changes, isotope data provide information on residential moves by single individuals and help in deciphering predominant residential customs or shedding light on long-distance mobility. Strontium ($^{87}Sr/^{86}Sr$) and oxygen ($\delta^{18}O$) isotope analyses are key methods for exploring human mobility (Knipper 2004; Bentley 2006; Evans *et al.* 2012). While strontium isotope

ratios reflect geological conditions, oxygen isotope ratios vary with climate, elevation, and distance from the sea. Both elements are incorporated during tooth enamel formation in childhood and adolescence and later remain unchanged.

To ensure comparability in the results obtained from these individuals, we opted for the sampling of second molars, whose crowns form between about three and seven years of age (Schroeder 2000; AlQahtani *et al.* 2010). Wherever these molars were not available, either because they had not yet formed in young children or were missing due to unfavourable preservation conditions, alternative teeth were sampled. If possible, deciduous and permanent front teeth and first molars were avoided, because their $\delta^{18}O$ values may be elevated as a result of breastfeeding (Wright and Schwarcz 1998; Britton *et al.* 2015). For individuals identified as non-local in the first place, an additional sample of a tooth forming later or earlier was taken to identify whether the residential change already occurred in childhood or later, after enamel formation was complete.

The dense accumulation of sites in the Lech Valley provides very favourable conditions for strontium and oxygen isotope analyses. The potential arable land and the

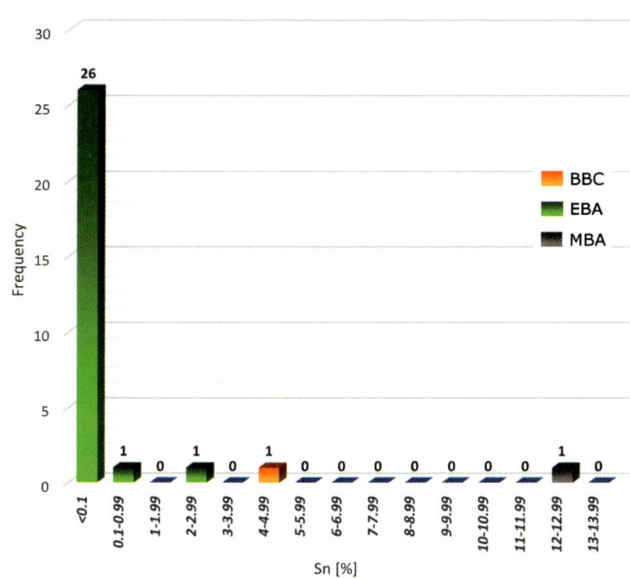

Figure 20.9: Sn contents in the analysed metal objects from H-UT 58-62 according to their chronological assignment.

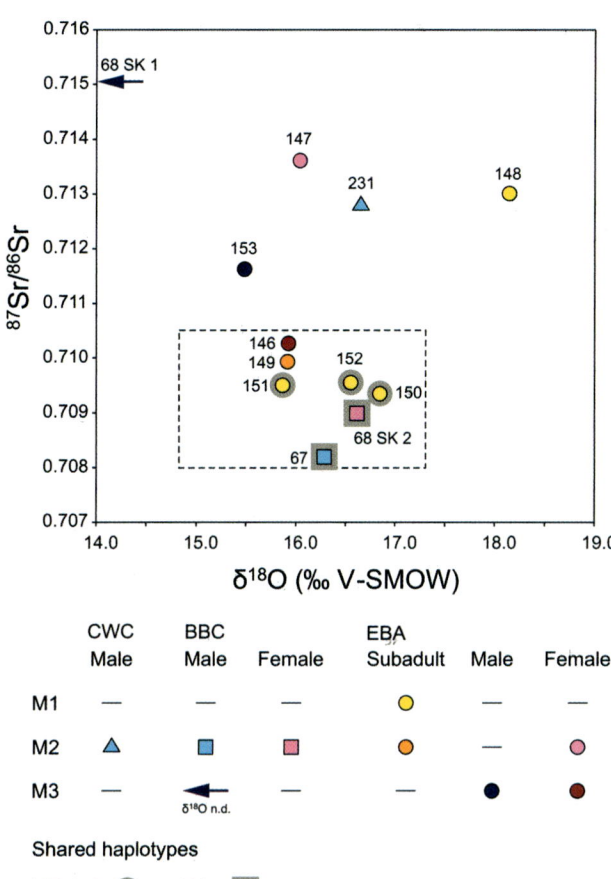

Figure 20.10: Scatter plot of strontium and oxygen isotope ratios of enamel samples from H-UT 58–62 and H-UT 89. The box indicates the baseline ranges for both isotope systems.

habitation areas of all the sites are dominated by loess and gravel and also shaped by the same climatic conditions. Accordingly, comparability between the analytic data from the different locations is high. Moreover, southern Bavaria and especially BBC burials have been the subject of earlier Sr isotope studies providing valuable comparative data on regional baseline values and mobility patterns (Price *et al.* 1998; Schweissing and Grupe 2003; Price *et al.* 2004; Bentley and Knipper 2005; Bickle *et al.* 2011; Bertemes and Heyd 2015).

In addition to Sr and O isotope analyses focusing on mobility, we also determined the stable isotope compositions of carbon ($\delta^{13}C$) and nitrogen ($\delta^{15}N$) in bone collagen to explore dietary habits. Carbon isotopes primarily reflect the photosynthetic path of the plants at the bottom of the food webs or the part played by marine diet components (Katzenberg 2000), but they also indicate smaller variations in the environmental conditions of arable land, including leaf coverage and humidity (Drucker *et al.* 2008; Kohn 2010; Mörseburg *et al.* 2015). Nitrogen isotope data primarily vary with trophic levels and can therefore indicate the importance of animal-based foods in a human diet (Hedges and Reynard 2007). They also react to the manuring of arable land (Bogaard *et al.* 2013), which makes them highly informative on farming practice (Styring *et al.* 2016) but may also interfere with data interpretation in connection with human diet composition.

The specific research issues at stake in connection with the stable isotope analyses in the Lech Valley project focused on the investigation of sex, age, site, and status-specific patterns in dietary habits and mobility, including residential customs and long-distance contacts and continuity and

change in these factors from the Final Neolithic to the Early Bronze Age.

Haunstetten – Unterer Talweg 58–62 and 89: Implications from a case study

The small burial communities of H-UT 58–62 and H-UT 89 are presented here as case examples illustrating the nature and scope of the information that can be gleaned from the complete dataset produced by the project (Knipper *et al.*, forthcoming). Sample preparation and analysis of the strontium and oxygen isotope composition of the phosphate component of the enamel were undertaken in line with the methods described elsewhere (Knipper *et al.* 2012; 2014; *cf.* Knipper *et al.* forthcoming), while carbon and nitrogen isotope composition was determined on the basis of bone collagen remains after extraction for radiocarbon dating (Stockhammer *et al.* 2015a). Based on archaeological faunal enamel, data compilation from the literature, and modern precipitation data, $^{87}Sr/^{86}Sr$ ratios lower than 0.7080 and higher then 0.7105 and $\delta^{18}O$ values beyond the range of about 14.7 to 17.4‰ VSMOW are indicators of non-local individuals from outside the Lech

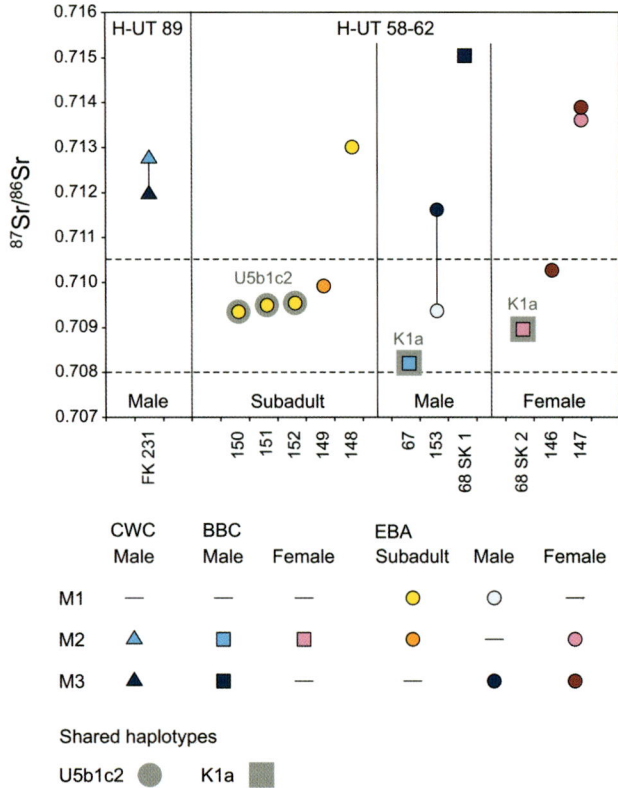

Figure 20.11: Strontium isotope ratios of enamel samples from H-UT 58-62 and H-UT 89. The data of subsequently formed teeth from the same individuals are plotted on top of each other and are connected with a line.

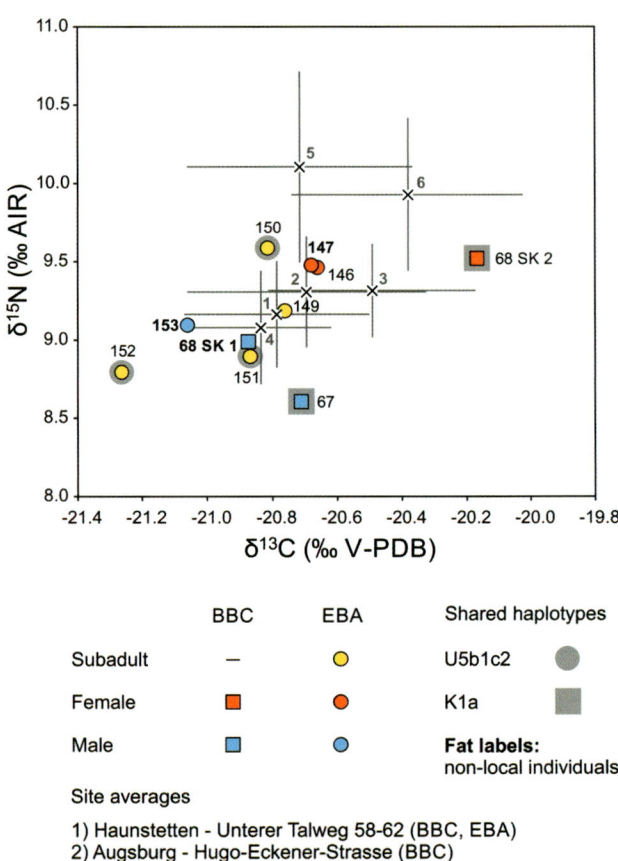

Figure 20.12: Scatter plot of carbon and nitrogen isotope ratios of bone collagen of human individuals from H-UT 58-62 and H-UT 89 as well as site averages with one standard deviation of other BBC and EBA sites from the Lech valley.

Valley (Knipper *et al.* forthcoming). Isotope ratios within these ranges are consistent with local origins but may also occur in non-local individuals from localities with similar geological and climatic conditions (*i.e.* most of southern Germany south of the Danube, excluding the Black Forest and the Bavarian Forest).

The ^{87}Sr/^{86}Sr ratios in the enamel of the individuals investigated varied widely from 0.70820 to 0.71505, while the δ^{18}O values ranged from 15.49 to 18.15‰ VSMOW (Fig. 20.10). The co-existing strontium and oxygen isotope signals of the BBC male (H-UT 58–62 feat. 67) and the female (H-UT 58–62 feat. 68 SK 2) were consistent with the local ranges for the stable isotope ratios of both elements. The same also applied to four EBA sub-adult individuals (feat. 149, 150, 151, and 152) and the EBA female from feature 146. More radiogenic ^{87}Sr/^{86}Sr ratios than were to be expected for individuals growing up in the Lech Valley were found in the teeth of five individuals. Among them were the CWC male (H-UT 89 feat. 231), the BBC male (H-UT 58–62 feat. 68 SK 1), and three individuals from of the EBA graves of H-UT 58–62, including the sub-adult (feat. 148), the adult female (feat. 147), and the adult male (feat. 153). The first molar (tooth 46) of the approx. five-year-old individual (feat. 148) also yielded an elevated δ^{18}O value of

18.15‰. However, because crown formation in first molars starts around birth and proceeds over the first three years, this measurement may reflect breast-feeding as breastmilk has higher δ^{18}O values than directly imbibed water (Wright and Schwarcz 1998; Britton *et al.* 2015). Accordingly, this δ^{18}O score does not necessarily point to origins in a region that was warmer or situated closer to the sea, *i.e.* to a long-distance change of residence.

To establish when residential changes took place, we conducted strontium-isotope analysis of a second tooth with a different mineralisation period in three of the non-local individuals. In the cases of the CWC male (H-UT 89 feat. 231) and the EBA female (H-UT 58–62 feat. 147), both the second and third molars yielded similar ^{87}Sr/^{86}Sr ratios above the local range (Fig. 20.11). This suggests that relocation to the Lech Valley took place after adolescence, when the crown formation of the wisdom teeth analysed was already complete. By contrast, the ^{87}Sr/^{86}Sr ratios of the male (H-UT 58–62 feat. 153) indicate a move to a new residence in childhood already. While the first molar yielded an $^{87}Sr/^{86}Sr$ ratio within the local range, the isotope ratio of the third molar was more radiogenic. These data imply that the male originated from the Lech Valley or a location with similar baseline values, moved away in late childhood or adolescence, and then returned in early adulthood or later to the Lech valley, where he was buried. The nearest geographic area with radiogenic, biologically available strontium as detected in the enamel of these non-local individuals is the Nördlinger Ries, some 60 km NNW. Parts of Bohemia (Scheeres *et al.* 2014) and glacially influenced areas of central Germany (Knipper *et al.* 2016a) are also possible areas of origin, because they boast geological conditions that produce both radiogenic, biologically available strontium and a rich archaeological record of the Final Neolithic and the Early Bronze Age. From a geochemical perspective, low mountain ranges, such as the Bavarian or Black Forests (Bentley and Knipper 2005) and parts of the Alpine region (Grupe *et al.* 2015) are also possible candidates, but the archaeological record of the relevant periods in these areas is anything but rich.

This small dataset already gives some indications of sex- and time-specific similarities and differences that are confirmed or differentiated in the complete dataset of the project (Knipper *et al.* forthcoming). Non-local individuals appeared in all three cultural complexes investigated, the CWC, the BBC, and the EBA, implying that human mobility played a crucial role not only in the EBA but also in the Final Neolithic. There were also non-local individuals of both sexes, adult and sub-adult, and residential changes by the specific individuals took place at different ages. This indicates fairly complex mobility patterns and individual reasons for resettlement. However, the dataset of the case study is also too small to reveal overarching trends. These are naturally easier to see in the study as a whole (Knipper *et al.* forthcoming).

The individuals from H-UT 58–62 were also subjected to carbon and nitrogen isotope analysis. The δ¹³C scores for their bone collagen varied between -21.3 and -20.2‰ V-PDB (average: -20.8 ± 0.3‰), and the δ¹⁵N values ranged from 8.6 to 9.6‰ AIR (average: 9.2 ± 0.3‰) (Fig. 20.12). Females appear to have slightly higher δ¹⁵N values than males, a result that has however to be regarded with caution due to the very small sample size of only three males and females each. No obviously different δ¹³C or δ¹⁵N values were found between age categories (sub-adults vs. adults), cultural complexes (BBC vs. EBA), or personal origins (local vs. non-local individuals). The average stable isotope data were also very similar to those at the nearby sites of Augsburg – Hugo-Eckener-Strasse (BBC), Haunstetten – Postillionstrasse (EBA), and Haunstetten – Unterer Talweg 85 (BBC, EBA), while the EBA burials of Königsbrunn – Obere Kreuzstrasse and Wehringen – Hochfeld (the only site studied that is situated on the other side of the loess ridge) appear to have higher average δ¹⁵N values.

Overall, the collagen stable isotope data are consistent with a mixed diet from plant and animal sources with C_3 plants at the base of the food chain. There is no evidence of C_4 plant consumption. This agrees with previous studies indicating that millet, the most important dietary C_4 plant in prehistoric Europe, became important as a staple crop in the Late Bronze Age and is clearly visible isotopically from the Hallstatt period onwards (Oelze *et al.* 2012; Moghaddam *et al.* 2016, Knipper and Maus 2016; Knipper *et al.* 2016b). As for δ¹⁵N, it generally tends to increase from Early Neolithic *Linearbandkeramik* contexts to the Early Bronze Age. This is well seen in Central Germany (Siebert *et al.* 2016) but is also identifiable in southern Bavaria (Asam *et al.* 2006). It can best be explained by the increasing importance of meat and dairy products in the human diet. Changes in manuring practices for arable land may also have played a role. The EBA datasets from southern Bavaria are more heterogeneous than those from Central Germany (Asam *et al.* 2006; Koch and Kupke 2012; Knipper *et al.* 2016a), and this diversity is also visible on a very local scale among the sites in the Lech Valley.

Integration and interpretation of the data
Combining multiple bio-archaeometric and archaeological data enables us to decipher very specific and detailed histories, right down to the individual or family level.

With regard to the single CWC individual, all we can do is establish that his origins were elsewhere and that he arrived at the Lech Valley in his adulthood. We do not know whether he was involved in large-scale population movements from the east or whether he came to the valley with his family or as an individual migrant.

The three BBC burials of H-UT 58–62 seem likely to represent a single family – possibly a couple with their son (buried at a later point in time) or the woman's brother (buried before or after the couple). The biological relationship between the two individuals is expressed in the close position of the burials and the construction of a tumulus on top of each of them. The similar equipment given to the two men in the graves could also be interpreted as the material expression of their close relationship. While the 'mother' (or sister) and the 'son' (or brother) lived in the Lech Valley (or a geologically similar area) in their early childhood, the putative father spent some time elsewhere. His third molar – mineralised approx. between his 7th and 17th year of life – has a high strontium isotope value that does not match to the soil of the Lech Valley or adjacent southern Germany (Fig. 20.10 and 20.11). Together with what we know about the distribution of fertile radiogenic soils and densities of settlement during the 3rd millennium BC, the strontium isotope ratio points to the Mittelelbe-Saale region or Bohemia as possible places where he spent at least his late childhood and his teenage years. More isotope analyses of different molars will clarify whether this man was indeed born outside the Lech Valley and whether the woman and the 'son' also moved during their childhoods. Possibly due to his mobility, the 'father' was able to acquire a dagger with a surprisingly high tin content, a much more common feature of Bell Beaker metal objects in the Mittelelbe-Saale region than in southern Germany (Junghans *et al.* 1960; 1968; Krause 2003). We are unable to say whether the three BBC individuals were also infected by the plague (*Yersinia pestis*) that raged in the Lech Valley during this period (Andrades Valtueña *et al.* 2017). The absence of the bacterium in the dentine of the three individuals can easily be explained by the very selective preservation of pathogens in human individuals from the past (Schuenemann *et al.* 2011).

The neighboring EBA cemetery indicates continuous settlement at H-UT 58–62 in the transition from the BBC to the EBA. Continuous settlement activity at this time (*i.e.* around 2150 BC) is clearly evidenced by radiocarbon dating (Stockhammer *et al.* 2015a; 2015b). However, settlement continuity within the valley went hand in hand with the relocation of single farmsteads in the course of time. The gap of at least 150 years between the different burial places might also indicate a settlement hiatus at H-UT 58–62. These relocations were probably not due to an absence of fertile ground, as the adjacent loess terrace offered enough space to shift fields. Other reasons, such as the extinction of the family or unknown societal or ritual issues, need to be taken into consideration.

After the gap between the BBC and the Early Bronze Age, a new farmstead was established at the very same place or at least very close to its Final Neolithic predecessor. The people now buried at H-UT 58–62 were probably not biologically related to the BBC individuals, as neither the mtDNA nor the Y chromosomes nor the full genomic evidence point to any closer relationship between the two groups of individuals.

Within the EBA cemetery, four children were buried close to each other and three of them (with sufficient bone preservation) have been identified as siblings (they shared the same mtDNA). Again, the close vicinity of graves mirrors a close biological relationship during their lifetimes as well as in their burial equipment. All metal objects of the EBA individuals were made of copper; none were identified as tin-bronze. The inhabitants of the EBA farmstead had a different exchange and barter network from their BBC forerunners and either no desire or no opportunity to obtain tin from faraway places. Nevertheless, the EBA individuals also had far-reaching networks, as several of the individuals on this site were non-local. One woman only arrived at the Lech Valley after she had turned 17 and thus probably after (and because of) reaching fertility. Her place of origin had similar radiogenic soils to those where the BBC 'father' originated or at least spent part of his youth. One man was probably born in the Lech Valley but

left his home at the age of at least seven and spent several years at a place with radiogenic soil. Was he sent by his parents to the Mittelelbe-Saale region or Bohemia – *i.e.* the region of the Únětice culture – to learn some particular craft and then return with his wife? If so, either she did not have children after her arrival in the Lech Valley or her children were sent back to her place of origin at some point in time or were buried outside the H-UT 58–62 site in a cemetery where we have not sampled all individuals and have thus missed the children. So far, her mtDNA has not been found in any other individual studied by us in the Lech Valley.

Zooming out

Combining the archaeological and scientific evidence has provided completely new insights into biological relationships, individual mobility, and the relation of mobility to the archaeological evidence. We are confronted with a society that combined individual long-distance mobility (already or at least or especially during childhood) with continuous settlement in the Lech Valley. Settlements in the valley took the form of single farmsteads whose positions changed continuously over time. A burial place was always located to the east of each farmstead. The relevance of the eastern location for the deceased was also expressed in their positioning within the grave and the alignment of posts extending from several graves to the east. The plague hit the Lech Valley once or several times, and generations of inhabitants suffered from it. The societal changes all around Europe during the 22nd century BC, especially the establishment of the Únětice culture and El Argar (Meller *et al.* 2015) may have been responsible for the transformations in the networks of barter and exchange. Tin was not used for alloying copper any more and gold, silver, and amber did not arrive in the Lech Valley during the EBA, although they were obtainable for BBC individuals before that time (Mahnkopf and Nitsch 2002). In the Lech Valley, a Final Neolithic 'Bronze Age' was succeeded by an Early Bronze 'Copper Age'.

Acknowledgements

We want to thank Michaela Hermann (Stadtarchäologie Augsburg), Sebastian Gairhos (Stadtarchäologie Augsburg), Hanns Dietrich (Bayerisches Landesamt für Denkmalpflege), Catharina Kociumaka, Rainer Linke (Arbeitskreis für Vor- und Frühgeschichte, Gruppe Augsburg Süd), Jürgen Hald (Kreisarchäologie Konstanz) and Joachim Wahl (Landesamt für Denkmalpflege, Baden-Württemberg) for their permissions to study and sample the burials as well as their great support of our work. Anja Staskiewicz and Elizabeth Nelson carried out anthropological age and sex determinations of the human skeletal remains.

We are indebted to Gerlinde Borngässer, Melanie Gottschalk, Bernd Höppner, Sigrid Klaus, and Sandra Pagacs and for help with sample preparation and strontium isotope analysis at the Curt-Engelhorn-Centre for Archaeometry Mannheim, Germany. Willi Dindorf and Michael Maus analysed the carbon, nitrogen, and oxygen isotope samples at the Institute for Organic Chemistry and the Institute of Geosciences, Department of Applied Geochemistry at the University of Mainz. Susanne Lindauer, Robin von Gysegheim, Ronny Friedrich, and Ute Blach conducted radiocarbon dating at the Curt-Engelhorn-Centre for Archaeometry Mannheim, Germany. Isil Kucukkalipci provided support in the lab and Franziska Göhringer documented the samples for the DNA analysis.

Notes

1 This article presents the results of the collaborative research project 'Times of Upheaval: Transformations of Societies and Landscapes at the Onset of the Bronze Age' under the direction of Philipp W. Stockhammer and (first) Johannes Krause (2012–2016) and (later) Alissa Mittnik (as of 2016). The Heidelberg Academy of Sciences and Humanities has kindly financed the project in the funding line of its WIN projects.

2 In the following, Final Neolithic (German: *Endneolithikum*) will be used synonymously for the Beaker Complexes, *i.e.* for CWC and BBC.

3 So far, all excavated hamlets seem to date to the Early Bronze Age. In line with the general lack of BBC settlement evidence in Central Europe (Nadler 2001), BBC hamlets have not been found yet.

4 Ruckdeschel 1978, 286. Archaeological analysis of the EBA burials we know of is part of Ken Massy's dissertation (Massy forthcoming); Catharina Kociumaka is investigating their BBC counterparts.

5 Müller (2000, 71–74) proposes such an overlap between the CWC, BBC, and the Early Bronze Age for the Mittelelbe-Saale region.

6 For the sampling strategy, see Stockhammer *et al.* 2015a. The number of samples has grown since this publication.

7 For a description of the respective methodologies and analytical protocols, *cf.* Stockhammer *et al.* 2015a (for radiocarbon dating) and Knipper *et al.* forthcoming (for Sr and O isotope ratio analyses and mtDNA). The comprehensive evaluation of the full genomic data has yet to be completed.

8 For the traditional dating of Singen, *cf.* Becker *et al.* 1989. Volker Heyd (2000) has already argued in favor of a slightly later transition between the two phases around 2150 BC.

9 With a concentration of 4.6% tin the dagger belongs to a small group of BBC metal objects with high tin contents in Central Europe, the others being a flange-hilted dagger from Altenmarkt (Bavaria) with 3.6% tin and a dagger from Mörigen (Switzerland) with 2.9% tin. Most BBC tin bronzes have been found in Thuringia and Saxony, *i.e.* regions where the Únětice culture was soon to appear (Junghans *et al.* 1960; Junghans *et al.* 1968; Krause 2003).

References

Allentoft, M. E *et al.* (2015) Population Genomics of Bronze Age Eurasia. *Nature* 522, 167–172.

Alqahtani, S. J., Hector, M. P. and Liversidge, H. M. (2010) Brief Communication: The London Atlas of Human Tooth Development and Eruption. *American Journal of Physical Anthropology* 142, 481–490.

Andrades Valtueña, A., Mittnik, A., Massy, K., Allmae, R., Daubaras, M., Jankauskas, R., Torv, M., Pfrengle, S., Spyrou, M. A., Feldman, M., Haak, W., Bos, K. I., Stockhammer, P. W., Herbig, A. and Krause, J. (2017) The Stone Age Plague: 1000 Years of Persistence in Eurasia. *bioRxiv* 094243. DOI: https://doi.org/10.1101/094243.

Asam, T., Grupe, G. and Peters, J. (2006) Menschliche Subsistenzstrategien im Neolithikum: Eine Isotopenanalyse bayerischer Skelettfunde. *Anthropologischer Anzeiger* 64, 1–23.

Bartelheim, M. (1998) *Studien zur böhmischen Aunjetitzer Kultur. Chronologische und chorologische Untersuchungen.* Universitätsforschungen zur prähistorischen Archäologie 46. Bonn, Habelt.

Bartelheim, M. (2010) Schmiedefürsten oder Großbauern? Eliten und Metalle in der Frühbronzezeit Mitteleuropas. In H. Meller and F. Bertemes (eds), *Der Griff nach den Sternen. Wie Europas Eliten zu Macht und Reichtum kamen. Internationales Symposium in Halle (Saale) vom 16.–21. Februar 2005.* Tagungen des Landesmuseums für Vorgeschichte Halle 5/2, 865–880. Halle (Saale), Landesamt für Denkmalpflege und Archäologie Sachsen-Anhalt und Landesmuseum für Vorgeschichte.

Becker, B., Krause, R., and Kromer, B. (1989) Zur absoluten Chronologie der Frühen Bronzezeit. *Germania* 67, 421–442.

Bentley, R. A. and Knipper, C. (2005) Geographical Patterns in Biologically Available Strontium, Carbon and Oxygen Isotope Signatures in Prehistoric SW Germany. *Archaeometry* 47, 629–644.

Bentley, R. A. (2006) Strontium Isotopes from the Earth to the Archaeological Skeleton: A Review. *Journal of Archaeological Method and Theory* 13, 135–187.

Bernbeck, R., Kaiser, E., Parzinger, H., Pollock, S. and Schier, W. (2011) Spatial Effects of Technological Innovations and Changing Ways of Life. *eTopoi. Journal for Ancient Studies* 1, 1–16.

Bertemes, F. and Heyd, V. (2015) 2200 BC – Innovation or Evolution? The Genesis of the Danubian Early Bronze Age. In H. Meller, R. Risch, R. Jung and H.-W. Arz (eds), *2200 BC – A Climatic Breakdown as a Cause for the Collapse of the Old World? 7th Archaeological Conference of Central Germany October 23–26, 2014 in Halle (Saale).* Tagungen des Landesmuseums für Vorgeschichte Halle 12/2, 561–578. Halle (Saale), Landesamt für Denkmalpflege und Archäologie Sachsen-Anhalt und Landesmuseum für Vorgeschichte.

Bickle, P., Hofmann, D., Bentley, R. A., Hedges, R. E. M., Hamilton, J., Laiginhas, F., Nowell, G., Pearson, D. G., Grupe, G. and Whittle, A. (2011) Roots of Diversity in a Linearbandkeramik Community: Isotope Evidence at Aiterhofen (Bavaria, Germany). *Antiquity* 85, 1243–1253.

Bogaard, A., Fraser, R., Heaton, T. H. E., Wallace, M., Vaiglova, P., Charles, M., Jones, G., Evershed, R., Styring, A. K., Anderson, N. H., Arbogast, R.-M., Bartosiewicz, L., Gardeisen,

A., Kanstrup, M., Maier, U., Marinova, E., Ninov, L., Schäfer, M. and Stephan, E. (2013) Crop Manuring and Intensive Land Management by Europe's First Farmers. *Proceedings of the National Academy of Sciences* 110, 12589–12594.

Bosch, T. L. (2009) *Archäologische Untersuchungen zur Frage von Sozialstrukturen in der Ostgruppe des Glockenbecherphänomens anhand des Fundgutes.* Unpublished thesis, Universität Regensburg, Regensburg. Latest update 21.04.2014, <urn:nbn:de:bvb:355-opus-12922> (24.01.2017).

Bramanti, B., Thomas, M. G., Haak, W., Unterlaender, M., Jores, P., Tambets, K., Antanaitis-Jacobs, I., Haidle, M. N., Jankauskas, R., Kind, C.-J., Lueth, F.,Terberger, T., Hiller, J., Matsumura, S., Forster, P. and Burger, J. (2009) Genetic Discontinuity between Local Hunter-Gatherers and Central Europe's First Farmers. *Science* 326, 137–140. DOI: 10.1126/science.1176869.

Brandt, G., Haak, W., Adler, C. J., Roth, C., Szécsényi-Nagy, A., Karimnia, S., Möller-Rieker, S., Meller, H., Ganslmeier, R., Friederich, S., Dresely, V., Nicklisch, N., Pickrell, J. K., Sirocko, F., Reich, D., Cooper, A. and Alt, K. W. (2013) Ancient DNA Reveals Key Stages in the Formation of Central European Mitochondrial Genetic Diversity. *Science* 342, 257–261. DOI: 10.1126/science.1241844.

Braun-Thürmann, H. (2005) *Innovation.* Bielefeld, transcript.

Britton, K., Fuller, B. T., Tütken, T., Mays, S. and Richards, M. P. (2015) Oxygen Isotope Analysis of Human Bone Phosphate Evidences Weaning Age in Archaeological Populations. *American Journal of Physical Anthropology* 157, 226–241.

Brumlich, M. (2014) Alte Thesen und neue Forschungen zur Eisenproduktion in der Jastorfkultur. In J. Brandt and B. Rauchfuß (eds), *Das Jastorf-Konzept und die vorrömische Eisenzeit im nördlichen Mitteleuropa. Beiträge der Internationalen Tagung zum einhundertjährigen Jubiläum der Veröffentlichung der Ältesten Urnenfriedhöfe bei Uelzen und Lüneburg durch Gustav Schwantes 18.–22. Mai 2011 in Bad Bevensen.* Veröffentlichungen des Archäologischen Museums Hamburg 105, 155–168. Hamburg, Archäologisches Museum Hamburg.

Buchvaldek, M. (1967) *Die Schnurkeramik in Böhmen.* Prague, Universita Karlova.

Burmeister, S. (2011) Innovationswege – Wege der Kommunikation. Erkenntnisprobleme am Beispiel des Wagens im 4. Jt. v. Chr. In S. Hansen and J. Müller (eds), *Sozialarchäologische Perspektiven. Gesellschaftlicher Wandel 5000–1500 v. Chr. zwischen Atlantik und Kaukasus, Internationale Tagung 15.–18. Oktober 2007, Kiel,* 211–240. Mainz, Philipp von Zabern.

Burmeister, S. and Müller-Scheeßel, N. (2013) Innovation as a Multi-Faceted Social Process: An Outline. In S. Burmeister, S. Hansen, M. Kunst and N. Müller-Scheeßel (eds), *Metal Matters. Innovative Technologies and Social Change in Prehistory and Antiquity.* Mensch – Kulturen – Traditionen. Studentisches Forschungscluster DAI 12, 1–12. Rahden/Westf., Marie Leidorf.

David, W. (1998) Zu früh- und ältermittelbronzezeitlichen Grabfunden in Ostbayern. In J. Michálek, K. Schmotz and M. Zápotocká (eds), *Archäologische Arbeitsgemeinschaft Ostbayern/West- und Südböhmen. 7. Treffen in Landau (Isar) 11.–14. Juni 1997,* 108–129. Rahden/Westf., Marie Leidorf.

Dresely, V. (2004) *Schnurkeramik und Schnurkeramiker im Taubertal.* Stuttgart, Theiss.

Drucker, D. G., Bridault, A., Hobson, K. A., Szuma, E. and Bocherens, H. (2008) Can Carbon-13 in Large Herbivores Reflect the Canopy Effect in Temperate and Boreal Ecosystems? Evidence from Modern and Ancient Ungulates. *Palaeogeography, Palaeoclimatology, Palaeoecology* 266, 69–82.

Dumler, M. and Wirth, S. (2001) Gräber und Siedlungen aus vier Jahrtausenden – Neue Ausgrabungen am Unteren Talweg in Haunstetten. *Das archäologische Jahr in Bayern* 2001, 47–50.

Evans, J. A., Chenery, C. A. and Montgomery, J. (2012) A Summary of Strontium and Oxygen Isotope Variation in Archaeological Human Tooth Enamel Excavated from Britain. *Journal of Analytical Atomic Spectrometry* 27, 754–764.

Fu, Q., Meyer, M., Gao, X., Stenzel, U., Burbano, H. A., Kelso, J. and Pääbo, S. (2013) DNA Analysis of an Early Modern Human from Tianyuan Cave, China. *Proceedings of the National Academy of Sciences of the United States of America* 110, 2223–2227. DOI: 10.1073/pnas.1221359110.

Furholt, M. (2003) *Die absolutchronologische Datierung der Schnurkeramik in Mitteleuropa und Südskandinavien.* Bonn, Habelt.

Gairhos, S. (2007) Von der Schlitzgrube zur Abseitsfalle – Archäologie unter dem neuen Fußballstadion in Göggingen. Stadt Augsburg, Schwaben. *Das archäologische Jahr in Bayern* 2007, 86–88.

Gebers, W. (1978) *Endneolithikum und Frühbronzezeit im Mittelrheingebiet. Katalog.* Bonn, Habelt.

Grupe, G., Price, T. D., Schröter, P., Söllner, F., Johnson, C. M. and Beard, B. L. (1997) Mobility of Bell Beaker People Revealed by Strontium Isotope Ratios of Tooth and Bone: A Study of Southern Bavarian Skeletal Remains. *Applied Geochemistry* 12, 517–525.

Grupe, G., Grünewald, M., Gschwind, M., Hölzl, S., Kocsis, B., Kröger, P., Lang, A., Mauder, M., Mayr, C., McGlynn, G., Metzner-Nebelsick, C., Ntoutsi, E., Peters, J., Renz, M., Reuß, S., Schmahl, W. W., Söllner, F., Sommer, C. S., Steidl, B., Toncala, A., Trixl, S. and Wycisk, D. (2015) Networking in Bioarchaeology: The Example of the DFG Research Group FOR 1670 'Transalpine Mobility and Culture Transfer'. In G. Grupe, G. McGlynn and J. Peters (eds), *Bioarchaeology Beyond Osteology.* Documenta Archaeobiologiae 12, 13–51. Rahden/Westf., Marie Leidorf.

Haak, W., Brandt, G., de Jong, H. N.,Meyer, C., Ganslmeier, R., Heyd, V., Hawkesworth, C., Pike, A. W. G., Meller, H., and Alt, K. W. (2008) Ancient DNA, Strontium Isotopes, and Osteological Analyses Shed Light on Social and Kinship Organization of the Later Stone Age. *Proceedings of the National Academy of Sciences of the United States of America* 105, 18226–18231. DOI: 10.1073/pnas.0807592105.

Haak, W., Lazaridis, I., Patterson, N., Rohland, N., Mallick, S., Llamas, B., Brandt, G., Nordenfelt, S., Harney, E., Stewardson, K., Fu, Q., Mittnik, A., Bánffy, E., Economou, C., Francken, M., Friederich, S., Garrido Pena, R., Hallgren, F., Khartanovich, V., Khokhlov, A., Kunst, M., Kuznetsov, P., Meller, H., Mochalov, O., Moiseyev, V., Nicklisch, N., Pichler, S. L., Risch, R., Rojo Guerra, M. A., Roth, C., Szécsényi-Nagy, A., Wahl, J., Meyer, M., Krause, J., Brown, D., Anthony, D., Cooper, A., Alt, K. W. and Reich, D. (2015) Massive Migration from the Steppe Was a Source for Indo-European Languages in Europe. *Nature* 522/7555, 207–211.

Hájek, L. (1968) *Die Glockenbecherkultur in Böhmen.* Prag, Archeologický Ústav Čsav.

Hedges, R. E. M. and Reynard, L. M. (2007) Nitrogen Isotopes and the Trophic Level of Humans in Archaeology. *Journal of Archaeological Science* 34, 1240–1251.

Heyd, V. (2000) *Die Spätkupferzeit in Süddeutschland. Untersuchungen zur Chronologie von der ausgehenden Mittelkupferzeit bis zum Beginn der Frühbronzezeit im süddeutschen Donaueinzugsgebiet und den benachbarten Regionen bei besonderer Berücksichtigung der keramischen Funde.* Saarbrücker Beiträge zur Altertumskunde 73. Bonn, Habelt.

Heyd, V., Winterholler, B., Böhm, K. and Pernicka, E. (2002–2003) Mobilität, Strontiumisotopie und Subsistenz in der süddeutschen Glockenbecherkultur. *Bericht der Bayerischen Bodendenkmalpflege* 43–44, 109–135.

Hofmann, K. P. (2012) Kontinuität trotz Diskontinuität? Der Wechsel von der Körper- zur Brandbestattung im Elbe-Weser-Dreieck und die semiotische Bedeutungsebene 'Raum'. In D. Bérenger, J. Bourgeois, M. Talon and S. Wirth (eds), *Gräberlandschaften der Bronzezeit. Paysages funéraires de l'âge du Bronze. Internationales Kolloquium zur Bronzezeit, Herne, 15.–18. Oktober 2008.* Bodenaltertümer Westfalens 51, 355–373. Darmstadt, Philipp von Zabern.

Hofmann, K. P. and Patzke, S. (2012) Von Athen nach Etrurien. Zum Diffusionsprozess der entlehnten Innovation 'ceramica sovraddipinta'. In A. Kern, J. K. Koch, I. Balzer, J. Fries-Knoblach, K. Kowarik, C. Later, P. C. Ramsl, P. Trebsche and J. Wiethold (eds), *Technologieentwicklung und -transfer in der Hallstatt- und Latènezeit.* Beiträge zur Ur- und Frühgeschichte Mitteleuropas 65, 83–101. Langenweißbach, Beier & Beran.

Innerhofer, F. (2000) *Die mittelbronzezeitlichen Nadeln zwischen Vogesen und Karpaten. Studien zur Chronologie, Typologie und regionalen Gliederung der Hügelgräberkultur.* UPA 71. Bonn, Habelt.

Jiménez, A. C. (2005) Changing Scales and the Scales of Change: Ethnography and Political Economy in Antofagasta, Chile. *Critique of Anthropology* 25/2, 157–176.

Junghans, S., Sangmeister, E. and Schröder, M. (1960) *Metallanalysen kupferzeitlicher und frühbronzezeitlicher Bodenfunde aus Europa.* Berlin, Mann.

Junghans, S., Sangmeister, E. and Schröder, M. (1968) *Kupfer und Bronze in der frühen Metallzeit Europas. Band 1: Die Materialgruppen beim Stand von 12000 Analysen.* Berlin, Mann.

Katzenberg, M. A. (2000) Stable Isotope Analysis: A Tool for Studying Past Diet, Demography, and Life History. In A. M. Katzenberg and S. R. Saunders (eds), *Biological Anthropology of the Human Skeleton.* New York and Chichester, Wiley-Liss.

Kienlin, T. L. (2006a), Frühbronzezeitliche Randleistenbeile von Böhringen-Rickelshausen und Hindelwangen. Ergebnisse einer metallographischen Untersuchung. *Prähistorische Zeitschrift* 81, 97–120.

Kienlin, T. L. (2006b) Waffe, Werkzeug, Barren. Zur Deutung frühbronzezeitlicher Randleistenbeile in Depotfunden des nordalpinen Raums. In H.-P. Wotzka (ed.), *Grundlegungen. Beiträge zur europäischen und afrikanischen Archäologie für Manfred K. H. Eggert,* 461–476. Tübingen, Francke.

Kienlin, T. L. (2008) *Frühes Metall im nordalpinen Raum: Eine Untersuchung zu technologischen und kognitiven Aspekten früher Metallurgie anhand der Gefüge frühbronzezeitlicher Beile.* UPA 162. Bonn, Habelt.

Kienlin, T. L. (2010) *Traditions and Transformations. Approaches to Eneolithic (Copper Age) and Bronze Age Metalworking and Society in Eastern Central Europe and the Carpathian Basin.* British Archaeological Reports, International Series 2184. Oxford, Archaeopress.

Kienlin, T. L. (2014) Aspects of Metalworking and Society from the Black Sea to the Baltic Sea from the fifth to the second Millennium BC. In B. W. Roberts and C. P. Thornton (eds), *Archaeometallurgy in Global Perspective. Methods and Syntheses,* 447–472. New York, Springer.

Knipper, C. (2004) Die Strontiumisotopenanalyse: Eine naturwissenschaftliche Methode zur Erfassung von Mobilität in der Ur- und Frühgeschichte. *Jahrbuch des Römisch-Germanischen Zentralmuseums Mainz* 51, 589–685.

Knipper, C., Maurer, A.-F., Peters, D., Meyer, C., Brauns, M., Galer, S. G., von Freeden, U., Schöne, B., Meller H. and Alt, K. W. (2012) Mobility in Thuringia or Mobile Thuringians: A Strontium Isotope Study from Early Medieval Central Germany. In E. Kaiser, J. Burger and W. Schier (eds), *Population Dynamics in Prehistory and Early History. New Approaches Using Stable Isotopes and Genetics,* 287–310. Berlin, de Gruyter.

Knipper, C., Meyer, C., Jacobi, F., Roth, C., Fecher, M., Schatz, K., Stephan, E., Hansen, L., Posluschny, A. G., Pare, C. F. E. and Alt, K. W. (2014) Social Differentiation and Land Use at an Early Iron Age 'Princely Seat': Bioarchaeological Investigations at the Glauberg (Germany). *Journal of Archaeological Science* 41, 818–835.

Knipper, C., Fragata, M., Nicklisch, N., Siebert, A., Szécsényi-Nagy, A., Hubernsack, V., Metzner-Nebelsick, C., Meller, H. and Alt, K. W. (2016a) A Distinct Section of the Early Bronze Age Society? Stable Isotope Investigations of Burials in Settlement Pits and Multiple Inhumations of the Únětice Culture in Central Germany. *American Journal of Physical Anthropology* 159, 496–516.

Knipper, C., Pichler, S., Rissanen, H., Stopp, B., Kühn, M., Spichtig, N., Röder, B., Schibler, J., Lassau, G. and Alt, K. W. (2016b) What is on the Menu in a Celtic Town? Iron Age Diet Reconstructed in Remains from Settlement Features and two Celsies at Basel-Gasfabrik, Switzerland. *Archaeological and Anthropological Sciences.* DOI: 10.1007/s12520-016-0362-8.

Knipper, C. and Maus, M. (2016) *Isotopenanalysen zur Rekonstruktion von Mobilität und Ernährungsweise der Bestatteten der hallstattzeitlichen Nekropole von Mauenheim.* In L. Wamser (ed.), Mauenheim und Bargen. Zwei Grabhügelfelder der Hallstatt- und Frühlatènezeit aus dem nördlichen Hegau. Forschungen und Berichte zur Archäologie in Baden-Württemberg 2, 461–486. Wiesbaden, Dr. Ludwig Reichert Verlag.

Knipper, C., Mittnik, A., Massy, K., Kociumaka, C., Kucukkalipci, I., Maus, M., Metz, S., Wittenborn, F., Krause, J. and Stockhammer, P. W. (forthcoming) *Female Exogamy and Gene Pool Diversification at the Transition from the Final Neolithic to the Early Bronze Age in Southern Germany.*

Koch, J. K. and Kupke, K. (2012) Life-Course Reconstruction for Mobile Individuals in an Early Bronze Age Society in

Central Europe: Concept of the Project and first Results for the Cemetery of Singen (Germany). In E. Kaiser, J. Burger and W. Schier (eds), *Population Dynamics in Prehistory and Early History. New Approaches Using Stable Isotopes and Genetics*, 225–240. Berlin, de Gruyter.

Kohn, M. J. (2010) Carbon Isotope Compositions of Terrestrial C3 Plants as Indicators of (Paleo)Ecology and (Paleo)Climate. *PNAS* 107, 19691–19695.

Krause, R. (2003) *Studien zur kupfer- und frühbronzezeitlichen Metallurgie zwischen Karpatenbecken und Ostsee*. Rahden/Westf., Marie Leidorf.

Mahnkopf, G. and Nitsch, G. (2002) Bestattungsplatz der Glockenbecherkultur, Lehmgrube Markt. *Heimatverein Landkreis Augsburg. 28. Jahresbericht 2001–2002*, 14–23.

Maran, J. (2004a) Die Badener Kultur und ihre Räderfahrzeuge. In M. Fansa and S. Burmeister (eds), *Rad und Wagen. Der Ursprung einer Innovation. Wagen im Vorderen Orient und Europa. Wissenschaftliche Begleitschrift zur Sonderausstellung 'Rad und Wagen. Der Ursprung einer Innovation. Wagen im Vorderen Orient und Europa', Landesmuseum Oldenburg vom 28. März bis 11. Juli 2004*. Archäologische Mitteilungen Nordwestdeutschland Beiheft 40, 265–282. Mainz am Rhein, Philipp von Zabern.

Maran, J. (2004b) Kulturkontakte und Wege der Ausbreitung der Wagentechnologie im 4. Jahrtausend v. Chr. In M. Fansa and S. Burmeister (eds), *Rad und Wagen. Der Ursprung einer Innovation. Wagen im Vorderen Orient und Europa. Wissenschaftliche Begleitschrift zur Sonderausstellung 'Rad und Wagen. Der Ursprung einer Innovation. Wagen im Vorderen Orient und Europa', Landesmuseum Oldenburg vom 28. März bis 11. Juli 2004*. Archäologische Mitteilungen Nordwestdeutschland Beiheft 40, 429–442. Mainz, Philipp von Zabern.

Maricic, T., Whitten, M. and Pääbo, S. (2010) Multiplexed DNA Sequence Capture of Mitochondrial Genomes Using PCR Products. *PLOS ONE* 5. DOI: 10.1371/journal.pone.0014004.

Massy, K. (forthcoming) Die frühbronzezeitlichen Gräber zwischen Nördlinger Ries, Lech und Alpen.

Mathieson, I. *et al.* (2015) Genome-Wide Patterns of Selection in 230 Ancient Eurasians. *Nature* 528, 499–503. DOI: 10.1038/nature16152 (2015).

Matthias, W. (1974) *Kataloge zur mitteldeutschen Schnurkeramik IV: Südharz-Unstrut-Gebiet*. Berlin, Deutscher Verlag der Wissenschaften.

Meller, H., Arz, H. W., Jung, R. and Risch, R. (eds) (2015) *2200 BC – Ein Klimasturz als Ursache für den Zerfall der Alten Welt? 7. Mitteldeutscher Archäologentag, 23.–26. Oktober 2014, Halle (Saale)*. Halle (Saale), Landesamt für Denkmalpflege.

Moghaddam, N., Müller, F., Hafner, A. and Lösch, S. (2016) Social Stratigraphy in Late Iron Age Switzerland: Stable Carbon, Nitrogen and Sulphur Isotope Analysis of Human Remains from Münsingen. *Archaeological and Anthropological Sciences* 8, 149–160.

Mörseburg, A., Alt, K. W. and Knipper, C. (2015) Same Old in Middle Neolithic Diets? A Stable Isotope Study of Bone Collagen from the Burial Community of Jechtingen, Southwest Germany. *Journal of Anthropological Archaeology* 39, 210–221.

Müller, J. (2000) Radiokarbonchronologie, Keramiktechnologie, Osteologie, Anthropologie, Raumanalysen. Beiträge zum Neolithikum und zur Frühbronzezeit im Mittelelbe-Saale-Gebiet (eds). *Bericht der Römisch-Germanischen Kommission* 80, 1999 (2000) 25–211.

Nadler, M. (2001) Einzelhof oder Häuptlingshaus? – Gedanken zu den Langhäusern der Frühbronzezeit. In B. Eberschweiler, J. Köninger, H. Schlichtherle and C. Strahm (eds), *Aktuelles zur Frühbronzezeit und frühen Mittelbronzezeit im nördlichen Alpenvorland. Rundgespräch in Hemmenhofen vom 6. Mai 2000*. Hemmenhofener Skripte 2, 39–46. Freiburg, Janus.

Oelze, V. M., Koch, J. K., Kupke, K., Nehlich, O., Zäuner, S., Wahl, J., Weise, S. M., Rieckhoff, S. and Richards, M. P. (2012) Multi-Isotopic Analysis Reveals Individual Mobility and Diet at the Early Iron Age Monumental Tumulus of Magdalenenberg, Germany. *American Journal of Physical Anthropology* 148, 406–421.

Price, T. D., Grupe, G. and Schröter, P. (1998) Migration in the Bell Beaker Period of Central Europe. *Antiquity* 72, 405–411.

Price, T. D., Knipper, C., Grupe, G. and Smrcka, V. (2004) Strontium Isotopes and Prehistoric Human Migration: The Bell Beaker Period in Central Europe. *European Journal of Archaeology* 7, 9–40.

Rasmussen, S. *et al.* (2015) Early Divergent Strains of Yersinia pestis in Eurasia 5.000 Years Ago. *Cell* 163/3, 571–582.

Reinecke, P. (1902) Beiträge zur Kenntnis der frühen Bronzezeit Mitteleuropas. *Mitteilungen der Anthropologischen Gesellschaft Wien* 32, 104–154.

Reinecke, P. (1924) Zur chronologischen Gliederung der süddeutschen Bronzezeit. *Germania* 8/2, 43–44.

Ruckdeschel, W. (1978) *Die frühbronzezeitlichen Gräber Südbayerns. Ein Beitrag zur Kenntnis der Straubinger Kultur*. Antiquitas II. Bonn, Habelt.

Scheeres, M., Knipper, C., Hauschild, M., Schönfelder, M., Siebel, W., Pare, C. and Alt, K. W. (2014) 'Celtic Migrations' – Fact or Fiction? Strontium and Oxygen Isotope Analysis of the Czech Cemeteries of Radovesice and Kutná Hora in Bohemia. *American Journal of Physical Anthropology* 155, 496–512.

Schefzik, M. (2001) *Die bronze- und eisenzeitliche Besiedlungsgeschichte der Münchner Ebene. Eine Untersuchung zu Gebäude- und Siedlungsformen im süddeutschen Raum*. Internationale Archäologie 68. Rahden/Westf., Marie Leidorf.

Schickler, H. (1981) 'Neolithische' Zinnbronzen. In H. Lorenz (ed.), *Studien zur Bronzezeit. Festschrift für Wilhelm Albert v. Brunn*, 419–445. Mainz am Rhein, Philipp von Zabern.

Schroeder, H. E. (2000) *Orale Strukturbiologie: Entwicklungsgeschichte, Struktur und Funktion normaler Hart- und Weichgewebe der Mundhöhle und des Kiefergelenks*. Stuttgart and New York, Thieme.

Schuenemann, V. J., Bos, K., DeWitte, S., Schmedes, S., Jamieson, J., Mittnik, A., Forrest, S., Coombes, B. K., Wood, J. W., Earn, D. J. D., White, W., Krause, J., Poinar, H. N. (2011). Targeted Enrichment of Ancient Pathogens Yielding the pPCP1 Plasmid of Yersinia Pestis from Victims of the Black Death. *PNAS* 108, 746–752. DOI: 10.1073/pnas.1105107108.

Schweissing, M. M. and Grupe, G. (2003) Stable Strontium Isotopes in Human Teeth and Bone: A Key to Migration Events of the Late Roman Period in Bavaria. *Journal of Archaeological Science* 30, 1373–1383.

Siebert, A., Knipper, C., Nicklisch, N., Friedrich, S. and Alt, K. W. (2016) Wandel der Ernährungsweise in Mitteldeutschland zwischen 5500 und 1600 v. Chr. *Archäologie in Deutschland* 2016/4, 24–25.

Sjögren, K.-G., Price, T. D. and Kristiansen, K. (2016) Diet and Mobility in the Corded Ware of Central Europe. *PLOS ONE* 11. DOI: 10.1371/journal.pone.0155083.

Stockhammer, P. W. (2008) *Kontinuität und Wandel – Die Keramik der Nachpalastzeit aus der Unterstadt von Tiryns.* Heidelberg, Universität Heidelberg.

Stockhammer, P. W. (2012) Performing the Practice Turn in Archaeology. *Transcultural Studies* 1, 7–42.

Stockhammer, P. W. (2015) Die Wirkungsmacht des Identischen: Zur Wahrnehmung von Metallobjekten am Beginn der Bronzezeit. *Germania* 93, 77–96.

Stockhammer, P. W., Massy, K., Knippe, C., Friedrich, R., Kromer, B., Lindauer, S., Radosavljević, J., Wittenborn, F. and Krause, J. (2015a) Rewriting the Central European Early Bronze Age Chronology: Evidence from Large-Scale Radiocarbon Dating. *PLOS ONE* 10. DOI: 10.1371/journal.pone.0139705.

Stockhammer, P. W., Massy, K., Knipper, C., Friedrich, R., Kromer, B., Lindauer, S., Radosavljević, J., Pernicka, E. and Krause, J. (2015b) Kontinuität und Wandel vom Endneolithikum zur frühen Bronzezeit in der Region Augsburg. In H. Meller, H. W. Arz, R. Jung and R. Risch (eds), *2200 BC – Ein Klimasturz als Ursache für den Zerfall der Alten Welt? 7. Mitteldeutscher Archäologentag in Halle (Saale), vom 23.–26. Oktober 2014.* Tagungen des Landesmuseums für Vorgeschichte Halle 12,2, 617–641. Halle (Saale), Landesamt für Denkmalpflege und Archäologie Sachsen-Anhalt.

Strahm, C. (1982) Zu den Begriffen Chalkolithikum und Metallikum. In *Atti X Simposio internazionale sulla fine del Neolitico e gli inizi dell'Etá del Bronzo in Europa,* 13–26. Verona, Museo Civico di Storia Naturale.

Styring, A. K., Maier, U., Stephan, E., Schlichtherle, H. and Bogaard, A. (2016) Cultivation of Choice: New Insights into Farming Practices at Neolithic Lakeshore Sites. *Antiquity* 90, 95–110.

Wright, L. E. and Schwarcz, H. P. (1998) Stable Carbon and Oxygen Isotopes in Human Tooth Enamel: Identifying Breastfeeding and Weaning in Prehistory. *American Journal of Physical Anthropology* 106, 1–18.

Zich, B. (1996) *Studien zur regionalen und chronologischen Gliederung der nördlichen Aunjetitzer Kultur.* Vorgeschichtliche Forschungen 20. Berlin, de Gruyter.

Chapter 21

Yet Another Revolution? Weapon Technology and Use Wear in Late Neolithic and Early Bronze Age Southern Scandinavia

Christian Horn

Introduction

Archaeology tends to emphasise the new and revolutionary aspects of past changes. In Bronze Age (BA) research, bronze is always the new material that brings about fundamental changes by extending exchange networks and the range of mobility (Kristiansen and Larsson, 2005). The sword introduces a new ideal of the warrior complete with new bronze weapons, toilet sets and international networks (Kristiansen 1984; Treherne 1995; Harding 2007). Although this picture is not wrong, it neglects the fact that there is a long history of metallurgy and specialised weaponry in what is somewhat arbitrarily called the Late Neolithic (LN; see Vandkilde 1996). The emphasis on constant progress plays down one important issue in the development of Nordic Early Bronze Age (EBA) weaponry: What is the contribution of the past?

The aim of this article is to highlight the multi-layered and interwoven nature of the development of specialised weaponry. To address combat and weapon design, it is important to observe actual traces of use. Too often research relies on inductive reasoning based on the form of weapons and on modern expectations: shorter spears are for throwing, longer ones for slashing, *etc.* (Tarot 2000, 41–45; Davis 2012, 57) . To rectify this view, I intend to discuss the use wear of 15 LN halberds and 204 weapons from period I of the Nordic EBA. By sketching a theoretical framework based on studies of the diffusion of innovation and the development of weapon technology, I shall investigate the scale on which change took place and the role of the LN in the adoption of innovations in weapon technology, fighting style and combat motions.

Chronology of early specialised weaponry in southern Scandinavia

In general, the use of halberds seems to end at different times across Europe (detailed discussion in Horn 2014). The final stage of the halberd development in Scandinavia is linked to the central European Únětice culture. Recently, attempts have been made to push the end of halberd use back to approximately 2000 BC for the Únětice region (Lorenz 2010; Rassmann 2010). However, this research is based on the dating of metal compositions. The authors argue that the main halberd production phase is earlier than previously assumed. This is supported by contextual and typo-chronological evidence (Horn 2014, 139).

Nonetheless, this finding should not be conflated with the end of the use of halberds in the Únětice culture. There is evidence to the contrary. The famous 'princely' grave at Leubingen, GER was constructed between 1950 and 1900 BC (Becker *et al.* 1989). Metal of the 'Bresinchen type' is assumed to be younger than the halberds (2000–1800 BC; Lorenz 2010), but the eponymous hoards from Bresinchen, GER contain two halberds (Breddin 1969). Lastly, there are halberds with wooden handles that may consist of tin bronze and are associated with late finds, such as the unique and very early socketed axe found in a hoard near Drobitz, GER (Bethge 1925). For all these reasons, and in accordance with the earliest dates proposed by other researchers, it has been suggested that the use of halberds in the Únětice region ended between 1900 and 1800 BC (Zich 1996; Schuhmacher 2002; Brandherm 2004; Horn 2014, 139). Recent radiocarbon dating has demonstrated the serious problems besetting a strict separation of phases in the central European Únětice culture (Stockhammer *et al.,* 2015). Accordingly, based on the typo-chronological evidence alone, there is nothing prompting us to assume that the use of halberds ends in the Únětice region or in typologically linked southern Scandinavia in 2000 BC (Vandkilde 1996, 193–199).

Since the initial periodisation of the Nordic EBA by Oskar Montelius, a great deal of work has been done to refine his system. For period I, this has resulted in a subdivision into periods Ia and Ib (*cf.* Vandkilde 1996 with older literature). In terms of specialised weaponry, spearheads of the Torsted type, Tinsdahl type, and central European type, including the Renderzhausen type, as well as short swords or daggers of the Virring type are traditionally dated to period Ia (Becker 1964). One of the grounds for this dating is the multi-type hoard at the eponymous Virring, DK containing a short sword of the Virring type, three spearheads of the Torsted type, one spearhead of the Renderzhausen type, and two flanged axes (Vandkilde 1996, fig. 83 no. 302). This period is usually thought to extend from 1700 to 1600 BC. In period Ib, a wide variety of weapon forms is assumed to have emerged within the space of 100 years (1600–1500 BC), spears of the Bagterp and Valsømagle types and swords of the Apa-derived, Sögel, and Wohlde types (Lomborg 1965; Vandkilde 1996, 223–256; Randsborg and Christensen 2006). Apa-derived swords are usually viewed as the result of direct influence by swords of the so-called Apa-Hajdúsámson type, with the swords, for example, from Torupgårde and Stensgård (both DK) being interpreted as direct imports (Lomborg 1960, 96; Bader 1991, 49; Wincentz Rasmussen and Boas 2006; Bergerbrant 2013).

This chronology has been very useful in addressing the Nordic material. However, rigid typo-chronology cannot account for the fluidity of reality and the development in weapon design. Furthermore, there are indicators suggesting that metalwork tends to cross the boundaries set by periodisation. Radiocarbon dates are a way of establishing a dating independently of find combinations. On this basis, Per Ethelberg has suggested that the Sögel-Wohlde complex emerged around 1800 BC (Ethelberg 2000). In Luttum (barrow 39, grave 2), GER, a Sögel-type blade was associated with a radiocarbon date that can be calibrated between 1897 and 1692 cal. BC (OxCal 4.2; KN-I. 2082 3480 ± 80 BP; Schwabedissen, 1978, 112; Bergerbrant, 2007, 25). The second important date stems from a grave with a short sword of the Wohlde type (grave 6) in a barrow near Rastorf, GER (Bokelmann, 1972, 1977). The date was published by Ethelberg (2000, 265) using the wrong standard deviation. In the archives of Schloß Gottorf castle, the author had access to the original letter by J. Lanting to K. Bokelmann, in which he specified the date as 3340±70 BP (GrN-10755). Calibrated with OxCal 4.2, the grave may have been built between 1730 and 1530 cal. BC (two sigma).

Bokelmann discovered an Apa-derived sword in grave 5. This grave can only be dated indirectly, because the only other another radiocarbon dating came from a charcoal sample from the preceding grave 4. The date for this grave is 1936–1752 cal. BC in the 2-sigma range (OxCal 4.2; GrN-10754; Bokelmann 1977). Given its stratigraphic position and taking due account of a possible old-wood effect, this would make the construction of grave 5 around 1750–1600 BC conceivable.

Ethelberg's proposal is potentially too radical and the radiocarbon dates are imprecise. However, since they are not obviously wrong, the dating should be taken seriously. Considering the problems and the long-established association-based typo-chronology (Vandkilde, 1996), we can legitimately suggest that Apa-derived swords may emerge in find contexts at the end of period Ia or in the transition to Ib and swords of the Sögel type shortly before or at the beginning of the second half of period Ib. Earlier emergence is conceivable (see Horn 2015) but we need new radiocarbon dates to address that issue. A grave or hoard find from Örebro, SWE contained a spear of the Ödeshög type and two flanged axes parallel to the Virring hoard axes, which suggests that Ödeshög-type spears emerge in period Ia (Oldeberg 1974: No. 2703; Vandkilde 1996, 102–103). Based on find associations in Lower Saxony, Friedrich Laux recently suggested that spears of the Bagterp type emerge parallel to Sögel swords, thus putting them in a similar temporal frame (2011, 9). Valsømagle spears may well be the last innovation of period Ib. That fits in well with the link-up of swords of the Valsømagle type to the central European swords of the Au-Spatzenhausen-Zajta-Livada group (Hachmann 1957, 132–146; Bader 1991, 53; Kemenczei 1991, 53).

Use-wear

Use-wear has been analysed on spears of the Torsted (51), Ödeshög (9), Bagterp (50), and Valsømagle types (41). In addition, two spears of the Renderzhausen type and the spear from the Tinsdahl, GER, hoard have been studied. One spear was not identifiable. Analyses were carried out on swords of various types: Apa-derived (11), Virring (3), Sögel (28), and Wohlde (7). One sword could not be identified. As a chronological compromise and in order to include the small number of swords of the Virring (3) and Wohlde (7) types analysed, the swords are divided into groups. Swords of the Virring and Apa-derived types form the group 'early swords', while the Sögel-Wohlde complex will be referred to as 'later swords'. Overall, the number of objects investigated is, of course, low. However, the material studied represents approximately one third of known period I material. Fifteen of the 33 known metal halberds from southern Scandinavia (45%) have been investigated. Therefore, the samples are significant. Future discoveries may be expected to alter the results and interpretations presented here.

We turn our attention to combat motions such as slashing and cutting as well as stabbing, thrusting, and throwing. The direction of the strike is highly significant for the kind of damage potentially incurred by the weapon. A slash or cut exposes more of the cutting edge, given that here the direction of the motion is at an oblique angle to the longitudinal axis of the weapon. It follows that weapons used for slashing and cutting are more likely to incur edge damage such as notches and indentations (Fig. 21.1a–c). Conversely, a stab, thrust or throw, the motion follows along the longitudinal axis of the weapon. Usually, the tip first strikes either the enemy, their armour or possibly the ground, if a combatant misses. Therefore, these kinds of motion expose the tip to higher risk of damage (Fig. 21.1d–f).

Before we look at the data, some problems should be mentioned. I will keep my remarks short, because a detailed discussion has been published elsewhere (Horn 2013). One eventuality is that the observable damage may have been caused by ritual acts of deliberate destruction. This is well known from the Chalcolithic, the Late Bronze Age, and the Latène periods (Nebelsick 2000; Sievers 2010; Horn 2011). In the material at hand, the large hoards from Torsted and Dystrup, DK, showed some evidence that tips may have been snapped off as a kind of ritual destruction (Melheim and Horn 2014). Sometimes it is impossible to deduce the origin of a particular instance of damage. Taphonomic processes are another problematic aspect. Graves represent an environment that is particularly corrosive and since in the timeframe under discussion swords are more often deposited in graves, they are sometimes too severely corroded for any use wear to be recognised. Experiments have shown that bronze weapons are remarkably resilient to damage if they are used properly (O'Flaherty 2007; Anderson 2011). Therefore, we may safely assume that damage is usually coincidental and fighters normally tried to protect weapons from damage, always depending of course on the combat situation.

In the following, we compare tip and edge damage (Fig. 21.2). Since each weapon may theoretically display both kinds, the absolute number of data points will be higher than the number of individual objects analysed. The overall margin of difference is narrow, working out at about 15%. To put the results into perspective, another chart will be provided summarising the absolute numbers of the damage categories (Fig. 21.2). The comparison of edge and tip damage indicates that spears of the Torsted type display the highest amount of edge wear (61%, 31) and that damage along the edges outweighs tip damage (39%, 20). At the other end of the scale are spears of the Bagterp type exhibiting a substantially higher amount of tip (54%, 31) than edge damage (46%, 26). Spears of the Ödeshög type seem to be very similar to the Bagterp type (tip: 50%, 6; edge: 50%, 6). Spears of the Valsømagle type (tip: 47%,

25; edge: 53%, 28) and both early and later swords take up similar positions in the middle of the range. If anything can be said about this close field, it would be that later swords display slightly more edge damage (tip: 46%, 13; edge: 54%, 15) than the earlier ones (tip: 50%, 5; edge: 50%, 5).

Comparing these findings to the earlier halberds, it is interesting to note that their tip (45%, 10) and edge damage (55%, 12) is almost equal. Accordingly, halberds do not substantially differ in terms of tip and edge use from spears of the Ödeshög and Valsømagle types or from later swords.

Discussion

The presence of edge wear on all spear types suggests that none of these spears were exclusively thrown during combat, *i.e.* there were no javelins. Secondly, both in southern Scandinavia and in Europe halberds have been underestimated in terms of their fighting capability and importance. The density of halberds in Denmark and Scania is akin to that of important halberd centres, for example the north-western Iberian Peninsula or Great Britain (Horn 2014, pl. 114). The presence of many notches indicates the possibility that many more weapons with metal blades were present. To create such notches, considerable force needs to be generated, making it unlikely that metal daggers caused this damage (for comparison, see O'Flaherty *et al.* 2008). As such, we may infer indirectly that other metal halberds are likely to have caused the notches on halberd blades, although these have not survived in the archaeological record. Thirdly, the combat differences between the various kinds of weapon are purely tendential. This is remarkable because in terms of design, halberds, spears, and swords represent quite a range of different forms. Thus, fighting with bladed weaponry in a fencing style (including slashing, cutting, stabbing and thrusting) may arguably have enjoyed a long tradition in southern Scandinavia.

It is unlikely that weapon design was coincidental. Instead, different weapon forms were perhaps intended to abet motions attuned to preconceptions about the way a fighter should fight. Judging from use wear, reality was messier and the approach to combat seems often to have been to win 'by whatever means required'. However, the small differences in tip and edge damage may tell us something about the kinds of fighting style particular weapons were designed to support. Spears of the Torsted type seem to have been used more often for slashing and cutting than stabbing and thrusting. The chronologically later spears of the Bagterp type reverse this emphasis in favour of thrusting and stabbing. The earlier spears of the Ödeshög type and early (short) swords may have initiated this development. Although they differ hardly at all in terms of edge and tip damage, they move away from the strong emphasis on slashing and cutting typical of the Torsted-type spears. After the Bagterp spears, fighting styles level

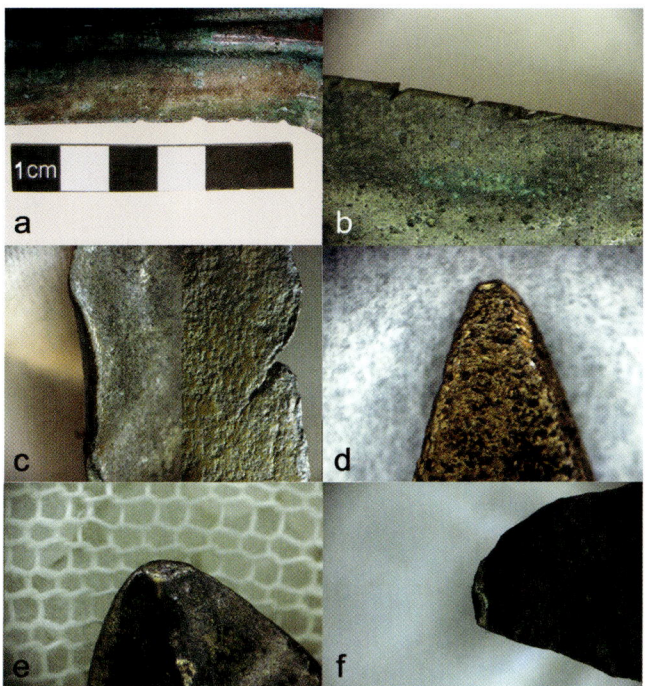

Figure 21.1: (a) Notches and indentations on a sword (Hohenlockstedt, Germany; LMSH KS 10802); (b) Notches and indentations on a spear (context unknown, Sweden; LUHM 8985; ×60); (c) Indentation and notch with material displacement on a halberd (context unknown, Denmark; NMK B 14623; ×60); (d) Tip pressure on a sword (Torupgårde, Denmark; NMK B 10600; ×60); (e) Tip pressure on a spear (context unknown, Sweden; SHM 2899; ×60); (f) Tip pressure on a halberd (Hvornum, Denmark; NMK B 6469; ×60).

out. The later swords encompass all motions, with a slight emphasis on the slashes and cuts typical of a fencing style of combat. Spears of the Valsømagle type may take up this trend and continue with it. Surprisingly, the LN halberds also favoured this style despite their very different construction and the chronological distance involved.

Innovation and adoption of weapon technology

For the ongoing exploration of these processes, a theoretical framework is provided by the studies by Pierre Lemonnier (1992), Everett Rogers (2010), and Robert L. O'Connell (1989). Lemonnier and Rogers investigate the preconditions for the uptake and spread of change by looking at unsuccessful innovations, for example new airplane designs (Lemonnier 1992) or water boiling as a health measure (Rogers 2010). Their investigations indicate that innovations may encounter major opposition even if they are objectively better in many or all ways than older designs, technologies or methods. To be successful, innovations have to preserve or relate to traditional technological aspects, social ideals or perceptions. This is what integrates innovations into societies, relates them to a given context and makes them a basis for future

developments (see also Hägerstrand 1967; Shils 1981). In that sense, innovations are never absolutely new.

O'Connell has studied the long history of weapons and their relation to the fighters using them from prehistoric to modern times. He identifies two processes in the development of weapon technology. Both are based on the urgent need to win. One process is diversification. That means that the people who design weapons, for example, fighters, war leaders or metalworkers, all seek to diversify existing weapon technology via innovation. The aim is to gain an advantage in combat by improving weapons or changing techniques. The other process takes place when there is an imbalance in weapon design or technique that gives one party a considerable advantage over another. The party in the inferior position will attempt to offset their disadvantage by adopting the innovations that provided the advantage in the first place. During adoption, innovations are potentially transformed to fit local needs. O'Connell called this process the drive towards 'symmetry' in weapon technology (O'Connell 1989). It means that warfare has the power to homogenise.

Interpretation

From the LN to the EBA, fighting was basically synonymous with an all-out form of fencing. In the course of time, one technique was emphasised over another as a form of diversification. However, if the cutting edge position and the theoretical motion necessary to achieve a stab, slash, or cut is considered, then the motions are remarkably consistent (Fig. 21.3). Due to familiarity with such motions, changes were quickly adopted and eventually homogenised in period Ib with the later swords and spears of the Valsømagle type. What this hypothetical reconstruction of motions may indicate is that EBA weapons are more specialised in the sense that they narrow down the range of motions necessary to achieve, say, a cut (Fig. 21.3). Familiarity with some motions may have been a precondition for the adoption of new weapons.

In the LN, fighting with halberds may have been replaced by fighting with flint and metal daggers and perhaps also flint spearheads (Montelius 1917: No. 450 484 488–490; Apel 2001; Schwenzer 2004). These weapons may have emphasised thrusting, and new use-wear studies address this issue. These weapons potentially facilitated the adoption of an influx of new designs like the Apa-derived swords. Full metal hilts on halberds and daggers were already known in the LN. Further support for this notion comes from the similarities in decoration and typology between early short swords and full-hilted daggers (*cf.* Schwenzer 2004; Laux 2009, pl. 1.1, 1.3–4; Engedal 2010). Differentiation between swords and daggers is misleading, because they could both be similar in length.

Torsted-type spears with their emphasis on slashing and cutting may have diversified combat techniques. Spearheads

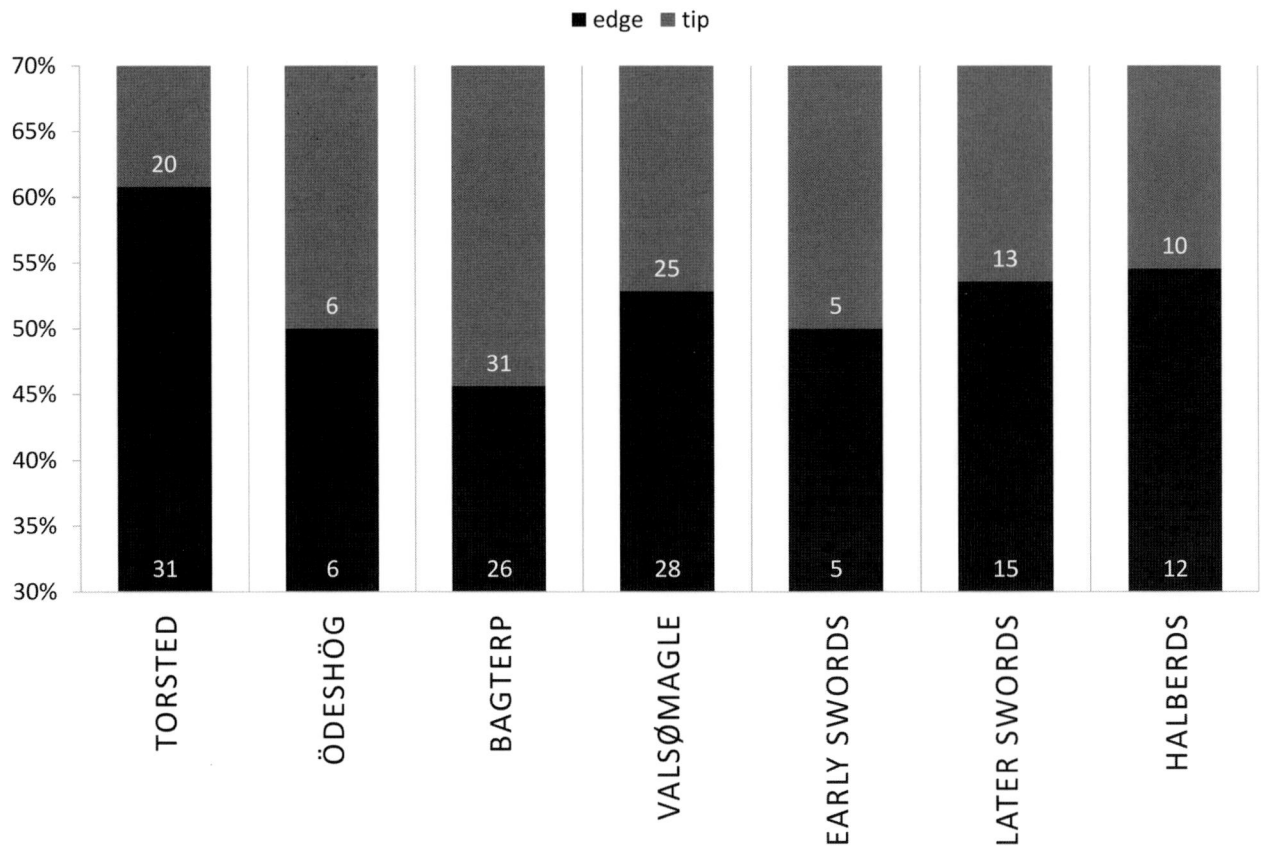

Figure 21.2: Chart comparing edge and tip separated by object type.

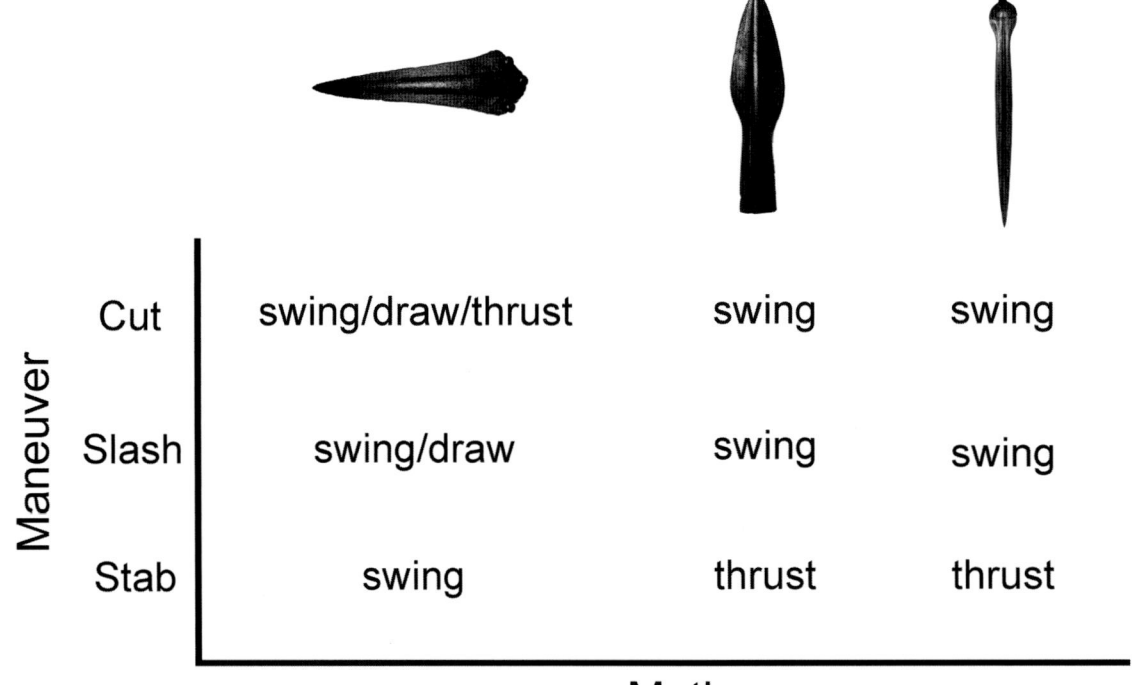

Figure 21.3: Chart comparing motions necessary to conduct attacks with halberds, spears and swords.

of the Bagterp type may represent a counter-innovation that was the first to spread throughout the region, demonstrating that diversification and homogenisation are not necessarily contradictory. Early and later swords arguably equalise the fighting techniques again. This and the wide spread of the Bagterp-type spear may have enabled the adoption of the Valsømagle-type spear. Its distribution in southern Sweden and the emphasis on combat techniques familiar from swords may have been what facilitated the wide introduction of the sword in Sweden with the onset of period II (Kristiansen 2000, fig. 34). In summary, period Ib and early II may possibly be understood as phases of symmetry in diversity, with the same types pervading southern Scandinavia and used in a similar fencing style, but with a new diversity of material forms. This development may have had its roots in LN fighting with halberds. However, the transformation was not achieved through revolutions and radical breaks but by a fluid succession of small changes.

Conclusion

A detailed chronological model of early specialised weapons suggests that there is no contradiction between the extended contemporaneity of sword and spear types in period Ib and a difference in their emergence. This demonstrates the intermesh between all sword and spear types in southern Scandinavia. It also points to longevity of specialised weapons in the LN already, *i.e.* halberds.

Use-wear analyses have been presented here as a comparison between edge and tip damage to investigate the role of fighting techniques such as slashing and stabbing. This helps to assess interpretations of their use based on their form. We may safely rule out the notion that spears were solely used for throwing. The differences between all these weapons, including the much older halberds, were relative.

I have proposed a theoretical framework emphasising the relatedness of successful innovations to the past and the interdependency of weapons in the development of weapon technology. On this basis, I have argued that change was instituted in weapon technology with the onset of the EBA, but that it was deeply rooted in the LN. Consistency in combat style permitted the adoption of new kinds of weapon in bronze.

References

Anderson, K. (2011) Slashing and Thrusting with Late Bronze Age Spears: Analysis and Experiment. *Antiquity* 85/328, 599–612.

Apel, J. (2001) *Daggers, Knowledge & Power*. Uppsala, Uppsala University.

Bader, T. (1991) *Die Schwerter in Rumänien*. Stuttgart, Franz Steiner Verlag.

Becker, B., Jäger K. D., Kaufmann, D. and Litt, T. (1989) Dendrochronologische Datierungen von Eichenhölzern aus den frühbronzezeitlichen Hügelgräbern bei Helmsdorf und Leubingen (Aunjetitzer Kultur) und an bronzezeitlichen Flußeichen bei Merseburg. *Jahresschrift für mitteldeutsche Vorgeschichte* 72, 299–312.

Becker, C. J. (1964) Neue Hortfunde aus Dänemark mit frühbronzezeitlichen Lanzenspitzen. *Acta Archaeologica* 35, 115–152.

Bergerbrant, S. (2007) *Bronze Age Identities: Costume, Conflict and Contact in Northern Europe 1600–1300 BC*. Lindome, Bricoleur Press.

Bergerbrant, S. (2013) Migration, Innovation and Meaning. Sword Depositions on Lolland 1600–1100 BC. In M. E. Alberti and S. Sabatini (eds), *Exchange Networks and Local Transformations. Interaction and Local Change in Europe and the Mediterranean from the Bronze Age to the Iron Age*, 146–155. Oxford, Oxbow Books.

Bethge, W. (1925) Vorgeschichtliche Funde der letzten Jahre aus dem Kreis Cöthen. *Mannus Ergänzungsband* 4, 35–51.

Bokelmann, K. (1972) Ein mehrphasiger Grabhügel der Stein- und Bronzezeit bei Rastorf, Kreis Plön. *Archäologisches Korrespondenzblatt* 2, 33–35.

Bokelmann, K. (1977) Ein Grabhügel der Stein-und Bronzezeit bei Rastorf, Kreis Plön. *Offa* 34, 90–99.

Brandherm, D. (2004) Porteurs de hallebardes? Überlegungen zur Herkunft, Entwicklung und Funktion der bronzezeitlichen Stabklingen. *Varia neolithica III. Beitrage zur Ur–und Frühgeschichte Mitteleuropas* 37, 279–334.

Breddin, R. (1969) Der Aunjetitzer Bronzehortfund von Bresinchen, Kr. Guben. *Veröffentlichungen des Museums für Ur- und Frühgeschichte Potsdam* 5, 15–41.

Davis, R. (2012) *The Early and Middle Bronze Age Spearheads of Britain*. Stuttgart, Franz Steiner Verlag.

Engedal, Ø. (2010) The Bronze Age of Northwestern Scandinavia. Unpublished Ph.D. thesis, University of Bergen, Bergen.

Ethelberg, P. (2000) Bronzealderen. In P. Ethelberg, E. Jørgensen, D. Meier and D. Robinson (eds), *Det Sønderjyske landbrugs historie: Sten- og bronzealder*, 135–280. Haderslev, Haderslev Museum og Historisk Samfund for Sønderjylland.

Hachmann, R. (1957) *Die frühe Bronzezeit im westlichen Ostseegebiet und ihre mittel- und südosteuropäischen Beziehungen*. Hamburg, Flemming.

Hägerstrand, T. (1967) *Innovation Diffusion as a Spatial Process*. Chicago, IL, University of Chicago Press.

Harding, A. F. (2007) *Warriors and Weapons in Bronze Age Europe*. Budapest, Archaeolingua alapítvány.

Horn, C. (2011) Deliberate Destruction of Halberds. In M. Uckelmann and M. Mödlinger (eds), *Bronze Age Warfare: Manufacture and Use of Weaponry*. British Archaeological Reports, International Series 2255, 53–65. Oxford, Archaeopress.

Horn, C. (2013) Weapons, Fighters and Combat: Spears and Swords in Early Bronze Age Scandinavia. *Danish Journal of Archaeology* 2/1, 20–44.

Horn, C. (2014) *Studien zu den europäischen Stabdolchen*. Bonn, Habelt.

Horn, C. (2015) Combat and Change: Remarks on Early Bronze Age Spears from Sweden. In P. Suchowska-Ducke, S. S. Reiter and H. Vandkilde (eds), *Forging Identities: The Mobility of Culture in Bronze Age Europe*. BAR, International Series 2772, 201–212. Oxford, BAR Publishing.

Kemenczei, T. (1991) _Die Schwerter in Ungarn II (Vollgriffschwerter)_. Munich, Beck.

Kristiansen, K. (1984) Krieger und Häuptlinge in der Bronzezeit Dänemarks. Ein Beitrag zur Geschichte des bronzezeitlichen Schwertes, _Jahrbuch des Römisch-Germanischen Zentralmuseums Mainz_ 31, 187–208.

Kristiansen, K. (2000) _Europe Before History_. Cambridge, Cambridge University Press.

Kristiansen, K. and Larsson, T. B. (2005) _The Rise of Bronze Age Society: Travels, Transmissions and Transformations_. Cambridge, Cambridge University Press.

Laux, F. (2009) _Die Schwerter in Niedersachsen_. Stuttgart, Franz Steiner Verlag.

Laux, F. (2011) _Die Dolche in Niedersachsen_. Stuttgart, Franz Steiner Verlag.

Lemonnier, P. (1992) _Elements for an Anthropology of Technology_. Ann Arbor, MI, Museum of Anthropology, University of Michigan.

Lomborg, E. (1959) Donauländische Kulturbeziehungen und die relative Chronologie der frühen nordischen Bronzezeit. _Acta Archaeologica_ 30, 51–146.

Lomborg, E. (1965) Valsømagle und die Frühe Nordische Spiralornamentik. _Acta Archaeologica_ 36, 223–232.

Lorenz, L. (2010) _Typologisch-chronologische Studien zu Deponierungen der nordwestlichen Aunjetitzer Kultur_. Bonn, Habelt.

Melheim, L. and Horn, C. (2014) Tales of Hoards and Swordfighters in Early Bronze Age Scandinavia: The Brand New and the Broken. _Norwegian Archaeological Review_ 47/1, 18–41.

Montelius, O. (1917) _Minnen frän var forntid. I_. Stockholm, Norstedt & Söners.

Nebelsick, L. (2000) Rent Asunder: Ritual Violence in Late Bronze Age Hoards. In C. F. E. Pare (ed.), _Metals Make the World Go Round: The Supply and Circulation of Metals in Bronze Age Europe. Proceedings of a Conference Held at the University of Birmingham in June 1997_, 160–175. Oxford, Oxbow Books.

O'Connell, R. L. (1989) _Of Arms and Men: A History of War, Weapons, and Aggression_. New York, Oxford University Press.

O'Flaherty, R. (2007) A Weapon of Choice–Experiments with a Replica Irish Early Bronze Age Halberd. _Antiquity_ 81/312, 423–434.

O'Flaherty, R., Bright, P., Gahan, J. and Gilchrist, M. D. (2008) Up Close and Personal. _Archaeology Ireland_, 22–25.

Oldeberg, A. (1974) _Die ältere Metallzeit in Schweden I_. Stockholm, Almqvist & Wiksell.

Randsborg, K. and Christensen, K. (2006) Bronze Age Oak-Coffin Graves: Archaeology & Dendrodating. _Acta Archaeologica_ 77/1. DOI: 10.1111/j.1600-0390.2006.00049.x.

Rassmann, K. (2010) Die frühbronzezeitlichen Stabdolche Ostmitteleuropas: Anmerkungen zu Chronologie, Typologie, Technik und Archäometallurgie. In H. Meller and F. Bertemes (eds), _Der Griff nach den Sternen: Wie Europas Eliten zu Macht und Reichtum kamen_, 807–821. Langenweißbach: Beier & Beran.

Rogers, E. M. (2010) _Diffusion of Innovations_. New York, Free Press.

Schuhmacher, T. X. (2002) Some Remarks on the Origin and Chronology of Halberds in Europe. _Oxford Journal of Archaeology_ 21/3, 263–288.

Schwabedissen, H. (1978) Konventionelle oder kalibrierte C14-Daten?: Argumente auf Grund der archäologisch-historischen Chronologie des zweiten vorchristlichen Jahrtausends. _Archäologische Informationen_ 4, 110–117.

Schwenzer, S. (2004) _Frühbronzezeitliche Vollgriffdolche: Typologische, chronologische und technische Studien_. Bonn, Habelt.

Shils, E. (1981) _Tradition_. Chicago, IL, University of Chicago Press.

Sievers, S. (2010) _Die Waffen aus dem Oppidum von Manching_. Wiesbaden, Reichert.

Stockhammer, P. W., Massy, K., Knipper, C., Friedrich, R., Kromer, B., Lindauer, S., Radosavljevic, J., Wittenborn, F. and Krause, J. (2015) Rewriting the Central European Early Bronze Age Chronology: Evidence from Large-Scale Radiocarbon Dating. _PlOS ONE_ 10/10: e0139705. DOI: 10.1371/journal.pone.0139705.

Tarot, J. (2000) _Die bronzezeitlichen Lanzenspitzen der Schweiz_. Bonn, Habelt.

Treherne, P. (1995) The warrior's beauty: the masculine body and self-identity in Bronze-Age Europe, _Journal of European archaeology_ 3, 105–144.

Vandkilde, H. (1996) _From Stone to Bronze: The Metalwork of the Late Neolithic and Earliest Bronze Age in Denmark_. Aarhus, Aarhus University Press.

Wincentz Rasmussen, L. and Boas, N. A. (2006) The Dystrup Swords: A Hoard with Eight Short Swords from the Early Bronze Age. _Journal of Danish Archaeology_ 14/1, 87–108.

Zich, B. (1996) _Studien zur regionalen und chronologischen Gliederung der nördlichen Aunjetitzer Kultur_. Berlin and New York, Walter de Gruyter.